Astronomy!
A Brief Edition

JAMES B. KALER
University of Illinois, Urbana-Champaign

▲▲ ADDISON-WESLEY

An imprint of Addison Wesley Longman, Inc.

Reading, Massachusetts • Menlo Park, California • New York • Harlow, England
Don Mills, Ontario • Sydney • Mexico City • Madrid • Amsterdam

For Aaron, Allison, Danielle, Josh,
Marissa, Natalie, and Sierra

Sponsoring Editor: Julia Berrisford
Developmental Editor: Rebecca Strehlow
Project Editor: Cathy Wacaser
Design Administrator: Jess Schaal
Text and Cover Design: Lesiak/Crampton Design Inc: Cynthia Crampton
Photo Researcher: Michelle E. Ryan
Production Administrator: Randee Wire
Compositor: Interactive Composition Corporation
Printer and Binder: R.R. Donnelley & Sons Company
Cover Printer: Phoenix Color Corporation

Cover Photo: NGC 604 in Galaxy M33.
Source: Hui Yang (University of Illinois), Jeff J. Hester (University of Arizona), NASA, STScI.

For permission to use copyrighted material, grateful acknowledgment is made to the copyright holders on pp. A-27–A-32, which are hereby made part of this copyright page.

Astronomy! A Brief Edition

Library of Congress Cataloging-in-Publication Data
Kaler, James B.
    Astronomy!/James B. Kaler.—A brief ed.
        p.   cm.
    Includes bibliographical references and index.
    ISBN 0-673-98561-X
    1. Astronomy.   I. Title
QB45.K35 1996
520—dc20   96-43503
                            CIP

ISBN 0-673-98561-X

96 97 98 99—DOW—9 8 7 6 5 4 3 2 1

CONTENTS IN BRIEF

CONTENTS

v

PART II   PLANETARY ASTRONOMY   151

TABLES

PREFACE

Welcome to *Astronomy! A Brief Edition*. As we look to the next millennium, astronomical knowledge is increasing at an accelerating pace, with exciting discoveries announced regularly in the daily news. Astronomy is without question a subject of the future—a science with no bounds as we probe ever deeper into the Universe in our quest for its origins and fate and, ever closer to home, to understand the nature of our own Solar System. At the same time, astronomy is a subject of the past—the most ancient of sciences—new knowledge being built upon old foundations.

This text is an amalgam of past, present, and future. We begin with the motions of the Earth that can be seen easily from the backyard, move outward through the planets to the stars, and finish with an examination of galaxies and the Universe that can be provided only by the most sophisticated instrumentation. This progression builds naturally upon prior knowledge to launch the student into unfamiliar territory.

Throughout the text, astronomy is associated with the reader's world. The sky is conceptually related to Earth by a substantial initial focus on classical astronomy; a unique chapter on stars and constellations allows the reader to relate the science to visible aspects of the sky; and a series of 13 Background essays ties astronomy to the familiar world, allowing students to experience days on the planets and presenting aspects of both the ancient and modern practice of science.

Coverage of each subject area moves from observational results to explanation and theory. Controversy and uncertainty are highlighted to show the workings of an active science.

Features

Astronomy! A Brief Edition keeps mathematics to a quiet minimum. Five separate discussions explain and refresh mathematical concepts. The history of astronomy, including the contributions of ancient astronomers, is made an integral part of the flow of astronomical discovery. Forty-five detailed tables set within the text summarize planetary conditions, list eclipse predictions, classify a variety of astronomical objects, and describe the different types of stars. A set of six star maps in Appendix 1 (which includes five more tables useful for telescopic observation) locates many of the stars and non-stellar objects discussed in the text. The brighter stars are colored according to their classes, aiding in the visualization of their distribution. A detailed map allows both naked-eye and telescopic exploration of the Moon.

Each chapter ends with a summary of key concepts and relationships. An average of 30 chapter exercises includes comparisons, numerical questions, thought and discussion questions, research ideas, and activities. Each chapter concludes with a pair of scientific writing exercises.

Organization

Astronomy! A Brief Edition is divided into four parts. Part I, Classical Astronomy, demonstrates the motions of the Earth and how they appear as motions in the sky. It includes the chapter on constellations, explains the motion of the Moon and nature of eclipses, and discusses the motions of the planets and the work of Copernicus, Kepler, and Galileo. Part I also provides a foundation for the rest of the text by exploring the nature of gravity (through both Newton and Einstein) and the properties of light, atoms, and telescopes.

Part II, Planetary Astronomy, begins with the Earth, then moves the reader through the characteristics of the planets with a strong emphasis on comparative planetology and how studying the planets allows us to understand our own world. It concludes with a chapter on the formation of the planets and the debris left behind.

The second half of the book concentrates on stars and their assemblies. Part III, Stellar Astronomy, begins with the Sun as a turning point, treating it as both the central controlling body of the Solar System and as an example of a star. The natures of other stars are first explained, and then the reader is taken along the entire course of stellar evolution, from the interstellar medium and star birth, through the evolutionary cycles of stars, to the quiet and violent forms of stellar death. It ends with a discussion of the Galaxy that demonstrates the integral natures of stellar and galactic evolution. Life in the Universe is not treated separately, but as a possible part of the study of star and planet formation. Throughout, Part III emphasizes the examination of our own planetary origins.

Finally, Part IV, Galaxies and the Universe, opens with an empirical discussion of galaxies, examines the expanding Universe and its meaning, and concludes with overviews of the Universe, its origin, and evolution. By the end, we see the Earth's place in the Universe and the unity of the Universe's construction.

Supplements

The text is accompanied by a supplement program that includes a transparency set; a test bank, available in print or on disk, offering 1,200 multiple-choice questions; and an instructor's guide, which provides a detailed tour through the text and answers to most of the end-of-chapter exercises.

Acknowledgments

I would like to thank the numerous individuals, especially Becky Strehlow, who, with Doug Humphrey, was instrumental in guiding the book through to completion. Nancy Brooks provided her usual superb copyediting, picture editor Michelle Ryan helped acquire the text's many images, and Cathy Wacaser furnished excellent coordination. Thanks also go to the talented crew at Precision Graphics.

I would like to thank my Illinois colleagues for helpful discussions, and particularly Joel M. Levine from Orange Coast College, who faithfully stayed with the project through several reviews. Special thanks go to Andrei Linde, who helped guide me through modern cosmology. I also give deep thanks to the other many reviewers who kept me on straight and accurate pathways:

Timothy C. Beers, Michigan State University

Jeffery A. Brown, University of Washington

John H. Campbell, Metropolitan State College of Denver

R. Kent Clark, University of South Alabama

George J. Corso, DePaul University

Duane R. Doty, California State University, Northridge

T. Stephen Eastmond, Rancho Santiago College

Philip Goode, New Jersey Institute of Technology

Ronald M. Haybron, Cleveland State University

Lon Clay Hill, Broward Community College

Thomas Hockey, The University of Northern Iowa

Terry Jay Jones, University of Minnesota

William C. Keel, The University of Alabama

Claud H. Sandberg Lacy, University of Arkansas

David Lamp, Texas Tech University

Kristine Larson, Central Connecticut State University

Joel M. Levine, Orange Coast College

John E. Littleton, West Virginia University

Douglas McCarty, Mt. Hood Community College

Thomas A. McDonough, Chemeketa Community College

J. Ward Moody, Brigham Young University

Edward W. Olszewski, University of Arizona

B.E. Powell, West Georgia College

William Romanishin, University of Oklahoma

Kenneth L. Russell, Houston Community College

John L. Safko, University of South Carolina, Columbia

John R. Sievers, Mesa College

Michael Sitko, University of Cincinnati

Thomas S. Statler, The University of North Carolina at Chapel Hill

Ronald Stoner, Bowling Green State University

Connie L. Sutton, Indiana University of Pennsylvania

Finally, I thank my wife Maxine for her continuing support and patience.

CLASSICAL ASTRONOMY

FROM EARTH TO UNIVERSE

A survey of the Universe to place the Earth in its surroundings

1

Come with me. We stand on a country hilltop, watching the late afternoon summer Sun glide slowly down a sapphire sky toward the west. It drops below the distant landscape, and the first stars begin to punctuate the on-rushing darkness overhead. Before long, the sky is filled with sparkling lights. We have picked a night when the stars surround a slim, bright, crescent Moon, and as we watch, they too all creep westward. What are they all? Where are they, how far away? Why do they move? What is their role in our lives? What else is out there?

They—and we—are all part of the **Universe,** defined as the totality of everything in existence. Before we proceed to examine the Universe, however, we need a framework, a survey of its contents and structure. We will then begin an exploration to reveal the extent of our knowledge, the depth of our ignorance, and what our search might mean to us as residents of Earth.

1.1 NUMBERS

The Sun is 93,000,000 miles away, the nearest star nearly 25,000,000,000,000. We need a sense of the significance of such huge numbers. Figure 1.1 shows 10,000 points, each new color demonstrating multiplication by 10. One lonely individual, singled out in yellow, represents the number 1; orange and yellow together make 10; and yellow, orange, and red constitute 100. These form one small block in the upper left. Multiply by 10 again. The total of 1,000 now stretches out in the row of yellow, orange, red, and green across the top. Another multiplication by 10 gives 10,000, represented by yellow, orange, red, green, and blue. Ten grids like this one will give you 100,000 points. A *million* is 10 times that. Imagine that the grid of 10,000 colored points fills the page; it would take 100 pages, about one-fifth the length of this book, to print 1,000,000 points.

Now you can have respect for a *billion,* which is a thousand million—1,000,000,000. To show a billion points would require a book 100,000 pages thick. One hundred or so copies of Tolstoy's *War and Peace* (each about 1,000 pages long) might serve. To print a *trillion* points (a thousand billion), we would need 100,000 copies and a library shelf nearly four miles

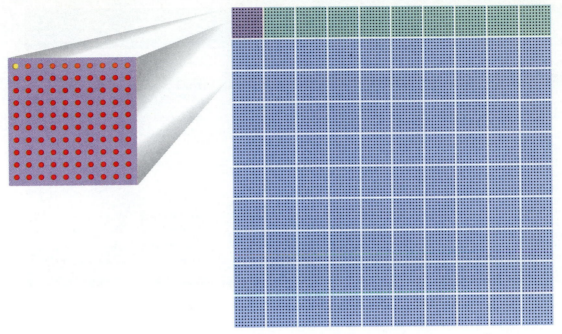

Figure 1.1
Each color change represents a multiplication by 10, progressing from 1 (yellow) to 10 (yellow plus orange) to 100 (yellow, orange, and red) to 1,000 (yellow, orange, red, and green) to 10,000 (all colors including blue).

long. A *quadrillion*, 1,000 times that, would require a row of books stretching from New York through Los Angeles and 1,000 miles out into the Pacific. *The number of stars in the observable Universe is over 10 million times more.*

1.2 A QUICK TOUR OF THE UNIVERSE

We now embark on a rapid tour of space, starting with Earth and looking outward as far as our instruments allow.

1.2.1 Earth, Moon, and Sun

Begin with yourself, a person most likely somewhat short of 2 meters tall. The *meter* (m), the fundamental unit of length in the **metric system** (Table 1.1), equals 39.4 inches or 1.09 yards. It is commonly multiplied and divided by powers of 10: a *kilometer* (km) is 1,000 meters (0.622 miles), a *centimeter* (cm) a hundredth of a meter (there are 2.54 cm per inch), and a *millimeter* (mm) a thousandth of a meter.

TABLE 1.1
Distance Units

Units	Examples
1 centimeter (cm) ⊢———⊣ = 10 millimeters (mm) = 0.394 inches	Thickness of pencil: 7 mm Diameter of golf ball: 4 cm
1 meter (m) = 100 cm = 1,000 mm = 39.4 inches = 1.09 yards	Basketball player: 2 m Football field: 90 m
1 kilometer (km) = 1,000 m = 1.609 miles	United States: 4,900 km across Distance to Moon: 384,000 km
1 Astronomical Unit (AU) = 1.50×10^8 km = 9.30×10^7 miles	Earth to Sun: 1 AU Sun to Pluto: 39 AU
1 light-year (ly) = 63,270 AU = 9.5×10^{12} km	Nearest star: 4 ly Diameter of Galaxy: 80,000 ly
1 parsec (pc) = 3.26 ly = 206,265 AU = 3.09×10^{13} km = 3.09×10^{18} cm	Nearest star: 1.3 pc Nearest large galaxy: 690,000 pc

Figure 1.2
The Earth is 12,700 km across. South America is seen in the lower half of this *Apollo 8* photograph. North America, toward the upper left, is swathed in clouds. The bulge of west Africa appears at the right.

You stand on your world, the Earth (Figure 1.2), a sphere 12,700 km (7,900 miles) in diameter, some 7 million (7×10^6: see MathHelp 1.1) times your length. If the Earth were a basketball, you would be only 4 millionths of a millimeter high, and no ordinary microscope could detect you; even Mount Everest, at the limit of the breathable atmosphere, would project upward less than two-tenths of a millimeter.

MATHHELP 1.1 Scientific Notation

To handle huge numbers reasonably, we must use *scientific notation*, or exponential numbers: 100 is 10×10, which can be written as 10^2; a million—1,000,000—is six tens multiplied in succession, $10 \times 10 \times 10 \times 10 \times 10 \times 10$, or 10^6; 200 is 2×100 or $2 \times 10 \times 10$, or 2×10^2. Numbers less than 1 can be expressed the same way: $1/10 = 0.1 = 10^{-1}$; $1/1,000,000$ is 10^{-6}; and $2/1,000,000$ is two times $1/1,000,000$, or 2×10^{-6}. Table 1.2 provides additional examples. Numbers are commonly rounded, as there is no reason to carry decimal places beyond our needs. For example, the average distance to the Sun is 149,597,706.1 km, which for most purposes can be written as 1.496×10^8 km or even as 1.5×10^8 km.

Multiplication and division with scientific notation is easier than with ordinary numbers. In multiplication, you multiply the prefixes and add the exponents. In division, you divide the prefixes and subtract the exponents. The rules are: $(a \times 10^x)$ times $(b \times 10^y) = (a \times b) \times 10^{x+y}$; $(a \times 10^x)$ divided by $(b \times 10^y)$ is $(a/b \times 10^{x-y})$. As an example, to find the length of a light-year (the distance light travels in a year) in kilometers, multiply the speed of light, 299,792 km per second (km/s), by the number of seconds in a year, 31,556,925. Without significant loss of accuracy, we can round 299,792 to 300,000 or 3.00×10^5, rewrite 31,556,925 as 3.16×10^7, and write the product as $3.00 \times 10^5 \times 3.16 \times 10^7$ to find 9.48×10^{12} (nearly 10 trillion) km.

TABLE 1.2
Scientific Notation and Powers of Ten

Number Name	Number	Scientific (Exponential) Form	Prefix[a]
trillion	1,000,000,000,000	10^{12}	(tera)
billion	1,000,000,000	10^9	(giga)
million	1,000,000	10^6	mega
thousand	1,000	10^3	kilo
hundred	100	10^2	(hecto)
ten	10	10^1	deca
one	1	10^0	—
one-tenth	1/10	10^{-1}	(deci)
one-hundredth	1/100	10^{-2}	centi
one-thousandth	1/1,000	10^{-3}	milli
one-millionth	1/1,000,000	10^{-6}	micro
one-billionth	1/1,000,000,000	10^{-9}	(nano)

[a]Indicates a multiplying factor. For example, 1,000 meters = 1 kilometer. Prefixes not used in this book are in parentheses.

The nearest of all astronomical bodies is an old friend, figuring in the stories and songs of many cultures. The **Moon** (Figure 1.3) is our lone natural **satellite**, a body that circles, or orbits, the Earth about once a month under the action of *gravity*, the attractive force that holds you to the ground. The Moon's diameter is a quarter that of Earth, its distance from us is 384,000 km (238,000 miles or 30 Earth diameters), and it is the only body in space to which human beings have traveled. With the Earth a basketball, the Moon is roughly the size of a baseball 7 m (6.5 yards) distant.

With the Earth a basketball, the Moon is roughly the size of a baseball 7 m distant, and the Sun is 30 m across and 3 km away.

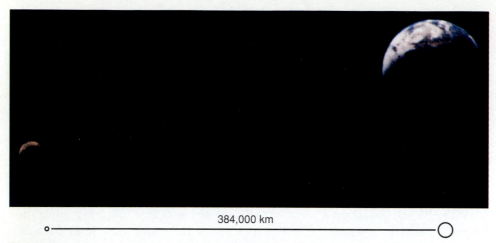

384,000 km

Figure 1.3

The Earth and Moon were photographed together for the first time by *Voyager 1* as it began its great flight in 1977 to the outer reaches of the Solar System. We see only slivers of the daylight sides of the two bodies. Because of the position of the spacecraft, the Earth and Moon seem relatively closer together than they really are. A true scale drawing appears below.

Now proceed to the **Sun** (Figure 1.4), which warms the day and keeps us all alive, and about which the Earth orbits once a year. The Sun, a vast gaseous ball 1.4×10^6 km (109 Earth diameters) across, is the nearest star. It appears small in the sky because it is so far away, 1.5×10^8 km (93 mil-

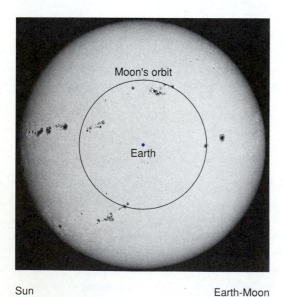

Moon's orbit

Earth

Sun

Earth-Moon system

Figure 1.4

The Sun, a typical star, is 1.4 million km across. The Moon's orbit and the Earth are seen projected against its surface. The solar diameter could hold 109 Earths side by side and almost two lunar orbits. The dark spots on the Sun, some of which dwarf the Earth, are regions of intense magnetism. Across the bottom is a scale drawing of the Sun relative to the Earth's orbit. The entire Earth-Moon system would fit inside the dot at the right, and the Earth itself would be a mere pinpoint.

lion miles), a fundamental distance in astronomy called the **Astronomical Unit** (AU). The AU is roughly 100 times the Sun's diameter (which makes the Sun about 10^{-2} AU across) and about 10,000 times the diameter of Earth. On our Earth-as-a-basketball scale, the Sun would be nearly 30 m (98 feet) wide and 3 km (almost 2 miles) away.

The Moon and Sun have profoundly different physical characteristics. The Moon is made of rock like the Earth and is cold and solid. The Sun is hot; it is mostly hydrogen, the simplest of all chemical elements, and it is gaseous throughout. The Moon, like the Earth, shines by sunlight reflected from its surface. The Sun, however, is self-luminous because of its great internal energy, and it shines as brightly as 4×10^{24}—more than a trillion trillion—standard 100-watt light bulbs. The Earth and the Sun, which were born together, are nearly 5 billion years old, and the Sun will stay alight for over 5 billion years more.

1.2.2 The Solar System

Many celestial bodies (*celestial* means "pertaining to the sky") orbit the Sun, the whole collection making up the **Solar System.** Among them is a family of nine traditional **planets** (Table 1.3 and Figure 1.5), of which the Earth is the third one out. All the planets orbit in the same direction, in

The Sun shines as brightly as 4×10^{24}—more than a trillion trillion—standard 100-watt light bulbs.

TABLE 1.3 The Planets		
Planet	*Average Distance from Sun (AU)*	*Characteristics*
Terrestrial[a]		
Mercury	0.4	Second smallest
Venus	0.7	Brightest; just smaller than Earth
Earth	1.0	Carries life
Mars	1.4	Red color
Jovian[b]		
Jupiter	5.2	Largest; prominent cloud belts; four bright satellites
Saturn	9.5	Surrounded by bright rings
Uranus	19.2	Tipped on its side; nearly featureless
Neptune	30.1	Blue-green clouds
Neither Class		
Pluto	39.4	Smallest

[a]Like Earth.

[b]Like Jupiter.

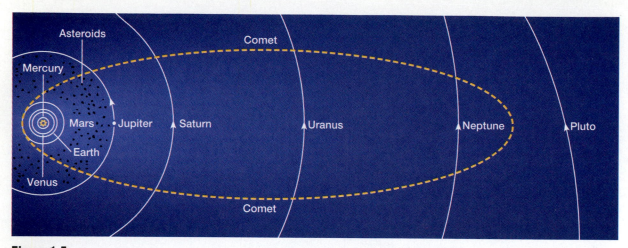

Figure 1.5

The Solar System and portions of the orbital paths of the planets around the Sun are drawn to scale. The Sun, on this scale only 0.04 mm in diameter, is at the center of the orbits. Jupiter is 10 times smaller and the Earth 10 times smaller yet. The dot that represents Jupiter encompasses the diameter of its whole satellite system. Most of the asteroids orbit between Mars and Jupiter. The dashed curve shows the path of a comet; most comets have orbits tilted out of the Solar System's plane.

nearly circular paths, and in nearly the same plane. Six of the nine spin in the same direction. All shine by reflected sunlight. Yet each has its own special characteristics, and each is a different world to explore.

The planets are divided into two main groups. The inner four, all within 1.5 AU of the Sun, are called the **terrestrial planets** because they are constructed of rock like the Earth, the largest of the four. (*Terrestrial*, from *Terra*, the Roman goddess of Earth, means "Earthlike.") The next four, orbiting between 5 and 30 AU from the Sun, are much larger than Earth, are made partly or mostly of hydrogen and helium, and are commonly called the **Jovian planets** after Jupiter, the biggest of them (Figure 1.6). Jupiter, however, is still only 0.001 AU across, and on the scale of Figure 1.5 would be invisible to the eye. The last planet is tiny Pluto. Only 2,400 km across and on average 39 AU from the Sun, it fits into neither of the two categories, a modern view removing it from the class of planets altogether. On the scale of a basketball Earth 3 km out from the Sun, Pluto would be about 110 km (75 miles) away.

The terrestrial planets have few satellites: Mercury and Venus have none, the Earth one, and Mars only two tiny ones. In contrast, the Jovian planets have extensive satellite systems (see Figure 1.6). Jupiter has 16 known satellites, 4 of them easily visible in binoculars. Saturn has at least 18, Uranus 15, and Neptune 8. Pluto has a companion half as big as itself.

Minor bodies throng the Solar System. The **asteroids** are small chunks of rock or metal that orbit the Sun in paths that lie largely, but not exclu-

Figure 1.6
Jupiter, the greatest of the planets, has a diameter 11 times that of Earth. Two of Jupiter's satellites, each about the size of our Moon, can be seen against the background of Jupiter's cloud tops.

Figure 1.7
Comet Bennett passed close to the Sun in 1970 and generated a long tail.

sively, between Mars and Jupiter (see Figure 1.5). Thousands have been catalogued, and we know there are countless more. The largest, Ceres, has a diameter only a quarter that of the Moon, and the smallest are mere pebbles. **Comets,** made of dusty ice, are typically a few kilometers in diameter and orbit the Sun on highly elongated paths. As a comet approaches the warmth of the Sun, the ice turns into gas that streams out in one or more *tails* that can be millions of kilometers long (Figure 1.7). We believe there are trillions of comets in clouds that extend perhaps as far as 100,000 AU from the Sun. Only a tiny fraction ever comes close enough to the Sun for us to see them.

Comets steadily disintegrate under the action of sunlight. As the Earth orbits, it continually collides with pieces of cometary debris and stray as-

Figure 1.8
This iron meteorite is a piece of an asteroid that collided with the Earth.

teroids that heat up in our atmosphere and streak brilliantly across the sky as **meteors.** The cometary dust all burns up in the air, but asteroids more than about a centimeter or so across can survive to strike the ground to become **meteorites** (Figure 1.8).

1.2.3 Stars

In the nighttime sky, the **stars** and planets seem to be at the same distance from Earth, but that is an illusion. The stars are distant suns. Even the nearest of them (called Alpha Centauri) is an astonishing 271,000 AU away, vastly farther than any of the Sun's planets. At such distances, the stars, no matter their size, appear as no more than points of light. In our scale model with a basketball Earth, Alpha Centauri would be 780,000 km (480,000 miles) distant, twice the actual distance to the Moon.

Such large distances require the use of a large unit. The **light-year** (ly) is the distance a ray of light will travel in a year at a speed of 299,792 km per second (km/s) (see Table 1.1). The light-year is 63,300 AU—almost 10^{13} km or 6×10^{12} miles (see MathHelp 1.1). Therefore, Alpha Centauri is 4.3 ly away. The light we see actually left the star 4.3 years ago; as we look out into space, we are looking back into the past. Professional astronomers more commonly use an even larger unit, the parsec (pc), which equals 206,265 AU or 3.26 ly, making Alpha Centauri 1.3 pc away. As great as this distance is, most stars seen with the naked eye are vastly farther from us—hundreds, even thousands, of light-years away.

We find great diversity among stars. Some are much hotter than the Sun, others are considerably cooler. The smallest are no larger than a small

The smallest stars are no larger than a small city; the biggest would nearly fill the orbit of Saturn.

city, and the biggest, if placed at the position of the Sun, would nearly fill the orbit of Saturn. A few stars are bright enough to be seen with the naked eye over distances of thousands of light-years, whereas the Sun would be invisible if only 70 ly away. Some other stars would barely be visible to the naked eye even if they were a mere 0.01 ly distant. The varied properties are caused by differences in the amounts of matter the stars contain—which range from about a tenth that of the Sun to a hundred times as much—and by the aging process. Stellar aging, however, is very slow—life spans are typically measured in millions and billions of years—and as a result the stars appear unchangeable.

1.2.4 Our Galaxy

In the grand design, all stars are arranged in individual **galaxies,** vast distinct collections of matter tied together by gravity (Figure 1.9). Some are huge, others very small. The one we live in, called simply **the Galaxy,** is quite large, with most of its 200 billion stars occupying a volume roughly 80,000 ly or 25,000 pc across. If the Sun were a marble, our Galaxy would be 5 million km in diameter, 13 times the actual distance to the Moon!

> If the Sun were a marble, our Galaxy would be 5 million km in diameter, 13 times the actual distance to the Moon.

We can obtain a good idea of the structure of our Galaxy by looking at others that are thought to be similar. The most important component of galaxies like our own is a flat *disk* that bulges in the middle (Figure 1.9a). The disk of our Galaxy contains the Sun, which is located about two-thirds of the way out from the center. A face-on view of this disk would reveal a set of spiral arms something like those in Figure 1.9b. The spiral arms are where stars are created, new stars replacing others that are dying.

As we look out into the sky from the Earth we see the disk of the Galaxy surrounding our heads in a great diffuse band of light that we call the **Milky Way** (Figure 1.10), which is made of the collected light of billions of stars that individually are too faint to be seen with the naked eye. From the dark countryside, the Milky Way can be spectacular. To under-

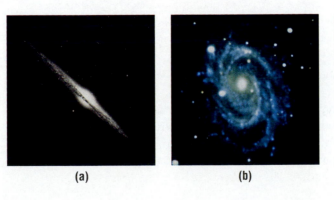

(a)   (b)

Figure 1.9
(a) A galaxy called NGC 4565 is a flat collection of billions of stars about 100,000 ly across. **(b)** A similar system called NGC 4603, seen face-on, displays spiral arms.

Figure 1.10
The Milky Way, the combined light of the stars of the galactic disk, stretches upward from the horizon.

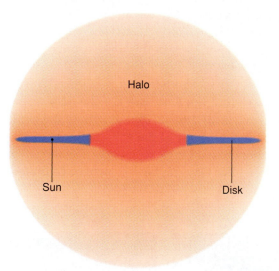

Figure 1.11
The disk of our Galaxy, depicted edge-on, is surrounded by a dim halo of stars.

stand the nature of the Milky Way, sit in a crowded roomful of people. The heads of the people in the room, including yours, collectively form a disk. As you survey the room, you see a band of heads surrounding you, just as the Milky Way surrounds the Earth.

A second component of our Galaxy is a vast spherical *halo* of stars that entirely surrounds and encloses the disk (Figure 1.11). Slice a basketball in two with a flat dinner plate. The plate represents the disk and the ball the halo. The disk is thickly populated with stars of all brightnesses, but those in the halo are relatively few and far between. As a consequence, this component is hard to see in other galaxies like those in Figure 1.9.

1.2.5 Other Galaxies

The nearest neighboring galaxy of a size comparable to our own is the *Andromeda Galaxy* or M 31 (Figure 1.12), 2.3 million ly (700,000 pc) away. This distance is about 30 times the diameter of our Galaxy. With the Sun a marble, M 31 would be 140 million km away, or about as far as the real Sun. Galaxies exhibit a great variety of sizes, shapes, and structures. Our Galaxy and M 31 are both large and have spiral arms. Others, such as those that accompany the Andromeda Galaxy (see Figure 1.12), have smooth elliptical shapes and lack spiral arms. Though most of these elliptical galaxies are small, some dwarf even M 31. Still other galaxies are sprawls of stars with little obvious structure at all.

M 31

The Galaxy

Figure 1.12

The Andromeda Galaxy, also called M 31, is the closest galaxy comparable in size and structure to our own. Though its disk is tilted, we can still make out its spiral arms. It is accompanied by a pair of small elliptical galaxies (arrows) that have no spiral arms. Our Galaxy and M 31 are drawn to scale below. On that scale the separation between the Sun and the nearest star, Alpha Centauri, is only 1/100,000 cm.

Figure 1.13
The central part of a large cluster of galaxies contains over a dozen bright members and many more fainter ones.

Galaxies tend to group in multiples (as seen in Figure 1.12) and clusters (Figure 1.13). Our Galaxy and the Andromeda Galaxy both belong to a sparse cluster with about two dozen members called the *Local Group*. Other clusters contain thousands of galaxies. There are even higher levels of organization, in which clusters team together in complex ways that we are only beginning to understand.

1.3 THE UNIVERSE ITSELF

We can see galaxies and their clusters as far away as our technology will allow, out to distances of billions of light-years. Had we the time, we could count more than 10 billion galaxies, each containing billions of stars. With improved scientific instruments, a trillion or more may be within our grasp.

But its size, immense as it is, is not the Universe's most remarkable property. With the exception of a few galaxies within the Local Group, every galaxy in the Universe is receding from us at a speed directly proportional to its distance; that is, if the distance is doubled, so is the speed. We in fact infer that all clusters of galaxies are moving away from all other clusters of galaxies with speeds that depend on the distances separating them. Our conclusion is that the Universe as a whole is expanding.

Because the speed is increasing in direct proportion to distance, it appears that the matter in the Universe was once enormously compressed and that the Universe began its expansion suddenly in an event called the **Big Bang.** From the speed of recession at a given distance, we can estimate how long it has been since that moment; that is, we can find the age of the

Matter in the Universe was apparently once enormously compressed and began its expansion suddenly in an event called the Big Bang.

Universe, which seems to fall between about 10 and 20 billion years. The Sun, only 5 billion years old, is a relative newcomer.

What were the discoveries and what was the progression of human thought that led us to our understanding of the structures of stars and the constitution of the Universe? What do we need yet to learn? To find out, we return to Earth and begin the story.

▶ KEY CONCEPTS

Asteroids: Rocky or metallic bodies, smaller than planets, that orbit the Sun between Mars and Jupiter.

Astronomical Unit (AU): The average distance between the Earth and the Sun.

Big Bang: The event that began the sudden expansion of the Universe some 10 to 20 billion years ago.

Comets: Small icy bodies that move around the Sun on highly elongated orbits.

Galaxies: Basic units of the Universe in which stars are collected.

Galaxy, The: The collection of stars in which we live.

Light-year (ly): The distance a ray of light travels in a year at 299,792 km/s.

Meteorites: Pieces of rock or metal from space that hit the surface of the Earth.

Meteors: Streaks of light in the sky caused by rocks and dust from space that heat up in the Earth's atmosphere.

Metric system: A system of measures in which the units differ by multiples of 10.

Milky Way: The band of light around the sky caused by the billions of stars in the disk of our Galaxy.

Moon: The Earth's satellite, which orbits the Earth under the force of gravity.

Planets: The larger bodies of the Solar System orbiting the Sun. The **terrestrial planets** are those like (and including) the Earth; the **Jovian planets** are those like (and including) Jupiter.

Satellite: A body that orbits a planet.

Scientific notation: A means of expressing numbers by using powers of 10.

Solar System: The Sun and its collection of orbiting bodies.

Stars: Gaseous, self-luminous bodies similar in nature to the Sun but with diverse properties.

Sun: The gaseous, self-luminous body that dominates the Solar System; the nearest star.

Universe: The all-encompassing structure that contains everything.

▶ KEY RELATIONSHIPS

$1 \text{ AU} = 1.5 \times 10^8 \text{ km} = 9.3 \times 10^7 \text{ miles}$
$1 \text{ ly} = 63,300 \text{ AU}$
$1 \text{ pc} = 3.26 \text{ ly} = 206,265 \text{ AU}$

▶ EXERCISES

Comparisons

1. What are the differences between terrestrial and Jovian planets?
2. What are the differences between the disk and the halo of our Galaxy?
3. What are the differences between comets and asteroids?
4. What is the difference between a meteor and a meteorite?
5. How does the Sun differ from a planet?
6. Compare the size of the Sun with the sizes of other stars.

Numerical Problems

7. Write the numbers 100, 1,000,000, 126, 0.335, and 0.000036 in scientific notation.

8. Write 22×10^2, 3.62×10^6, 2.34×10^{-3}, and 6.8456×10^{-5} in normal numerical form.

9. What are **(a)** $(3 \times 10^2) \times (2 \times 10^5)$; **(b)** $(6 \times 10^{-2}) \times (3 \times 10^{-7})$; **(c)** $(1.5 \times 10^5) \times (1.4 \times 10 \times 10^{-17})$?

10. If the Earth and Sun were separated by 100 meters, **(a)** what would be the diameters of the Earth and Sun in centimeters? **(b)** How far would the Moon be from the Earth in centimeters? **(c)** How big would the Solar System be in meters (out to the orbit of the most distant planet)? **(d)** How far away would the nearest star be in kilometers?

11. The star Vega is 24 light-years away. How far is that in parsecs, astronomical units, and meters?

12. Roughly how far is the Sun from the center of our Galaxy in parsecs and light-years?

13. **(a)** How many of our Suns could you line up from the Earth to Alpha Centauri? **(b)** How many of our own Galaxies could you line up from the Earth to the nearest large spiral galaxy, M 31?

Thought and Discussion

14. Describe the relationships of planets, comets, and asteroids to the Sun.

15. How does the structure of the Galaxy give rise to the Milky Way?

Research Problem

16. What countries still use distance units like the inch, foot, and mile in daily life? Look in old astronomy books to see when the metric system began to be used as the astronomical standard.

Activity

17. Using ordinary household items, make scale models of the Solar System and the Galaxy.

Scientific Writing

18. Write a letter to a grade school student in which you explain and demonstrate how far even the nearest star is from the Earth.

19. Describe in a single typewritten page how the Earth fits into the Solar System, how the Solar System fits into the Galaxy, how the Galaxy fits into the Local Group, and how the Local Group fits into the Universe as a whole.

EARTH, SUN, AND SKY

The rotation and revolution of the Earth and apparent motions in the sky

2

As you stand outdoors on a sunny day, the sky appears as a hemispherical blue bowl inverted over your head. The countryside stretches outward to the **horizon,** where land seems to meet the sky. You watch the Sun move slowly toward the west and eventually drop below the horizon, or *set*. When the sky darkens and the stars come out, they too are seen to be moving steadily, some *rising* over the eastern horizon, others setting below the western. Gradually the sky becomes light, the stars fade, and the Sun reappears in the east. The performance repeats itself day after day. The Earth looks flat, and it appears that the Sun, the stars, and the whole sky are circling around you. The truth is very different.

2.1 SHAPE, SIZE, AND SCIENCE

The concept that the Earth is a sphere can be traced back 2,500 years to the Greek mathematician Pythagoras. But how do we really *know* the Earth is round? The scientist does not accept authority but looks at evidence, endeavoring to see what **theory,** that is, what logical or mathematical description of nature, best fits and explains the observations. A theory allows the scientist to make predictions about how nature might behave and suggests new observations. The experimentalist or observer examines these predictions to see if they are true. If they are, the theory stands; if not, it falls and must be modified, or a new theory must be developed to accommodate the fresh data. Although science commonly does not work in such a methodical fashion—many discoveries are quite accidental—this *scientific method* provides a foundation on which we can build.

Examine the evidence for the Earth's shape. The sharp horizon at first suggests a flat Earth with an edge. But as you walk toward the horizon, the "edge" recedes ahead of you, implying that the Earth's surface is curved. Look also at a ship sailing out to sea (Figure 2.1a). It does not stay permanently visible, nor does it drop instantly away. First the hull disap-

The concept that the Earth is a sphere can be traced back 2,500 years to the Greek mathematician Pythagoras.

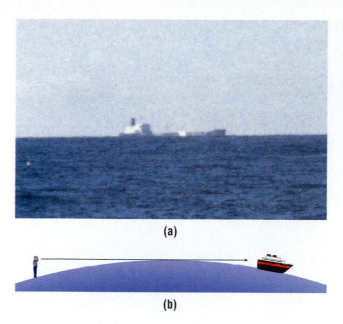

(a)

(b)

Figure 2.1
(a) A photograph of a ship at sea shows it "hull down," its lower parts below the horizon. **(b)** The drawing shows how the curved Earth hides the lower hull.

pears, then the masts. When you see only the top of the ship, you know that the bottom must be still resting on an ocean that is below your line of sight (Figure 2.1b).

Additional evidence is found in the sky. As long ago as the fourth century B.C., the Greek philosopher Aristotle discussed how the stars and the Sun change their positions as you move on Earth. As you walk north, the stars in front of you climb higher above the horizon and those to your back drop toward it, an effect difficult to explain unless either the Earth is curved or the stars are very close.

Still, these phenomena do not demonstrate that our home is a complete sphere, closing back upon itself. Aristotle pointed out more telling evidence. The orbiting Moon is eclipsed when it passes through the shadow the Earth casts into space. The shadow always has a circular outline (Figure 2.2), no matter where in the sky the Moon appears. The only solid figure whose outline is always a circle is a sphere. Confirming evidence is furnished by circumnavigation and by spaceflight, which allows us to see our planet from a distance. From any point of view in space the Earth has a circular outline, again demonstrating sphericity.

No scientist, therefore, doubts the accuracy of the theory that the Earth is at least approximately spherical, as the combination of evidence is overwhelming. We will encounter other theories, however, that rest on far less evidence, and then we must remember that no theory can actually be proved: it can only not be disproved. One contradictory observation shows a theory to be incorrect and in need of revision or even replacement.

Figure 2.2
The shadow of the Earth cast upon the Moon during an eclipse (seen here in a multiple exposure) is always circular.

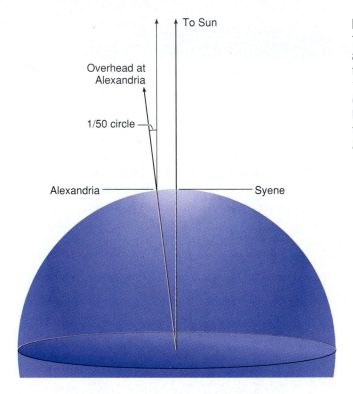

Figure 2.3
The overhead point for an observer is always on a line directed outward from the Earth's center. Alexandria is 1/50 of the way around the Earth (7°) from Syene. If the Sun is overhead at Syene, it will be 1/50 of a full circle away from overhead at Alexandria.

Once the ancient astronomers realized the Earth's shape, they could measure its size. In the third century B.C., Eratosthenes of Cyrene knew that in the town of Syene in southern Egypt, the Sun shone directly down a well at noon on the first day of summer, and therefore it had to be directly overhead. However, to the north, in Alexandria, the Sun was 1/50 of a whole circle south of the overhead point at noon (Figure 2.3). If the Earth is a sphere (and the Sun very far away), the distance between Syene and Alexandria must be 1/50 of the globe's circumference.

From the time it took the king's messenger to run between the two towns, Eratosthenes estimated the distance at 5,000 *stadia*. The Earth's cir-

Eratosthenes' measurement of the Earth's circumference was off by only 15%.

cumference must then be $50 \times 5{,}000$, or 250,000 stadia. The length of the Greek stadium (the singular of stadia) is about 185 m, so Eratosthenes' result corresponds to a circumference of 46,000 km. Modern measurement with the same technique gives 40,000 km (25,000 miles). Eratosthenes was off by only 15%.

2.2 ROTATION

Aristotle, whose opinions swayed human thought until the seventeenth century, argued forcefully that the Earth was too heavy to move. Yet it was clear to others of his time that the daily movement of the Sun, Moon, and stars is more easily explained by having the Earth rotate daily from west to east, in the direction opposite the motion seen in the sky. Stand in the middle of a room and begin to turn counterclockwise. *You* are spinning, but the room and its furnishings seem to go in the opposite direction, clockwise. The chairs and tables first "rise" into your view from the left and then disappear or "set" to the right when you turn away. The Earth is so large and the motion so smooth that you just do not feel it. You have to look outside the Earth to see that the movement is there at all.

The Earth (Figure 2.4) rotates about an imaginary **axis** that runs through its center and exits through its **poles.** If you place two fingers on

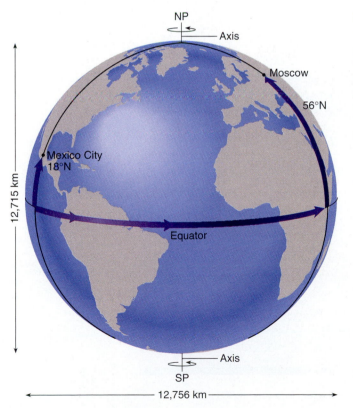

Figure 2.4

The Earth spins around an axis that projects outward from the north and south poles (NP and SP). The equator is a great circle equidistant from both poles. Latitude is measured north and south of the equator. That of Mexico City is 18°N, that of Moscow 56°N. The Earth's equatorial diameter of 12,756 km is 41 km greater than the polar diameter because of the effect of rotation.

the opposite sides of a basketball and spin it, they become the poles of rotation. We discriminate between the two poles by calling one north and the other south. The **equator** of the Earth is a circle that is everywhere equidistant between the poles. The equator is a special kind of circle called a **great circle,** one whose center is the same as the center of the sphere and that divides the sphere into two equal hemispheres. The Earth spins counterclockwise from the point of view of an observer looking down from above the north pole. If you were standing there you would spin to your left, a motion that makes the stars appear to move to your right.

Since the equator is farther from the axis than any place else, a person there must move the fastest, 1,670 km/h (1,030 miles/h), to make the full circuit in a day. No one is aware of this great velocity because everything there, including the air, is moving at the same speed. The great speed at the equator causes it to bulge outward, slightly distorting the Earth from a sphere into a flattened *oblate spheroid*. The Earth is actually about 40 km larger along its equatorial diameter than along the polar.

Distance north and south of the Earth's equator is measured by an angle called **latitude** (ϕ) in degrees, minutes, and seconds of arc (see MathHelp 2.1; also, note the Greek alphabet given in Table 2.1). Select a place in Figure 2.4, perhaps Mexico City. Draw a great circle through it and the poles. Then measure the arc from the equator to Mexico City along this circle to find a latitude of 18°N. The latitude of the equator is 0° and the latitudes of the north and south poles are respectively 90°N and 90°S.

TABLE 2.1
The Greek Alphabet[a]

Letter	Upper Case	Lower Case	Letter	Upper Case	Lower Case
Alpha	A	α	Nu	N	ν
Beta	B	β	Xi	Ξ	ξ
Gamma	Γ	γ	Omicron	O	o
Delta	Δ	δ	Pi	Π	π
Epsilon	E	ε	Rho	P	ρ
Zeta	Z	ζ	Sigma	Σ	σ
Eta	H	η	Tau	T	τ
Theta	Θ	θ	Upsilon	Y	υ
Iota	I	ι	Phi	Φ	ϕ
Kappa	K	κ	Chi	X	χ
Lambda	Λ	λ	Psi	Ψ	ψ
Mu	M	μ	Omega	Ω	ω

[a]Greek letters are commonly employed to denote scientific quantities and are used in the naming of stars.

MATHHELP 2.1 Angles

Distances or separations are commonly measured in terms of some metric unit or, in astronomy, in light-years or parsecs. However, astronomers also specify separations between two bodies in terms of *angles*. An angle is measured from the point of intersection of two lines in degrees (°) from 0° to 360° (Figure 2.5a). Each degree is subdivided into 60 parts called minutes ($1° = 60'$) and each minute into 60 seconds ($1' = 60''$). (These terms must not be confused with minutes and seconds of time measurement, which are specified by m and s.)

The angle can also be specified by the arc of the surrounding circle isolated by the two lines. The length of the arc in degrees is the fraction of the circle isolated by the lines times 360°. Arc and angular measurement are numerically identical. Although the physical length of the arc (in centimeters, meters, or some other unit) depends on the radius of the circle, the length of the arc in degrees does not.

The **angular diameter** (α) of an astronomical body is the angle formed by lines that project from the eye to either side of the body (Figure 2.5b). The body is then said to *subtend* that angle or arc. At a given distance, the larger the physical size or diameter of the body (in meters or kilometers), the larger its angular diameter. For a given physical size, the greater the distance the *smaller* the angular diameter. If we know the distance of a body and measure its angular diameter, we can find its physical diameter from the laws of trigonometry. (Or if we know the angle between two equidistant points of known distance, we can find their physical separation.) Conversely, if we measure the angular diameter and can infer the physical size, we can calculate the distance.

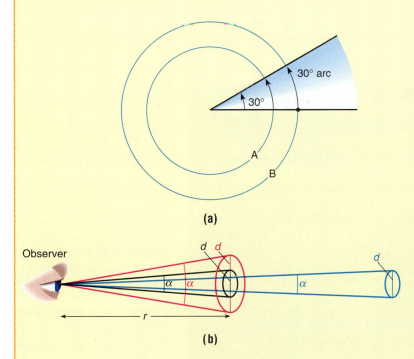

(a)

(b)

Figure 2.5

(a) Angles are formed at the intersection of two lines and are measured in degrees; a whole circle is 360°. In the example shown, the lines come together at 30°. The lines intercept arcs on circles that are drawn around the point of intersection. The arcs have the same length in degrees as the angle at the center, independent of the circle's size. **(b)** A body of physical dimension *d* is seen to subtend an angle α at distance *r* (black); that is, the body is seen to have an angular diameter of α. If *r* stays the same and *d* increases (red), the angular diameter increases (here they both double). If *d* stays the same and *r* increases (blue), the angular diameter decreases (here *r* is doubled and α is halved).

2.3 THE CELESTIAL SPHERE

Now turn your eyes to the sky. Imagine the Moon, Sun, planets, and stars to be affixed to a huge **celestial sphere** with the Earth at its center (Figure 2.6). Assume you are observing from the northern United States, at a latitude of 45°N, maybe from Minneapolis, Minnesota, or Portland, Oregon. Because gravity pulls you downward toward the center of the Earth, wherever you are you feel that you are standing at the top of our planet. Since you are actually standing halfway between the north pole and the equator, the Earth's axis must be drawn tipped by the same 45°. The point on the celestial sphere directly above your head is the **zenith.** Extend a plane at your feet outward in all directions perpendicular to the line to the zenith. This plane will intersect the celestial sphere at the **astronomical horizon,** a great circle that divides the sky into visible and invisible hemispheres and is approximated by the visible horizon at sea, where there are no foreground obstructions.

Next extend the Earth's rotational axis outward through the north and south poles. It will pierce the celestial sphere at the **north celestial pole** (NCP) and the **south celestial pole** (SCP). Then extend the plane of the Earth's equator outward into space. The plane will slice through the celestial sphere at the **celestial equator,** another great circle that is everywhere 90° from the NCP and SCP and that divides the heavens into equal northern and southern hemispheres centered on the NCP and SCP. Since the NCP is above the horizon (at your latitude of 45°N), it is visible, but the SCP, opposite the NCP and below the horizon, is not. However, you can still see a considerable portion of the southern celestial hemisphere.

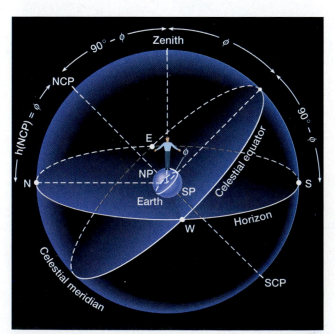

Figure 2.6
The celestial sphere surrounds the Earth. NP and SP are the north and south poles of the Earth. The zenith is over the observer's head 90° above the horizon. The axis of the Earth intersects the sphere at the north and south celestial poles (NCP and SCP). The celestial meridian runs through the poles and the zenith, and the celestial equator is parallel to the Earth's equator. North (N), south (S), east (E), and west (W) are defined by the intersections of the three circles. The observer is at 45°N latitude (ϕ), the latitude always equal to the altitude of the pole above the horizon.

Finally, your **celestial meridian** is defined as the great circle that passes through the NCP, SCP, and the zenith, splitting the sky into its eastern and western hemispheres. These celestial circles allow us to find our way on Earth: the intersections between the horizon and the celestial meridian define the directions north and south and those between the horizon and the celestial equator east and west.

Locate the celestial pole and you know your latitude.

We can even use the sky to find where we are. In Figure 2.6 we draw two lines from the center of the Earth. One passes through the observer to the zenith. The other goes through the Earth's equator to the celestial equator at the point where it crosses the celestial meridian. The two lines intersect at an angle ϕ equal to the observer's latitude and cut off an arc on the meridian also equal to ϕ. Since the celestial equator is 90° from the NCP, the arc from the zenith to the NCP must be (90° − ϕ). However, the zenith is 90° from the horizon, so the arc from the horizon to the NCP is equal to 90° − (90° − ϕ), or just the latitude, ϕ. An arc in the sky measured upward from (and perpendicular to) the horizon is called an **altitude** (*h*). Therefore, the altitude of the pole is equal to the latitude of the observer, or $h(\text{NCP}) = \phi$. *Locate the celestial pole and you know your latitude.*

A modestly bright star called Polaris, or the North Star, lies close to the NCP (Figure 2.7). To find it, look up your latitude in an atlas, then face

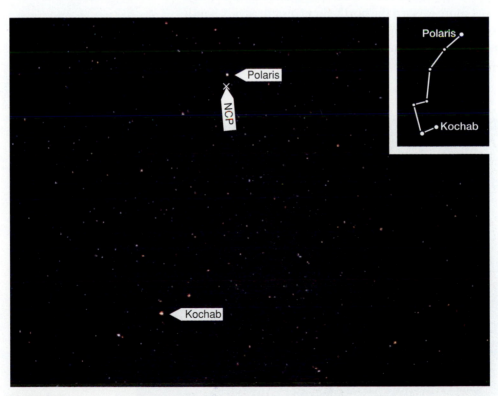

Figure 2.7
The star Polaris, at the end of the Little Dipper's handle, lies close to the north celestial pole, indicated by an "X."

north and look upward from the horizon by that many degrees (if you live near 45°N, look halfway up, if near 30°N, a third of the way). Once you have the star, point one arm toward it, position the other 90° away, and trace out the celestial equator. Then point north and trace out the circle through the NCP and the zenith to see your celestial meridian.

2.4 MOTIONS IN THE SKY

As the Earth turns counterclockwise, or eastward (as viewed from above the north pole), the stars seem to move oppositely, clockwise, or to the west, tracing out **daily paths** (Figure 2.8) that are parallel to the celestial equator. Figure 2.9 shows the sky and several stars at different locations north and south of the celestial equator. Because the apparent motions of stars are parallel to the celestial equator, they will stay at the same angles north or south of it as they track their daily paths.

A star exactly on the celestial equator (see Figure 2.9) will rise precisely at the east point, cross (or *transit*) the meridian at an altitude of $90° - \phi$, and set exactly in the west. If a star is north of the equator but not too close to the celestial pole, like Arcturus, it will rise in the northeast (that part of the horizon between north and east), transit higher, and set in the northwest. Stars yet farther north of the celestial equator (look at Vega in Figure 2.9) rise and set more toward the north. Stars progressively south of the celestial equator (for example Spica and Antares) transit ever lower and rise and set more and more in the southeast and southwest. Go outside and watch.

Figure 2.8
The photographer set the camera and allowed the stars to trail along their daily paths as the Earth rotated.

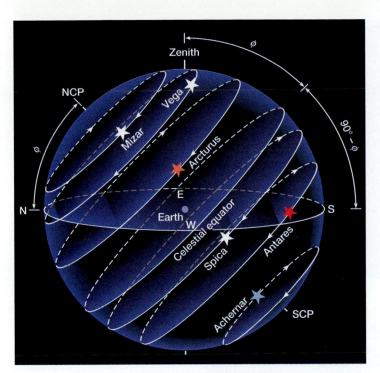

Figure 2.9
Daily paths of stars are parallel to the celestial equator, and the motions are always to the west. A star on the celestial equator will rise exactly in the east and set exactly in the west. Arcturus and Vega, progressively farther north of the celestial equator, transit the meridian higher in the sky and rise and set progressively closer to the north point, whereas Spica and Antares, south of the celestial equator, transit lower and rise and set more toward the south point. For the observer at 45°N latitude, Mizar is so far north that it cannot reach the horizon and is circumpolar; Achernar is so far south that it cannot come above the horizon and remains invisible.

Stars sufficiently north of the celestial equator transit so close to the NCP that they cannot reach the horizon, and like Mizar in Figure 2.9 cannot set (Figure 2.10). Such stars are called **circumpolar**, and in middle northern latitudes include many famous sky figures such as the Big and Little Dippers. If not for daylight, these stars would be perpetually visible as they go around the pole. Conversely, for a northern observer there is also a zone around the SCP in which the stars (like Achernar in Figure 2.9) never rise and remain perpetually unseen.

Now from 45°N (Figure 2.11a) head toward the north pole, following the sequence of Figures 2.11a through 2.11c. The NCP climbs the sky and daily paths begin to flatten out, as exemplified by those of Castor and Sirius. At 70°N latitude (Figure 2.11b), we see many more circumpolar stars and have lost more within the southern celestial hemisphere. When we stand at the north pole (Figure 2.11c), the NCP is in the zenith and the celestial equator falls on the horizon. The entire northern hemisphere of the sky is circumpolar and none of the southern is visible. Since daily paths are parallel to the equator, they are also parallel to the horizon. All stars move perpetually to the right, none ever rising or setting.

Next, travel south from 45°N by following the sequence of Figures 2.11a, 2.11d, 2.11e, and 2.11f. When you arrive at the Earth's equator (Figure 2.11d), latitude 0°, the NCP must lie on the horizon and have an altitude of 0°. If the NCP is on the northern horizon, the SCP must be on the southern. Here there are no circumpolar stars. Everything rises and sets

At the north pole all stars move perpetually to the right, none ever rising or setting.

(a)

(b)

Figure 2.10

Perpetually visible circumpolar stars tread their daily paths around the celestial pole, which is at the point of zero rotation. These pictures show the SCP rather than the NCP, but the effect is the same. The upper picture **(a)** was taken from Botswana in Africa at a latitude of 19°S, the lower **(b)** from an observatory in Australia at latitude 31°S. As you move toward either of the Earth's rotation poles, the corresponding celestial pole rises in the sky (the altitude of the pole equals the latitude of the observer), and you see more circumpolar stars.

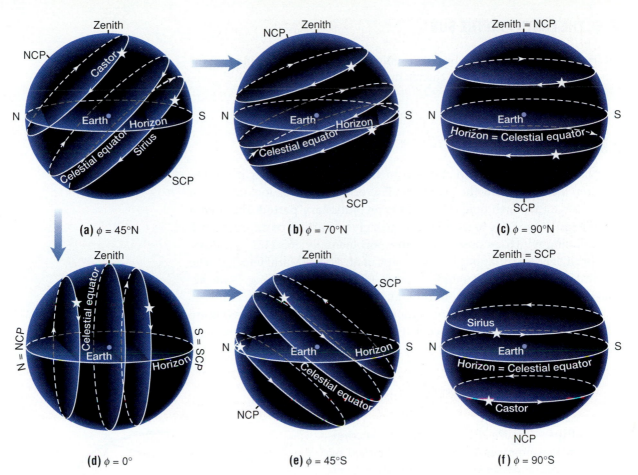

(a) $\phi = 45°N$ **(b)** $\phi = 70°N$ **(c)** $\phi = 90°N$

(d) $\phi = 0°$ **(e)** $\phi = 45°S$ **(f)** $\phi = 90°S$

Figure 2.11
In the sequence **(a)**, **(b)**, **(c)** you walk from 45°N to the north pole and watch the daily paths of Castor and Sirius flatten out and their rising and setting points move more toward the north and south points. At the pole, the NCP is in the zenith, the equator is on the horizon, the daily paths are parallel to the ground, and everything in the northern hemisphere is circumpolar. In the sequence **(a)**, **(d)**, **(e)**, **(f)** you travel past the equator to the south pole and watch the daily paths first become perpendicular to the horizon and then tilt the other way. At the south pole, the SCP is in your zenith, all the southern stars are circumpolar, and they now move to your left rather than to your right.

perpendicular to the horizon, and over the course of the day, everything (if not for daylight) would be visible.

As you step across the equator, the NCP disappears below the horizon and the SCP rises in the sky (Figure 2.11e). Now your southern latitude is given by the altitude of the SCP, some southern stars are circumpolar (see Figure 2.10), and some northern ones become perpetually invisible. When you finally arrive at the Earth's south pole (Figure 2.11f), it is the SCP that is in your zenith. The celestial equator again lies on the horizon, all the stars in the southern celestial hemisphere are circumpolar, and none of the northern hemisphere can be seen.

2.5 THE EARTH AND THE SUN

Of importance equal to the Earth's rotation is the annual movement of the Earth around the Sun, which causes the changes of the seasons and gives great variety to human life.

2.5.1 Annual Revolution

The Earth revolves around the Sun in the same direction as it rotates, counterclockwise from the point of view of an observer looking down from above the north pole. The Earth's revolutionary **period** (the time it takes to go around the Sun and return to a given starting position) is a full year, during which an observer counts $365\frac{1}{4}$ (actually 365.2422 . . .) days. The orbit is not circular but has the shape of an *ellipse* (MathHelp 2.2). The semimajor axis of Earth's elliptical orbit, 149.6 million km long, is the actual definition of the Astronomical Unit.

   The Sun is not located at the center of the ellipse but at a focus (Figure 2.13), and as a result, the distance between the Earth and Sun continuously

---

MATHHELP 2.2   The Ellipse

The **ellipse** (Figure 2.12), commonly encountered in astronomy, is a closed curve defined by two *foci* (plural of *focus*). Along the path of an ellipse the sum of the distances to the foci is a constant. You can easily draw an ellipse by placing two tacks on a board and tying a loose string between them. Pull the string taut with a pencil and trace out the curve. The line through the foci is called the *major axis*, and that perpendicular to it at the center is the *minor axis*. The size of the ellipse is characterized by the length of half the major axis, the **semimajor axis** (a). The shape of the ellipse, or the **eccentricity** (e), is the distance from the center to one of the foci divided by the length of the semimajor axis ($e = CF_1/a = CF_2/a$). If the two foci are brought together at the center, $e = 0$, and the ellipse becomes a *circle*. If they are separated while the semiminor axis remains the same, the curve becomes more and more elongated and e approaches (but never equals) 1.

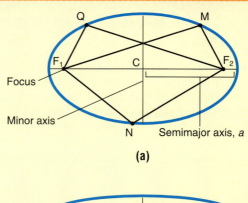

(a)

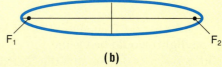

(b)

Figure 2.12
(a) In any ellipse, the sum of the distances of any points along the curve to each of two focus points (F_1 and F_2) is a constant: $MF_1 + MF_2 = QF_1 + QF_2 = NF_1 + NF_2$ and so on. The semimajor axis (a) is half the major axis. The eccentricity $e = CF_2/a = 0.84$.
(b) In this flatter ellipse, $e = 0.95$.

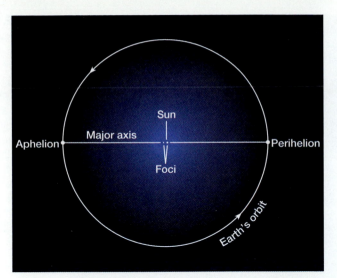

Figure 2.13
The elliptical orbit of the Earth and the Sun are drawn to scale and the Sun placed at the right-hand focus. The center of the orbit is between the two foci. The Earth is closest to the Sun at perihelion and farthest at aphelion. (The dot representing the Sun is drawn to the correct scale.)

changes. However, because the eccentricity is only 0.017, the Sun is just 2.5 million km from the center of the ellipse and the variation in distance is small. When the Earth and Sun are closest, at a point in the orbit called **perihelion** (from the Greek *peri*, "closest," and *helios*, "Sun"), the solar distance is 147.1 million km; when they are farthest apart, at **aphelion**, the distance is 152.1 million km.

The Earth's average orbital speed is 29.8 km/s, but we are unaware of the motion because everything on Earth is moving together. Instead, it *looks* as if the Sun is orbiting the Earth counterclockwise—to the east—across a background of stars (invisible in daylight) at an average rate of 360°/365 days, or just under 1° per day (Figure 2.14). If you walk around a chair placed in the middle of the room, the chair will appear to change its position in relation to the walls and will seem to be revolving around you. The apparent annual path of the Sun is a great circle on the celestial sphere called the **ecliptic.** The ecliptic passes through 12 ancient **constellations** (named patterns formed by stars) collectively called the **zodiac** (Table 2.2).

Which stars we see depends on the direction we face at night, roughly opposite the direction of the Sun. In late June, the Sun is in Gemini, and that constellation is invisible because of daylight. When we rotate away from the Sun, we see Sagittarius at night. As the Earth moves in its orbit and the Sun progresses along the ecliptic, we continuously face in a different direction. As a result, the appearance of the nighttime sky changes with the time of year. In December, the Sun is in Sagittarius, and Gemini appears high in the sky at midnight.

Imagine a star on the celestial meridian at midnight. By tomorrow night the Earth will have moved 1/365 of its orbit, or a little less than 1°. At midnight, you will be facing in a slightly different direction, and you will see the same star about 1° to the west. The following night at midnight the

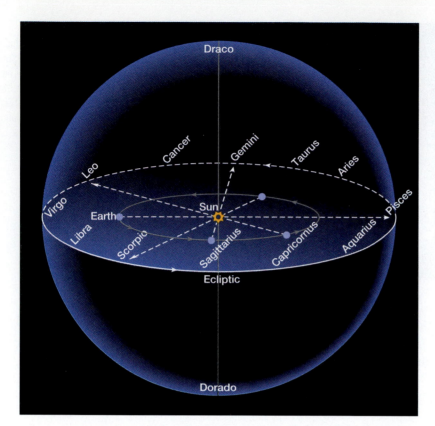

Figure 2.14
The Earth's orbit is set into the celestial sphere. As the Earth moves counterclockwise around the Sun (symbolized by ✿), the Sun appears to move counterclockwise around the Earth at the rate of about 1°/day, following the ecliptic, a great circle through the stars that passes through the constellations of the zodiac. At night, when you are facing away from the Sun, you see the stars in the opposite direction. The directions that are perpendicular to the ecliptic plane point to the constellations Draco and Dorado.

TABLE 2.2
Solar Passage Through the Constellations of the Zodiac

Constellation	Dates of Current Solar Passage
Aries	April 19 — May 14
Taurus	May 15 — June 20
Gemini	June 21 — July 20
Cancer	July 21 — August 10
Leo	August 11 — September 16
Virgo	September 17 — October 30
Libra	November 1 — November 24
Scorpius[a]	November 25 — December 17
Sagittarius	December 18 — January 19
Capricornus	January 20 — February 16
Aquarius	February 17 — March 11
Pisces	March 12 — April 18

[a]Scorpius is combined here with the nonzodiacal constellation Ophiuchus, through which the Sun passes between November 30 and December 17.

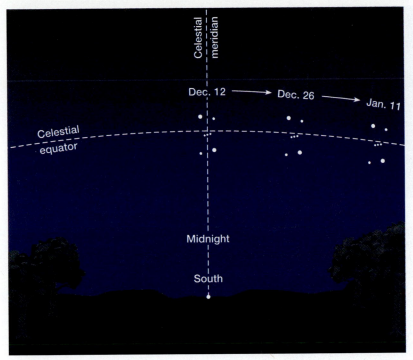

Figure 2.15
The constellation Orion is seen on the celestial meridian to the south at midnight on December 12. The Earth orbits 1° around the Sun each day. As a result, on each successive midnight Orion is seen to shift 1° to the west. In two weeks (December 26) it will have shifted west by 14° and in a month (January 12) by 31°.

star will be another degree farther yet. In a month it will have moved 30° (Figure 2.15), in 3 months 90°, and in a full year, 360°. Since Earth turns through 360° in 24 hours, it spins through 1° in 4 minutes. You must therefore look some 4 minutes progressively earlier each night to see the stars in the same position.

2.5.2 The Seasons

Opinion polls show that the majority of people believe the change of seasons to be caused by the varying distance between the Earth and Sun as the Earth moves on its elliptical orbit. However, the Earth passes perihelion about January 2, in the dead of northern hemisphere winter, and aphelion about July 4 during summer's heat. Moreover, if the varying solar distance were the cause of seasonal change, both the northern and southern hemispheres of the Earth would suffer summer or winter at the same time, but in fact they alternate: when it is summer in the north, it is winter in the south.

The only cause of seasonal change is the 23.5° tilt of the Earth's axis.

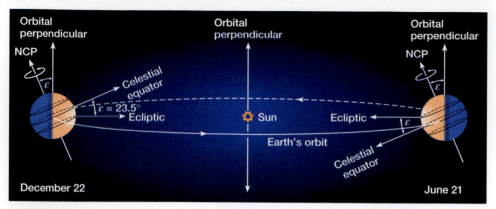

Figure 2.16
The Earth's rotational axis is tilted by $\varepsilon = 23.5°$ relative to the perpendicular to its orbit. The celestial equator is therefore tilted by 23.5° relative to the ecliptic plane. On the first day of northern winter on December 22 (shown on the left), the Sun (✿) appears 23.5° south of the celestial equator, low in the sky from the Earth's northern hemisphere and high from the southern. Six months later (June 21, the first day of northern summer, shown on the right), the Earth has moved through half its orbit. The axis maintains the same orientation in space, the Sun appears 23.5° north of the celestial equator, and appears high in the sky from the Earth's northern hemisphere and low from the southern.

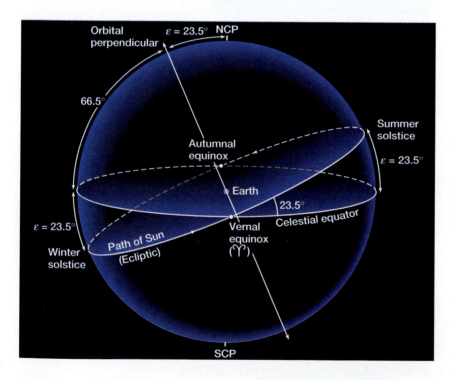

Figure 2.17
The apparent path of the Sun, the ecliptic, is inclined by $\varepsilon = 23.5°$ to the celestial equator. From the northern hemisphere, the Sun appears to move counterclockwise, crossing the ecliptic at the equinoxes, the vernal where it goes north and the autumnal where it passes south. Its farthest point north of the equator is at the summer solstice, 23.5° north of the celestial equator, and the farthest point south is at the winter solstice, 23.5° south of the celestial equator.

The only cause of seasonal change is the 23.5° (23°27′) tilt of the Earth's axis relative to the perpendicular to its orbit (Figure 2.16). The tilt causes the ecliptic to be inclined to the celestial equator by the same angle (Figure 2.17), which is therefore called the **obliquity of the ecliptic** (ε). In

Figure 2.16, the Earth's rotational axis maintains a constant direction as the Earth orbits the Sun. As a result, the Sun can be seen as far as 23.5° below the celestial equator (as on the left in Figure 2.16) or 23.5° above it (as on the right). As the Sun glides along its apparent path it must move alternately north of the celestial equator and then south of it, crossing it twice a year.

The Sun crosses the celestial equator (Figure 2.17) on its way north on March 20 or 21 at the **vernal equinox** (symbolized by ♈, the zodiacal symbol for the constellation Aries) to define the first day of northern-hemisphere *spring*. The vernal equinox is located in the direction of the constellation Pisces. When the Sun is at the vernal equinox, its daily path is along the celestial equator, and it must rise and set exactly east and west (Figure 2.18). Days and nights are then approximately equal, each (ignoring twilight) about 12 hours long.

As the Sun moves along the ecliptic and farther north of the celestial equator, it transits the meridian higher and higher in the sky, rises and sets progressively farther into the northeast and northwest, and is above the horizon for a longer and longer period of time. About June 21, the Sun reaches its extreme position 23.5° north of the celestial equator at a point (in Gemini) called the **summer solstice,** which marks the beginning of northern-hemisphere *summer* (see the right-hand Earth in Figure 2.16). The Sun now transits the celestial meridian as high (and rises and sets as far north) as possible, as seen from the diagram of the daily path of the summer solstice in Figure 2.18. Daylight hours are now at a maximum.

From the summer solstice, the Sun must move to the south. It crosses the celestial equator again at the **autumnal equinox,** in Virgo, about Sep-

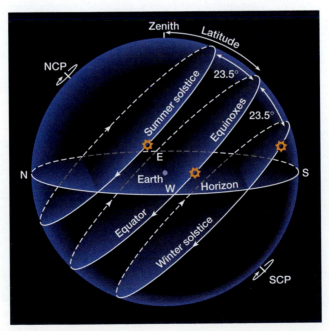

Figure 2.18
The daily paths of the two equinoxes must be the celestial equator, so on March 20 and September 23 the Sun (✿) rises and sets exactly east and west. The daily paths of the solstices are parallel to and 23.5° north and south of the equator. On June 21, the Sun must rise and set well north of east and west. On December 21, the rising and setting points are to the south of east and west.

BACKGROUND 2.1 Ancient Monuments to the Sun

On the Salisbury Plain of southern England stands an extraordinary partly ruined structure called *Stonehenge* (Figure 2.19a). Built between 3000 and 1000 B.C., its chief characteristic is a ring of vertical stones about 3 m high originally topped all the way around by capstones. Inside the ring were five huge stones joined at the top by more capstones. Outside the ring to the northeast is a single, smaller stone called the heelstone. As seen from the center of the ring, the summer solstice Sun rises exactly above this rock, demonstrating that Stonehenge was built at least in part as some sort of calendrical device, marking the day summer begins. Other lines of sight involve the Moon.

Stonehenge is only one of many such constructions scattered about the world. The Incas of Peru established astronomical lines of sight

(a)

Figure 2.19

(a) Stonehenge, in southern England, was used as a calendar; alignments of its stones mark the rising of the Sun on Midsummer Day, June 21. **(b)** The Big Horn Medicine Wheel, constructed by Native Americans in Wyoming, is also astronomically aligned.

(b)

Background 2.1 continued

near Cuzco, and the Mayans of Mexico built an observatory a thousand years ago that contained lines of sight for the rising of the planet Venus and that allowed their priests to place the position of the rising equinox Sun. Closer to home is the 400-year-old Big Horn Medicine Wheel in the Rocky Mountains of Wyoming (Figure 2.19b), built by Native Americans. This construction directs the observer to the rising and setting of the summer solstice Sun and to the risings of several stars.

Clearly, it was important to people in earlier cultures to know the sky. Positions of celestial objects were used to tell the dates on which planting and harvesting should take place. The sky served as a clock, as a calendar, and as you will see in the next chapter, as a storyboard. The knowledge and the engineering capability of these and many other ancient cultures were remarkable indeed. As impressive as our scientific feats are today, we must remember that they are firmly rooted deeply in the past.

tember 23, marking the beginning of northern-hemisphere *autumn*. Once again it rides the celestial equator and rises and sets due east and west. It then plunges smoothly into the southern hemisphere, transiting the meridian lower and lower, rising and setting progressively more toward the south point, and coming up later and going down earlier. Finally, the Sun bottoms out at the **winter solstice** in Sagittarius about December 22 (see the left-hand Earth in Figure 2.16), when it is 23.5° south of the celestial equator, and on this first day of northern-hemisphere *winter* begins to move back north.

The temperature of the ground and the air above it depends on how much warming sunlight the ground absorbs. At a latitude of 45°N, a beam of summer sunlight shines down from nearly overhead (see Figures 2.18 and 2.20a). In winter, however, when the Sun is low in the sky, a similar beam must spread itself over more ground (see Figures 2.18 and 2.20b), with the result that the temperature cannot climb as high. Therefore, the weather turns cold. The effect is intensified by the shortening of the day. However, when the Sun is low in the northern hemisphere, it is high in the southern (see Figure 2.11), so it is summer in Argentina when it is winter in Nebraska. The actual times of maximum high and low temperature at our northern latitudes do not coincide with the technical beginning of summer or winter but occur about a month later because of the time it takes the Earth's land and waters to heat and cool.

2.5.3 Special Places

Since the arc on the celestial equator between the celestial equator and the zenith equals the observer's latitude, and since the Sun can travel no more than 23.5° north of the equator, you can never see the Sun overhead at 45°N latitude (see Figure 2.18). To have the Sun in your zenith on the first

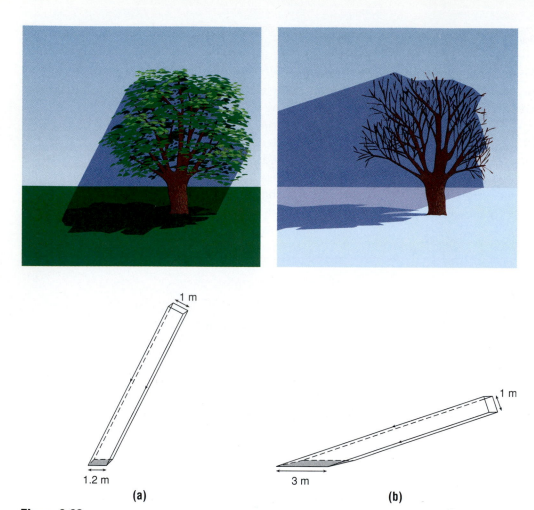

Figure 2.20
(a) Shadows are short as a square shaft of sunlight 1 m across shines nearly straight down and illuminates a 1 × 1.2 m rectangle and an area of 1.2 square meters. **(b)** When it strikes at a higher angle,
shadows are long, the shaft must cover a larger area (now 3 square meters), and the ground cannot get
as warm.

day of northern summer you must travel south to a latitude of 23.5°N, perhaps near the southern tip of Baja California in Mexico or into the Ba-
hamas (Figure 2.21a). A circle around the Earth parallel to the equator at
this special latitude is called the **tropic of Cancer** because when the science of astronomy was developing, the summer solstice was in that constellation. Farther south, the Sun will pass overhead twice during the year,
once on its way north and then on its way south. When you reach the
equator, $\phi = 0°$ in Figure 2.21b (maybe near Quito, Ecuador), the Sun
crosses the zenith at the time of its equinox passages. Here, solar daily

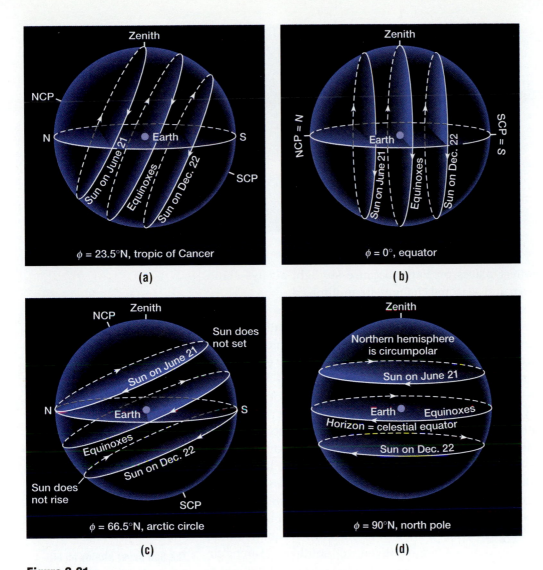

Figure 2.21

(a) At the tropic of Cancer, latitude 23.5°N, the Sun is overhead only on the first day of northern-hemisphere summer. **(b)** At the equator, latitude 0°, the Sun sets vertically and passes through the zenith twice per year. **(c)** At the arctic circle, latitude 66.5°N, the Sun is circumpolar on the first day of summer and does not rise on the first day of winter. **(d)** At the north pole, latitude 0°N, the Sun is up between March 20 and September 23.

paths are perpendicular to the horizon, and the days and nights are always 12 hours long no matter what the time of year. At 23.5°S latitude, the Sun will cross the zenith only when it reaches its southerly extreme of 23.5° south of the equator at the winter solstice, and once you are below 23.5°S latitude, the Sun can no longer be overhead. This latitude is the **tropic of Capricorn.** The lands between the two limits, loosely called the

Figure 2.22
The circumpolar Sun skims the horizon just above the arctic circle on June 21.

tropics, are the hottest places on Earth because sunlight always shines nearly straight down and seasonal changes are minimized.

Now walk north from 45°N latitude. The Sun transits ever lower and the weather becomes cooler. Eventually, you arrive at latitude 90° − 23.5° = 66.5°N (Figure 2.21c), possibly at Fort Yukon, Alaska, north of Fairbanks. On the first day of summer the Sun is 23.5° north of the equator and therefore 66.5° from the NCP; but the NCP is 66.5° from the horizon, so that on June 21 the Sun is circumpolar; that is, the Sun cannot set (Figure 2.22). Alternatively, on the first day of winter, the Sun will not rise, but will just graze the southern horizon. Latitude 66.5° girdles the globe as the **arctic circle** (from the Greek *arktos*, "bear"; the constellation known as the Great Bear passes overhead in these regions). The analogue to the arctic circle in the south is the **antarctic circle** (which encompasses most of Antarctica) at latitude 66.5°S. As you proceed above the arctic circle you will encounter more days of the circumpolar Sun during summer (that is, the Sun can be circumpolar when it is closer to the celestial equator) and a longer period of darkness during winter. (Even well south of the arctic circle the effect is noticed in summer in long extended periods of twilight as the Sun passes beneath the NCP not far below the horizon.) Since the Sun never gets very high, the weather is always cold.

Finally, you arrive at the north pole in the Arctic Ocean (Figure 2.21d): the NCP is overhead, daily paths are parallel to the ground, and only the northern celestial hemisphere is seen. The Sun is now visible for six months straight, from the time of its vernal equinox passage on March 20 to its crossing of the autumnal equinox on September 23. It then sets and cannot be seen for the next six months. When it is dark at the north pole, it is daylight at the south. At these extreme latitudes the Sun never attains an altitude greater than 23.5°, the temperature rarely rises above freezing, and the poles are covered with thick sheets of ice.

The Sun never attains an altitude greater than 23.5° at the poles, and they are covered with thick sheets of ice.

2.6 TIME AND THE CALENDAR

The motions of the Sun along its daily path and the ecliptic are used to reckon the passage of time. The Sun transits the celestial meridian at 12 hours on the clock, or *noon*. The Sun moves along its daily path at the rate of 15° per hour, so for every 15° the Sun moves to the west, we add another hour to the clock. At 15° west of the meridian the time is 13 hours, or 1 P.M. (*post meridiem*, Latin for "after noon"); at 45°, 3 P.M. For every 15° the Sun is east of the meridian, we subtract an hour: if the Sun is 30°E of the meridian, it is 10 hours or 10 A.M. (*ante meridiem*, "before noon"). Twelve hours after noon, the Sun crosses the meridian below the pole and it is $12^h + 12^h = 24^h$, and we start the new day at 0 hours or midnight.

In practice, timekeeping is much more complicated. The Sun does not move uniformly along the ecliptic because of the eccentricity of the Earth's orbit, and even if it did, it would not move uniformly to the east through the stars because of the ecliptic's tilt; the consequent effects must be averaged out. Worse, as you move east on the Earth, the Sun appears to move west of the celestial meridian, and the time of day increases, causing (for example) clocks to read later in New York than in California. *Standard time* relieves part of the problem by establishing zones roughly every 15° in the east-west direction. Since 15° corresponds to 1 hour of time, you change your clock by 1 hour as you pass from one time zone to the next (earlier to the west, later to the east). The time at Greenwich, England, is adopted as a worldwide standard called *Universal Time*.

We must keep track not only of hours, but also of days. Our year is 365.2422 . . . days long. If we adopt a calendar year of 365 days, the first day of spring would advance at the rate of about one day every four years. To remove this difficulty, two millennia ago Julius Caesar's astronomer established a calendar with a four-year cycle in which three years have 365 days and the fourth, a *leap year*, has 366 (the extra day added in February). He also placed the first day of spring on March 25. However, the average number of days in this *Julian calendar* is 365.25, slightly longer than the duration of the true year, so the date of the vernal equinox creeps slowly backward. By late in the sixteenth century it had shifted to March 12, and Pope Gregory XIII ordered a calendar reform.

Gregory's astronomers slightly modified the Julian calendar by declaring that leap years were to be skipped in century years not evenly divisible by 400. Thus in the *Gregorian calendar*, which we use today, the years 1700, 1800, and 1900 were not leap years despite their position in the four-year cycle, but the year 2000 will be a leap year. This scheme slightly shortens the average length of the calendar year, making it 365.2425 days long over a 400-year period—nearly perfect. Gregory's astronomers also skipped 10 days to make spring begin on March 21, the date on which it fell in the year 325 at the time of a church assembly at which the rules for calculating the date of Easter were established. England and its colonies finally adopted the system in 1752.

2.7 HOW DO WE KNOW THE EARTH MOVES?

Rotation is simple to demonstrate. At the north pole erect a pendulum, a heavy weight mounted on a long wire fastened to a support with a nearly frictionless bearing (Figure 2.23). Swing the pendulum toward a star, perhaps bright Betelgeuse, which you will find near the horizon. As Betelgeuse marches around the sky, the pendulum moves to follow it; that is, the plane of the swing rotates to the right with a period of a day. It looks at first as if there is some force making the pendulum's swing-plane move. The explanation, however, is that the weight is swinging freely in space. The Earth, with you on it, is actually turning beneath the pendulum, revealing that it is we who are moving, not the sky.

The first such pendulum was constructed in 1851 in Paris by Léon Foucault, who used an iron weight swinging from a 60-m-long cable. At locations other than the poles the pendulum's rotation period is greater than 24 hours, increasing with decreasing latitude, and at 45°N it is about 34 hours. Many museums around the world have a *Foucault pendulum,* by which you can easily watch the rotation of the Earth.

The Earth's revolution is not so obvious. Stand outdoors as snow is falling on a windless day. The snow falls directly on you from overhead. But if you walk or drive (Figure 2.24a), the snow appears to be coming from a point in front of you, hitting your face or the windshield. The faster you move, the greater the apparent shift from the overhead point. The same effect is produced when you move relative to a beam of light. In Fig-

Many museums around the world have a Foucault pendulum, by which you can easily watch the rotation of the Earth.

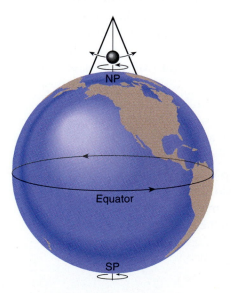

Betelgeuse

Figure 2.23

A Foucault pendulum is erected at the north pole and swung toward the star Betelgeuse. As the Earth turns under it, the pendulum's plane stays fixed in space, always directed at the star moving on its daily path. To the observer on Earth, it is the plane of the pendulum that seems to turn.

NP

Equator

SP

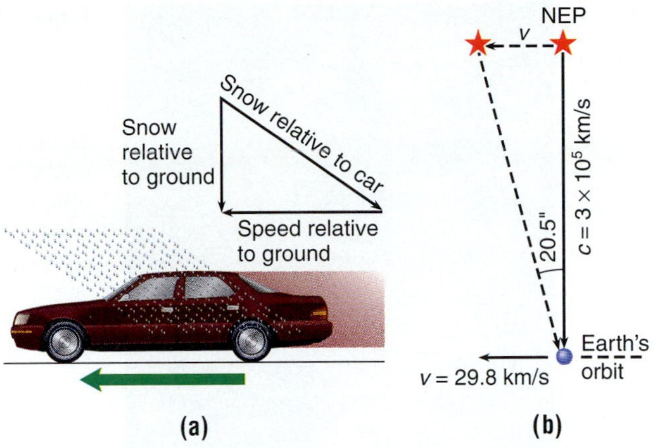

Figure 2.24
(a) If you drive into vertically falling snow, the flakes appear to come from the forward direction. The faster you move, the more the direction of their fall seems to come from a point in front of you. **(b)** A ray of light and the apparent position of a star are affected the same way. The observed shift of 20.5 seconds of arc is defined by the relative speeds of the Earth (v) and light (c).

ure 2.24b, the Earth is moving to the left in the ecliptic plane at speed v of 29.8 km/s and is intercepted by a ray of light moving at speed c (300,000 km/s) from a star that is in the perpendicular direction. From a triangle constructed with sides v and c (greatly out of scale in the diagram), we can calculate that the star is shifted by 20.5 seconds of arc (see MathHelp 2.1). And that is exactly what we see. As the Earth goes around the Sun, a star in the direction perpendicular to the Earth's orbital plane will move in a little loop 20.5" in radius (an effect called the *aberration of starlight*), providing a wonderful demonstration that it is the Earth that moves, not the Sun.

2.8 PRECESSION

You can think of our planet as a sphere of matter with a "spare tire" at the equator, an *equatorial bulge* (see Section 2.2). However, both the Sun and (on the average) our nearby Moon are found in the plane of the ecliptic rather than in the plane of the celestial equator. As a consequence, they exert a gravitational pull on the equatorial bulge (Figure 2.25). The result is a 26,000-year wobble, or **precession,** of the Earth's rotational axis about its orbital axis in which the obliquity of the ecliptic stays the same. The effect is similar to the wobbling of a spinning top.

Precession produces several interrelated effects. Looking upward from the Earth in Figure 2.25, you would see the NCP move counterclockwise in a circle of radius 23.5° (outlined in the star maps in Appendix 1). You, however, are observing from the moving body. The NCP always has an altitude equal to your latitude, and therefore it seems that the stars are moving past the pole rather than that the pole is moving through the stars. The NCP is currently pointed nearly in the direction of the star Polaris in the Little Dipper; for the ancient Egyptians of 2700 B.C., however, the pole star was Thuban, in the constellation Draco; in 13,000 years, the NCP will be on the other side of the precessional circle, near the bright star Vega.

For the ancient Egyptians of 2700 B.C., the pole star was Thuban, in the constellation Draco.

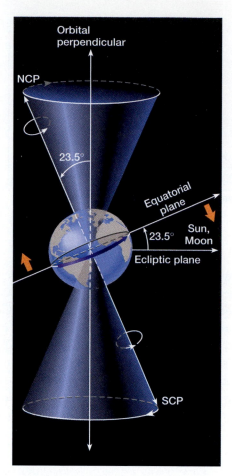

Figure 2.25
The Sun and, on the average, the Moon are in the ecliptic plane and exert gravitational forces on the Earth's equatorial bulge (highly exaggerated here) in the direction of the orange arrows. Because the Earth is rotating, the spin axis wobbles, causing the NCP to trace out a circle of radius 23.5° about the perpendicular to the orbit with a period of nearly 26,000 years.

    If the celestial poles move through space, so must the celestial equator (which is halfway between them), and so too must the points of intersection between it and the ecliptic. The effect is to move the equinoxes and the solstices to the west through the constellations of the zodiac at the rate of 50″ per year (a full degree in 70 years). Because we are on the moving body, it appears that the constellations are sliding in the opposite direction, past the equinoxes and solstices, as if on a moving belt. In 13,000 years, Gemini, which now appears high in the northern-hemisphere sky in winter, will have changed places with Sagittarius and will be low in the south in summer. The progress of the vernal equinox is evident from its symbol, ♈, which represents the horns of Aries, the Ram, even though the vernal equinox is now one constellation of the zodiac to the west, in Pisces. The vernal equinox *was* in Aries when the symbol was assigned; about the year 2700 it will move into Aquarius. Similarly, the tropics of Cancer and Capricorn bear these names even though the summer and winter solstices are now respectively in Gemini and Sagittarius, each one constellation to the west.

Twenty-six thousand years seems like a long time, but in fact modern astronomers can monitor the effects of precession from one night to the next. The motion was actually discovered with naked-eye observations by Hipparchus of Nicaea more than 2,000 years ago.

▶ KEY CONCEPTS

Altitude (h): The angular elevation of a body above the horizon.

Angular diameter : The angle formed by lines projecting from the viewing point to opposite sides of a body.

Arctic and **antarctic circles:** The latitudes (respectively 66.5°N and 66.5°S) above and below which it is possible to observe the Sun at midnight.

Axis: An imaginary line about which a body rotates or moves.

Celestial equator: The great circle on the celestial sphere above the Earth's equator, equidistant from the celestial poles.

Celestial meridian: The great circle through the celestial poles and the zenith.

Celestial sphere: The apparent sphere of the sky.

Circumpolar stars: Stars that do not set on their daily paths.

Constellations: Named patterns of stars.

Daily paths: The apparent paths taken by celestial bodies as the Earth rotates.

Eccentricity (e): The degree of flattening of an ellipse, ranging from 0 for a circle up to (but not including) 1.

Ecliptic: The apparent path of the Sun through the stars.

Ellipse: A curve defined by two foci such that sum of the distance from any point on the curve to each of the foci is constant.

Equator (terrestrial): The great circle on the Earth that is equidistant from the poles.

Great circle: A circle on a sphere whose center is the same as the center of the sphere.

Horizon: The line where the land seems to meet the sky; the **astronomical horizon** is the great circle defined by the intersection between the celestial sphere and a plane at the observer's feet perpendicular to the line to the zenith.

Latitude (ϕ): Arc measurement north and south of the Earth's equator.

North and **south celestial poles** (NCP and SCP): The points of rotation of the celestial sphere that lie above the north and south poles of the Earth.

Obliquity of the ecliptic (ε): The tilt of the Earth's axis relative to the perpendicular to the Earth's orbit, or the angle between the celestial equator and the ecliptic, equal to 23°27'.

Perihelion: The point of closest approach of a planet to the Sun; **aphelion** is the point of greatest distance between them.

Period: The time required for a body to return to a given starting point or position.

Poles (terrestrial): The points on the Earth where its rotation axis emerges.

Precession: The motion of the celestial poles and equinoxes; caused by a 26,000-year wobble of the Earth's axis.

Semimajor axis (a): Half the major axis of an ellipse; the measurement that characterizes the ellipse's size.

Summer and **winter solstices:** The points on the ecliptic farthest north and south (23.5°N and 23.5°S, respectively) of the celestial equator, which the Sun crosses on June 21 and December 22.

Theory: A logical or mathematical description of nature that embraces and explains observational or experimental data.

Tropics of Cancer and **Capricorn:** The latitudes (respectively 23.5°N and 23.5°S) at which the Sun passes overhead when it crosses the summer solstice (June 21) and winter solstice (December 22).

Vernal and **autumnal equinoxes:** The points at which the Sun and the ecliptic cross the celestial equator, the Sun moving respectively north (March 20) and south (September 23).

Zenith: The point on the celestial sphere over your head.

Zodiac: The band of constellations that contains the ecliptic.

 KEY RELATIONSHIP

Latitude:
$$\phi = h \text{ (NCP).}$$

 EXERCISES

Comparisons

1. How do the celestial poles relate to the terrestrial poles?

2. What is the difference between perihelion and aphelion?

3. Compare the visibility of the Sun at the south and north poles of the Earth.

Numerical Problems

4. If Eratosthenes had observed a 10° change in the position of the Sun in the sky between Syene and Alexandria, what would be the circumference of the Earth?

5. What is the latitude of a city **(a)** one-third the way from the Earth's equator to the north pole; **(b)** one-quarter of the way from the equator to the south pole? Give your answer to a minute of arc.

6. How many kilometers are there in a degree of latitude?

7. What are the altitudes of stars when they are seen **(a)** halfway up the sky; **(b)** setting; **(c)** two-thirds the way from the horizon to the zenith?

8. What is the observer's latitude when the north celestial pole is **(a)** 27° above the horizon; **(b)** 33° below the horizon?

9. You live at 33°N latitude. What is the minimum angular distance north of the celestial equator of a circumpolar star?

10. You live at 42°N latitude. What is the minimum angular distance south of the celestial equator of a star that does not rise?

11. If the axis of the Earth were tilted by 33° relative to the perpendicular to the ecliptic plane, what would be the latitudes of **(a)** the tropic of Capricorn; **(b)** the antarctic circle?

12. Draw an ellipse with a semimajor axis of 10 cm and an eccentricity of 0.5. Note that the ends of the semimajor axis lie on the ellipse, and therefore you can find the sum of the distances from these points to the foci.

13. Tonight at 8 P.M. you see a certain star transiting the celestial meridian. How far west of the meridian will that star be seen at 8 P.M. two days, a week, and two months from now?

14. A sphere 200 m away has a diameter of 1 cm. What is its angular diameter in seconds of arc? (See MathHelp 2.1 and Figure 2.24b.)

15. In what constellations of the zodiac will the vernal and autumnal equinoxes be in the year 7000?

Thought and Discussion

16. How does an oblate spheroid differ from a sphere?

17. What is the significance of **(a)** the tropic of Cancer; **(b)** the arctic circle?

18. Which latitudes on Earth receive the greatest solar heating; which receive the least?

19. How can the coldest days of the northern hemisphere coincide with the Earth's perihelion?

20. At what latitude will the Sun be seen overhead on March 20?

21. Over what range of latitudes will the Sun be circumpolar on **(a)** June 21; **(b)** March 20; **(c)** December 22?

22. What century years in the next millennium will *not* be leap years?

23. What is the evidence for the rotation and revolution of the Earth?

Research Problems

24. What are the latitudes of **(a)** Quito, Ecuador; **(b)** Melbourne, Australia; **(c)** Sacramento, California; **(d)** Tokyo, Japan; **(e)** your hometown?

25. Find a city in another country that has the same latitude as your hometown.

Activities

26. Draw and properly label celestial spheres for observers with latitudes of 60°N, 10°N, 30°S, and 90°S. On each place a star 10°N north of the celestial equator as it transits the meridian and show its daily path.

27. Fasten two straight sticks together at their ends with a single nail. Find Polaris in the sky and turn the sticks until one points at Polaris and the other points at the horizon below Polaris. Measure the angle with a protractor. You now have a measurement of your latitude. How close did you come to the latitude given in an atlas?

28. Draw celestial spheres, like those in Figure 2.21, showing the daily paths of the Sun during the year at the tropic of Capricorn, the antarctic circle, and the south pole.

Scientific Writing

29. A magazine for a travel club describes a trip from New York to Buenos Aires. Write a sidebar (a short one-column essay of about 200 words that relates to a main article) in which you describe the changes the voyagers will see in the sky. One of your goals is to interest your readers in sky-watching.

30. Write an article for a gardening magazine in which you explain the cause of the change of the seasons and the reason that growing conditions at lower latitudes are different from those at higher latitudes.

THE FACE OF THE SKY
Stars and constellations and the organization of the heavens

3

On a clear, dark night, stars of different brightnesses and colors shimmer everywhere, so many that they shed a pale embracing light. They scatter into patterns, much like a handful of thrown sand. Here you see a circle of them, there a square, somewhere else a figure of a man and a meandering river. Night after night they repeat their display, friendly, familiar, reigning supremely and untouchably over the Earth. As you learn to recognize stars, you begin to name them and the patterns they seem to form, translating to the sky images from your experience. Such is the meaning of the constellations and the myths they illustrate as they carry forward the ancient lore that allows you to reach back and touch your ancestors of unrecorded time.

3.1 THE ANCIENT CONSTELLATIONS

Everywhere, in every land, people have named the stars. In China, India, Africa, North and South America, and elsewhere, the nightly patterns have been organized, though in different ways. Our Western constellations began to be developed some 4,000 years ago in the cradle of our civilization in the Middle East, which was centered around ancient Mesopotamia. The stars were used as a calendar and to tell stories, honor gods and heroes, and serve as reminders of days gone by (Figure 3.1). Few of these constellations resembled what they were supposed to be. How could they? After all, the stars are not really fastened to a celestial sphere, but are all at different distances and distributed largely at random. The constellations were—and still are—meant to *represent*, not *portray*.

The ancient Greeks adopted the early constellations and identified them with their own fables and religious beliefs. The old star patterns were formally codified by the Greek mathematician Eudoxus of Cnidos about 375 B.C., lavishly embraced by Aratos (about 270 B.C.) in a famous poem called *On the Phaenomena*, and finally cast into their present form about A.D. 150 by the last of the great Greek astronomers, Ptolemy, in his *Syntaxis* (meaning "composition"), in which he set down the astronomical knowledge of the time.

The constellations are meant to represent, not portray.

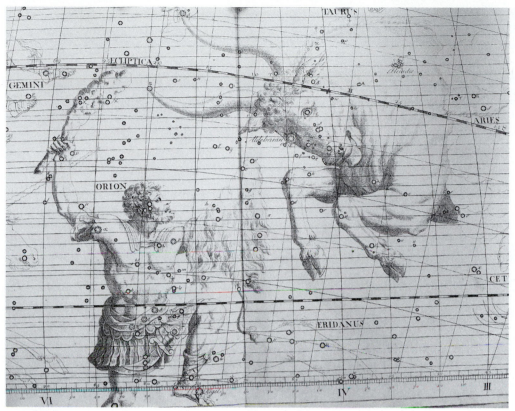

Figure 3.1

The sky is populated with the imagination of humankind, its stories, dreams, and inventions. Here, as depicted in John Flamsteed's *Atlas Coelestis* of 1729, the hunter Orion prepares to strike the zodiacal bull Taurus: they are surrounded by a river, a whale, a ram, a pair of twins, and a unicorn.

The Romans accepted these celestial figures whole, and after the gradual collapse of their great empire, the stellar patterns passed on to the Arabian lands, where they were further embellished and the stars graced with new names. Finally they returned to Europe in much the form we see them today, 48 **ancient constellations** engraved into the sky (Table 3.1). As the story unfolds, find the constellations on the star maps in Appendix 1, and then at night try to locate the more prominent figures.

Most ancient constellations do not stand alone but are parts of groups of patterns that commonly tell the stories of one or more myths. The most famous group is the *zodiac* (Figure 3.2), which was used in Chapter 2 to illustrate the ecliptic and precession. The zodiacal constellations are among the oldest known, and to the ancients were certainly the most important as they contain all the moving bodies of the Solar System that were closely identified with the gods—the Sun, Moon, and (as we will see in Chapter 4) the planets. There are 12 of these constellations because the Moon runs

The origins of the zodiacal constellations lie deep in our agrarian past.

TABLE 3.1
The Ancient Constellations

Name	Meaning	α^a	δ^a	Possessive	Abbreviation
The Zodiac					
Aries	Ram	3	+20	Arietis	Ari
Taurus	Bull	5	+20	Tauri	Tau
Gemini	Twins	7	+20	Geminorum	Gem
Cancer	Crab	8.5	+15	Cancri	Cnc
Leo	Lion	11	+15	Leonis	Leo
Virgo	Virgin	13	0	Virginis	Vir
Libra	Scales	15	−15	Librae	Lib
Scorpius	Scorpion	17	−30	Scorpii	Sco
Sagittarius[b]	Archer	19	−25	Sagittarii	Sgr
Capricornus	Water goat	21	−20	Capricorni	Cap
Aquarius	Water-carrier	22	−10	Aquarii	Aqr
Pisces	Fishes	1	+10	Piscium	Psc
Ursa Major					
Ursa Major	Greater bear	11	+60	Ursae Majoris	UMa
Ursa Minor[c]	Lesser bear	16	+80	Ursae Minoris	UMi
Boötes	Herdsman	15	+30	Boötis	Boo
Orion					
Orion	Proper name: hunter, giant	6	0	Orionis	Ori
Canis Major	Greater dog	7	−20	Canis Majoris	CMa
Canis Minor	Lesser dog	8	+5	Canis Minoris	CMi
Lepus	Hare	6	−20	Leporis	Lep
Perseus					
Perseus	Proper name: hero	3	+45	Persei	Per
Andromeda	Proper name: princess	1	+40	Andromedae	And
Cassiopeia	Proper name: queen	1	+60	Cassiopeiae	Cas
Cepheus	Proper name: king	22	+65	Cephei	Cep
Cetus	Whale	2	−10	Ceti	Cet
Pegasus	Proper name: winged horse	23	+20	Pegasi	Peg
*The Ship **Argo**: Jason and the Argonauts*					
Carina[d]	Keel	9	−60	Carinae	Car
Puppis[d]	Stern	8	−30	Puppis	Pup
Vela[d]	Sails	10	−45	Velorum	Vel
Hercules	Hero	17	+30	Herculis	Her
Hydra	Water serpent	12	−25	Hydrae	Hya
Centaurus					
Centaurus	Centaur	13	−45	Centauri	Cen
Lupus	Wolf	15	−45	Lupi	Lup
Ara	Altar	17	−55	Arae	Ara
Ophiuchus					
Ophiuchus	Serpent-bearer	17	0	Ophiuchi	Oph
Serpens	Serpent	17	0	Serpentis	Ser

TABLE 3.1
Continued

Name	Meaning	α[a]	δ[a]	Possessive	Abbreviation
		Single Constellations			
Aquila	Eagle	20	+15	Aquilae	Aql
Auriga	Charioteer	6	+40	Aurigae	Aur
Corona Australis[e]	Southern crown	19	+40	Coronae Australis	CrA
Corona Borealis[f]	Northern crown	16	+30	Coronae Borealis	CrB
Corvus[g]	Crow, raven	12	−20	Corvi	Crv
Crater	Cup	11	−15	Crateris	Crt
Cygnus	Swan	21	+40	Cygni	Cyg
Delphinus	Dolphin	21	+10	Delphini	Del
Draco	Dragon	15	+60	Draconis	Dra
Equuleus	Little horse	21	+10	Equulei	Equ
Eridanus	River	4	−30	Eridani	Eri
Lyra[g]	Lyre	19	+35	Lyrae	Lyr
Piscis Austrinus	Southern fish	22	−30	Piscis Austrini	PsA
Sagitta	Arrow	20	+20	Sagittae	Sge
Triangulum	Triangle	2	+30	Trianguli	Tri

[a]Use to locate the constellations on the star maps in Appendix 1, where right ascension (α) and declination (δ) are explained.

[b]Contains the galactic center.

[c]Contains the north celestial pole.

[d]Carina, Puppis, and Vela are modern subdivisions of the original constellation Argo and together make one of the ancient 48.

[e]Sometimes considered as Sagittarius' crown.

[f]Ariadne's crown.

[g]Lyra was Orpheus' harp, Corvus his companion.

Figure 3.2

The zodiac's deadly scorpion, Scorpius, with red Antares at its heart, presides over the summer southern landscape.

through its phases (see Chapter 4) somewhat over 12 times a year. As a consequence, the Sun is roughly in a different constellation every month.

The origins of the zodiacal constellations lie deep in our agrarian past. Aries, the rutting ram that held the Sun on the first day of spring 2,500 years ago, is clearly a fertility symbol associated with planting. Taurus, the Bull (see Figure 3.1), now seen in northern winter, had similar significance even earlier in our history. And Virgo, the Virgin, assumed much the same kind of symbolism when she embraced the Sun at the time of the late summer harvest.

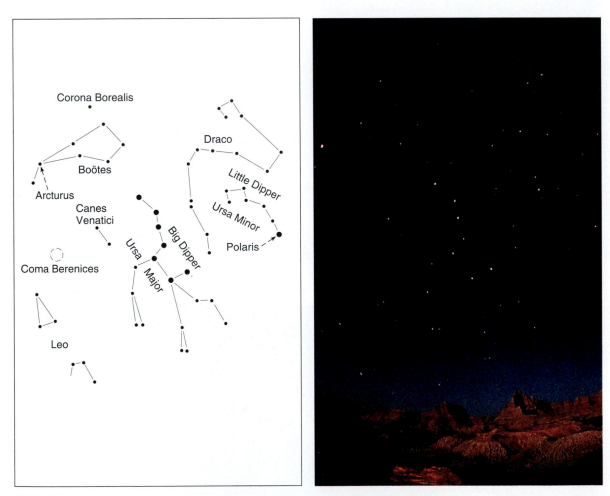

Figure 3.3

The Big Dipper, part of the constellation Ursa Major, the Greater Bear, falls silently in the northwest toward the Black Hills of South Dakota. The paws of the bear are below the Dipper, whose front bowl stars point directly at Polaris, the star at the end of the handle of the Little Dipper, which stands upright. Draco coils between the two. The arc of the Big Dipper's handle points to orange Arcturus in Boötes. Below the curve is the pair that make the modern constellation Canes Venatici. Setting at lower left are the hindquarters and part of the head of Leo.

There is an important distinction to be made between the astrological *signs* of the zodiac and the astronomical *constellations* of the zodiac. The constellations are real star groups that can be seen on any clear night; the signs are symbolic concepts that are locked to the vernal equinox. The signs were once aligned with the constellations, but are no longer because precession has moved them westward. To the superstitious the signs possess magical powers that derive from the names of the parent constellations and that provide the means to foretell the future.

The next constellation groups are listed in Table 3.1 more or less in order of renown, beginning with the bears. Ursa Major and Ursa Minor, the Greater and Lesser Bears, contain respectively the Big and Little Dippers (Figure 3.3). Almost everyone can find the Big Dipper, its high northerly location giving it almost year-round visibility. The Dippers are not formal constellations but are **asterisms** (Table 3.2), which are either smaller portions of constellations or larger figures that can embrace several. One legend has it that the bear was hurled into the sky by her tail, which stretched out into the string of stars that makes the Dipper's handle. The third member of the group, Boötes, the celestial herdsman, is recognizable by the kitelike shape that terminates in the great orange star Arcturus, which watches over the bear as she plods around the pole.

In northern winter, Orion and his companions (Figures 3.1 and 3.4) stalk the sky. Three bright stars near the celestial equator stud his belt, bright reddish Betelgeuse and bluish Rigel respectively mark his right shoulder and left foot, and another triplet of stars makes the sword hanging from his belt. This constellation is such an imposing figure that the ancient Arabs called it Al Jauza, "the central one"—a mysterious woman.

TABLE 3.2
Prominent Asterisms

Asterism	Constellation	Description
Big Dipper	Ursa Major	Seven bright stars ($\alpha - \eta$ UMa)
Circlet	Pisces	Head of the western fish
Great Square	Pegasus	Central figure of Peg (α And, α, β, γ Peg)
Hyades	Taurus	Star cluster around Aldebaran
Keystone	Hercules	Four stars in north Her (η, π, ε, ζ Her)
Kids	Auriga	Triangle south of Capella (ε, ζ, η Aur)
Little Dipper	Ursa Minor	Seven stars beginning at Polaris
Little Milk Dipper	Sagittarius	Five-star dipper (λ, τ, σ, ϕ, ξ Sgr)
Northern Cross	Cygnus	Deneb at top of cross
Pleiades	Taurus	Seven Sisters star cluster
Sickle	Leo	Terminates in Regulus
Summer Triangle	——	Deneb, Vega, Altair
Teapot	Sagittarius	Little Milk Dipper plus stars to west
Urn, (water jar)	Aquarius	Four stars in Y in northern Aqr
Winter Triangle	——	Betelgeuse, Procyon, Sirius

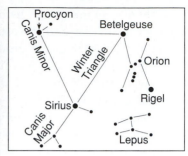

Figure 3.4

Great Orion draws attention in the winter sky with his prominent belt and the brilliant stars Betelgeuse and Rigel. His hunting dogs, Canis Major and Canis Minor, follow him across the sky, and beneath his feet is Lepus, the Hare. Betelgeuse, Procyon, and Sirius form the Winter Triangle. The winter Milky Way runs downward to the left. Compare this view with the map in Figure 3.1.

Orion's belt points down and to the left to brilliant Sirius in Canis Major (see Figure 3.4), Orion's larger dog, who stands on his hind legs ready to do his master's bidding. East of Orion is bright Procyon, the brightest star of Canis Minor, one of a pair that make Orion's smaller dog. In one story Orion met his end when he was stung by Scorpius, and when the gods honored the hero by placing him forever in the sky they put him opposite the celestial scorpion so that he need never look upon his killer: as one rises, the other sets.

The Perseus myth ties together a number of major autumn constellations. The queen *Cassiopeia*, wife of *Cepheus*, boasted that her daughter *Andromeda* was more beautiful even than the sea nymphs. In anger, Neptune had Andromeda chained to a rock by the sea as a sacrifice to *Cetus*, the sea monster or whale. But as Cetus bears down upon her, *Perseus* (Figure 3.5), riding his flying horse *Pegasus*, comes dramatically to the rescue. He has just slain the dreaded Medusa, a woman with hair of snakes who is so horrifying that one look at her will turn the viewer to stone. Perseus bran-

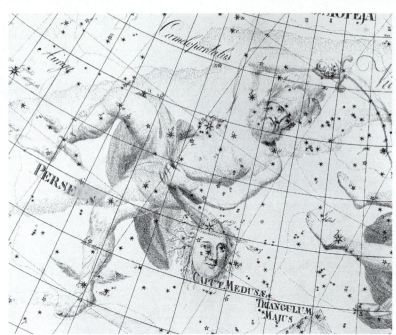

Figure 3.5
Bold Perseus, the rescuer of Andromeda, holds the Medusa in John Bevis's
Uranographia Britannica of 1786 as
he turns Cetus to stone. Constellation
boundaries are drawn as curved dot-
ted lines. The other lines are grids for
measuring star positions.

dishes her severed head, petrifies the monster, and rescues Andromeda.
(Figure 3.5 at first may appear to be merely a fanciful drawing. It is actually a scientifically accurate star map from the year 1786. The engraving is
an artistic holdover from an earlier time in which the drawings served to
identify the constellations and aided in the location and naming of stars.
Modern star maps use straight lines to outline constellation boundaries or
do away with them altogether. Some plot over a million stars.)

Argo, or *Argo Navis,* the ship in which Jason sailed to find the golden
fleece, is the largest constellation in the sky, so big that in more recent
times it was broken into three more-manageable parts—Carina, the Keel;
Puppis, the Stern (or poop); and Vela, the Sails—thereby raising the number of ancient constellations from 48 to 50. Much of the ship, which contains the second brightest star, Canopus, is hidden below the southern
horizon from most of the United States. Jason's quest was plagued by dis-
asters epitomized by Hydra, the Water Serpent, which wraps one-third of
the way around the sky. Among Jason's crew was Hercules, the strongest
man of ancient, or any other, times. The Argo group overlaps that associated with Centaurus, the Centaur (Jason's foster father and tutor). This
half man-half horse holds the closest and third brightest star, Alpha Centauri (Figure 3.6), and is sometimes shown sacrificing Lupus, the Wolf,
upon Ara, the Altar.

The modern boundaries of another powerful figure, Ophiuchus, cut
across the ecliptic. He is depicted as fighting a massive serpent that, although considered one constellation, is divided in two: Serpens Caput,
the head, seen to the west of Ophiuchus, and Serpens Cauda, the tail, to

**Ophiuchus' serpent-wrapped
body is the model for the modern physician's symbol.**

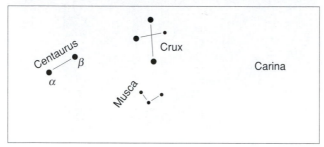

Figure 3.6
Alpha and Beta Centauri lie in the Milky Way to the east of Crux, the famous Southern Cross. A dark patch called the Coalsack is seen down and to the left of the Cross. Musca, the Fly, a modern constellation, is just below the Cross.

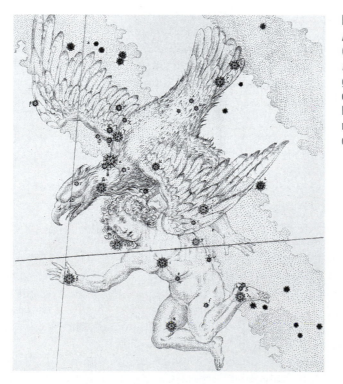

Figure 3.7
Aquila soars the Milky Way south of Cygnus in Bayer's famous *Uranometria* of 1603, which includes the engravings of Jacobo de Gehyn. Aquila carries Antinoüs, companion to the Roman emperor Hadrian. Antinoüs is no longer officially recognized as a constellation.

the east. Ophiuchus is identified with Asclepius, the mythic physician of the ancient Greeks. His serpent-wrapped body is the model for the caduceus, the staff entwined with snakes that is the modern physician's symbol—the sky brought back to Earth.

Finally, several prominent ancient constellations stand by themselves. During northern summer months, two great birds, Cygnus, the Swan, and Aquila, the Eagle (Figure 3.7) fly through the Milky Way. Cygnus, its tail lit by the bright star Deneb, transits near the zenith in mid-northern latitudes. Aquila is formed by bright Altair and two flanking stars that give the impression of outstretched wings. West of Cygnus we find Lyra, the lyre played by Orpheus, who sang so beautifully he could make the rocks applaud. It is marked by the brilliant white star Vega, from which hangs a small, near-perfect parallelogram of faint stars.

High in northern winter, Auriga, the Charioteer, adds to the luster created by Orion, Canis Major, Canis Minor, and Gemini. To the west of Orion springs Eridanus, the celestial depiction of Ocean Stream that in ancient times was thought to girdle the world. From most of the United States, Eridanus flows below the southern horizon before ending in bright Achernar. And there are more: fish, fowl, royal crowns, and an array of ancient artifacts all waiting in the stars for you to find.

3.2 THE MODERN CONSTELLATIONS

The ancient star patterns do not fill the heavens, since there are broad areas of sky with few bright stars and the northern inventors of the constellations could not see the stars of the deep southern hemisphere around the south celestial pole. So beginning with the scientific renaissance of the late sixteenth century, and continuing for some 200 years, European astronomers created fresh constellations to fill in the blanks. Like their ancient forebears, they designed these constellations around their own civilization, placing a hundred or more familiar objects, animals, kings, and heroes into the sky. Most of the figures did not survive the passage of time, and the astronomers of our own century finally officially adopted 38 **modern constellations** (Table 3.3), which when added to the ancient constellations yield a total of 88. This full set is significant to modern astronomy because it parcels the sky into manageable, easily recognizable sections.

The large majority of the recognized modern constellations were created by a handful of people. The seventeenth-century German astronomer Hevelius filled the spaces of the northern hemisphere with animals like Canes Venatici, the Hunting Dogs (see Figure 3.3) and Lacerta, the Lizard. However, the deep southern hemisphere provided the greatest treasury of unnamed stars and patterns. Constellations invented around the year 1600 by the Dutch travelers Pieter Keyser and Frederik de Houtman were immortalized in Johannes Bayer's great atlas, his *Uranometria* of 1603, its

The constellations parcel the sky into manageable, easily recognizable sections.

TABLE 3.3
The Modern Constellations

Name	Meaning	α^a	δ^a	Possessive	Abbreviation
Antlia	Air pump	10	−35	Antliae	Ant
Apus	Bee	16	−75	Apodis	Aps
Caelum	Graving tool	5	−40	Caeli	Cae
Camelopardalis	Giraffe	6	+70	Camelopardalis	Cam
Canes Venatici	Hunting dogs	13	+40	Canum Venaticorum	CVn
Chamaeleon	Chameleon	10	−80	Chamaeleontis	Cha
Circinus	Compasses	15	−65	Circini	Cir
Columba	Dove	6	−35	Columbae	Col
Coma Berenices[b]	Berenice's hair	13	+20	Comae Berenices	Com
Crux	Southern cross	12	−60	Crucis	Cru
Dorado	Swordfish	6	−55	Doradus	Dor
Fornax	Furnace	3	−30	Fornacis	For
Grus	Crane	22	−45	Gruis	Gru
Horologium	Clock	3	−55	Horologii	Hor
Hydrus	Water snake	2	−70	Hydri	Hyi
Indus	Indian	22	−70	Indi	Ind
Lacerta	Lizard	22	+45	Lacertae	Lac
Leo Minor	Lesser lion	10	+35	Leonis Minoris	LMi
Lynx	Lynx	8	+45	Lyncis	Lyn
Mensa	Table	6	−75	Mensae	Men
Microscopium	Microscope	21	−40	Microscopii	Mic
Monoceros	Unicorn	7	0	Monocerotis	Mon
Musca (Australis)	(Southern) fly	12	−70	Muscae	Mus
Norma	Carpenter's square	16	−50	Normae	Nor
Octans[c]	Octant	—	−90	Octantis	Oct
Pavo	Peacock	20	−70	Pavonis	Pav
Phoenix	Phoenix	1	−50	Phoenicis	Phe
Pictor	Painter's easel	6	−55	Pictoris	Pic
Pyxis[d]	Compass	9	−30	Pyxidis	Pyx
Reticulum	Net	4	−60	Reticuli	Ret
Sculptor	Sculptor's workshop	1	−30	Sculptoris	Scl
Scutum[e]	Shield	19	−10	Scuti	Sct
Sextans	Sextant	10	0	Sextantis	Sex
Telescopium	Telescope	19	−50	Telescopii	Tel
Triangulum Australe	Southern triangle	16	−65	Trianguli Australis	TrA
Tucana	Toucan	0	−65	Tucanae	Tuc
Volans	Flying fish	8	−70	Volantis	Vol
Vulpecula	Fox	20	+25	Vulpeculae	Vul

[a]See footnote a, Table 3.1.

[b]Star cluster with many old references, but still not considered one of the ancient 48.

[c]Contains the south celestial pole.

[d]Grouped with argo.

[e]Shield of the Polish king and hero John Sobieski.

style exemplified by Figure 3.7. Over a century later, the Abbé Nicolas de Lacaille honored the artifacts of a burgeoning industrial revolution, including a furnace (Fornax), an air pump (Antlia), and a microscope (Microscopium), mixed in with a bird of paradise (Apus), a chameleon (Chamaeleon), and a peacock (Pavo).

Two modern constellations are special. Coma Berenices, Berenice's Hair, is a lovely sprinkling of faint stars south of the Big Dipper's handle (see Figure 3.3). The figure goes back perhaps to Eratosthenes, and ought to take its place among the ancient constellations, but it was not included in Ptolemy's definitive list. Berenice was an Egyptian queen of about 250 B.C. who gave a lock of her hair to the gods in thanks for bringing her husband safely home from war. And the gods gave it to the sky.

The other is that gem of the southern hemisphere, Crux, the Southern Cross (see Figure 3.6). From the southern hemisphere it rises just ahead of Centaurus' two bright stars, and as they all proceed across the celestial sphere they present a sight never to be forgotten.

3.3 STARS AND THEIR NAMES

Roughly 6,000 to 8,000 stars can be seen with the naked eye. Some are so luminous that they can be viewed even from the centers of big cities, while others are only just visible on the blackest of nights. About 130 B.C., Hipparchus, the discoverer of precession, devised a simple scheme to order stellar brightness. He arranged the naked-eye stars into six categories or **magnitudes,** with first magnitude the brightest and sixth the faintest. The modern magnitude system (to be explained in Chapter 13) uses decimal places. Regulus, a first-magnitude star in Leo, has a modern magnitude of 1.35. "First magnitude" now ranges from 0.50 to 1.49, second from 1.50 to 2.49, and so on. The magnitude system has also been extended with higher numbers to fainter stars that can be observed only through telescopes. In addition, the stars of Hipparchus' first magnitude have such a big range that the brightest stretch into the modern magnitude classes 0 and even −1. However, by tradition the brightest 21 stars are still informally lumped together as "first magnitude."

About a thousand of the brighter stars carry **proper names** that come from a variety of languages and that commonly relate to the stars' properties or locations. Brilliant Sirius, for example, derives from a Greek word that means "scorching." The majority of proper names, however, are derived from Arabic and are a heritage of the time after the fall of the Roman Empire when the great Arabian civilizations cultivated the knowledge of Greece and Rome (Figure 3.8). The Arabians had their own constellation lore, but they adopted the Greek figures and little of it survived. The Arabians' principal contribution was to apply a large number of names that generally describe the positions of the stars within the Greek constellations. Deneb, for example, means "tail." The name is given to the first-magnitude star in the tail of Cygnus, the Swan, and appears as Denebola in Leo, Deneb Kaitos in Cetus, and Deneb Algedi in Capricornus.

When the Arabic names were translated into Latin in the Middle Ages and later, many were so corrupted that their present form and meaning might be unintelligible to a modern Arab. The name of the great reddish

About a thousand of the brighter stars carry proper names.

Figure 3.8
Arabian astronomers, depicted here with a variety of astronomical instruments, carried forward the work of the Greeks and made their own original contributions.

star in Orion, Betelgeuse, comes from the phrase "yad al-jauza," meaning "the hand of the central one," but it has been mangled into the "armpit" of the hunter. A list of 31 proper names in Table 3.4 includes the first-magnitude stars plus several others of interest.

Proper names can be hard to remember and may be ambiguous, as one name may have been given to several stars. In the early 1600s, Johannes Bayer applied Greek letters to stars within a constellation more or less in order of brightness (there are several exceptions), and then added an ending to the constellation name that in Latin signifies possession. Vega, the brightest star in Lyra, thus becomes α Lyrae (abbreviated α Lyr), meaning "α of Lyra"; the second brightest becomes β Lyrae, and so on. Tables 3.1 and 3.3 list the possessive forms and official abbreviations for all the constellations. **Bayer Greek letters** are seen in Figures 3.1, 3.5, and 3.7.

TABLE 3.4
Proper Names of Selected Stars

Proper Name	Meaning[a]	Greek-Letter Name	Magnitude[b]
Achernar	End of the river	α Eri	0.46
Aldebaran	Follower	α Tau	0.85
Algol	Demon's head	β Per	2.12v
Alkaid	Chief of mourners	η UMa	1.86
Altair	Flying eagle	α Aql	0.77
Arcturus	Bear watcher (Gk)	α Boo	−0.04
Antares	Like Mars (Gk)	α Sco	0.96
Bellatrix	Female warrior (Lat)	γ Ori	1.64
Betelgeuse	Hand of the central one	α Ori	0.50v
Canopus	Proper name (Gk)	α Car	−0.72
Capella	She-goat (Lat)	α Aur	0.08
Castor	Proper name (Gk)	α Gem	1.58
Deneb	Tail	α Cyg	1.25
Denebola	Tail	β Leo	2.14
Dubhe	Bear	α UMa	1.79
Fomalhaut	Fish's mouth	α PsA	1.16
Megrez	Root of the tail	δ UMa	3.31
Merak	Loin of the bear	β UMa	2.37
Mizar	Groin	ζ UMa	2.06
Phecda	Thigh	γ UMa	2.44
Polaris	Pole star (Lat)	α UMi	2.02v
Pollux	Proper name (Lat)	β Gem	1.14
Procyon	Before the dog (Gk)	β CMi	0.38
Regulus	Little king (Lat)	α Leo	1.35
Rigel	Foot of the central one	β Ori	0.12
Sirius	Scorching (Gk)	α CMa	−1.46
Spica	Ear of wheat (Lat)	α Vir	0.98
Thuban	Serpent (corr)	α Dra	3.65
Vega	Swooping eagle	α Lyr	0.03
Zubenelgenubi	Southern claw of the scorpion	α^2 Lib	2.75
Zubeneschamali	Northern claw of the scorpion	β Lib	2.61

[a] "Lat" and "Gk" denote Latin and Greek names; other names are of Arabic origin. Serious corruption is indicated by "corr."

[b] "v" indicates that the magnitude is variable.

Then, in the eighteenth century, the Englishman John Flamsteed (who created the atlas from which Figure 3.1 is taken) organized the stars to roughly fifth magnitude within each constellation, working eastward from its western boundary. Shortly thereafter other astronomers assigned numbers. By this system, Vega, α Lyrae, is also 3 Lyrae. Astronomers generally use proper or Greek-letter names for the first-magnitude stars, Greek-letter names for all the others when available, and finally employ these **Flamsteed numbers.** Fainter stars are assigned a variety of catalogue numbers that ignore constellation boundaries.

3.4 THE MILKY WAY

The Milky Way, the visible manifestation of the disk of our Galaxy (see Section 1.2.4 and the star maps in Appendix 1), appears as a luminous circular band with us at the center. It has a mythology equal to that of the constellations. In a dark sky, the billions of stars that compose it blend into a near-continuous stream, with islands of darkness caused by the patchy distribution of light-absorbing dust in the vast spaces between the stars (Figure 3.9). The Greeks named this glowing ribbon the Milky Circle, and our word *galaxy* comes from the Greek word for "milk." To some cultures, the Milky Way is the street of souls on their way to heaven, to others a great river that brings water from the ocean to the irrigating streams.

Because of the absorbing interstellar dust and the Sun's off-center position in the Galaxy, the Milky Way varies considerably in brightness around its circle. During the northern winter months, when we are facing away from the Galaxy's center, the Milky Way is weak and hard to see through Auriga, Taurus, and Orion. It brightens as the white band passes south to the horizon from Canis Major (see Figure 3.4) to Carina and Puppis, only to be lost from sight. Its beauty is revealed in the northern summer when we look in the other direction. It brightens as it comes out of Cassiopeia into Cygnus, then forms massive star clouds as it plunges through Scutum and down to Sagittarius, where our view is directed through the thickest part of the system toward the center of the Galaxy. Then it is lost to northerners as it passes below the horizon. It is in the southern hemisphere that the Milky Way shines in full glory. In middle

In a dark sky, the billions of stars that compose the Milky Way blend into a near-continuous stream.

Figure 3.9
The Milky Way, the combined light of the stars of the disk of our Galaxy, cascades through Scutum and Sagittarius.

southern latitudes, the Galaxy's center passes overhead, the great silver stream brilliant as it extends through Centaurus and Crux and back north to float the mighty ship *Argo*.

 KEY CONCEPTS

Ancient constellations: The 48 constellations handed down by the ancient Greeks.

Asterisms: Small, named portions of constellations, or stellar groupings that extend over constellation boundaries.

Bayer Greek letters: Greek letters assigned to stars, usually in order of brightness within a constellation.

Flamsteed numbers: Numbers assigned to stars west-to-east within a constellation.

Magnitudes: Classes of star brightnesses.

Modern constellations: Constellations generally invented since about 1600.

Proper names: Individual names assigned to the brighter stars, usually reflecting the stars' properties or positions; most are of Arabic origin.

 EXERCISES

Use the star maps in Appendix 1 where necessary.

Comparisons

1. What are the differences between the ancient and the modern constellations?

2. Discriminate between constellations and asterisms.

Numerical Problems

3. What is the highest northern latitude at which you can see all of **(a)** Crux; **(b)** Centaurus; **(c)** Grus?

4. What is the most southerly latitude at which you can see all of **(a)** Ursa Major; **(b)** Aquila; **(c)** Pegasus?

Thought and Discussion

5. To what constellations do the Big and Little Dippers belong?

6. What are the proper names of **(a)** the two bears; **(b)** the two constellations involving fish; **(c)** the three constellations involving dogs; **(d)** the two crowns?

7. What first-magnitude stars are found in the zodiac?

8. What constellation is **(a)** divided into two noncontiguous parts; **(b)** broken into three subdivisions? Name the subdivisions of each.

9. In what principal northern seasons do you find the following constellations in the evening sky: **(a)** Orion; **(b)** Leo; **(c)** Sagittarius; **(d)** Ursa Minor; **(e)** Octans?

10. What constellations surround **(a)** Aquila; **(b)** Scorpius; **(c)** Virgo; **(d)** Monoceros; **(e)** Hydra; **(f)** Dorado?

11. In what celestial hemispheres do you find **(a)** Lynx; **(b)** Fornax; **(c)** Lacerta; **(d)** Microscopium?

12. What constellations represent **(a)** a clock; **(b)** a table; **(c)** an air pump; **(d)** an arrow; **(e)** a cup; **(f)** a dove?

13. To what constellations do the following stars belong: **(a)** Arcturus; **(b)** Achernar; **(c)** Rigel; **(d)** Antares; **(e)** Deneb?

14. What are the prominent asterisms in **(a)** Taurus; **(b)** Auriga; **(c)** Sagittarius (name two); **(d)** Pisces; **(e)** Aquarius?

15. What are two asterisms that cross over constellation boundaries?

16. What are three ways in which the naked-eye stars are named?

17. Name five constellations that fall within the Milky Way.

Research Problems

18. Try to find an old star atlas or globe in a library or museum. Note the date and any constellations that are not in the accepted lists in this chapter. What constellations now contain these groupings?

19. Use a library to research Native American or Chinese constellations. Compare the constellations of the Chinese zodiac with those of the ancient Greek zodiac.

Activities

20. Some clear night go outside and make your own list of constellations. Find a few prominent groupings, outline them, and give them names. Compare your results with those of the ancients.

21. Select a first-magnitude or otherwise well-known star. If your library has the materials, find its name in as many different star catalogues as you can.

22. Look in the sky for the star you chose in the previous exercise and draw its surrounding constellation.

Scientific Writing

23. Write a commentary for beginners explaining the significance and importance of the constellations in the past and at present.

24. Write a three-page article for a nature magazine in which you explain how a teacher or parent could begin to teach the stars and constellations to children.

EARTH, MOON, AND PLANETS
The view of the Solar System from Earth and the explanations
of lunar and planetary motions

4

With the starry sky as our background, we now turn to the Moon and the
planets to see how they behave and to explore the roles they have played
in astronomy and in the development of human thought.

4.1 THE MOON

Continuously changing both its apparent shape and its location in the sky,
the Moon is a source of endless fascination.

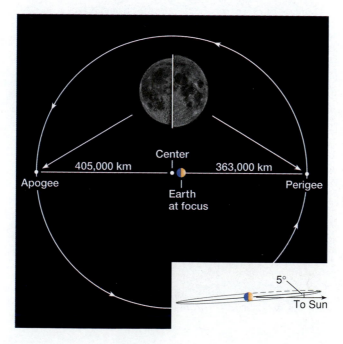

Figure 4.1
As we look down from the north, the
Moon revolves counterclockwise
around the Earth, its orbit an ellipse of
eccentricity 0.055 with the Earth at
one focus. At perigee, the Moon is
363,000 km away; at apogee, 405,000
km. The 11% difference in the Moon's
angular diameter at apogee and
perigee is readily seen from the photographs. The inset shows the orbit
nearly on edge and its 5° tilt relative
to the ecliptic plane.

4.1.1 The Lunar Orbit

The Moon, our nearest neighbor, takes just under a month to orbit the Earth, moving counterclockwise, in the same direction as the Earth's rotation and revolution (Figure 4.1). The plane of its orbit is tilted from that of the Earth's by only 5°; as a result, the Moon is always found within 5° of the ecliptic (and thus within the constellations of the zodiac), and its daily paths across the sky are similar to those of the Sun.

The distance to the Moon is easy to find. If you look at a nearby object from two different places it will appear to be in different directions. Hold your finger in front of you and then look alternately at it with one eye and then the other (Figure 4.2a). It will seem to jump back and forth against the background of the room, an effect called **parallax**, which is the basis of

The Moon is always found within the constellations of the zodiac.

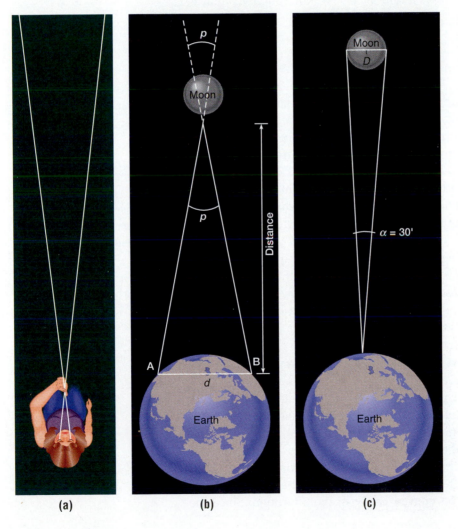

(a)   (b)   (c)

Figure 4.2
(a) Each eye sees an outstretched finger in a different direction.
(b) Like the eyes in **(a)**, observers A and B each see the Moon in a different direction. The lines of sight from the two locations intersect in the parallax angle *p*. If the separation of the observers (*d*) is known, they can calculate the lunar distance from the triangle made of A, B, and the Moon.
(c) The Moon subtends an angle α equal to 30 minutes of arc. With the distance of the Moon known, we can calculate the lunar diameter (*D*) in kilometers.

three-dimensional vision. The angle through which your finger jumps depends on how far it is from your face: the farther you hold it away, the less it appears to move. In Figure 4.2b, we observe the parallax angle p through which the Moon seems to shift upon observing it from points A and B on the Earth. As seen from the Moon, A and B subtend the same angle p. Since we know p and the physical distance in kilometers between A and B we can find the lunar distance (see MathHelp 2.1), which averages 384,000 km (1/400 AU).

The Moon's orbit is close to the shape of an ellipse with the Earth at one focus (see MathHelp 2.2 and Figure 4.1). The eccentricity is 0.055, triple that of the terrestrial orbit. The minimum distance of 363,000 km takes place at **perigee** (Greek *peri-*, "near," and *ge*, for "Earth"). The maximum of 405,000 km is at **apogee** (Greek *apo*, "away from"). Given the distance and the measured angular lunar diameter of 30 minutes of arc (Figure 4.2b), the physical diameter is found to be 3,500 km, roughly a quarter that of the Earth.

4.1.2 Phases

Over the course of its orbit, the appearance of the Moon changes radically. Its different apparent shapes, or **phases** (Figure 4.3), are correlated with the Moon's rising and setting times. The Moon produces no light of its own but shines purely by sunlight reflected from its surface. As with the Earth, one full hemisphere is bright with daylight, the other dark with night. The phases are commonly and incorrectly thought to be produced by shadows thrown by the Earth. Instead, the apparent shape of the Moon depends upon how much of its daylight side is visible to us on Earth at different orbital positions.

The phases of the Moon depend upon how much of its daylight side is visible to us on Earth.

In Figure 4.3, the rotating Earth is in the center, ringed by the lunar orbit. The illuminating Sun is far off the page to the right. If the orbiting Moon is between the Earth and the Sun (position 1), the Moon and Sun rise and set at the same time. The Moon's sunlit portion faces away from us, we look at the unilluminated nighttime side, and the Moon is not visible. In this position the Moon is said to be *new*.

Now let the Moon proceed a quarter of the way along its orbit to position 3, where it is 90° to the east of the Sun. From Earth, we now see half of each of the daylight and nighttime sides. This *first-quarter* phase (its name reflecting the amount of the orbit the Moon has traversed, not its appearance) looks like a half moon in the sky. When the observer (who is being carried on the rotating Earth) passes from day into night, the Sun is setting on the horizon and the time (on the average) must be 6 P.M. or 18 hours. The first-quarter Moon will therefore be seen on the celestial meridian at sunset and can also be viewed easily in the afternoon. Since the

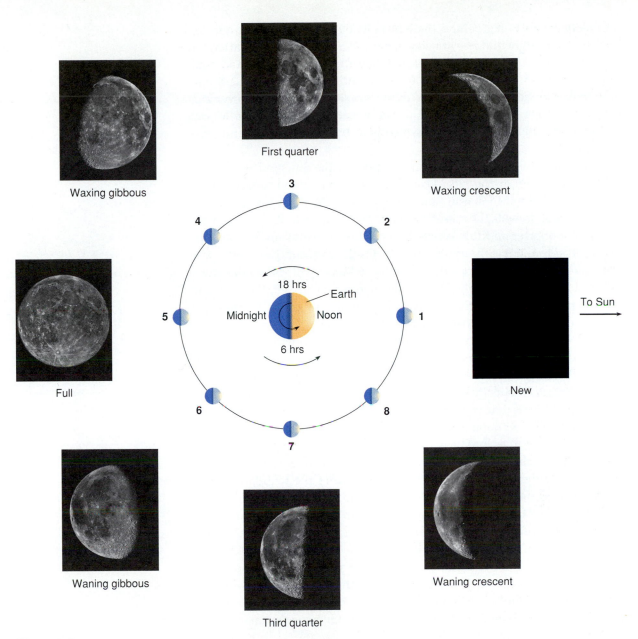

First quarter

Waxing gibbous

Waxing crescent

18 hrs

Earth

Midnight   Noon

6 hrs

To Sun

Full

New

Waning gibbous

Waning crescent

Third quarter

Figure 4.3

The phases of the Moon are produced by the amount of daylight that is seen from Earth as our satellite revolves about us. At position 1, the new phase, we look only at the nighttime side, and the Moon is invisible as it rises and sets with the Sun. Opposite, at position 5, the full phase, we see the entire sunlit face, which will rise at sunset and set at sunrise. In between are the quarters (positions 3 and 7) where only half the daytime side can be viewed. First quarter will be on the celestial meridian at sunset, third at sunrise. Between the quarters and new we see only a crescent, and between the quarters and full the Moon appears gibbous. As seen from the Moon, the Earth would go through similar, but opposite, phases.

Moon continues to lag behind the Sun in its daily path, the first quarter will set at midnight and rise at noon. In between new and first quarter, for example at position 2, we see less than half the daylight side, and consequently the Moon takes on the appearance of a *crescent*. As the Moon now makes less than a 90° angle with the Sun, it must set in the evening, before midnight. At this point you can actually see the whole Moon, including the nighttime portion, which is lit by light reflected from the bright Earth.

At position 5, the Moon is directly opposite the Sun, and we see the full daytime side, or the *full Moon*. The nighttime side now faces away from us. Since the Moon and Sun are opposite in the sky, the full Moon rises at sunset, transits the meridian at midnight, and sets at sunrise. In between first quarter and full, typified by position 4, we see more than half the sunlit side and it is the *dark* portion that is the crescent. The visible part now takes on a *gibbous* appearance, and the Moon will rise in the afternoon and set after midnight. Except for the full Moon itself, these are the *waxing* phases (from the Anglo-Saxon *weaxen*, "to grow"), since the amount of visible sunlit surface increases with time.

Following full, the sequence is reversed, the Moon now in its *waning* phases as the amount of visible sunlit surface decreases. After full (see position 6) the Moon becomes gibbous again, now rising *after* sunset, and at position 7 it reaches its *third quarter*, where we again see only half the daylight face (the half that was dark at first quarter). Now the Moon *rises* at midnight and *sets* at noon, just the opposite of its behavior at first quarter. The Moon then passes into its waning crescent phase (as in position 8) and is seen in the early morning hours before sunrise, again with its nighttime portion lit by the Earth. And finally, 29.5 days after the sequence began, the Moon is again new and disappears from the sky.

You can see from the photographs in Figure 4.3 that only one side of the Moon is ever visible from the Earth. As our satellite orbits, it also slowly rotates on its own axis so that one side faces us at all times. Place a chair in the center of the room. Walk around it while facing it and notice that you have to rotate to do so. At new Moon the invisible *farside* is in full sunlight and its opposite, the *nearside*, the side we see, is in darkness. At full Moon, the nearside is fully lit and night has fallen on the farside.

4.1.3 Synodic and Sidereal Periods

The 29.5-day period of the phases, that relative to the Sun, is called the **synodic period** (from the Greek *synodos*, "coming together" or "meeting"). The Moon's **sidereal period,** which relates to the stars, is shorter, 27.3 days. The difference is the result of the Earth's going about the Sun while the Moon goes about the Earth.

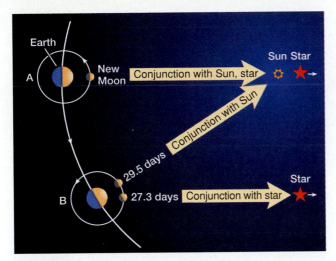

Figure 4.4
The lunar sidereal period of 27.3 days is the time it takes the Moon to travel from a star in the zodiac (A) back to that same star (B). The cycle of the phases, however, the synodic period, is geared to the Sun, and takes longer, 29.5 days, because the Sun is continuously moving to the east of the star at the rate of about 1° per day.

Begin with the Moon, the Sun, and a star all aligned at position A in Figure 4.4. As the Moon circles the Earth, our own motion will cause the Sun to appear to move steadily to the east of the star. The Moon will then first return to the star at position B as it completes a sidereal period, by which time the Sun has shifted (at the rate of 1° per day) about 27° to the east of where it started. To go about the Earth in 27.3 days requires that the Moon slide through the zodiac at a rate of about 13° per day. It will then take 2.2 more days (taking into account the additional motion of the Sun over that time) to catch up with the Sun and return to its new phase. Watch for yourself, and note the position of the Moon against the constellations at any particular phase. It will return to that constellation about two days before it completes its cycle of phases.

4.1.4 Eclipses

Figure 4.5a shows the Sun shining on the Earth and the Earth casting a shadow in space. The diagram is very much out of scale, as it must be to show the Earth, Moon, and Sun on a page. All shadows have two parts: an *umbra*, in which no direct light falls, and a *penumbra*, a region of partial shade. If you were to stand in the umbra of the terrestrial shadow you could see none of the Sun; if you were in the penumbra you would see that

When the Moon is caught in the Earth's shadow, sunlight is cut off, and we see an eclipse of the Moon.

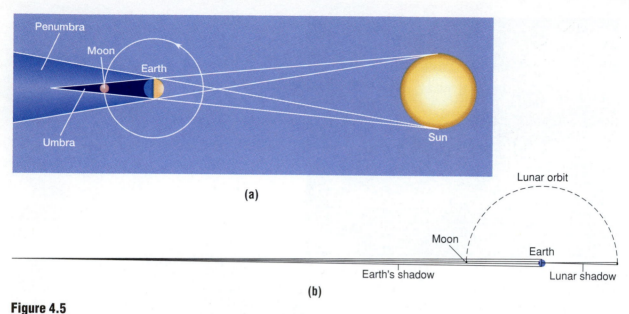

(a)

(b)

Figure 4.5

(a) In the dark part of the Earth's shadow, the umbra, the Sun is completely hidden. In the penumbra, the lightly shaded area, an observer would see the Earth block only part of the Sun. The diagram is out of scale so as to show all three bodies. **(b)** The Earth's 1.38 million-km-long shadow, drawn here to scale, can completely swallow the Moon. The Moon's shadow, however, can just barely reach the Earth.

the Earth cuts off only a part of the solar disk. The closer to the umbra you get, the darker the penumbra.

Figure 4.5b shows the shadows of both the Earth and Moon drawn to scale. The Earth's umbra is very long, projecting 1.38 million km into space. When the Moon is caught in the Earth's shadow, sunlight is cut off, and we see an **eclipse of the Moon.** At the Moon's average distance of only 384,000 km, the umbral shadow is 9,700 km in diameter, 2.8 times the size of the lunar globe. Consequently the Moon can be entirely engulfed by the umbra to produce a **total eclipse of the Moon.** If only part of the Moon passes through the umbra, the event is called a **partial eclipse of the Moon.**

From Figure 4.5 you see that the Moon can be eclipsed only when it is opposite the Sun, that is, when it is full. But experience tells you that most full Moons are *not* eclipsed. The reason is the 5° tilt of the lunar orbit, which carries the Moon as much as 5° above or below the ecliptic. The Moon therefore usually misses the umbral shadow, which at the lunar distance is only 1.4° across. The lunar orbit, a great circle in the sky, must cross the great circle of the ecliptic at two opposite points. Total eclipses of the Moon can take place only if the full Moon happens to be within 5° of one of these crossing points (if between 5° and 11° we see a partial eclipse). Table 4.1 gives a list of lunar eclipses through the year 2020. We typically see one from any given location about once every three years.

TABLE 4.1
Eclipses of the Moon, 1996–2020[a]

Greenwich	Date and	Hour	Type	Moon Overhead at Mid-Eclipse[b]
1996	Apr. 4	0	T	1°W, 6°S; Gulf of Guinea
1996	Sept. 27	3	T	46°W, 1°N; Brazil
1997	Mar. 24	5	P	69°W, 1°S; Colombia
1997	Sept. 16	19	T	77°E, 3°S; Indian Ocean
1999	July 28	12	P	172°W, 19°S; Samoa
2000	Jan. 21	5	T	68°W, 20°N; Haiti
2000	July 16	14	T	153°E, 21°S; Coral Sea
2001	Jan. 9	20	T	57°E, 22°N; Oman
2001	July 5	15	P	136°E, 23°S; Australia
2003	May 16	4	T	56°W, 19°S; Brazil
2003	Nov. 9	1	T	23°W, 16°N; Cape Verde Islands
2004	May 4	21	T	52°E, 16°S; Madagascar
2004	Oct. 28	3	T	50°W, 13°N; West Indies
2005	Oct. 17	12	P	176°E, 10°N; Central Pacific
2006	Sept. 7	19	P	76°E, 6°S; Indian Ocean
2007	Mar. 3	23	T	13°E, 7°N; E. Africa coast
2007	Aug. 28	11	T	158°W, 10°S; Central Pacific
2008	Feb. 21	3	T	48°W, 11°N; W. Central Atlantic
2008	Aug. 16	21	P	44°E, 14°S; Madagascar
2009	Dec. 31	19	P	69°E, 23°N; W. India
2010	Jun. 26	12	P	174°W, 23°S; Samoa
2010	Dec. 21	8	T	125°W, 23°N; W. Mexican coast
2011	Jun. 15	20	T	57°E, 23°S; Madagascar
2011	Dec. 10	15	T	140°E, 23°N; Marianas
2012	Jun. 4	11	P	166°W, 22°S; Cook Islands
2013	Apr. 25	20	P	57°E, 13°S; Madagascar
2014	Apr. 15	8	T	117°W, 9°S; Central Pacific
2014	Oct. 8	11	T	166°W, 6°N; Central Pacific
2015	Apr. 4	12	T	180°, 6°S; Central Pacific
2015	Sept. 28	3	T	44°W, 2°N; N. Brazil coast
2017	Aug. 7	18	P	87°E, 16°S; Indian Ocean
2018	Jan. 31	14	T	161°E, 17°N; Central Pacific
2018	July 27	20	T	56°E, 19°S; Mauritania
2019	Jan. 21	5	T	75°W, 20°N; Cuba
2019	Jul. 16	22	P	39°E, 21°S; Madagascar

[a]Twenty-three total (T) and 12 partial (P) lunar eclipses are scheduled to occur in the 25-year-interval 1996 through 2020. Roughly, each is visible over the terrestrial hemisphere whose pole is located at the indicated latitude and longitude. There is no eclipse in 2020.

[b]The first number is longitude, an angular measure of position east or west of Greenwich, England; the second number is latitude.

BACKGROUND 4.1 Ancient Measures of Distance

In the third century B.C., 2,300 years ago, Aristarchus of Samos actually made an estimate of the ratio of the distance of the Sun to that of the Moon by observing the angle between the Moon and the Sun at the time of the quarters. If the Sun is infinitely far away, the angle should be exactly 90°. The closer the Sun, the smaller the angle (Figure 4.6). Aristarchus found an angle of 87° and announced that the Sun is 20 times farther from the Earth than is the Moon. Unfortunately, his method is impossible to apply with any accuracy because the Sun is so far away—20 times more distant than he thought—and because the true angle is so close to 90° that it cannot be distinguished from a right angle. His result was produced by simple (and understandable) observational error. Nevertheless, the idea is ingenious, and even if his measurement was wrong, his conclusion of enormous distance was correct.

A century later, the great Hipparchus measured the distance of the Moon relative to the diameter of the Earth. By watching a total lunar eclipse, he could easily determine that the angular size of the Earth's shadow at the Moon's distance is 1.4°, from which he deduced that the Moon is at a distance of 59 times the terrestrial radius: this figure is very close to the actual value of 60.3. In the second century A.D., Ptolemy found essentially the same ratio as had Hipparchus from the lunar parallax. Eratosthenes' measure of the physical size of the Earth gives the true distance of the Moon, and Aristarchus' determination of the ratio of the distances of the Moon and Sun gives at least a measure of the true solar distance! The intellectual achievements of the ancient thinkers would be impressive in any age.

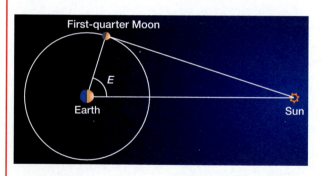

First-quarter Moon

E

Earth   **Sun**

Figure 4.6

If the Sun were infinitely far away, the first-quarter Moon would appear perpendicular to the solar direction; but if the Sun is nearby, the angle *E* between the Moon and the Sun at that phase must be significantly less than 90°.

Total lunar eclipses are fun to watch. The sequence of an eclipse is diagrammed in Figure 4.7. The Moon is moving to the left. Its passage through the penumbral shadow, which starts at P_1, is barely noticeable. The visible event starts at position 1, where we see the Earth's umbra begin to take a dark bite out of the lunar disk. The eclipse will be partial until position 2, when the whole Moon is in shadow. Until position 3 the eclipse will be total; then the partial phase resumes and continues until position 4. The eclipse is finally over when the Moon leaves the penumbra

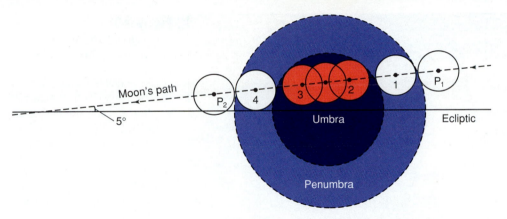

Figure 4.7

The Moon's orbit crosses the ecliptic at an angle of 5°. The Earth's shadow, opposite the Sun, is centered on the ecliptic, and if the shadow is close enough to a crossing point, the passing full Moon can be eclipsed. The Moon first strikes the penumbra at P_1 and contacts the umbra, where the eclipse becomes partial, at position 1. Between positions 2 and 3 it is totally immersed in the umbra. A glimpse of sunlight is seen at position 3; the Moon leaves the umbra at position 4 and the penumbra at P_2.

at P_2. Since the Moon moves at an angular rate of about half a degree per hour, the total portion (positions 2 to 3) can last as long as $1\frac{1}{2}$ hours.

The Moon does not usually disappear during totality, because the Earth's atmosphere, bright with sunlight, partially illuminates the dark shadow. However, absorption of sunlight by the air also causes the light to be considerably reddened (the reason that the Sun often looks red on the horizon), and consequently the eclipsed Moon takes on a dull, dark brick-red color (see Figure 2.2). The brightness of the eclipsed Moon depends upon the atmosphere's transparency and thus on terrestrial volcanic activity, which can make the air dusty and opaque.

If *new* Moon occurs when the Moon is close to crossing the ecliptic, the Moon can cover the solar disk and produce an **eclipse of the Sun.** The Moon is only a quarter the diameter of the Earth and its shadow is correspondingly shorter than Earth's (see Figure 4.5b). At the average lunar distance, the umbral shadow cannot quite reach us, and we see an **annular eclipse** (Figure 4.8), one in which the darkened Moon is surrounded by a thin ring (or annulus) of sunlight.

Figure 4.8

A multiple exposure (viewed from right to left) shows the development of the annular eclipse of May 10, 1994, whose path went directly across the United States.

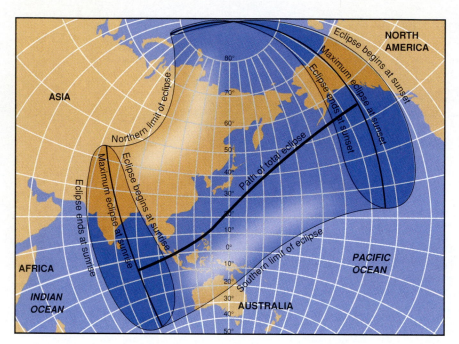

Figure 4.9

The umbra of the Moon's shadow is so small that as it flies across the Earth (black path) only a relatively few people see totality. The penumbra (shaded in darker blue), in which only a portion of the solar disk is covered, is much larger. This eclipse, which took place on March 18, 1988, crossed Sumatra and the Philippines and just missed the Aleutian Islands of Alaska. The apparent curvature of the eclipse path is caused by the distortion of the flat map.

For the new Moon to cover the Sun fully so that the umbra strikes the Earth to produce a **total eclipse of the Sun,** the Moon must be near perigee. At best, the shadow on the ground directly below the Moon is a spot no more than 269 km across (Figure 4.9). Since this shadow flies along across the Earth's surface at high speed—the speed of the orbiting Moon minus that of the spinning Earth—an individual can be immersed in the lunar umbra for at most only about seven minutes! The shadow spot is so small that a given location will see an eclipse only on the average of once every 300 or so years. However, whether the central eclipse is annular or total, the lunar penumbra is thousands of kilometers across, so that at any one time a large portion of the population can see a **partial eclipse of the Sun,** in which only a portion of the Sun is covered. There must be at least two eclipses of the Sun somewhere on Earth each year, and there can be as many as five. Predictions of total or annular solar eclipses through the year 2020 are given in Table 4.2.

A total solar eclipse presents a number of remarkable phenomena (Figure 4.10 on page 78). *Note, however, that only during totality is it safe to look directly at the Sun without an adequate solar filter. Any direct sunlight is bright enough to cause partial blindness.* During the partial phases use pinhole projection. Make a small hole in a piece of cardboard and project the image against paper or even onto the ground. At first you see only a tiny piece of the Moon over the Sun, but then the dark section gets larger, the Sun becomes a crescent (Figure 4.10a), and the sky begins to take on a cool deep-blue

At first you see a fiery red ring surrounding the darkened lunar disk and then the glorious corona.

TABLE 4.2
Eclipses of the Sun, 1996–2020[a]

Date	Type	Path
1997 Mar. 9	T	Siberia, Arctic Ocean
1998 Feb. 26	T	Central Pacific, N. South America, N. Atlantic
1998 Aug. 22	A	Indonesia, S. Pacific
1999 Feb. 16	A	S. Africa, Indian Ocean, Australia
1999 Aug. 11	T	North Atlantic, Europe, India, Thailand
2001 June 21	T	S. Atlantic, S. Africa, Madagascar
2001 Dec. 14	A	Pacific Ocean, Central America
2002 June 10	A	Northern Pacific Ocean
2002 Dec. 4	T	S. Africa, Indian Ocean, Australia
2003 May 31	A	Greenland, North Atlantic
2003 Nov. 23	T	S. Indian Ocean, Antarctica
2005 Apr. 8	A-T	S. Pacific, N. South America
2005 Oct. 3	A	N. Atlantic, N. Africa, Indian Ocean
2006 Mar. 29	T	Atlantic, N. Africa, Central Asia
2006 Sept.22	A	Indian Ocean, S. Atlantic, N. South America
2008 Feb. 7	A	Antarctica, S. Pacific
2008 Aug. 1	T	N. Canada, Arctic, Russia
2009 Jan. 26	A	Atlantic, Indian Ocean, Indonesia
2009 Jul. 22	T	India, China, Central Pacific
2010 Jan. 15	A	Africa, Indian Ocean, China
2012 May 20	A	China, N. Pacific, United States
2012 Nov. 13	T	N. Australia, S. Pacific
2013 May 10	A	N. Australia, New Guinea, Central Pacific
2013 Nov. 3	A	Atlantic, Central Africa
2014 Apr. 29	A	Antarctica
2015 Mar. 20	T	N. Atlantic, Arctic
2016 Mar. 9	T	Indian Ocean, Borneo, Central Pacific
2016 Sept. 1	A	Atlantic, Africa, Indian Ocean
2017 Feb. 26	A	S. South America, Atlantic, Central Africa
2017 Aug. 21	T	Pacific, United States, Atlantic
2019 Jul. 2	T	Pacific, S. South America
2019 Dec. 26	A	Arabia, Indian Ocean, Borneo
2020 Jun. 21	A	Africa, China, Central Pacific
2020 Dec. 14	T	Pacific, S. America, Atlantic

[a]Eighteen annular (A), 16 total (T), and 1 annular-total (A-T) eclipses are scheduled for the 25-year interval 1996 through 2020. The path of each eclipse begins in the first-listed geographic area, sweeps sequentially across other regions indicated, and ends in the last-listed area. In addition, there are partial-only eclipses not listed here that can be seen only at higher latitudes, generally near the poles.

hue. Look beneath trees shortly before totality: you will see hundreds of little crescents on the ground, the holes between the leaves acting like little pinhole projectors. In the distance you can see the onrushing shadow of the Moon high in the Earth's atmosphere. The last bit of sunlight then comes shining through the valleys at the lunar edge (Figure 4.10b), and finally, during totality, the Sun is covered. At first you see a fiery red ring

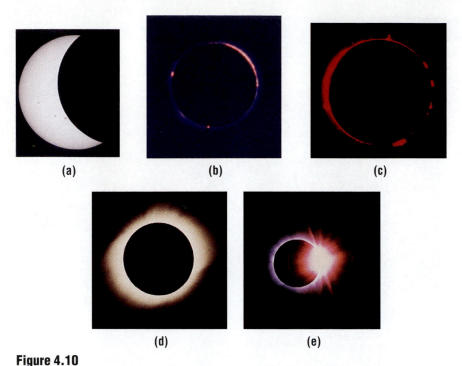

(a)   (b)   (c)

(d)   (e)

Figure 4.10
(a) The Moon takes a small bite out of the Sun. **(b)** Just before totality, the last bit of sunlight shines through the lunar valleys. Suddenly, the Sun is covered and the red chromosphere **(c)** and the corona **(d)** pop into view. **(e)** At the end of the total eclipse, the first bit of sunlight coupled with the fading corona creates the beautiful "diamond ring."

surrounding the darkened lunar disk (Figure 4.10c)—the *chromosphere*—and then the glorious *corona* (Figure 4.10d), a pearly-white envelope that extends many solar diameters outward (the natures of these layers will be explained in Chapter 12). The corona can be seen by eye only during totality—the blue sky hides it during the day. After only a few minutes, the first rays of brilliant sunlight break out, the corona disappears, and for a fraction of a second you see the stunning "diamond ring" (Figure 4.10e) surrounding the blackened disk of the new Moon. The partial phases then repeat in reverse. Once seen, the event will never be forgotten.

4.2 THE PLANETS

If you watch the constellations of the zodiac for any length of time you will see that five of their brightest "stars" do not stay fixed, but move (Figure 4.11). These bodies long ago received the name **planets,** from the Greek *planetai*, "wanderers." Because of their brilliance and their motions through the seemingly mystic constellations of the zodiac, the *ancient plan-*

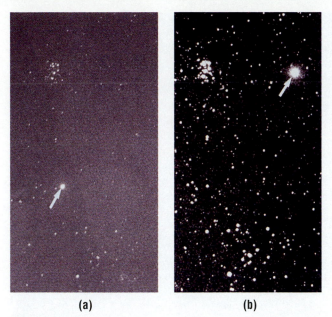

(a)   **(b)**

Figure 4.11
Jupiter moves westerly against the stars of the constellation Taurus between **(a)** October 1988 and
(b) January 1989. In both photographs the Pleiades star cluster is seen at upper left and the Hyades
cluster at lower center.

ets (those known since antiquity) were long ago assigned the names and
personalities of gods. The search for the laws that actually govern their apparent wanderings lasted more than 2,000 years.

4.2.1 Orbital Characteristics

The nine planets, with their orbital characteristics, are listed in Table 4.3.
Like our Moon, the planets shine by reflected sunlight, their apparent
brightnesses (given in column 2 by their magnitudes at maximum) depending on their physical sizes, reflectivities, and distances. The ancient
planets are bright and easily visible. Of those discovered in modern times
(Uranus, Neptune, and Pluto), only Uranus can be seen with the naked
eye, and then just barely.

Like the orbit of the Earth (see Section 2.5.1), all planetary orbits are ellipses with the Sun at one focus (Figure 4.12). The semimajor axes, or average distances, are given in column 3 of Table 4.3 and range from only
0.4 AU to 39 AU. Column 4 gives the eccentricities: only the paths of Mercury, Mars, and Pluto deviate much from circles. Pluto's orbit is so eccentric that near perihelion it can come closer to the Sun than does Neptune
(where it will be until 1999). Except for Pluto, the planets have low tilts rel-

TABLE 4.3
Planetary Orbital Data

1	2	3	4	5	6	7	8	9	10
				P	P				
Planet	m	a (AU)	e	$i°$	(sidereal)	(synodic)	°/day	v	R
Mercury	−1.8	0.387	0.206	7.00	88d	116d	4.09	48	23
Venus	−4.7	0.723	0.007	3.39	225d	584d	1.60	35	42
Earth	—	1.000	0.017	0.00	1.000y	—	0.99	30	—
Mars	−2.8[a]	1.524	0.093	1.85	1.88y	780d	0.52	24	73
Ceres[b]	+7.9	2.77	0.077	10.6	4.61y	466d	0.21	18	85
Jupiter	−2.9	5.20	0.048	1.31	11.9y	399d	0.083	13	121
Saturn	+0.0	9.54	0.056	2.49	29.5y	378d	0.033	10	138
Uranus	+5.5	19.2	0.047	0.77	84.0y	370d	0.012	7	152
Neptune	+7.7	30.1	0.009	1.77	165y	367d	0.006	5	158
Pluto	+14	39.4	0.249	17.1	248y	367d	0.004	5	162

The columns are 1, names; 2, maximum brightnesses in magnitudes (m); 3, semimajor axes (a) in AU; 4, orbital eccentricities (e); 5, tilts of the orbits to the plane of the ecliptic (i); 6 and 7, sidereal and synodic periods (P); 8, mean rates of motion along the orbit in degrees per day; 9, mean orbital velocities in km/s (v); 10, average number of days each planet spends in retrograde motion each year as viewed from Earth (R).

[a]At closest approach in its elliptical orbit.

[b]The largest asteroid (see Chapter 11).

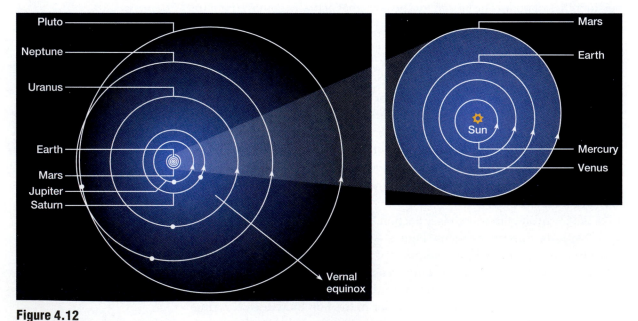

Figure 4.12
All the planets orbit the Sun and, except for Mercury, Mars, and Pluto, on paths that are almost circular. The positions of Jupiter, Saturn, Uranus, Neptune, and Pluto are shown for the year 1997 (the others move too quickly to place). On this scale the actual sizes of the planets are microscopic; even huge Jupiter would be only 0.001 mm across.

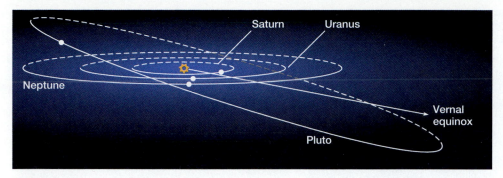

Figure 4.13
A side view of the outer four planets clearly indicates the plane of the Solar System and Pluto's 17° orbital tilt. The positions of the planets are shown for 1997.

ative to the plane of the Earth's orbit (given by i in column 5), and so are found within the zodiac (Figure 4.13).

The sidereal period of a planet (column 6 of Table 4.3) is the time required for it to orbit the Sun, that is, to return to a given point in its orbit. The farther a planet is from the Sun, the slower it moves and the greater distance it must travel. Consequently, more distant planets have longer sidereal periods. Average orbital velocities in degrees per day and km/s are given in columns 8 and 9.

4.2.2 Apparent Motions

As the planets orbit the Sun, they change their apparent positions relative to the Sun and to one another. The **inferior planets** (Mercury and Venus, those closer to the Sun than is the Earth) behave somewhat differently from the **superior planets** (those farther away). The alignments of the superior planets are typified in Figure 4.14 by Mars. If a planet is directly lined up with the Sun (as viewed from Earth), it is said to be in **conjunction** with the Sun. It then rises and sets with the Sun and cannot be seen. If the Sun and a planet are in opposite directions in the sky (like the Moon at full phase), they are in **opposition** to each other. Now, also like the full Moon, the planet rises at sunset, sets at sunrise, transits the celestial meridian at midnight, and we see it all night.

The planets have synodic as well as sidereal periods, defined (as for the Moon) as the interval between successive oppositions or conjunctions. The synodic periods are the lapping times of the speedier Earth relative to the more slowly moving outer planets (or the lapping times of even faster moving Mercury and Venus relative to Earth), and they express the intervals between successive times of best observability. In Figure 4.15, start with the Earth and Jupiter each in position 1, where Jupiter is in opposition to the Sun. After half a year has elapsed, by position 2, the Earth has moved through half its orbit and Jupiter has moved through only a small

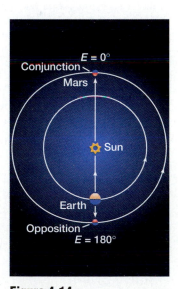

Figure 4.14
As seen from Earth, when a superior planet like Mars is lined up with the Sun, it is in conjunction; when it is opposite the Sun in the sky, it is in opposition.

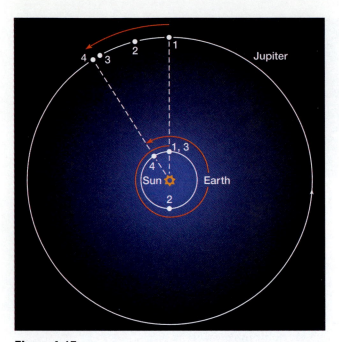

Figure 4.15
The sidereal period of a planet (Jupiter is depicted here) is the time it takes to revolve entirely about its orbit, from position 1 back to position 1 again. The synodic period is the interval between successive oppositions or conjunctions, here the time between position 1, at which Jupiter is in opposition to the Sun, through 2 and 3, and then 4.

arc. After a full terrestrial orbit, position 3, the Earth has returned to its starting point and Jupiter, with a sidereal period of 12 years, has traced out 1/12 of its path. It will take the Earth about another month to catch up with it, during which time Jupiter will have shifted a bit more, to position 4. Finally, after 399 days, or 1.09 years, Jupiter is back to opposition with the Sun. The farther the superior planet is from the Sun, the less it moves in orbit in a year, and the shorter the synodic period.

As opposition approaches, a superior planet slowly grinds to a halt and then appears to go backward.

Because the planets move counterclockwise around the Sun, they will generally appear to be traveling eastward through the zodiac. But as opposition approaches, a superior planet slowly grinds to a halt and then begins **retrograde motion** (Figure 4.16), in which it appears to go *backward*, to the west. The planet reaches maximum angular westward velocity at opposition, and then after a time it again slows, stops, and resumes traveling east. The phenomenon occurs because we are observing from a moving body (Figure 4.17). As you pass a car on the highway while you both are driving north, the other car seems to be in reverse, moving south. Similarly, as the Earth overtakes a superior planet and passes between it and the Sun, the outer one appears to be moving in the reverse direction.

Since the inferior planets are closer to the Sun than is the Earth, they can never be in solar opposition; instead, they pass through two different kinds of conjunctions (Figure 4.18). **Inferior conjunction** occurs when an

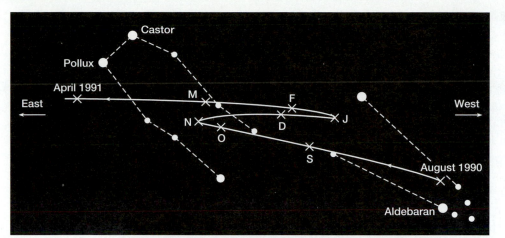

Figure 4.16
The path traced out by Mars in 1990–91 is shown against the background of Taurus and Gemini. The letters indicate the Martian positions at the beginnings of the indicated months. Opposition occurred on November 27, 1990.

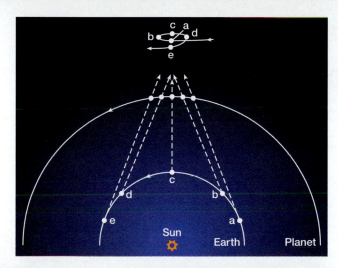

Figure 4.17
Retrograde motion occurs when the Earth overtakes a superior planet (or is overtaken by one of the inferiors). The planet is orbiting counterclockwise and, if the Earth were stationary, would appear to be moving easterly through the stars. The dashed arrows show the direction of the planet as seen from Earth from points a to e: the loop above shows the appearance of the motion in the sky. From a to b the planet still moves easterly, but from b through c (opposition) to d it moves westerly. At d it resumes easterly movement.

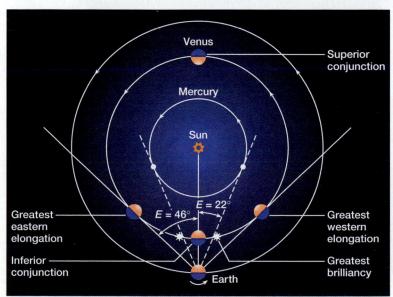

Figure 4.18
Mercury and Venus go through two kinds of conjunctions, one with the planet between the Earth and Sun (inferior) and again with the planet on the far side of its orbit (superior). For simplicity, Mercury's orbit is shown as a circle. Venus's average greatest elongation from the Sun (solid lines) is 46° and Mercury's (dashed lines) is only 22°. (It can actually reach 28° as a result of its orbital eccentricity.) The positions of Venus's greatest brilliancy are indicated by asterisks.

inferior planet is between the Earth and Sun, and **superior conjunction** when the Sun is in the middle, between the Earth and the planet. The synodic period is the interval between successive inferior or superior conjunctions. The inferior planets undergo retrograde motion when they swing past the Earth near inferior conjunction.

An inferior planet's greatest angular separation from the Sun, or **greatest elongation,** will occur at the two points where our line of sight just grazes the orbit. At greatest elongation, Venus averages 46° from the Sun, so it can be seen in a dark sky well after sunset or before sunrise. Poor Mercury, however, is so close to the Sun that its greatest elongation averages only 22°. It therefore rises or sets close upon the Sun and is never seen in a dark sky, only in twilight, and so is difficult to locate.

Like the Moon, the inferior planets go through phases (Figure 4.19). If you could turn a telescope onto Venus or Mercury at superior conjunction, you would see the full sunlit face, or full phase. But at greatest

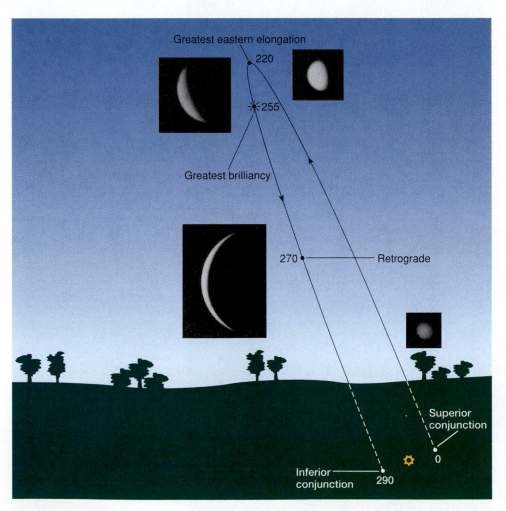

Figure 4.19

Venus is marvelous to watch as it moves relative to the Sun. The numbers indicate the number of days since superior conjunction, and the photographs show the change of phase (changes in direction along the horizon are ignored). Following greatest elongation, the thinning crescent quickly reaches greatest brilliancy as it descends to the horizon, enters retrograde, and approaches inferior conjunction. The planet will then appear to move in the reverse direction in the morning sky.

elongation you see half the daylight and half the nighttime sides, and consequently the planet will look like the Moon at one of its quarters. At inferior conjunction, when you look toward the nighttime side it will be new, and invisible. Between greatest elongation and inferior conjunction the planet takes on the appearance of a crescent. Since its distance from Earth is continuously decreasing between superior and inferior conjunctions, its angular size will steadily grow and correlate with the phases and with angular separation from the Sun.

The brightness of an inferior planet depends on the amount of the visible illuminated angular area (the number of visible square seconds of arc that are in daylight), which in turn depends on a combination of distance and phase. Venus's *greatest brilliancy* actually occurs in the crescent phase between elongation and inferior conjunction. It is then a glorious sight at night, casts shadows in a dark location, and can be seen in full daylight. Mercury's greatest brilliancy is at superior conjunction, when it cannot be seen because of the Sun.

4.3 THEORIES

Although the great astronomer Aristarchus of Samos, who lived between 310 and 250 B.C., had introduced a **heliocentric theory** of the Solar System, one in which the planets go around the Sun, the common view nearly to modern times was **geocentric**, centered on the Earth. But if the Earth is immobile, how do you explain retrograde motions?

4.3.1 Ptolemy

Early ideas about the Solar System were based on the ingrained idea of the circle and the sphere as the perfect figures and on the concept of uniform circular motion. A variety of geocentric systems, some quite ingenious and complex, culminated in the work of Claudius Ptolemaeus, or Ptolemy, an Alexandrian Greek, who in A.D. 140 produced his great *Syntaxis* or, as it became known to the eighth-century Arabs who translated and preserved it, the *Almagest* (from *al Magisti*, "the greatest").

In the *Ptolemaic system* (Figure 4.20), each planet moves counterclockwise once every synodic period on a small circular orbit, an *epicycle*, taking a synodic period to complete its round. Each epicycle rides a large circular orbit centered on the Earth that takes the epicycle around us in a *sidereal* period. As the planet swings between the Earth and the epicycle's center, it appears to go backward, or retrograde. The whole system, including an outer starry sphere, rotates about the Earth once a day, causing the stars to rise and set. The Ptolemaic system allowed approximate predictions of planetary positions and was regarded as the absolute standard for the next 1,400 years.

Figure 4.20
Ptolemy's view of a geocentric Solar System had the planets moving on circular epicycles that in turn moved about the Earth on circular orbits. The planets appear to move backward when they pass between the Earth and the centers of their epicycles. The epicyclic centers of Venus and Mercury lie on a line to the Sun to constrain them within their observed greatest elongations. The Moon moves backward on its epicycle to account for observed irregularities. (The diagram is not to scale.)

4.3.2 Copernicus

Absolute standard or not, however, the Ptolemaic system was incorrect. The modern view was ushered in by Nicolaus Copernicus, a Polish astronomer born in 1473. He attended the University of Cracow, studied medicine as well as law in Italy, and worked as a church administrator in Prussia, where he pursued his real interest, astronomy. He was struck by the old Greek idea of a heliocentric Solar System and decided to examine it from a mathematical point of view. It was a labor of love that occupied his entire working life.

The **Copernican system** (summarized in Figure 4.21) appeared in a massive and complex work called *De revolutionibus orbium coelestium* ("On the Revolutions of the Celestial Spheres"). The volume was completed in the year of its author's death, 1543. Much of the book involves the development of the mathematics needed to explain the subject. Before Copernicus's time, the planets were placed at distances from the Earth according to their sidereal periods. Copernicus, however, determined accurate relative distances by calculation. Figure 4.22 shows Venus at greatest elonga-

The distances of planets from the Sun derived by Copernicus are essentially the modern values.

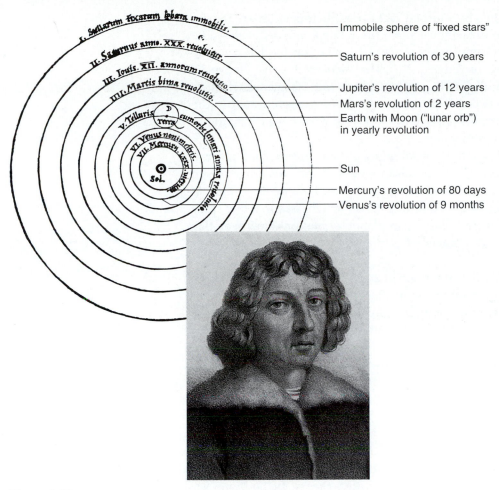

Immobile sphere of "fixed stars"

Saturn's revolution of 30 years

Jupiter's revolution of 12 years

Mars's revolution of 2 years

Earth with Moon ("lunar orb")
in yearly revolution

Sun

Mercury's revolution of 80 days

Venus's revolution of 9 months

Figure 4.21
The Copernican view of the Solar System (Copernicus himself at lower right), taken from *De revolutionibus*, is drawn deliberately out of scale to enable placement on a page. Names of the bodies are given in Latin. Telluris refers to Earth, about which the Moon is seen in orbit; Jovis refers to Jupiter. The largest circle is the celestial sphere, which contains the stars.

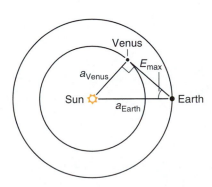

Figure 4.22
Copernicus's method for obtaining the distance to Venus (and to Mercury as well) involves simply determining the angle at which the planet is at maximum elongation, E_{max}. Since Earth-Venus-Sun is at that time a right triangle, and $a_{Earth} = 1$ AU, the distance of Venus (a_{Venus}) is easily found to be 0.72 AU.

tion. Since Earth-Venus-Sun form a right triangle (one with an angle of 90°), Copernicus could easily calculate the relative lengths of the sides, and the distance Sun-Venus in terms of the distance Sun-Earth (or, in modern terms, the distance of Venus in AU). The calculation of the relative distances of the superior planets involves observing the planet when the Earth would appear to be at maximum elongation as seen from the other body. Copernicus's results, all determined from naked-eye observations, are essentially the modern values.

There was as yet no way of knowing which theory, Copernican or Ptolemaic, was correct. The Copernican view had the advantage of simplicity. However, Copernicus still clung to the idea of circular orbits. Although his theory explained retrograde movement, he had to retain epicycles to explain smaller irregularities in planetary motions, and even then his system gave worse predictions of planetary positions than had the Ptolemaic.

Copernicus represents an extraordinary turning point in the history of science and of human thought. He displaced us forever from our centric view of ourselves, from the notion that the Universe was created with our planet as the focal point. Much of the subsequent history of astronomy displaces us ever farther, until we now know ourselves to be at the edge of a galaxy of billions of stars in a Universe of billions of galaxies, a Universe with no center at all.

4.3.3 Tycho

The Copernican system clearly needed improvement that could only be provided by careful observation and innovative theory. Tycho Brahe was a brilliant, contentious Danish nobleman, born only three years after the death of Copernicus. From his youth he had a strong inclination toward astronomy. As his prominence rose, he caught the interest of King Frederik II of Denmark, who provided funds for the construction and support of an observatory, which Tycho named Uraniborg, or Castle of the Heavens (Figure 4.23). The observatory contained large circles marked off in degrees for precise naked-eye measurements of the positions of the stars and planets.

Tycho produced a catalogue of star positions that was far more accurate than anything done before and compiled a continuous record of the movements of the planets among them, working to an astonishing precision of 1 or 2 minutes of arc (only 1/15 the angular diameter of the Moon). After the king's death, Tycho's funds were cut off and he departed for Prague, where he published his catalogue of 777 stars. He died there in 1601.

(a)

(b)

Figure 4.23

(a) Uraniborg, Tycho's observatory, is seen from overhead. **(b)** Tycho works within the curve of his great graduated quadrant, an instrument used to measure angles in the sky to an accuracy of one or two minutes of arc.

4.3.4 Johannes Kepler and the Laws of Planetary Motion

An interpreter of nature rather than an observer, Johannes Kepler was born in Württemberg (part of modern Germany) in 1571, nearly 30 years after the publication of *De revolutionibus.* After his student years he eventually became an instructor in mathematics in what is now Austria. Religious persecution forced him to leave in 1598, and in 1600 he was invited to Prague to be Tycho's assistant. Following the Danish astronomer's death, he succeeded to Tycho's academic position and, most important, acquired his data.

Kepler set out to find how the planets truly move about the Sun by a thought process that was radically new: he would directly use Tycho's observations, particularly those of Mars, to create a new theory, letting the data lead the way. Kepler knew the observations were accurate to within 2 or so minutes of arc, and any new extension of the Copernican system had to fit the observed planetary positions to within these limits.

He proceeded by setting up pairs of observations at intervals of a Martian sidereal year, at the beginning and end of which the Earth would be at two different places, as well as pairs at intervals of an *Earth* year, at the beginning and end of which *Mars* would be at two different places. He first tried circular orbits with centers offset from the Sun. However, he then found that to match Tycho's observations he had to flatten the circle. Rejecting the wisdom of the ages, as he wrote in his *New Astronomy,* "It is as if I awoke from sleep and saw a new light." After three years of crushing calculation, he saw that the orbit was an *ellipse.*

What went for Mars should go for the other planets as well, and in 1609 he finally announced his **first law of planetary motion** (introduced in Section 2.5.1):

1. The orbits of planets are ellipses with the Sun at one focus.

More profoundly, with this **law of ellipses** he had discovered the first real indication that the Sun *controls* the movements of its tiny companions.

At the same time, Kepler explained the variation of Mars's angular motion against the background of the constellations (Figure 4.24). Kepler's **second law of planetary motion** states:

2. The line that connects the planet to the Sun sweeps out equal areas in equal times.

According to this **equal areas law,** a planet must move farther at perihelion in a given amount of time (and therefore it must move faster) than it would at aphelion in order that the two swept areas be the same. The

Kepler saw that the orbit of Mars was an ellipse.

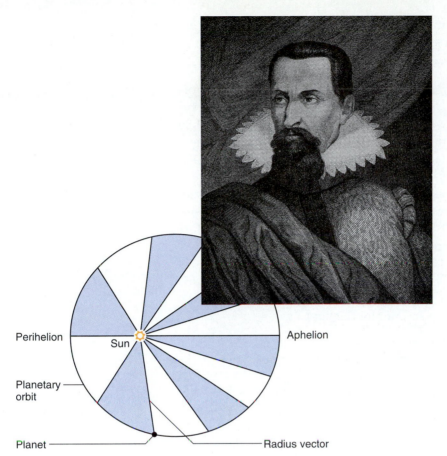

Figure 4.24
Kepler's second law (Kepler at upper right) states that the line connecting the planet to the Sun must sweep out equal areas in equal times. All the sectors have the same area. The planet takes the same time to move from one edge of each sector to the other edge. As it gets closer to the Sun, it therefore moves faster.

movement is such that the average distance of a planet from the Sun over the course of its year equals the length of its semimajor axis.

The effect of Kepler's second law is witnessed by the *inequality of the seasons*. Northern hemisphere spring and summer each have 93 days, but autumn has 90 and winter 89. The Earth is near its perihelion point only two weeks after the winter solstice and near aphelion two weeks past the beginning of summer. The Earth consequently moves faster during autumn and winter, making those pass quickly, and slower during spring and summer, stretching them out by a few days.

Kepler finally set out to examine the orbits of the other visible planets and to find the relations among them. Ten years after his publication of the first two laws, he established his **third law of planetary motion:**

3. The squares of the orbital periods of the planets about the Sun are proportional to the cubes of the orbital semimajor axes.

This **harmonic law**—one of the central ideas in astronomy—is usually expressed relative to the Earth's orbit by the formula

$$P^2 = a^3,$$

where P is the planet's sidereal period expressed in years, and a is its orbital semimajor axis in AU. Use Jupiter as an example. Its sidereal period is 11.86 years and its distance from the Sun is 5.20 AU. The square of 11.86 (11.86 × 11.86) is 141, as is the cube of 5.2 (5.2 × 5.2 × 5.2). (Differences found by carrying the calculation into tenths and hundredths are caused by round-off errors in the periods and semimajor axes.)

Look at the power of this gloriously simple law. You take your telescope outdoors and discover a new body in motion about the Sun. Continued observation shows it to have a sidereal period of 2.6 years. Just square 2.6 to get 4.74. Then find the number that when multiplied by itself twice (the cube root) gives 4.74 to find a semimajor axis of 1.89 AU. Together the three laws allow calculations of accurate planetary positions both into the future and back into the past. We now know how the planetary system works, and we can account for the things we see.

4.3.5 Galileo

At about the same time Kepler produced his epochal first two laws, the Italian Galileo Galilei was observationally confirming the Copernican system. He was born in Pisa in 1564, seven years before Kepler. Absorbed by science and astronomy, he eventually became a professor at the universities of Pisa and Padua. Although he lectured on the Ptolemaic system in his younger years, he was increasingly drawn toward the logic of the Copernican. He had an immense natural curiosity about the world around him. Although he did not invent the telescope (Figure 4.25), and may not even have been the first to turn it on the sky, he was unquestionably the first to use it in a systematic way to learn something of the nature of the heavens. He first examined the sky during 1609 and 1610, and in those marvelous years found that our Moon has mountains and valleys like the Earth, Jupiter has moons of its own, Venus exhibits phases, the Milky Way is made of faint stars, Saturn has "appendages" (later determined to be rings), and the Sun is spotted. Copernicus helped open our minds to the true nature of the Universe; Galileo now helped open our eyes, for the first time peering out into the darkness to see not just how things moved, but how they are *made*. He began our real study of the planets and the stars.

Galileo's greatness was that he continued to observe for years, thinking about what he saw and, most important, drawing conclusions. It was evident to him that his view of the Solar System amply confirmed Copernican thought. The lunar surface looks remarkably like that of the Earth, with mountains and basins, so the Earth is not unique. The Jovian satel-

Galileo helped open our eyes to see how things are made.

Figure 4.25
Galileo, who represents a scientific transition to modern times, is seen with one of his telescopes in this fresco by the Florentine painter L. Sabatelli.

lites go around Jupiter just as Copernicus said the planets go around the Sun; that is, a central body other than the Earth is surrounded by orbiting companions. Venus can pass through a full set of phases only if it has a path that takes it around the Sun, not the Earth. Galileo's astonishing labor helped put the Ptolemaic system to death. In his most famous work, the *Dialogue on Two Chief Systems of the World* (1632), he promoted Copernicanism by an imaginary, but not impartial, debate. For so vigorously setting out to destroy Aristotelian and Ptolemaic ideas, Galileo was called to the Inquisition at Rome. His fame protected him from harsh punishment, but he was forced to recant his positions and was confined to his home from 1633 until his death in 1642. The church ultimately cleared Galileo in 1992.

The combined work of these men told us *how* the planets move about the Sun, but not *why*. The discovery of the reason—gravity and how it works—remained for Isaac Newton and Albert Einstein. In Newton's hands, Kepler's third law became a way not just of probing the Solar System but of examining the structure of the entire Universe.

KEY CONCEPTS

Conjunction: The position in which a planet (as viewed from the Earth) is in the same direction as the Sun (or in which two planets are aligned with each other); in **opposition,** a planet is opposite the Sun as viewed from Earth (or two planets are opposite each other).

Copernican system: A system of circular heliocentric planetary orbits.

Eclipse of the Moon: The passage of the Moon through the Earth's shadow; in a **total eclipse of the Moon,** the entire Moon is immersed; in a **partial eclipse of the Moon,** only part of the Moon passes through shadow.

Eclipse of the Sun: The passage of the Moon across the Sun and the lunar shadow across the Earth; in an **annular eclipse,** the Moon is too far from the Earth to cover the Sun completely, leaving a ring of sunlight; in a **total eclipse of the Sun,** the Moon covers the entire Sun; in a **partial eclipse of the Sun,** the Moon blocks only part of the Sun.

Geocentric theory: A theory of the Solar System in which the planets orbit the Earth.

Greatest elongation: The maximum possible angular separation from the Sun for an inferior planet.

Heliocentric theory: A theory of the Solar System in which the planets orbit the Sun.

Inferior conjunction: A conjunction between Venus or Mercury and the Sun in which the planet lies between the Earth and the Sun; in a **superior conjunction** the planet is on the other side of the Sun.

Inferior planets: Mercury and Venus, the two inside Earth's orbit; the **superior planets** are Mars through Pluto, those outside the Earth's orbit.

Parallax: The apparent shift in the position of a body when viewed from different directions.

Perigee: The point in the lunar orbit (or in any other orbit around Earth) closest to the Earth; **apogee** is the most distant point.

Phases (of the Moon): The different apparent shapes of the Moon caused by viewing different segments of the lighted side as the Moon orbits the Earth.

Planets: The Sun's family of major orbiting bodies.

Retrograde motion: The apparent westward motion of a planet relative to the stars as a result of the Earth passing a superior planet in orbit or of the Earth being passed by an inferior planet.

Sidereal period: The orbital period of the Moon or a planet relative to the stars, or from one orbital point back to the same point.

Synodic period: The orbital period of the Moon or a planet relative to the Sun; for the Moon, the period of the phases.

KEY RELATIONSHIPS

Kepler's laws of planetary motion:

1. The **law of ellipses:** The orbits of planets are ellipses with the Sun at one focus.

2. The **equal areas law:** The line that connects a planet to the Sun sweeps out equal areas in equal times.

3. The **harmonic law:** $P^2 = a^3$, where P is the sidereal period in years and a is the semimajor axis of the orbit in AU.

EXERCISES

Comparisons

1. What is the difference between waning and waxing crescent Moons?

2. What is the difference, numerically and conceptually, between the Moon's sidereal and synodic periods?

3. What is the conceptual difference between a partial eclipse of the Sun and a partial eclipse of the Moon?

4. What is the difference between total and annular solar eclipses?

5. How do the inferior planets differ from the superior planets?

6. What are the differences between opposition and conjunction, and between inferior and superior conjunctions?

7. List the essential elements of the Ptolemaic and Copernican systems and describe the observational tests that can discriminate between them.

Numerical Problems

8. At approximately what time would you expect the Moon to set if it were a crescent halfway between new and first quarter? About what time would you expect it to rise if it were gibbous and halfway between full and third quarter?

9. How far in angle does the Moon move with respect to the background stars between two successive new Moons?

10. Assume that you live at a latitude of 40°N. What is the maximum altitude the Moon can attain and why is that altitude different from the maximum solar altitude?

11. What is the maximum deflection of a star caused by the aberration of starlight as seen from Mercury?

12. NASA launches a spacecraft into orbit about the Sun with a period of six years. What is the craft's semimajor axis? If the orbit is circular, how fast in km/s does the craft travel?

Thought and Discussion

13. If the Moon is full on March 12, what will be the phase on **(a)** March 19; **(b)** March 31?

14. If the Moon is in its third quarter, what is the phase of the Earth as viewed from the Moon?

15. How would the Moon appear to move over the month if viewed from the Earth's north pole?

16. Why can we see the full Moon during a total lunar eclipse?

17. What are the conditions under which you will see **(a)** a total lunar eclipse; **(b)** a total solar eclipse; **(c)** an annular eclipse?

18. Which three planets have the most eccentric orbits? Which planet has the most nearly circular orbit?

19. Pluto can come closer to the Sun than Neptune over a short portion of its orbit. Why do Pluto and Neptune not crash into each other?

20. Why is the synodic period of Saturn shorter than the synodic period of Jupiter?

21. At what orbital positions do Mars and Venus move at their maximum angular speeds (as seen from Earth) in the retrograde direction?

22. If Jupiter is in Capricornus at opposition, in what constellation will it appear at the next opposition?

23. Why is Mercury so hard to see even though it is quite bright?

24. Why is Venus at greatest brilliancy in a crescent phase rather than in its full phase?

25. What were Galileo's and Tycho's essential contributions toward proving the correctness of the Copernican system?

26. At what point in its orbit will a planet move most slowly in km/s?

Research Problems

27. Use popular astronomy magazines in the library to find information on the brightness of the Moon during the past six lunar eclipses. List the reported reasons for the differences among eclipses.

28. Fit the works of Copernicus, Tycho, Kepler, and Galileo into their times. Using library materials, list the explorations of the world

that were in progress and provide a brief summary on the state of Europe during that period.

Activities

29. Photocopy the equatorial star maps in Appendix 1. Plot the apparent path of the Moon over the course of a month. Draw the phases you see and note the visibility of reflected light from the Earth. Locate the points at which the Moon crosses the ecliptic.

30. If any of the planets are visible during the school term, locate them in the sky using information from a popular astronomy magazine. Using photocopied star maps, plot their motions, noting retrograde if you see it. Locate planetary oppositions if appropriate.

31. If the timing is right, observe a lunar eclipse. Note **(a)** the times of shadow contact; **(b)** the appearance of the Moon during the total phase, especially the variation in brightness across the lunar surface. Why should the brightness vary?

32. Make a chart in which you organize the planets by their three categories: ancient or modern (discovered by telescope); inferior or superior; terrestrial or Jovian (see Chapter 1).

33. If you have access to a telescope, use it to track the phases of Venus. Estimate the elongation of Venus from the Sun and, from a scale drawing of the orbits of the Earth and Venus, plot the position of Venus in orbit relative to the Earth, indicating the phase.

34. Use a popular astronomy magazine to find the time of the next maximum elongation of Mercury; then find the planet.

Scientific Writing

35. It is commonly believed that the phases of the Moon are somehow caused by the Earth's shadow. Write an illustrated article for a children's magazine that shows the difference between lunar phases and lunar eclipses.

36. Some years ago, a television production called "Meeting of Minds" assembled actors portraying historical figures from different eras to discuss a variety of topics. You are on the show with Aristotle. Write a monologue for yourself in which you tell him of discoveries made in the sixteenth and seventeenth centuries regarding the construction of the Solar System.

GRAVITY

The glue of the Universe and how it relates to celestial motions

Throw a ball in the air and it returns to Earth. Why? Nothing seems to pull on it, nothing you can see. The ball is subject to **gravity,** a force of nature that acts over a distance to draw things together. Our understanding of it begins with Kepler and Galileo, continues through the brilliant deductions of Isaac Newton, and culminates in theories by Albert Einstein that give us a clearer view of how the Universe actually works.

5.1 NEWTON AND THE LAW OF GRAVITY

Isaac Newton (Figure 5.1) was born in England in 1642, the year of Galileo's death. A founder of modern science, Newton applied rigorous mathematical treatments in search of the physical principles that underlie the Copernican and Keplerian discoveries. He studied at Cambridge University, where he became a professor of mathematics in 1669. For two

Figure 5.1
Isaac Newton discovered the natures of motion and gravity; he is seen here investigating properties of light. The reflecting telescope he invented is to the right of center.

decades he roamed through science, making discoveries in mechanics, optics, and mathematics. In his great work of 1687, *Philosophiae Naturalis Principia Mathematica* ("Mathematical Principles of Natural Philosophy," known as the *Principia*), he laid down the rules for how things move and presented a law that describes gravity. To do so, he (simultaneously with Leibniz in Germany) invented a branch of mathematics, the calculus. His labors took the discoveries of Copernicus, Kepler, and Galileo and changed our view of the world.

If a body is left to itself, it will move in a straight line forever (Newton's *first law of motion*), its motion described by its **velocity,** a term meaning both its speed (the rate of change of distance with time, expressed, for example, in km/s) and direction. An **acceleration** is any change in velocity, in either speed *or* direction. To accelerate a body you must apply some kind of **force.** You apply a force with your hand to start a ball rolling from rest (an acceleration because the speed is changed); the ball is then slowed by the force of friction against the floor. You accelerate around a curve as you drive at constant speed (a change of direction) because of the force imparted to the car by the tires against the road (Figure 5.2). In his *second law of motion*, Newton showed that the acceleration (A) given to a body is directly proportional to the force (F) applied, but is inversely proportional to the body's *mass* (M), or $A = F/M$ (see MathHelp 5.1). **Mass** is a fundamental property of any physical body and is commonly thought of as the amount of matter the body contains. Newton found how to define and measure mass properly, since if $A = F/M$, $M = F/A$ (and, for completeness, $F = MA$). That is, to find a body's mass, apply a known force and measure the acceleration, whether it be a change in speed, direction, or both. (Mass is measured in *grams*, where the gram is the mass of a cubic centimeter of water, or in *kilograms*, a thousand grams).

Falling bodies accelerate at the same rate independently of their natures.

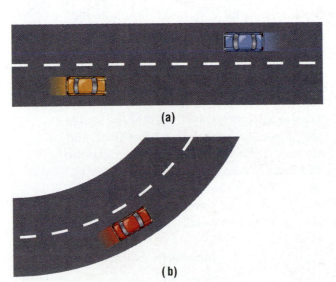

(a)

(b)

Figure 5.2
(a) Although two cars approach each other at identical speeds, their velocities are different because the directions of motion are not the same.
(b) A car passes through a curve at constant speed but is accelerating because it is changing direction.

MATHHELP 5.1 Direct and Inverse Relationships

Many rules in science, like Newton's second law of motion, make use of inverse relationships. If two numbers, D and N, are in direct proportion to each other, one varies directly with the other. If N doubles, so does D. The inverse or reciprocal (R) of a number N is the number 1 divided by N, or $R = 1/N$. We then say that R is proportional to $1/N$. For example, if $N = 100$, $R = 1/100$ or 0.01. N is also the reciprocal of R. Reciprocals behave oppositely to their corresponding numbers. As N increases, $R = 1/N$ decreases and vice versa. If $N = 200$, $R = 1/200 = 0.005$. When N is large, R is small; when N is small, R is large. If $N = 0.001$, $1/N = 1,000$.

Such relationships also commonly involve squares, cubes, and other powers, as they do in Newton's law of gravity. A number S may be the inverse square of N, or $S = 1/N^2$. In that case, we multiply N by itself before dividing into 1. For example, if $N = 100$, $S = 1/100^2 = 1/10,000 = 0.0001$. The rate of change in an inverse square relationship or inverse square law is much greater than it is for a simple reciprocal. As N goes from 2 to 4, $R = 1/N$ goes from 1/2 to 1/4, whereas $S = 1/N^2$ goes from 1/4 to 1/16. Inverse cubes, $C = 1/N^3$, will be more extreme.

Direct and inverse relationships can be combined. Z may be directly proportional to Y and inversely proportional to N squared, so that $Z = Y/N^2$. If Y is doubled, so is Z, but if N is doubled, Z is quartered. If both are doubled, Z is increased by a factor of 2 but decreased by a factor of 4, or is multiplied by a factor of $2/4 = 1/2$.

Gravity is a force—one of the four fundamental forces of nature—and as such can produce accelerations. If you drop a ball, it continuously increases its speed until it hits the ground. The **acceleration of gravity** (g) is measured near the surface of the Earth at 9.8 meters per second per second (abbreviated m/s^2) in the downward direction. After one second the ball will be moving at a speed of 9.8 m/s, after 2 seconds double that at 19.6 m/s, after 3 seconds triple, and so on. Neglecting air resistance, all falling bodies, as Galileo knew, accelerate at this same rate independently of their natures.

If there were not some force acting on the planets to accelerate them into curved paths around the Sun, they would fly off into space along straight lines (Figure 5.3). The Sun, as the central body, must be the source of the force. The more distant the planet, the less its orbital curvature, and the less its acceleration. Measurement of the relative accelerations of the planets showed Newton's friend Edmund Halley (and Newton as well) that the force behaves according to the inverse square of the distance from the Sun; that is, if you double the distance, the strength of the force goes *down* by a factor of two squared, or 4. The Earth clearly has the same effect on the Moon, accelerating it on its curved orbit. Newton realized that if gravity, the downward force felt at the surface of the Earth, extends as far

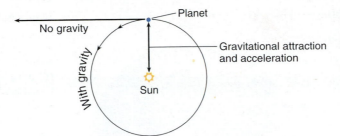

Planet

No gravity

With gravity

Gravitational attraction
and acceleration

Sun

Figure 5.3
A planet is accelerated in the direction of the Sun by a gravitational attraction that keeps the planet moving in a curved path. If the gravity could be switched off (which it cannot), then according to Newton's first law of motion, the planet would fly away in a straight line.

as the Moon, it probably causes the lunar orbital acceleration as well and should thus be the same force that causes planetary accelerations.

From this reasoning, Newton discovered how gravity works. Every mass is surrounded by a gravitational *field* through which it mutually attracts other masses. A force field is space within which the force has an influence. Two masses M_1 and M_2 have a gravitational force between them that is proportional to the product of the masses divided by the square of the distance between their centers (see MathHelp 5.1), or

$$F = G\frac{M_1 M_2}{R^2}.$$

(G, the *gravitational constant*, is found from the force with which two known masses attract each other and equals 6.67 x 10^{-11} when mass and length are in kilograms and in meters). No matter how large R, the force never can become zero. The gravitational field is therefore infinite, gravity acting over all of space. With Newton's **law of gravity** we can follow the motions of the planets and stars in exquisite detail.

Each of the two bodies M_1 and M_2 must accelerate the other. Since $F = MA$, the acceleration of mass M_2 toward mass M_1 is GM_1/R^2, and that of mass M_1 toward M_2 is GM_2/R^2. The acceleration of gravity (g) must then be proportional to the mass of the Earth divided by the distance to the center of the Earth: $g = GM_{Earth}/R_{Earth}^2$. If we measure g from the rate of the increase in the speed of a falling body and know the radius of the Earth, the only unknown quantity (after finding G in the laboratory) is the mass of the Earth, calculated to be 6×10^{24} grams.

When you stand on the Earth, not accelerating because of the solid ground, you are still subject to the force between you and the Earth, which you feel as **weight** (W). Your weight, or the weight of any body, is therefore

$$W = G\frac{M_{Earth} \times M_{body}}{R_{Earth}^2}.$$

The weight of a body thus equals its mass times the acceleration of gravity, or Mg. Mass and weight are entirely different concepts. (Mass is measured in grams or kilograms. Weight measured in these units is a social convention by which weight is set equal to the mass at the Earth's

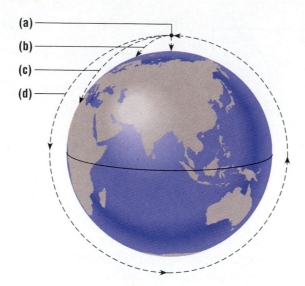

(a)
(b)
(c)
(d)

Figure 5.4

If you drop a rock **(a)** it will fall toward the center of the Earth and land at your feet. If the rock is thrown **(b)**, it still falls toward the Earth's center, but now it drops a good distance away because of its horizontal motion. The faster it is thrown **(c)**, the farther it goes, in part because the Earth drops away beneath it. If it could be thrown fast enough **(d)**, the Earth would curve away at the same rate at which the rock drops. The rock is now in orbit and cannot ever reach the Earth.

surface.) If you ride an elevator upward, your mass stays the same, but your weight decreases as you move away from the Earth's center. Your mass is the same on the Moon, but your weight would be different because the lunar mass and radius (substituted above for the Earth's mass and radius) are different.

Newton showed that orbiting bodies are falling bodies. The Moon is subject to Earth's gravity, and thus the actual direction of acceleration must be toward us. The Moon does not reach us because it is also moving in a direction that on the average is perpendicular to the line between the two bodies. Drop a rock and see how long it takes to strike the ground (Figure 5.4). Then throw the rock horizontally. As it moves away, it drops at the same rate as before and strikes the ground after the same amount of time. Now imagine throwing it very hard, at such a speed that as it falls, the Earth curves below it. The rock will now take longer before it strikes. If you hurl the rock hard enough that it travels at 28,400 km/h (17,500 miles/h), and ignore the resistance of the Earth's atmosphere, the rock drops at the same rate at which the Earth curves and, just like the Moon, is in orbit.

Newton showed that orbiting bodies are falling bodies.

5.2   KEPLER'S GENERALIZED LAWS OF PLANETARY MOTION

Newton calculated that a planet accelerated according to a law in which the force dropped by the inverse square of the distance would move in an elliptical path. He had reproduced Kepler's first law! Upon application of his laws of motion and the law of gravity, he then discovered a more general version:

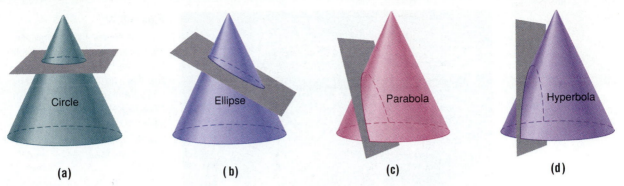

Figure 5.5

The conic sections are created by intersecting a cone with a plane. **(a)** If the plane is parallel to the cone's base, the intersection is a circle. **(b)** If you tip the plane but leave the pitch angle less than that of the cone's side, the intersection is an ellipse. **(c)** A parabola is generated by a plane that cuts the cone parallel to the side. **(d)** A hyperbola is made by a plane with a pitch angle greater than that of the side.

1. The paths of orbiting bodies are **conic sections** with the other body at one focus of the curve.

The conic sections (Figure 5.5) are curves created when a cone is intersected at different angles by a plane: the *circle, ellipse, parabola,* and *hyperbola.* When the plane is parallel to the cone's base, the intersection is a circle. If the plane is tilted, the intersection is an ellipse that becomes more elongated as the eccentricity (*e*) increases from 0 to 1. The parabola is the limiting case of the ellipse in which the plane is parallel to the cone's side, one end does not close, and *e* = 1. If the cone is cut at a steeper angle, *e* is greater than 1 and open-ended hyperbolas are created.

Imagine a body in uniform circular motion around the Sun, curve *A* in Figure 5.6. If you apply a force at *X* to slow the body, it goes into an elliptical orbit (curve *B*) with *X* at aphelion. It will follow Kepler's second law of motion, speeding up as it falls toward the Sun, slowing as it recedes. If you apply a force at *X* to speed up the body, you can place it into an elliptical orbit with *X* at perihelion (curve *C*). Then apply so much force that solar gravity cannot bring the outbound speed to zero. It is now in a one-way, open-ended hyperbolic orbit (*E*) *and can never return.* There is a special case in which the velocity decreases to zero, but only after an infinite time has passed. In that case, the path is that of a parabola (*D*).

Throw a ball into the air. The faster you throw it, the higher it will go. At a critical **escape velocity** the ball will go into a parabolic orbit and not come back. All bodies—the Earth, Sun, other planets—have escape velocities. The larger the mass of the planet, the larger the escape velocity; the farther you are from the center of the planet, the smaller the escape velocity (it is actually proportional to the square root of M_{planet}/R_{planet}). The escape velocity at the Earth's surface is 11.1 km/s or 39,960 km/h, the

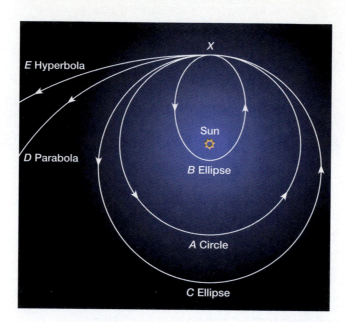

Figure 5.6
Orbit *A* about the Sun is a circle. An elliptical orbit can be created by either slowing the body at *X* (curve *B*) or by speeding it up (curve *C*). If the body is given enough speed, the orbit can be parabolic (*D*) or hyperbolic (*E*). Since these two curves, the parabola and the hyperbola, are open-ended, the body will leave, never to return.

minimum speed required to send a spacecraft toward an interplanetary mission.

To see how Newton generalized Kepler's second law, imagine tying a rock to a string and whirling it in a circle. The rock's **angular momentum** (*L*) is its mass (*M*) times its velocity (*v*) times the length of the string (the orbital radius, *R*), or $L = MvR$. The general restatement of Kepler's second law is:

2. In any closed system, angular momentum is conserved; that is, angular momentum does not change unless there is an action by an outside force.

The **conservation of angular momentum** is a powerfully important concept in astronomy. Its personal effects can be impressive. A skater starts her spin with outstretched arms, applying a force with her skates to establish her angular momentum. She then brings her arms inward to reduce her body's average radius, and her rotation speed increases dramatically. In an elliptical orbit, the distance of a planet from the Sun varies. As the distance goes down, the speed goes up to produce Kepler's equal areas law.

Kepler's third law states that $P^2 = a^3$, where *P* and *a* (the sidereal period and semimajor axis respectively) are expressed in years and Astronomical Units. To analyze orbits in terms of seconds and meters, however, we must write $P^2 = ka^3$. The problem is then to find the nature of k. The gravitational force depends on the separation between the two bodies and the product of their masses, so k must contain the masses. The law of gravity leads to a generalization of Kepler's third law:

3. $P^2 = \dfrac{\text{constant}}{(M_1 + M_2)} a^3,$

where the masses of the two bodies are given by M_1 and M_2 and k is seen to be a constant divided by $(M_1 + M_2)$. (The constant is actually $4\pi^2/G$.) Kepler's third law works in its original form for the Solar System because the mass of the Sun (M_1) is so much greater than the masses of any of the planets, M_2. Consequently, $M_1 + M_2$ is essentially the same for all the planets, and for all practical purposes (certainly for naked-eye observations) the sum is *also* a constant. As a result, we can compare all the other planets with the Earth and write $P^2 = a^3$ with P in years and a in AU. The generalized law, however (with P in seconds and a in meters), is applicable to *any* orbit that involves *any* two bodies.

Kepler's generalized third law provides us with great power in our quest to understand the Universe. The Earth's orbital period is 3.16×10^7 seconds, and its semimajor axis (the AU) is 1.50×10^8 km. The unknown quantity in the equation is the sum of the masses of the Earth and Sun, which is readily found to be 1.99×10^{30} kg. The mass of the Earth is negligible by comparison, and when subtracted, the result—the solar mass—is still 1.99×10^{30} kg. We can apply this law to *any* orbit to find the sum of the masses of the two bodies involved, a vital step toward knowledge of their natures.

5.3 DISCOVERY

Newton's theories let us use observations of planetary positions to calculate the properties of planetary orbits. In turn, knowledge of the orbital paths allows accurate calculation of future or past positions. However, Kepler's generalized laws apply to orbiting systems consisting solely of two bodies, of which there are no examples in our Solar System. The Earth is indeed dominated by the Sun, but it is also accelerated by Jupiter as well as by all the other planets. Furthermore, gravity is a mutual affair, and all the planets accelerate the Sun and consequently affect the solar location, and accelerate one another as well. Nine planets and the Sun produce a total of 45 interactions! As a result, the Earth cannot maintain a perfectly elliptical orbit, nor can any of the other planets. If we wish to calculate a correct planetary path, it is necessary to take these small forces, or gravitational **perturbations,** into account.

We first find the elliptical orbit a planet would have under the influence of the Sun alone. Then we calculate all the gravitational accelerations imposed by the other planets, enabling us to compute the way in which the orbital semimajor axis, eccentricity, tilt relative to the ecliptic, and orientations change with time, allowing the accurate prediction of planetary positions well into the future.

Theory could now play a role in the exploration of the Solar System. At the time of the American Revolution, only six planets were known. A modern era, however, was being ushered in by the work of William Herschel. Herschel was born in Germany in 1738 and earned his living in England as a musician. His interests also ranged over mathematics and astronomy. By 1774, he had built his first telescope and had begun to scan the skies in earnest, discovering and studying new celestial sights. In 1781, he came across an unusual object that did not appear in the telescope as a point, like a star, but as a small circular disk. Subsequent observations showed it to move like a planet, and a distance of 19 AU—well outside the orbit of Saturn—could be derived from Kepler's laws.

Herschel's discovery, the planet Uranus (in mythology, Uranus was Saturn's father), is surprisingly bright. At opposition it has a magnitude between 5 and 6 and is just visible to the naked eye; it had been seen and recorded as a faint star as early as 1690. Oddly, neither the old observations nor the new ones that were rapidly being made quite agreed with the positions expected from orbital calculations performed with Newton's laws. In the early 1840s, John Couch Adams in England and Urbain Leverrier in France decided that the culprit must be the gravitational pull of yet another planet even farther from the Sun than Uranus. From the deviation between the observed and expected positions of Uranus, they independently calculated the trans-Uranian planet's position, both men completing the work by 1846. Adams had little success in persuading the English astronomers to look for the body. Leverrier transmitted his results to the Berlin Observatory, where Johann Galle found the new planet—to be named Neptune, after the god of the sea—almost immediately. This discovery was rightly considered a triumph of Newtonian theory.

However, even with Neptune's perturbations added to the calculations, Uranus still did not quite behave as predicted. The American philanthropist and astronomer Percival Lowell, who had founded an observatory in Arizona, suggested the existence of yet another planet. In 1929, he hired a young observer, Clyde Tombaugh, to look for it. A year later Tombaugh came across dim Pluto (Figure 5.7), named after the god of Hades, the underworld. Pluto, however, is now known to have such a low mass that it could not have caused any significant changes in Uranus's orbit; its discovery had been accidental. Calculations of Uranus's position with modern planetary masses predict its position well. Though the Solar System beyond Neptune contains many small bodies of low mass (including Pluto), current observations indicate that there are no more large planets.

The discovery of Neptune was rightly considered a triumph of Newtonian theory.

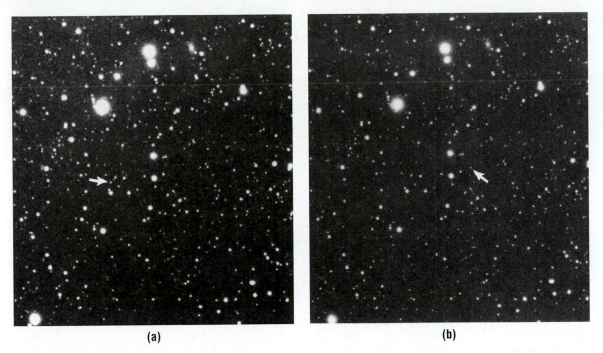

(a)   **(b)**

Figure 5.7
Portions of Clyde Tombaugh's original discovery photographs of Pluto, taken **(a)** January 23 and
(b) January 30, 1930, reveal the planet's motion (retrograde, since Pluto was near opposition).

5.4 SPACEFLIGHT

The space shuttle rises from the launching pad at Cape Canaveral revealing Newton's third and final law, that for every action there is an equal
and opposite reaction. A great blast of hot gas spews from the nozzles of
the rocket engines, and the spacecraft accelerates in the opposite direction.
As it climbs, it begins to move horizontally (Figure 5.8a). By the time it
reaches an altitude of 300 km and a speed of 27,700 km/h, it is moving
parallel to the Earth's surface and is in orbit. The engines shut down and
the shuttle coasts around the planet, taking 91 minutes to make the
journey.

The farther away a spacecraft is from Earth, the less is the acceleration
of gravity, the slower the craft moves, the greater the orbital radius (a), and
the longer the period (P) (in accordance with Kepler's third law, in which
P^2 is proportional to a^3). The orbital radius is the distance from the Earth's
center, the Earth's radius plus the spacecraft's altitude. At 1,000 km above
the Earth's surface the period climbs to 105 minutes. At a distance of
42,200 km from the Earth's center (26,200 miles), the period is 24 hours,

(a)   (b)

Figure 5.8

(a) The space shuttle is accelerating and its path is tilting parallel to the Earth's surface. (b) The astronauts are orbiting around the world with the same acceleration as the cabin and are therefore weightless.

equal to that of the Earth's rotation. If the satellite is above the Earth's equator, it will appear stationary from the ground. The satellite is now in *geosynchronous orbit*. Communications satellites are placed in such orbits so that receivers—TV satellite dishes, for example—can remain fixed, always pointing to the same place in the sky.

> **Astronauts are never out of the Earth's gravity; they simply feel no weight.**

The astronauts riding in the shuttle are accelerated toward the Earth along with the shuttle cabin, and all orbit at the same speed (Figure 5.8b). There is nothing to press the passengers to the shuttle walls or floor, and consequently they are weightless and float around the cabin bouncing off the walls. The astronauts are not "out of the Earth's gravity." One is *never* out of the Earth's gravity—it is this gravity that makes the craft and its inhabitants orbit. They simply feel no weight.

To reach another planet, a spacecraft is accelerated past the escape velocity into a hyperbolic orbit relative to the Earth and into an elliptical (or even hyperbolic) orbit relative to the Sun. Since the planet is also moving, the craft is directed ahead of it so that the two arrive at the same place at the same time. Planetary gravity is even used as an accelerating mechanism. The interplanetary probe *Voyager 2* was directed behind Jupiter in 1979, causing the craft to change direction and to speed up enough to be hurled to Saturn. The same trick was used there to throw it to Uranus and Neptune. Newton would have enjoyed it all immensely.

5.5 EINSTEIN AND RELATIVITY

About 1870, Urbain Leverrier, the co-discoverer of Neptune, found a discrepancy in Newtonian theory. Perturbations produced by the other planets should cause the perihelion point of Mercury's orbit to move counterclockwise around the Sun at the rate of 527 seconds of arc per century. The observations, however, showed the rate to be 38 seconds of arc larger (more precise observations eventually showed a discrepancy of 43 seconds). The difference between theory and observation defied explanation. The problem was ultimately resolved by a young man named Albert Einstein (Figure 5.9), who extended Newton's science of mechanics through the theories of **relativity.**

Einstein was born in 1879 in Germany and educated there and in Switzerland, where he made his living between 1901 and 1909 as a patent examiner. The first part of his great work, the **special theory of relativity,** was published in 1905; it concerns bodies in uniform constant relative motion, those with no accelerations. Imagine that a man on a flatbed truck moving at constant speed passes by a woman standing on the side of the road (Figure 5.10). Pretend that any observation or measurement is possible. To her, the truck is in motion; to him, the woman appears to be moving in the opposite direction. There is no such thing as absolute motion,

Figure 5.10
A truck moves past a woman standing by the roadside. A man on the truck shines a light in the direction of the truck's motion and measures a speed c. In spite of the truck's motion, the woman also sees the light move at c. The woman also sees lightning bolts strike the ground simultaneously. Because the man is moving, he sees them strike at different times.

Figure 5.9
Albert Einstein formulated the theories of relativity and gave us our modern view of space, time, and gravity.

only *relative* motion. He throws a ball in the direction of his motion at 20 m/s. If the truck is moving at 50 m/s, she will see the ball move at 70 m/s. She just adds the velocities, a basic Newtonian concept. Now say the truck is moving at half the speed of light and that he shines a light in the direction of his motion. He sees the beam move away from him at a speed (c) of 299,792 km/s (186,282 miles/s). You would intuitively say that the stationary woman should measure the beam moving forward at the sum of the two speeds, or 1.5c. However, contrary to expectation, she *too* measures the speed of the light at c! The speed of light is independent of the speeds of the source or the observer, facts documented by laboratory experiments.

Now the woman sees two lightning bolts strike the road simultaneously, one in front of the truck, the other behind it. Because the man is moving, the light from the forward stroke will reach him before the light from the rearward one arrives. To him, the strokes are *not* simultaneous: his sense of time and space must be different from hers. If the woman and the man are carrying identical clocks, each will see the other's clock appear to run slow relative to the one in hand, an effect called *time dilation*. Each person also holds a 100-g meter stick in the direction of motion. As they move past each other, each sees the other's stick as shorter than a meter, and each measures a mass greater than 100 g.

These effects are important only at speeds that approach c. Even at 0.1c, the distortion in time and space is only 1%. Consequently, Newtonian rules work very well at speeds encountered in everyday life. But as soon as the speed is high enough, close to that of light, relativistic rules must be used. If v = c, the measured length of the moving body becomes zero and the time interval and the mass become infinite. Since acceleration requires a force, it would take an infinitely strong force to accelerate a mass to the speed of light. Such a force does not exist. Consequently, no material body can attain the speed of light; it can only approach it. The speed of light is thus the absolute upper limit to speeds in the Universe.

Proofs of special relativity are ample. Time dilation has been measured with precise clocks in relative motion. Atomic accelerators can speed particles nearly to c, and observations of their masses agree perfectly with prediction. But special relativity does not solve Leverrier's problem with Mercury's perihelion. That was finally explained when Einstein expanded the theory to include accelerations, which led to the **general theory of relativity** (published in 1914) and to the modern theory of gravity.

"Space" has three dimensions, their directions easily seen where the walls and ceiling meet in the corner of a room. We are used to thinking of "time" as independent of space. In the theory of relativity, however, it is necessary to combine space and time into a single four-dimensional system called **spacetime.** We move through the Universe in a continuum of four dimensions, our location given by three coordinates in space, x, y, and z, and by one in time, t. The four cannot be separated, and the distor-

(a)　　　　　　　　(b)

Figure 5.11
(a) A woman rides an elevator in interstellar space and feels the upward acceleration as a downward weight. She holds a mass on a spring, which stretches to the floor under the acceleration. **(b)** A man stands on the Earth, holding an identical mass. The Earth's gravitational field pulls him downward and also stretches his spring. If the elevator accelerates at 9.8 m/s², each person sees and feels exactly the same effects.

tions of time and space that involved the truck experiment in Figure 5.10 cannot be separated in time and space either. We have to deal with distortions in spacetime as a unit.

A woman floats in a weightless state into an elevator placed deep in interstellar space away from any significant gravitational field. The doors close and she is accelerated upward at a steady rate (Figure 5.11). She is forced to the floor and acquires weight. To her, it feels exactly as if a gravitational field has been switched on. If she drops a ball, it will accelerate to the floor, and if the elevator is accelerating at 9.8 m/s², she feels as if she is back on Earth. There is no way for her to tell the difference. The identity of the experience of the acceleration of the elevator and the acceleration of gravity is called the *principle of equivalence*.

A ray of light now passes through the accelerating elevator. To the passenger the light will appear to move in a curved path. Because of the principle of equivalence, the same effect must be seen in a gravitational field. The path followed by a light ray, however, serves as our ultimate definition of a straight line. Therefore, it is *spacetime* that is curved. Einstein's great discovery is that you fall to Earth as a result of the curvature of spacetime caused by the very existence of mass. The effect is something like a well in spacetime in which bodies will be accelerated (Figure 5.12). The planets orbit around the Sun within this gravitational well. The greater the mass, the more spacetime curves, the greater the acceleration, and the greater the apparent force that is felt. As you move away from the mass and the curvature decreases, so do the force and acceleration, and the planets move more slowly.

Mercury is the fastest-moving planet, and although its average speed of 48 km/s is far below that of light, its proximity to the Sun is enough to affect its motion in the deep solar gravitational well. The elliptical orbit

You fall to Earth because of the curvature of spacetime.

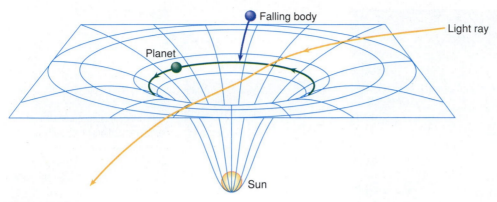

Figure 5.12

It is impossible to draw four-dimensional spacetime on a sheet of paper. Instead, imagine that space has only two dimensions, so that spacetime has a total of three. This perspective drawing shows how a mass causes a distortion in spacetime's fabric that results in a gravitational field. The central mass will make bodies fall toward it, will accelerate planets to keep them in orbit, and will bend light rays (greatly exaggerated here).

keeps the planet moving through different curvatures of spacetime, making Mercury's perihelion advance by the missing 43 seconds of arc per century, which does not happen under Newtonian gravitational theory. A vast body of other evidence now supports general relativity. When a ray of light passes the Sun, it must curve. As a result, the stars that surround the Sun in the daytime are slightly displaced outward from the solar center, the observed effect matching prediction. The rate of a clock should depend on the strength of the gravitational field in which it runs. Precise clocks on the ground are observed to run slower than those placed on top of tall towers, where the strength of the gravitational field is diminished. Relativity is not just an abstract theory. It is real, it is true, and it helps describe our world; it will be an important factor in chapters to come.

We must admit, however, that Einstein's discovery only pushes the unknown back a little farther. *Why* a mass curves spacetime is yet to be understood, and the mystery that is gravity remains.

▶ KEY CONCEPTS

Acceleration: Any change in velocity, that is, in speed or direction.

Angular momentum (*L*)**:** For an orbiting body, mass times velocity times orbital radius.

Conic sections: The curves (circle, ellipse, parabola, and hyperbola) defined by the intersections of a cone and a plane.

Conservation of angular momentum: The concept that in a closed system total angular momentum is always constant.

Escape velocity: The velocity a body needs to achieve a parabolic orbit from another body's surface.

Force (*F*)**:** That which produces an acceleration of a mass.

Gravity: An attractive force; the curvature of spacetime caused by the presence of mass.

Mass (*M*): The amount of matter in a body, or the degree to which a body's motion resists a force; usually measured in grams or kilograms.

Perturbations: Orbital changes induced by outside gravitational forces.

Relativity: The branch of mechanics, developed by Albert Einstein, that lets the speed of light be independent of the speeds of the source or observer; the **special theory of relativity** involves constant speed, the **general theory of relativity** involves accelerations.

Spacetime: A four-dimensional construction that consists of the three dimensions of space and the one of time.

Speed: The rate at which a body changes its distance with time (measured, for example, in m/s, km/s, or km/h).

Velocity (*v*): The combination of speed and direction.

Weight (*W*): The force with which a body is pressed to the surface of another body as a result of gravity.

KEY RELATIONSHIPS

Acceleration of gravity (g): The acceleration of a falling body at the Earth's surface is

$$g = G\,\frac{M_{\text{Earth}}}{R^2}.$$

Kepler's generalized laws of planetary motion:

1. The path of an orbiting body is a conic section with the other body at one focus of the curve.

2. In any closed system, angular momentum is conserved.

3. $P^2 = \dfrac{\text{constant}}{(M_1 + M_2)}a^3.$

Law of gravity:

$$F = G\,\frac{M_1\,M_2}{R^2}.$$

EXERCISES

Comparisons

1. Distinguish between velocity and speed.

2. Distinguish between mass and weight.

3. Distinguish a body in an elliptical orbit from one in a hyperbolic orbit.

4. What distinguishes special from general relativity?

Numerical Problems

5. What are the periods of Earth satellites in circular orbits 2,000 km and 10,000 km above the ground?

6. How many times more (or less) would you weigh if you could travel to a planet that has **(a)** the same radius as the Earth but one-third the mass; **(b)** the same mass as the Earth but one-third the radius; **(c)** one-third the mass of the Earth and one-third the radius?

7. The Moon has 1/82 the mass of Earth and a radius is 27% that of Earth. How much would your lunar weight be?

8. How fast must a body be launched from the surface of the Moon to escape?

9. Tie a string 2 m long to a 1-kg weight and whirl it at 10 m/s. Reel in the string to a radius of 1 m. How fast is the weight moving now?

10. Use the orbit of Mars to determine the mass of the Sun.

Thought and Discussion

11. What is meant by the term *acceleration*?

12. If you shut off the engine of a car and coast in neutral on a straight, level road, the car will come to a stop. Is this behavior a violation of Newton's first law of motion?

13. How might you use the acceleration of gravity to measure the shape of the Earth?

14. Write a formula for the acceleration of gravity at the surface of Mars.

15. If you could increase the mass of the Sun but keep the semimajor axis of the Earth a constant, what would happen to the Earth's sidereal period?

16. If you could decrease the mass of the Sun but hold the sidereal period of the Earth constant, what would happen to the semimajor axis of the Earth's orbit?

17. What role did astronomy play in the discovery of laws of motion and mechanics?

Research Problem

18. List the different important functions of Earth satellites.

Activities

19. Try Galileo's proposition regarding the acceleration of gravity for yourself. Use two different weights that are both heavy enough to be unaffected by air resistance.

Drop them at the same time and have someone with a stopwatch try to detect any difference in the times at which they strike. What might cause any such difference?

20. Using an annual map from a popular astronomy magazine like *Sky & Telescope* and *Astronomy*, find Uranus in the nighttime sky with or without binoculars.

Scientific Writing

21. You read in the newspaper that "the shuttle astronauts are floating around in their cabin because they have been launched out of the gravitational field of the Earth." Write a letter to the editor in which you set the matter straight.

22. Write a one- or two-page article on the importance of Kepler's laws to modern science, as he deduced them and as Newton derived them.

ATOMS, LIGHT, AND TELESCOPES

To observe and understand the very large we first examine
the very small

6

As early as 400 B.C., the Greek scholar Democritus suggested that matter is
composed of tiny particles called **atoms,** from the Greek word meaning
"uncuttable." We now know that these atoms are made of smaller particles and that they interact with light and other forms of radiation. Observation of this radiation with our telescopes allows us to understand the
natures and atomic constituencies of astronomical bodies that we cannot
reach out to touch.

6.1 ATOMS

All matter—the Sun, stars, grass, the air we breathe—is made of chemical
elements, each of which is composed of a different kind of atom. Of the
110 known elements (Table 6.1), 90 are present in the Earth. Many are
known to us in our daily lives, including the oxygen and nitrogen we
breathe, the carbon that helps compose living things, the metals—iron,
aluminum, copper—with which we build.

TABLE 6.1
The Chemical Elements

Name	Symbol	A	M	Name	Symbol	A	M
Hydrogen	H	1	1	Neon	Ne	10	20
Helium	He	2	4	Sodium	Na	11	23
Lithium	Li	3	7	Magnesium	Mg	12	24
Beryllium	Be	4	9	Aluminum	Al	13	27
Boron	B	5	11	Silicon	Si	14	28
Carbon	C	6	12	Phosphorus	P	15	31
Nitrogen	N	7	14	Sulfur	S	16	32
Oxygen	O	8	16	Chlorine	Cl	17	35
Fluorine	F	9	19	Argon	Ar	18	40

TABLE 6.1
continued

Name	Symbol	A	M		Name	Symbol	A	M
Potassium	K	19	39		Europium	Eu	63	153
Calcium	Ca	20	40		Gadolinium	Gd	64	158
Scandium	Sc	21	45		Terbium	Tb	65	159
Titanium	Ti	22	48		Dysprosium	Dy	66	164
Vanadium	V	23	51		Holmium	Ho	67	165
Chromium	Cr	24	52		Erbium	Er	68	166
Manganese	Mn	25	55		Thulium	Tm	69	169
Iron	Fe	26	56		Ytterbium	Yb	70	174
Cobalt	Co	27	59		Lutecium	Lu	71	175
Nickel	Ni	28	58		Hafnium	Hf	72	180
Copper	Cu	29	63		Tantalum	Ta	73	181
Zinc	Zn	30	64		Tungsten	W	74	184
Gallium	Ga	31	69		Rhenium	Re	75	187
Germanium	Ge	32	74		Osmium	Os	76	192
Arsenic	As	33	75		Iridium	Ir	77	193
Selenium	Se	34	80		Platinum	Pt	78	195
Bromine	Br	35	79		Gold	Au	79	197
Krypton	Kr	36	84		Mercury	Hg	80	202
Rubidium	Rb	37	85		Thallium	Tl	81	205
Strontium	Sr	38	88		Lead	Pb	82	208
Yttrium	Y	39	89		Bismuth	Bi	83	209
Zirconium	Zr	40	90		Polonium[b]	Po	84	210
Niobium	Nb	41	93		Astatine	At	85	210
Molybdenum	Mo	42	98		Radon	Rn	86	222
Technetium[a]	Tc	43	99		Francium	Fr	87	223
Ruthenium	Ru	44	102		Radium	Ra	88	226
Rhodium	Rh	45	103		Actinium	Ac	89	227
Palladium	Pd	46	106		Thorium	Th	90	232
Silver	Ag	47	107		Protactinium	Pa	91	231
Cadmium	Cd	48	114		Uranium	U	92	238
Indium	In	49	115		Neptunium[a]	Np	93	237
Tin	Sn	50	120		Plutonium[a]	Pu	94	242
Antimony	Sb	51	121		Americium[a]	Am	95	243
Tellurium	Te	52	130		Curium[a]	Cm	96	247
Iodine	I	53	127		Berkelium[a]	Bk	97	249
Xenon	Xe	54	132		Californium[a]	Cf	98	251
Cesium	Cs	55	133		Einsteinium[a]	Es	99	254
Barium	Ba	56	138		Fermium[a]	Fm	100	253
Lanthanum	La	57	139		Mendelevium[a]	Md	101	256
Cerium	Cs	58	140		Nobelium[a]	No	102	254
Praseodymium	Pr	59	141		Lawrencium[a]	Lw	103	257
Neodymium	Nd	60	142		Kurchatovium[a]	Ku	104	—
Promethium[a]	Pr	61	147		Hahnium[a]	Ha	105	—
Samarium	Sm	62	152					

The column headed A gives the atomic number. The column headed M gives the mass number or atomic weight (number of nucleons, or protons plus neutrons) of the most abundant isotope of each element, or in the case of radioactive elements, the weight of the most stable isotope.

[a] Radioactive elements that do not exist in the Earth and are made in the laboratory; others have been made above element 105.

[b] All elements heavier than bismuth are radioactive; M gives mass of most stable isotope.

Figure 6.1

Protons, neutrons, and electrons are the building blocks of atoms. The proton and electron have opposite electric charges of the same strength, whereas the neutron is neutral. The proton and neutron are about 10^{-13} cm across and have about the same mass. The electron has 1/1,800 the mass of the proton.

In modern times physicists discovered that all the different kinds of atoms are composed of three more-elementary particles that can be arranged in a variety of ways: **protons, neutrons,** and **electrons** (Figure 6.1). Protons and electrons carry the **electromagnetic force,** which has two manifestations, the **electric charge** and the **magnetic field.** (The concept of the field is defined in Section 5.1). Electromagnetism is another of the four fundamental forces of nature, forces that act over a distance, the first being gravity. The electromagnetic force, however, is vastly stronger. Moreover electric charge (which produces a surrounding *electric field*) has two forms, positive and negative: opposite charges attract, similar charges repel.

Atoms are mostly empty space.

The proton, which has a mass of only 1.67×10^{-24} g and a tiny diameter of 10^{-13} cm, carries the positive charge. The electron has an equal and opposite negative charge, but its mass is 1,800 times smaller and it has no measurable size. By contrast, the neutron, which has about the same mass as a proton, carries no charge at all. Since protons and electrons have opposite charges, they will attract each other. An electron attached to a central proton (the **nucleus**) creates the simplest kind of atom, hydrogen (Figure 6.2a). Atomic electrons are at great relative distances from their nuclei and atoms are therefore mostly empty space. What we feel as the solid surface of an object is not the atoms themselves but the electrical forces of their constituents.

Each kind of atom (and chemical element) has a different number of protons (the *atomic number*) in its nucleus. That of hydrogen is 1. Helium has a nucleus that contains two protons tightly bound together (Figure 6.2b) along with (usually) two neutrons. Since similar electric charges repel, the protons must be held together by yet another force, the most powerful of all, the **strong** or **nuclear** force, which is carried by both protons and neutrons. Unlike gravity or electromagnetism, the strong force extends over only the size of the nucleus. Within that realm, it overpowers the electric repulsion and keeps the protons and neutrons tied within its grip. The progression of elements continues with additional protons. Lithium has 3 protons, carbon 6 (Figure 6.2c), iron 26, and uranium 92 (Table 6.1). As the number of protons increases, so must the number of neutrons in order to add binding force. The sum of the two for any element is called the *atomic weight*.

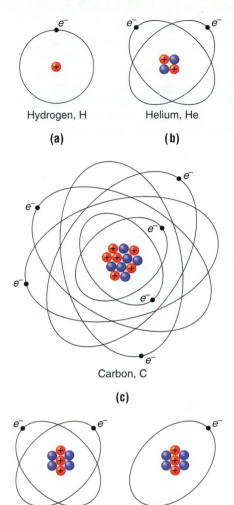

Hydrogen, H

(a)

Helium, He

(b)

Carbon, C

(c)

Ionized lithium, Li⁺

(d)

Doubly ionized
lithium, Li⁺²

(e)

Figure 6.2

(a) The simplest atom, hydrogen, has a single proton (red, +) surrounded by an electron (e^-). The size of the proton is about 1/100,000 the size of the atom. Electrons are sometimes said to orbit their nuclei, although the analogy with the planets and the Sun is only very rough. **(b)** The helium (He) atom consists of coupled pairs of protons and electrons. Neutrons (blue) attached to the nuclei help bind them together. **(c)** The carbon atom has 6 protons and 6 neutrons. **(d)** Ionized lithium Li^+ has three protons but only two electrons, so it carries a net positive charge of +1. **(e)** Doubly ionized lithium, Li^{+2}, has another electron removed and carries a charge of +2.

Electrons are not bound very tightly to their nuclei and can be removed with relative ease. Atoms with missing electrons take on positive electric charges, those with extra electrons negative charges. Both forms are called **ions** (Figures 6.2d and 6.2e). The removal of one electron creates an ion with an excess positive charge of +1, resulting in the singly ionized state. The removal of two electrons yields a net charge of +2, the doubly ionized state, and so on. A helium atom with a missing electron is called He^+, and if both electrons are gone, revealing a bare nucleus, the atom is He^{+2}. A lone proton is H^+.

Electrons can be also shared among atoms to produce **molecules,** which make the *chemical compounds* (Figure 6.3). Atoms can combine in a limitless number of ways. We count the number of individual atoms in a

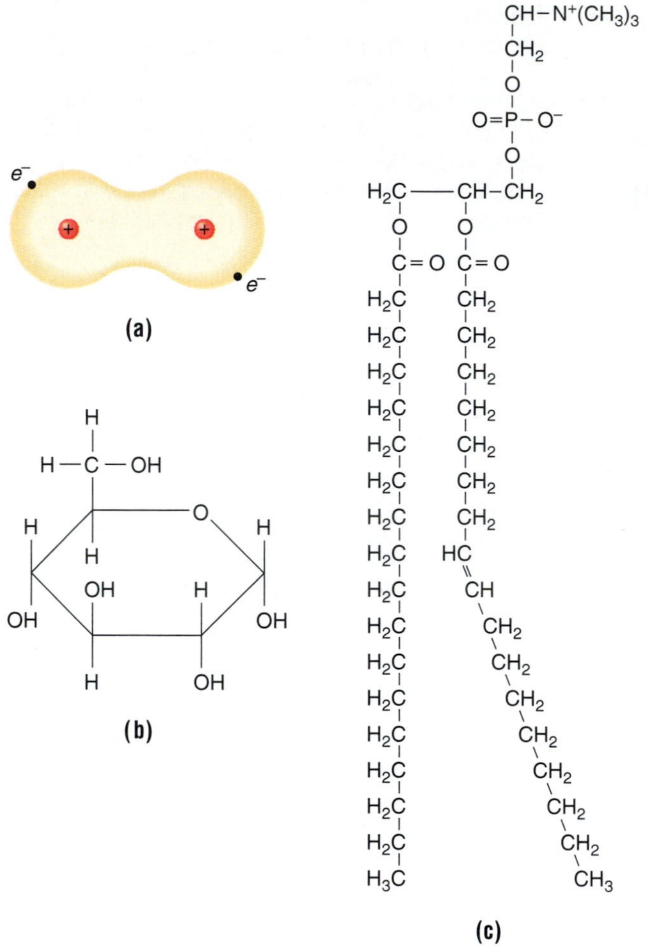

Figure 6.3
(a) A hydrogen molecule, H_2, is held together because the protons share their electrons. Molecules can be quite complicated, as indicated by **(b)** one of soap and **(c)** a complex organic (carbon-containing) one called phosphatidylcholine.

molecule and indicate the number by a subscript. The air you breathe is mostly nitrogen and oxygen in the molecular forms of N_2 (two nitrogen atoms joined together) and O_2: atomic oxygen is poisonous. You are made mostly of water, H_2O, and you exhale carbon dioxide, CO_2. More complex molecules, like those that make soap and gasoline, can consist of dozens of atoms. *Organic molecules*, those that have carbon in them and control the processes of life, can be very complicated and can contain thousands of atoms.

Together, atoms and molecules make three basic **states of matter.** In a **gas** (like the air you breathe) these particles are free to move past one another, interacting only electrically or by collisions. A gas has neither shape nor definite volume but can readily expand and contract. In a **solid** (like a rock), the particles are locked together, giving the substance both definite shape and specific volume. A **liquid** (like water from a tap) is an intermediate state. A liquid has specific volume, but the particles can still move past one another, allowing it to change its shape.

6.2 ELECTROMAGNETIC RADIATION

The electromagnetic force spreads outward from a source at a speed of 299,792 km/s (or 186,282 miles/s, the maximum possible speed for anything in the Universe) in the form of **electromagnetic radiation,** which is most familiar to us as visible *light*. Our basis for understanding its properties goes back to Newton, who passed a beam of sunlight through a glass prism and discovered that light could be split into a *spectrum* of myriad colors (Figures 5.1 and 6.4), those seen in the common rainbow. Light therefore can be decomposed.

The nature of light was revealed by the Scottish physicist James Clerk Maxwell, who in the 1860s formulated the equations that describe electric and magnetic fields. A moving electric charge produces a magnetic field. If an electric charge is accelerated, it produces waves in both its electric and magnetic fields. A solution to his complex equations showed that the waves must move in a vacuum at 3×10^8 m/s, the known speed of light. Maxwell then immediately identified light as an *electromagnetic wave*, as electromagnetic radiation.

A wave is most easily described by its **wavelength,** (λ), the distance in meters or centimeters between adjacent wave crests (Figure 6.5). We can also count the number of wave crests that pass a specific point per second, the **frequency** (ν) measured in cycles per second, or Hertz (Hz). As wavelength goes down, there are more crests per centimeter, and consequently more pass you per second; that is, wavelength and frequency are inversely

Figure 6.4
Sunlight is resolved by a glass prism into a spectrum of all the visible colors from red to violet.

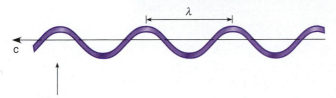

Figure 6.5
A wave moves to the left at velocity c. The wavelength (λ) is the distance between crests or troughs. The frequency of the wave (ν) is the number of wave crests passing a fixed point (vertical arrow) per second.

proportional to each other. They are related through the speed of the wave, here the velocity of light (c), where

$$\lambda\nu = c,$$

so that $\nu = c/\lambda$ and $\lambda = c/\nu$.

There are no restrictions on the lengths or frequencies of electromagnetic waves. The full set of waves of all wavelengths (or frequencies) is called the **electromagnetic spectrum** (Figure 6.6). Different names are assigned to waves with roughly similar wavelengths, but all waves move in a vacuum at the same speed, c. The human eye is sensitive to electromagnetic radiation with wavelengths between about 4 and 7×10^{-7} m. These numbers are so small that it is convenient to use a unit called the *Ångstrom* (Å), which is 10^{-10} m (or 10^{-8} cm), about the size of the hydrogen atom. The range of this *visual radiation* (ordinary light) is thus expressed as 4,000–7,000 Å. Within the visual realm, we see different-length waves as the different colors of Figures 6.4 and 6.6. At the long-wave end, light appears red, and at the short-wave end violet, with the other colors—orange, yellow, green, blue, and hundreds of recognizable intermediate shades—in between.

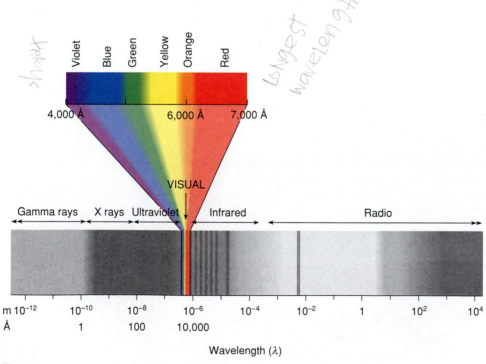

Figure 6.6

The electromagnetic spectrum is composed of different wavelength domains. The boundaries of the descriptive terms are not physically real, nor are they sharp: the X-ray section gradually overlaps the ultraviolet, and so on. The shading roughly indicates the degree to which light is blocked by the Earth's atmosphere. None of the darkly shaded parts of the spectrum can penetrate to the ground.

Waves with lengths greater than about 7,000 Å, called *infrared* (*IR*), cannot excite the eye but can be felt by the skin as heat. Near a tenth of a millimeter, we begin to enter the *radio* region. Used for communications, it extends through wavelengths of meters to over a kilometer. At wavelengths below 4,000 Å, the waves again cannot be detected by the eye. This *ultraviolet* (*UV*) *radiation* extends downward to about 100 Å, at which point it becomes known as *X rays*, which can penetrate the tissue of a human body and are indispensable in medicine. Finally, below about an Ångstrom are *gamma rays* produced by, among other things, atomic bomb explosions.

Although the wave theory of light successfully explains many experiments, in others it fails completely. An example is the *photoelectric effect*, in which a beam of light knocks electrons loose from certain substances to produce a flow of electricity. The phenomenon is exploited in "electric-eye" door openers and motion-picture sound systems. To produce the photoelectric effect, electromagnetic radiation must behave not as waves but as *particles*, called **photons**. Light has the remarkable property of acting like both a wave and a particle *at the same time*, a concept outside our everyday experience.

The photon is a packet of **energy** with an underlying wave nature. Energy is the ability of a body to impress a force on, and to accelerate, another. There are many kinds of energy. A moving body can strike another, accelerating it to some velocity or changing its path. The faster the first body is moving, the more energy it has, and the more it can impart to the second body. However, a body may also have *potential* energy as a result of its position or configuration. If you raise a bowling ball above the floor you give it potential energy within the Earth's gravitational field. When it drops, gravity exchanges the ball's potential energy for energy of motion.

All matter contains *heat energy*, a microscopic form of energy of motion related to the speeds at which particles—atoms and molecules—move in a gas or vibrate in a solid. Heat energy (the amount per gram) is measured by **temperature.** The worldwide standard is degrees centigrade (°C), defined by the freezing (0°C) and boiling (100°C) points of water (both at sea level). As you remove heat from a body, its particles slow, and its temperature goes down. At −273°C (−459° Fahrenheit), *absolute zero*, the body reaches minimum energy and can get no colder. To avoid negative temperatures, the Kelvin (K) scale starts at 0° K at absolute zero and then counts centigrade degrees upward. Water thus freezes at 273° K and boils at 373° K. In this scale the degree symbol is now usually dropped, and "degrees Kelvin" are referred to as "kelvins," or K.

Electromagnetic radiation is *radiant energy* that can be absorbed by a body's atoms to heat it and raise its temperature or even to accelerate it. Each photon carries an amount of energy (E), that depends upon its frequency (ν), where

$$E = h\nu$$

Light has the remarkable property of acting like both a wave and a particle at the same time.

and h is a laboratory constant (Planck's constant, named after the early-twentieth-century German physicist Max Planck). The higher the frequency and thus the shorter the wavelength (λ), the more the energy. A familiar unit, the *watt*, describes the rate at which energy is used or released per second. A 100-watt yellow light bulb radiates about 3×10^{20} photons per second and a powerful megawatt FM radio transmitter an astonishing 10^{31} radio photons per second. These enormous numbers show that the energies of individual visual and radio photons are *very* low, and except under unusual circumstances visual and radio radiation can cause little biological damage. Ultraviolet photons, however, have enough energy to produce severe burns. Sunburns are caused by ultraviolet radiation between 3,000 and 4,000 Å that penetrates the Earth's atmosphere; the air is opaque to shorter radiation as a result of absorption by ozone (O_3). Both X rays and gamma rays can enter the body, where they have the potential to ionize atoms and cause burns or even cancer. Gamma rays created by nuclear explosions were a principal cause of death in the atomic bombings of Japan in 1945.

6.3 RADIATION AND MATTER

Electromagnetic radiation is both emitted and absorbed by matter, specifically by charged particles—electrons, protons, and the atoms, ions, and molecules that contain them—that are accelerated or otherwise change their energies.

6.3.1 The Blackbody and Continuous Radiation

Radiation is *continuous* if its spectrum has no breaks or gaps nor any sudden changes in brightness. In the continuous spectrum in Figure 6.6, one color blends smoothly into another. Many physical processes produce continuous radiation. One kind of radiator, an idealization called the **blackbody,** is particularly important because the Sun, stars, and planets, bear a close similarity to it.

   A blackbody absorbs all the radiation falling upon it, reflecting none. The absorption of radiative energy will raise a blackbody to some particular temperature. To maintain that temperature even though the body is constantly receiving energy, the body must emit just as much energy as it receives. Consequently, a blackbody is not necessarily black, but can be bright to the eye. The greater the temperature of a blackbody, the more energetic the atomic particles within it, and the more radiation they can emit. The amount of radiation emitted per square meter of surface (*F*) by a blackbody depends on the fourth power of the temperature (*T*), or

$$F = 5.7 \times 10^{-8} T^4 \text{ watts/m}^2,$$

If you double a blackbody's temperature, it becomes 16 times brighter.

(the **Stefan-Boltzmann law**): that is, each square meter of surface releases energy at a rate of $5.7 \times 10^{-8}T^4$ watts. If you double a blackbody's temperature, it becomes 2^4, or $2 \times 2 \times 2 \times 2 = 16$, times brighter!

Look at the power of this law. A spherical blackbody has a radius of r meters and temperature T. The surface area (A) of the sphere is $A = 4\pi r^2 = 12.6r^2$ square meters (where π, the ratio of the circumference of a circle to its diameter, equals 3.14 . . .). Each square meter radiates at a rate of $F = 5.7 \times 10^{-8}T^4$ watts. The amount of *total* energy produced per second by a source of radiation is called its **luminosity** (L). The luminosity of a spherical blackbody, in watts, is thus the product of the surface area (A, the number of square meters) times the energy radiated per square meter per second (F) (Figure 6.7), or

$$A \times F = L \text{ (watts)} = 7.2 \times 10^{-7}r^2\,T^4.$$

The temperature is then just the fourth root of $L/7.2 \times 10^{-7}r^2$. Even though stars and planets are not perfect blackbodies, we can apply this formula to measure their *effective temperatures*, the temperatures they *would* have if they *were* perfect blackbodies. Effective temperatures provide a superb way of comparing astronomical bodies with one another.

As the temperature of a blackbody rises, so does the average atomic speed and energy and the average energy of the emitted photons. At very

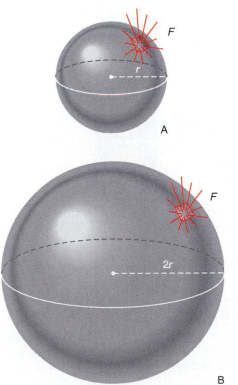

Figure 6.7

Spherical blackbody B has twice the radius of spherical blackbody A and therefore has *four* times the surface area. If the two bodies have the same temperature, the radiant energy emerging from a square meter of each (F, in red) is the same. Since B has four times as many square meters as A, it also radiates four times as much total energy, that is, its luminosity is four times as great.

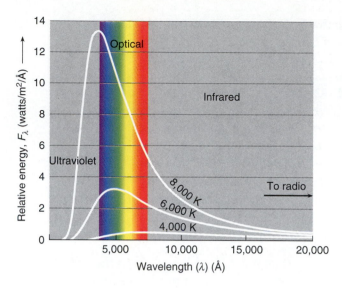

Figure 6.8

The graph shows the relative energy emitted by blackbodies per square meter per second per Ångstrom (F_λ) plotted against wavelength for three different temperatures. All the curves have the same shape, rising slowly to a peak, then dropping rapidly. As temperature (indicated on the curves) rises, the amount of radiation increases at all wavelengths and the peak of the curve shifts to shorter wavelengths. The 8,000-K blackbody produces 16 times as much energy per second as the one at 4,000 K.

low temperatures, from a few to a few tens of kelvins, the blackbody produces only low-energy, long-wavelength photons in the radio spectrum. As temperature climbs into the hundreds of kelvins, the blackbody begins to radiate infrared photons in addition to those in the radio, and when the temperature reaches into the thousands of kelvins, the energies are high enough to produce visual photons (in addition to infrared and radio), and we can see the body glow. If the temperature is elevated into the millions of kelvins, even X rays and gamma rays will be produced.

A graph of the energy radiated by a blackbody per unit area per Ångstrom (F_λ) plotted against wavelength is called a *blackbody curve*. Figure 6.8 shows blackbody curves at three different temperatures. In all cases, F_λ climbs slowly from the radio region of the spectrum (far off the graph to the right) to a sharp peak, then suddenly drops. The wavelength of the peak of the curve, where F_λ is at a maximum, or λ_{max}, depends inversely on the temperature; this relationship is stated by the **Wien law**:

$$\lambda_{max} = 2.898 \times 10^{-3}/T \text{ (meters)} = 2.898 \times 10^{7}/T \text{ (Ångstroms)}.$$

As the temperature of a blackbody climbs, the energy radiated per Ångstrom increases at all wavelengths, and λ_{max} shifts to shorter wavelengths, or to the left. At room temperature (295 K), $\lambda_{max} = 9.82 \times 10^{-6}$ m (about 0.01 mm), which lies far into the infrared. Such a blackbody would be invisible to the human eye, and therefore the walls of your room do not visibly glow in the dark. At 2,000 K, though, the body emits a small amount of red radiation. At 4,500 K you see the body as orange, at 6,000 K yellowish (like sunlight), at 10,000 K white, and above that there is so much blue and violet light that the body takes on a bluish cast (Figure 6.9). Star colors (see Figures 3.2, 3.4, and 3.6) are the result of different stellar

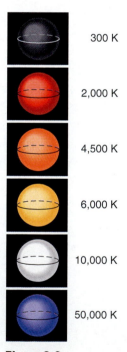

Figure 6.9

The color of a blackbody (exaggerated for clarity) as seen by the human eye depends on its temperature.

surface temperatures. If a blackbody's λ_{max} can be measured from observation, we can also find its temperature from the Wien law, where $T(K) = 2.898 \times 10^{-3}/\lambda_{max}$ and λ_{max} is in meters.

6.3.2 Spectrum Lines

Newton and his followers saw only the continuous spectra of Figure 6.6. However, in 1802, William Wollaston discovered narrow dark lines in the solar spectrum that ran perpendicular to the flow of colors. Within a dozen years, Joseph von Fraunhofer had catalogued over 300 such gaps (Figure 6.10) where light and color seemed to be missing, features that have since become known as **absorption lines.** It was then found that if light was made to pass through a substance, such as a rarefied or low-density gas, absorption lines could be made in the laboratory. (**Density** is mass per unit volume, measured in kilograms per cubic meter, kg/m^3, or, more commonly in astronomy, in g/cm^3). Furthermore, if a substance was made to burn so as to emit light from a hot but rarefied gas, the absorption lines (Figure 6.11a) were replaced by **emission lines** (Figure 6.11b), bright lines of color, at exactly the same wavelengths.

Every ion of every atom produces a unique pattern of spectrum lines.

It quickly became obvious that individual kinds of atoms could be identified by their **spectrum lines,** which could be either absorption or emission lines. Different materials always produce a variety of spectrum lines at different wavelengths. Hot rarefied hydrogen always produces emissions at 6,563 Å (a line called $H\alpha$), 4,861 Å ($H\beta$), and others that get closer together toward the short-wave end of the spectrum. The set of these lines, seen in Figures 6.11a and b, is called the *Balmer series*. A hot rarefied sodium gas, however, causes a pair of bright orange-yellow lines (Figure 6.11c) at 5,890 Å and 5,896 Å; other metals and hot gases produce quite different patterns. Some, like iron (Figure 6.11d), are extremely complex, with thousands—even millions—of lines. *Every ion of every atom produces a unique pattern*; there is no possibility of confusing one with another.

By 1859, the chemist Gustav Kirchhoff had determined basic rules, now known as **Kirchhoff's laws** of spectral analysis, under which the different types of spectra are produced. Spectra can be observed by a device called a *spectroscope* (to be examined in Section 6.4). In Figure 6.12,

Figure 6.10
Fraunhofer's spectrum of the Sun showed an enormous number of dark lines, each produced by a particular atom, ion, or molecule in the solar atmosphere.

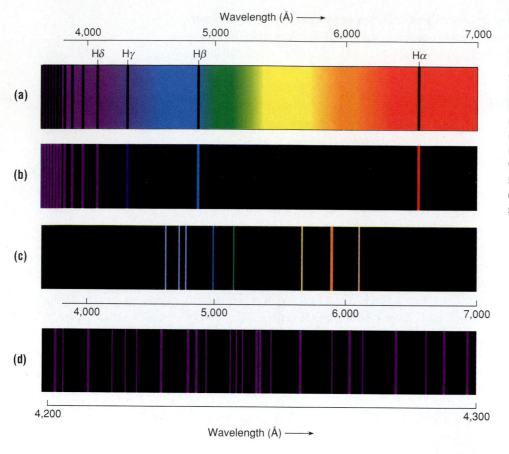

Figure 6.11

Hydrogen displays **(a)** absorption and **(b)** emission lines at exactly the same wavelengths (here, the lines of the Balmer series). Neutral sodium **(c)** produces a pair of bright orange-yellow emissions (seen here as one line) as well as several others. The spectrum of iron **(d)** is so complex that only a 100-Å segment is shown.

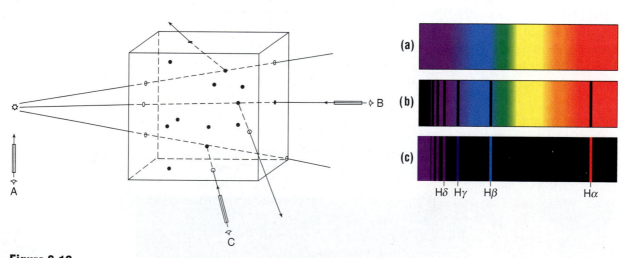

Figure 6.12

Light from a hot blackbody enters a transparent box filled with hydrogen and is observed through spectroscopes A, B, and C. **(a)** The blackbody seen through A has a continuous spectrum. **(b)** If the light is looked at through the box, through B, the continuous spectrum is seen to be crossed by absorption lines. **(c)** If the box alone is looked at, through C, the observer sees only emission lines.

spectroscope A looks only at a hot blackbody that produces a continuous spectrum. But spectroscope B looks at the blackbody through a transparent box filled with a rarefied hydrogen gas that is cooler than the blackbody. The observer now sees dark lines superimposed on the continuous spectrum as the hydrogen atoms absorb light at their characteristic wavelengths. Spectroscope C points directly at the box and avoids the blackbody radiation. Now the observer sees emission lines that create the exact reversal of the spectrum seen in spectroscope B.

Kirchhoff's laws are summarized as:

1. An incandescent solid or hot high-density gas produces a continuous spectrum (Figure 6.12a).
2. A hot low-density gas produces an emission-line spectrum (Figure 6.12c).
3. A source of continuous radiation viewed through a cooler low-density gas produces an absorption-line spectrum (Figure 6.12b).

Here is a key to learning the properties of astronomical objects, from planets to galaxies, over great distances. If we replace the laboratory apparatus with nature and look at a star, we find an absorption-line spectrum like that seen in spectroscope B, showing that a light from a source of continuous radiation is passing through a lower-density cooler gas, which in this case is the star's low-density atmosphere.

Understanding of how atoms create spectra requires knowledge of atomic structure, which began to be revealed by the work of the Danish scientist Niels Bohr in 1913. He adopted the common idea that electrons orbit their nuclei but made the critical assumption that the potential energies of the electrons (in terms of the electromagnetic force between the protons and the electrons) can have only specific values that are ultimately determined by multiples of Planck's constant. The radii of the orbits can therefore also take only specific values, and there must be an orbit with a minimum radius and energy, an innermost orbit called the *ground state* (Figure 6.13).

The electrons are free to jump between any two orbits. If an electron jumps downward between two orbits, it will give up its energy in the form of a photon that has an energy equal to the energy difference between the orbits and that will contribute to an emission line. Look at higher and lower energy orbits m and n in Figure 6.14a. The energies of these orbits are E_m and E_n. The radiated photon thus has energy $E = E_m - E_n$. The energy of a photon corresponds to a specific frequency through $E = h\nu$, so that $\nu_{mn} = E/h$. Once we know the frequency, we can calculate the wavelength λ_{mn} from $\lambda = c/\nu_{mn}$. Conversely, an electron in a lower orbit n in Figure 6.14b can *absorb* a photon of wavelength λ_{mn} and be raised upward into orbit m. The essence of the process, however, is that *all* the energy of the incoming photon must be absorbed, which means that the photon must

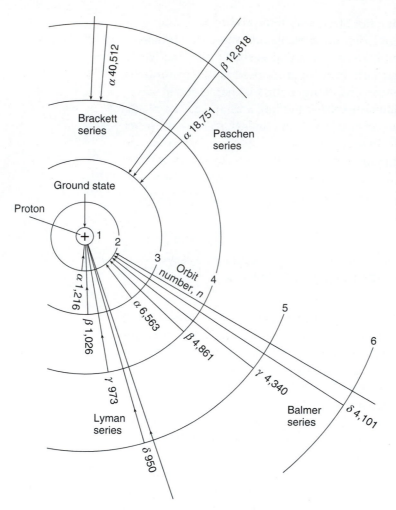

Figure 6.13

The first six orbits of the hydrogen atom are drawn to scale in this diagram. A downward-jumping electron (one that jumps toward the nucleus) gives up energy that is radiated outward as a photon, the wavelength given in Ångstroms. The whole array of jumps is arranged in various series in which the electron lands on a specific orbit, the Lyman series on orbit 1, the Balmer series on orbit 2, and so on. Upward jumps cause absorption lines.

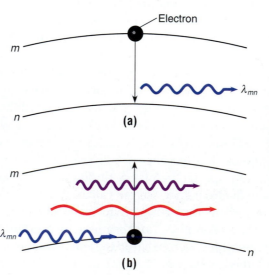

Figure 6.14

(a) An electron in a higher orbit m jumps downward to a lower orbit n and gives up its energy as a photon of energy $E_m - E_n$ and wavelength λ_{mn}. **(b)** Photons of various wavelengths pass an electron in orbit n. The electron can absorb only the photon with wavelength λ_{mn} (energy $E_m - E_n$), which elevates the electron to the upper orbit m.

have energy $E_m - E_n$ (and wavelength λ_{mn}) exactly. Hydrogen's $H\alpha$ emission line is created by downward electron jumps between orbits 3 and 2, the $H\beta$ emission line by jumps between 4 and 2, and so forth (see Figure 6.13), and the respective absorption lines by upward jumps between 2 and 3, 2 and 4, and so on.

Hydrogen's orbital structure is simple because there is only one electron; two or more electrons will interact with each other to create more complex orbital structures. With more orbits there are many more possible transitions. The complexity can be awesome, as seen for iron in Figure 6.11d. The champions of complexity, however, are molecules. Because a molecule can rotate and vibrate, electronic orbits are subdivided into sub-orbits and sub-suborbits. What would be a single spectrum line for an atom becomes a series of bands, each band consisting of hundreds or even thousands of individual lines (Figure 6.15).

This theory of the atom readily explains Kirchhoff's laws. The atoms in the box in Figure 6.12 are constantly colliding with one another. The collisions move electrons into outer, more energetic, orbits. These energetic or *excited* atoms (as well as those with electrons in the ground state) absorb the photons from the incoming beam of continuous radiation, but only those photons whose energies correspond to energy differences between the various orbits. Absorption lines are therefore seen in spectroscope B. In addition, if a photon's energy is sufficiently great, it can give the electron enough energy to escape the nucleus, to rip it away, and the atom becomes ionized. Electrons in upper orbits can also jump downward. When they do, they radiate the emission lines seen in spectroscope C.

The absorption lines seen in spectroscope B will not be completely dark, as some photons at the critical wavelengths will get through. By measuring the *strengths* of absorption lines (the amounts of energy extracted from the continuous spectrum), the temperature and density of the gas and the amount of the chemical element forming the lines can be determined. If we can detect lines of more than one element, we can also derive the relative numbers of the different kinds of atoms. Similar procedures can be applied to the brightnesses of emission lines seen in spectroscope C. When these concepts are applied to real astronomical objects, we can analyze their conditions and compositions and begin to understand the nature of the Universe.

Violet   Red

Figure 6.15
Molecular spectra, like that of C_2 shown here, consist of many series of bands, each band containing hundreds or thousands of individual lines.

6.4 TELESCOPES

Light from distant stars falls upon the Earth, carrying the coded messages that tell us what they are and how they are made. We collect and decipher the light with our **telescopes,** instruments whose purpose is to make distant objects appear nearer, larger, and brighter. The earliest telescopes used the principle of **refraction.** Figure 6.16 shows two light rays falling from the air into a denser substance, perhaps glass or water, at two different *angles of incidence* (*i*) between the light rays and the perpendicular to the surface. The rays slow as they enter the substance. As a result, they change direction and bend toward the perpendicular, so that the *angle of refraction* (*r*) (the angle between the bent ray and the perpendicular) is less than *i*. The degree to which the light is bent ($i - r$) increases with *i* and with the ratio of c to the velocity (*v*) of light in the substance: glass will cause more bending than water, diamond more than glass. The paths of the light rays in Figure 6.16 are independent of the direction in which they travel. If a ray starts *in* the denser substance (ray 3) and goes into the air, it bends *away* from the perpendicular, *i* and *r* are reversed, and the angle of refraction is larger than the angle of incidence.

The convex (or converging) *lens* in Figure 6.17 is a circular piece of glass whose surfaces are shaped into sections of spheres that face away from each other. The line through the center of the lens perpendicular to the surfaces points toward a star, a body so far away that all its rays are effectively parallel to one another. The ray of starlight on the central line is not refracted. But a ray hitting the lens away from the center strikes at an angle to the perpendicular to the lens's surface (line a) and is bent in the direction of the central line. When this ray leaves the glass it is now bent away from the perpendicular to the lens's surface (line b), *again* toward the central line.

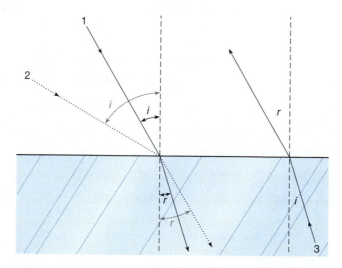

Figure 6.16
Light rays 1 and 2 enter a dense substance from air and are refracted toward the perpendicular to the surface. The angle of refraction (*r*) is less than the angle of incidence (*i*). If the light source is in the denser substance (ray 3) and the ray travels in the opposite direction, it is bent away from the perpendicular.

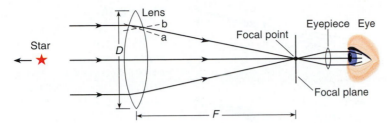

Figure 6.17

A simple refracting lens of diameter *D* bends starlight, the star far off the page to the left. Light rays that strike or leave the lens's surface at an angle to the perpendicular (lines a and b, respectively) are refracted toward the line through the center of the lens and come to a focus at the focal point a distance *F* (the focal length) from the lens. The result is an image of the distant star. To view the image, we place an eyepiece behind it at a distance equal to *its* focal length.

All the rays falling on the lens—the *objective*—are refracted to cross or focus at the *focal point* at a distance *F*, the *focal length*, from the lens. If a piece of paper is placed at the focal point, you will see a point of light, the *image* of the star. Substitute a photographic plate (a sheet of glass with a photographic emulsion spread on it) for the paper and you can take a picture of it. You now have a **refracting telescope,** or *refractor*, which acts like an ordinary camera. The telescope can image stars over a specific area of the sky, focusing their light on the *focal plane* in Figure 6.17.

The brightness of the star's image (the amount of energy falling into it per second) depends on the telescope's *light-gathering power*, the amount of light that the lens can capture, which in turn depends on the area of the lens or on the square of its diameter (*D*). If you double the size of the lens, the image becomes four times brighter, and you are able to record stars that are four times fainter (Figure 6.18a). Telescopes, like the "40-inch" at the Yerkes Observatory in Figure 6.19, are named by lens diameter.

If you wish to look at an object through the telescope, you can place an *eyepiece* behind the focus (at a distance equal to the eyepiece's focal length) to make the rays parallel again so that they can be focused onto the retina of the eye (see Figure 6.17). The eyepiece increases the angular separations of stars and the angular sizes of extended objects, thereby making them look bigger. The degree of magnification, or the *magnifying power*, is the focal length of the objective divided by the focal length of the eyepiece ($M = F_{objective}/F_{eyepiece}$). It is easy to change the power of a telescope just by switching eyepieces.

More important is the telescope's *resolving power*, which is its ability to reveal fine detail or to separate two sources of light that are close together. When any wave passes through an opening, it is sent outward in all directions, a process called *diffraction*. At our distance a star is effectively a point in the sky, even with telescopic magnification. However, waves diffracted

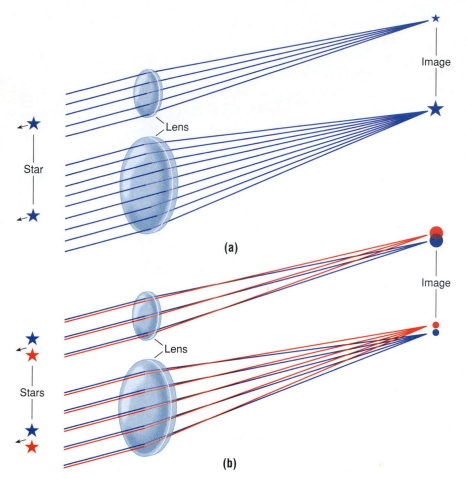

Figure 6.18
The stars are far off the page to the left. **(a)** The diameter of the bottom telescope lens is twice that of the one on the top. Its surface area is therefore four times as great, it collects four times as much light, and the image of the star is four times brighter. **(b)** The two telescopes now view a pair of stars close together (image brightness is ignored). The image of any star is not a point but a disk. The image formed by the smaller telescope has the larger disk, which smears the images of the two stars into one. The larger telescope, however, separates them.

Figure 6.19
The great nineteenth-century refractor of the Yerkes Observatory of the University of Chicago, located in Williams Bay, Wisconsin, has a lens 40 inches (about 1 m) in diameter.

through the different parts of the lens interfere with one another and smear the point-like image of the star into a disk, the *diffraction disk* (Figure 6.18b). If two stars are so close that their diffraction disks overlap, they will be seen as one. The diameter of the diffraction disk (d) is directly proportional to wavelength and inversely proportional to the diameter of the telescope's lens (that is, d is proportional to λ/D). The larger the diameter, the smaller the size of the diffraction disk. The ability of the telescope to resolve detail thus improves along with increasing lens diameter.

The resolving power of the naked human eye is at best about 1 minute of arc, 1/30 of the Moon's angular diameter. At typical visual wavelengths, a 10-cm (4-inch) telescope can separate two stars about 1.3" of arc apart, the angle between car headlights at a distance of 300 km! A 20-cm (8-inch) telescope will do twice as well. In practice, however, the resolving power is limited for most telescopes by variable refraction in different layers of the Earth's atmosphere, which smears a stellar image into a larger *seeing disk* that is typically a second of arc across. The same process produces the familiar twinkling of stars.

Lenses are expensive. Moreover, the size of a lens is limited because the weight of the glass causes it to sag, distorting it and reducing the quality of the image. We can overcome these deficiencies with mirrors and the **reflecting telescope** (Figure 6.20), which was invented by Isaac Newton. A mirror can be supported from the back and can be made much larger than a lens.

Figure 6.21a shows the principle of **reflection,** in which the angle of incidence (i) between the incoming ray and the perpendicular is equal to the *angle of reflection* (r). As a result, you always see a faithful (though

Figure 6.20
The primary (objective) mirror of the Palomar 200-inch (5-m) telescope is at the bottom of the open tube, and the secondary system is at the top.

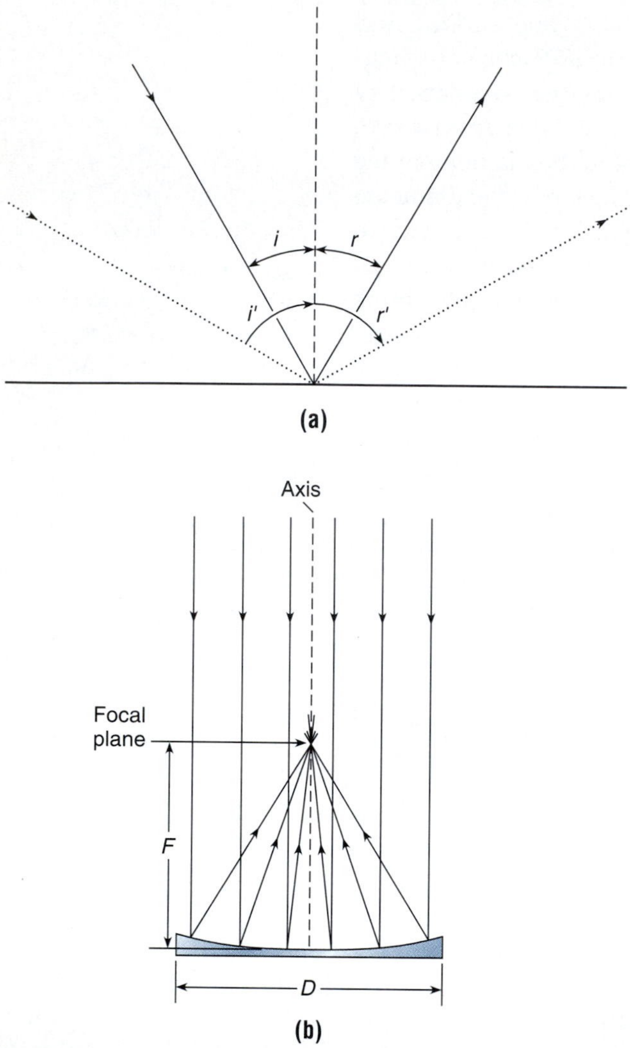

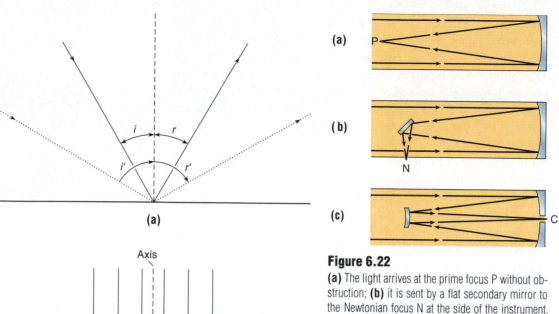

(a)

Axis

Focal
plane

F

D

(b)

Figure 6.21
(a) Rays fall downward from the left onto a smooth surface at angles i and i' (the angles of incidence) to the perpendicular to the surface. They are reflected up and to the right at the angles of reflection, r and r'. The angle of reflection equals the angle of incidence for any ray, so $i = r$ and $i' = r'$. **(b)** A mirror shaped like a paraboloid will bring all the rays to a focus at the same point.

(a) P

(b) N

(c) C

Figure 6.22
(a) The light arrives at the prime focus P without obstruction; **(b)** it is sent by a flat secondary mirror to the Newtonian focus N at the side of the instrument. **(c)** The Cassegrain focus uses a curved secondary to extend the focal length to C, shooting the light through a hole in the large primary mirror.

backward) image of yourself in a bathroom mirror. A reflecting telescope uses a mirror curved into the shape of a *paraboloid* (the surface generated when a parabola is spun about its axis) to focus the light (Figure 6.21b). The actual reflecting surface is a thin coat of aluminum. Other than in the way the light is collected, the reflector and refractor behave similarly, and all the terms defined above have the same meanings.

The focal point of a mirror, called the *prime focus* (Figure 6.22a), is in front of the light-collecting surface rather than behind it. Therefore, the act of viewing the sky will block incoming light. (The prime focus can still be used, but only if the mirror is significantly larger than the astronomer or the device used to detect the light.) Newton solved the problem by setting a small flat *secondary mirror* in the converging beam at a 45° angle to the big, or *primary*, mirror. The *Newtonian focus* (Figure 6.22b) is at the side of the telescope. The *Cassegrain focus* (Figure 6.22c), however, uses a curved secondary mirror to extend the focal length and to send the light directly back through a hole in the primary. Additional mirrors can send the light elsewhere, even to fixed positions within the observatory.

The light-gathering power of the Palomar 200-inch (5-m) telescope pictured in Figure 6.20 is staggering. A star seen through an eyepiece would be nearly a million times brighter than if viewed with the naked eye. The world's largest telescopes, the two 10-m Kecks atop Mauna Kea, Hawaii (Figure 6.23), feature 36 hexagonal 1.8-m mirrors, each a piece of a giant computer-controlled paraboloidal surface. Each of the pair has four times the light-gathering power of Palomar. Table 6.2 provides a list of the largest telescopes operating as of 1996; several additional large ones are planned or under construction.

Whether refractor or reflector, all telescopes are mounted on two axes that are set perpendicular to each other, an arrangement that allows them

A star seen through the Palomar 5-m telescope would be nearly a million times brighter than if viewed with the naked eye.

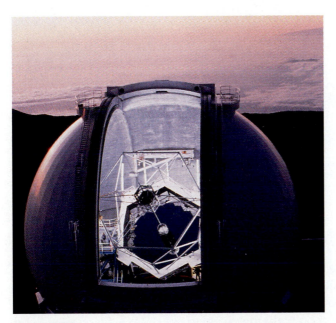

Figure 6.23
The Keck 10-m telescope consists of 36 separate mirrors, each 1.8 m across, arranged over a paraboloidal surface.

TABLE 6.2
The Largest Telescopes

Refractors

40-inch (1-m), Yerkes Observatory, University of Chicago, Williams Bay, Wisconsin, U.S.A.
36-inch (0.9-m), Lick Observatory, University of California, Mt. Hamilton, California, U.S.A.

Reflectors

10-m, Keck Telescope, California Institute of Technology and the University of California, Mauna Kea, Hawaii, U.S.A.
10-m Keck II, Mauna Kea, Hawaii
6-m, Bolshoi Teleskop Azimutal'nyia, Caucasus Mountains, Russia
5-m, Hale Observatory, Palomar Mountain, California, U.S.A.
4.5-m equivalent, Multiple Mirror Telescope, University of Arizona and the Smithsonian Center for Astrophysics, Mount Hopkins, Arizona, U.S.A.[a]
4.2-m, Royal Greenwich Observatory, Canary Islands, Spain
4-m, Kitt Peak, Cerro Tololo Interamerican Observatory, Chile[b]
4-m, Kitt Peak National Observatory, Arizona, U.S.A.[b,c]
3.9-m, Anglo-Australian Telescope, New South Wales, Australia

Radio[d]

8,000-km Very Long Baseline Array (VLBA), National Radio Astronomy Observatory; a VLBI that spans the distance from the Virgin Islands to Hawaii
36-km Very Large Array (VLA) interferometer, National Radio Astronomy Observatory, Socorro, New Mexico, U.S.A.
20-km Westerbork Radio Synthesis Observatory Interferometer, Westerbork, The Netherlands
300-m single fixed dish, Arecibo Observatory, Arecibo, Puerto Rico
100-m single dish, Max Planck Institute for Radio Astronomy, Bonn, Germany
100-m single dish, National Radio Astronomy Observatory, Green Bank, West Virginia, U.S.A.

[a] To be replaced by a single 6-m primary.

[b] Parts of the U.S. National Optical Astronomy Observatories; Cerro Tololo is operated in cooperation with the goverment of Chile.

[c] Mirror is masked down to 3.8 m to improve image quality.

[d] The largest interferometer had one component in geosynchronous orbit and others on the ground, the components separated by 42,000 km, about three Earth-diameters.

to be pointed to any part of the sky. One axis is geared to a clock that automatically drives the telescope to follow a star as the Earth's rotation moves it along its daily path. To reduce atmospheric effects, telescopes are placed on remote, dark, desert mountaintops that have minimal cloud cover and are as high as possible. From the top of a tall peak you see the stars with wonderful clarity, and on an especially good night they seem to stare at you, hardly twinkling at all.

BACKGROUND 6.1   Do It Yourself

Astronomy is a remarkably accessible science: you need only go outside and look. Even simple binoculars will provide marvelous views. Binoculars are described by a pair of numbers, for example, 7 × 35; the first gives the magnifying power, the second the lens diameter in millimeters. For nighttime use, 7 × 50 is excellent. Small telescopes, which are simple, affordable, and available from several mail-order companies as well as from camera stores, are better. The minimum is the 60-mm (2.4-inch) refractor. Classic Newtonian reflectors (Figure 6.24) will give the greatest diameter for the money. The minimum usable mirror diameter is 3 inches (75 to 80 mm). The bigger the telescope you can afford, the better: a 6-inch will make stars seem four times brighter than a 3-inch. At a more sophisticated

(and expensive) level, you might opt for a Cassegrain design, in which a long focal length is folded into a short tube. Cassegrains are therefore relatively easy to transport into a dark site. They start as small as 3 inches, and there is really not much of an upper limit except that imposed by the checkbook. Most amateur telescopes come with a variety of eyepieces, allowing the power to be changed. As a general rule, always use the lowest power needed, which will yield the widest field of view and the brightest images. The maximum that can generally be supported is about 50 power per inch of aperture.

Astronomical observing is a learned skill that requires practice. Use your new telescope first in the daytime to align the attached low-power finder telescope. When you do you will become aware of one of the astronomical telescope's most confounding properties: the view is inverted, up-for-down, left-for-right. Do not try to "erect" the image with a special eyepiece—you will only be introducing another lens that absorbs precious light. At night you will discover another problem: the telescope also magnifies the apparent speed of the Earth's (or the sky's) rotation. Therefore, to follow an object on its daily path, you have to move the telescope constantly, and, because of the field inversion, apparently in the "wrong" direction! Have patience, you get used to it. If you have invested enough, you may have a clock drive that will follow a star for you, but only if you have remembered to align the telescope's polar axis on the celestial pole. That means making a latitude adjustment and another adjustment at night to point the axis north (or south if you live in the southern hemisphere). A great deal of help—and fun—can be had from your local amateur astronomy club.

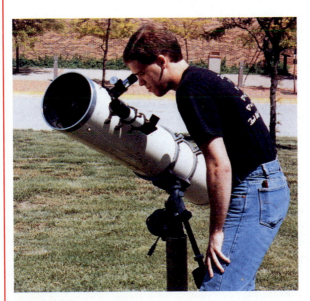

Figure 6.24
The simple, easy-to-use Newtonian reflector is the amateur's workhorse.

6.5 ANALYSIS AND INSTRUMENTATION

Until the late nineteenth century, the human eye was the only astronomical detector. The development of photography, with its ability for creating permanent records, revolutionized the science. An astronomical camera is simply a plateholder attached to the telescope's focus. During the time the plate is exposed—which may be minutes or hours—it will record photons, allowing the astronomer to build up images of things too faint to be seen by eye.

The photographic plate, however, is a slow collector of light. Modern detectors are electronic in nature. The most effective apparatus, developed since the 1970s, is the *charge-coupled device*, or CCD, a light-sensitive surface divided into rows and columns of individual cells or pixels. Each pixel can build up an electric charge that is proportional to the amount of light that falls on it. A computer records the charge of each pixel and stores the data on magnetic tape or disk. The data are in numerical, or digital, form, and can be reconstructed by computer to produce an image on a video terminal or be printed for reproduction. The brightness of every position in the picture can also be accurately measured. CCDs are so efficient that the exposure times needed to produce images are many times shorter than they are with photographic plates.

The real natures of celestial objects, however, can be revealed only through their spectra. The angle through which a substance refracts light depends on both the light-transmitting material and the wavelength. In Figure 6.25, a beam of white light (which is a mixture of colors) enters a glass prism from the air. The blue photons (the photons are not actually colored, the term descriptive only of wavelength) in the light are refracted more than the yellow, and the yellow more than the red. As a result, the white light is spread or *dispersed* into its component colors. It is dispersed even more when it leaves the prism. We can now let it fall on a screen and, like Newton (see Figures 5.1 and 6.4), observe a beautiful spectrum. Light

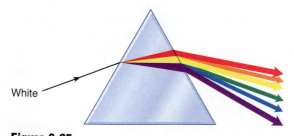

Figure 6.25
White light is dispersed as it enters the face of a triangular prism and is dispersed even more as it exits. The colors can now fall on a screen and be seen.

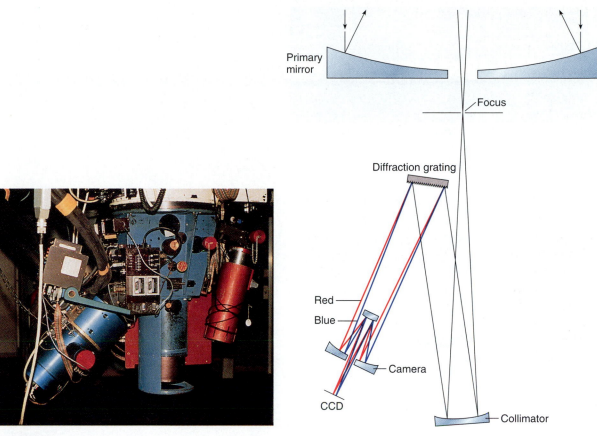

Figure 6.26

This spectrograph is attached to the Cassegrain focus of the 2.3-m telescope of the University of Arizona; its light path is diagrammed on the right. The light comes out of the telescope's focal plane and is collimated; that is, the rays are made parallel. The diffraction grating disperses the radiation and the camera mirror focuses it onto a detector, usually a CCD.

can also be dispersed by reflecting it from a *diffraction grating,* a plate ruled with narrow fine lines. Such spectra are easily seen reflected from a compact audio disk, which contains a fine spiral groove.

Astronomical spectra are created with the **spectrograph** (Figure 6.26). A narrow slit or aperture is placed at the telescope's focal plane to isolate the source of radiation (or a small portion thereof) to be observed. Light passes through the aperture to a *collimating mirror* that makes the incoming rays parallel and sends them on to a diffraction grating that disperses them. All rays of the same color are now parallel to one another, but each color is going in a slightly different direction. The spectrum is then focused by a *camera mirror* onto the focal plane, where it can be photographed (Figure 6.27a) or recorded by a CCD as a **spectrogram.** CCD recordings can be reconstructed to look like photographs, but are almost

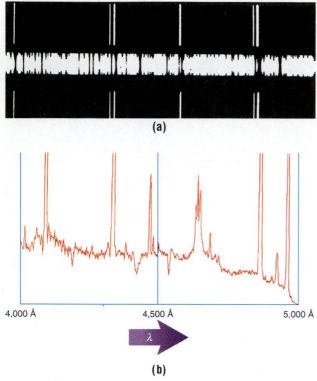

(a)

(b)

Figure 6.27

(a) A small section of the absorption spectrum of Arcturus, recorded with a photographic plate, is flanked by iron emission lines produced in the spectrograph. Their precisely known wavelengths allow those of the absorption lines to be measured. **(b)** A CCD detector can make a graphical spectrogram, here one of a cloud of interstellar gas. Its emission lines appear as spikes on top of a continuous spectrum.

always displayed graphically, with the brightness of the spectrum plotted against wavelength (Figure 6.27b). (The visual version of the spectrograph, shown in Figure 6.12 to illustrate Kirchhoff's laws, is called a *spectroscope*.)

Within the spectrograph is a device for generating an emission-line *comparison spectrum* from electrically vaporized iron or electrically excited helium, neon, or argon that can be directed into the spectrograph. The precise wavelengths of the emission lines are known from laboratory measurements, and they serve to calibrate the resulting spectrum to allow the measurement of accurate stellar wavelengths. In a photographic spectrogram, such as the one in Figure 6.27a, twin comparison spectra are placed alongside the stellar spectrum. If a CCD is used, a separate image of the emission spectrum generated by the spectrograph's calibration device allows the assignment of specific wavelengths to specific pixels.

6.6 THE INVISIBLE UNIVERSE

In the late 1700s, William Herschel opened the first nonvisual domain when he used thermometers to measure the energies of different portions of the solar spectrum. To his surprise, he found that the thermometers responded beyond the visible red end. He had discovered infrared radiation.

The Earth's atmosphere is not very transparent in the infrared (see Figure 6.6) because of strong absorption lines belonging to atmospheric carbon dioxide and water vapor. As a result, infrared astronomers will go to great lengths to reduce the amount of air over their heads. Two IR observatories are located high on Mauna Kea, and to observe in the submillimeter domain, where infrared and radio merge, there is an observatory at the south pole, which has some of the driest air on Earth. Astronomers even use telescopes mounted into jet aircraft that fly at high altitudes, where the amount of water vapor overhead is reduced by some 99%.

Like the infrared, radio waves from space were also discovered by accident. In 1933, Karl Jansky, an engineer for Bell Laboratories, built a large radio *antenna* in New Jersey to find the source of interference that was disturbing transatlantic radio communication. An antenna is merely a metal wire or rod (or a set of them). Because a radio wave or photon is an electromagnetic signal, it will produce an electric current in the antenna that can be amplified, detected, and measured. An antenna can be designed to have greater sensitivity in one direction than in another: that is, it can be pointed, even if crudely. Such a device is the simplest form of **radio telescope.** Jansky's interference was strongest when the Earth's rotation turned his telescope toward the Milky Way. The signals were coming from outside the Earth.

In 1936, an avid amateur scientist named Grote Reber built the first real reflecting radio telescope in his Wheaton, Illinois, backyard from spare and homemade parts. To increase the amount of radiation collected and to make the telescope more directional, he used a 28-foot-wide parabolic reflector, a metal radio mirror, to focus the radio waves onto the antenna (Figure 6.28). Reber could steer his telescope over the sky, and with it he made the first map of celestial radio sources. Today's astronomers have available dozens of fully steerable instruments with diameters up to 100 m, the size of a typical athletic field. The largest radio telescope is a fixed instrument 300 m (1,000 feet) in diameter built into a basin in Arecibo, Puerto Rico.

Radio telescopes "see" radio radiation just as optical ones see light; they "listen" to nothing. The purpose of the radio telescope is to measure the amount of radio radiation coming from celestial objects. The instrument can be made to scan over a source to map it and to produce computer-generated radio images that look very much like photographs made with optical telescopes. Some radio telescopes also have spectrographs

Figure 6.28
Grote Reber built the first paraboloidal reflecting radio telescope. The 28-foot dish focused radio radiation from space onto the antenna at the top.

that can be tuned to specific frequencies to search for absorption or emission lines.

Since the diameter of a telescope's diffraction disk is directly proportional to wavelength (see Section 6.4), the resolving power at long radio wavelengths is much worse than it is at short visual wavelengths. The problem is partially compensated for by a radio telescope's great size. However, even a 100-m radio telescope operating at a wavelength of 10 cm has a resolution of only four minutes of arc, worse than the unaided human eye. A solution would be to build bigger telescopes, but the practical engineering limit has already been reached. Instead of building a kilometer-wide radio telescope, we build *two* radio telescopes separated by an arbitrary distance and point them in the same direction. The electrical signals from the twin instruments are linked, and we observe their sum. With this **radio interferometer** we can reconstruct an image with a resolution equivalent to that of a single dish with a diameter equal to the separation between the two individual telescopes.

However, a two-telescope interferometer still provides limited information on the structure of a source. It is better to use three or more, or even to fill in the area of a hypothetical single dish by constructing an array of

Figure 6.29
Twenty-seven 25-m radio telescopes of the Very Large Array spread over 26 miles of the New Mexican desert. The telescopes can be moved on railroad tracks so that the interferometer diameter and resolving power can be changed.

instruments. The largest of these is the Very Large Array (VLA) of the National Radio Astronomy Observatory (Figure 6.29) in New Mexico. Its 27 radio telescopes are spread in a Y with arms 21 km long. The effect is that of a single-dish instrument 36 km in diameter, roughly the size of the beltway around Washington, D.C.! With it, astronomers can achieve a maximum resolution of a tenth of a second of arc, better than any ground-based optical telescope.

At greater sizes, direct linkage of the interferometer components becomes impractical. So we separate the instruments entirely, synchronize the observations with precise clocks, and mix the data by computer instead of by direct cabling. The concept is called a *very long baseline interferometer (VLBI)* and there is no limit to its size. The Very Long Baseline Array, a set of 10 radio telescopes that stretches across 8,000 km from the Virgin Islands to Hawaii, can produce an astonishing resolution of 0.0002 seconds of arc: car headlights separated at 1.5 million km, four times the distance of the Moon from the Earth.

The Very Long Baseline Array can produce an astonishing resolution of 0.0002 second of arc.

6.7 SPACE OBSERVATORIES

From space, the stars shine steadily, untwinkling, and we can see the infrared, ultraviolet, X-ray, and even gamma-ray spectral regions that are blocked by our atmosphere. A remarkable number of astronomical spacecraft have been flown. There are two broad varieties of such spacecraft: those that travel directly to the bodies of the Solar System and those that are placed in orbit around the Earth and are designed more for general

TABLE 6.3
Some Astronomical Spacecraft[a]

Name	Launch, Status	Function[b]	Region or Wavelength	Countries
High Energy Astronomical Observatory (HEAO2, Einstein)	1971–72	I, S	X ray	U.S.A.
International Ultraviolet Explorer (IUE)	1976, active	S	UV: 1,100-3,000 Å	U.S.A., U.K., ESA[c]
Infrared Astronomical Satellite (IRAS)	1983	I, S	IR: 12, 20, 50, 100 microns[d]	U.S.A.
Cosmic Background Explorer (COBE)	1989–91	I, S	Microwave, radio	U.S.A.
Röntgen Satellite (ROSAT)	1990, active	I, S	X ray	Germany, U.S.A.
Hubble Space Telescope (HST)	1990, active	I, S	UV through IR	U.S.A.
ASTRO	1990	I, S	UV	U.S.A.
Compton Gamma Ray Observatory (CGRO)	1991, active	S	Gamma ray	U.S.A., ESA
Extreme Ultraviolet Explorer (EUVE)	1992, active	I, S	UV: 70–760 Å	U.S.A.

[a] For planetary probes, see Part II.

[b] I denotes imaging capability; S, spectral capability.

[c] European Space Agency.

[d] A micron (μ) is 10^{-6} m = 10^{-3} mm = 10^4 Å; 1 μ = 10,000 Å.

observation. The first type will be examined in Part II; several in the latter category are listed in Table 6.3.

Though in principle a space telescope is just like one on Earth, there are some major differences. An orbiting space telescope orients itself by locking onto bright stars and is stabilized and pointed by gyroscopes (rapidly rotating wheels that resist turning) and small rocket motors. All observing is done by remote control and radio command from the ground, and the data are returned via a radio link. When its fuel supply is exhausted, or too many motors break down, the satellite dies. Space telescopes, such as the International Ultraviolet Explorer (IUE) (Figure 6.30a), the Infrared Astronomical Satellite (IRAS) (Figure 6.30b), and others (sampled in Table 6.3) have provided us with unparalleled views of the Universe.

The most versatile of astronomical satellites is the Hubble Space Telescope (HST) (Figure 6.30c). Conceived decades ago, this 2.4-m (90-inch) Cassegrain telescope finally flew in 1990. The originally flawed mirror was optically corrected in 1993. The HST contains two spectrographs and two cameras and can observe from the ultraviolet into the near infrared. Because it does not have to look through the Earth's atmosphere, Hubble's resolving power is limited only by the size of its small diffraction disk, allowing it to resolve objects 0.04 second of arc apart, some 10 times closer together than Earth-based instruments. With it and others we have finally broken our science loose from the constricting Earth, and in a sense have taken our telescopes to the stars themselves.

(a)

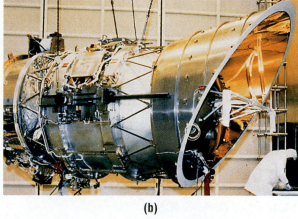

(b)

Figure 6.30

(a) The International Ultraviolet Explorer (IUE) observes from geosynchronous orbit. **(b)** The Infrared Astronomical Satellite (IRAS) is shown under construction. **(c)** The Hubble Space Telescope (HST) is a complete orbiting observatory able to make images as well as to do photometry and spectroscopy from the deep ultraviolet to the infrared. The "wings" on the spacecraft generate electricity from sunlight.

(c)

▶ KEY CONCEPTS

Atom: The basic unit that forms the chemical elements.

Blackbody: A body that absorbs all the radiation that falls upon it and radiates an energy spectrum that depends only on temperature.

Density: Mass per unit volume, measured in kg/m^3 or g/cm^3.

Electromagnetic force: The force of nature that combines electricity and magnetism; manifested by the **electric charge,** which produces an electric field, and the **magnetic field.**

Electromagnetic radiation: Energy transported by electric and magnetic fields (that is, electromagnetic waves or photons).

Electromagnetic spectrum: The array of the kinds of electromagnetic waves, categorized by wavelength; these include gamma rays, X rays, and ultraviolet, optical, infrared, and radio waves.

Electrons: Negatively charged atomic particles.

Elements: The basic constituents of common matter; each element is defined by a unique number of nuclear protons.

Energy: The ability of a body to impress a force on, and to accelerate, another.

Frequency (ν): The number of waves passing a location per second.

Ions: Atoms or molecules that are electrically charged because electrons are missing or added.

Luminosity (L): The amount of energy radiated by a body per second, measured in watts.

Molecules: Combinations of atoms.

Neutrons: Neutral atomic particles.

Nucleus: The combined protons and neutrons at the atomic center.

Photons: Particles of electromagnetic radiation that also incorporate wave motion.

Protons: Positively charged atomic particles.

Radio interferometer: Two or more linked radio telescopes that together have a resolving power given by their separation.

Radio telescope: A telescope that collects and detects radio radiation.

Reflecting telescope: A telescope that focuses radiation with a mirror (usually a paraboloid) and that has a variety of focal positions.

Reflection: The return of radiation from a surface at an angle equal to the angle of incidence.

Refracting telescope: A telescope that uses a lens to refract light to a focus.

Refraction: The bending of the path of radiation as it travels from one substance to another.

Spectrogram: A photographic or graphical recording of a spectrum.

Spectrograph: A device that separates electromagnetic radiation by wavelength.

Spectrum lines: Radiation at specific wavelengths (**emission lines**) or gaps in a continuous spectrum (**absorption lines**).

States of matter: The forms matter may take: **gases** do not have fixed shapes or volumes; **solids** have fixed shapes and volumes; and **liquids** have fixed volumes but not shapes.

Strong (nuclear) force: The force that binds atomic nuclei together.

Telescope: A device for gathering and focusing electromagnetic radiation.

Temperature: A measure of the amount of heat energy in a body.

Wavelength (λ): The distance between successive crests of a wave.

KEY RELATIONSHIPS

Frequency and energy:
$$E = h\nu.$$

Kirchhoff's laws:

1. An incandescent solid or hot high-density gas produces a continuous spectrum.
2. A hot low density gas produces an emission line spectrum.
3. A source of continuous radiation viewed through a cooler low density gas produces an absorption line spectrum.

Luminosity of a spherical blackbody:
$L = 7.2 \times 10^{-7} r^2 T^4$ watts, where r is the body's radius.

Stefan-Boltzmann law:
$$F = 5.7 \times 10^{-8} T^4 \text{ watts/m}^2.$$

Wavelength and frequency:
$$\lambda\nu = c.$$

Wien law:
$$\lambda_{max} = 2.898 \times 10^{-3}/T \text{ meters}$$
$$= 2.898 \times 10^7/T \text{ Ångstroms}.$$

EXERCISES

Comparisons

1. Compare the forces of nature introduced so far with regard to their strengths, their actions, and their ranges of effectiveness.
2. Compare the electric charges and masses of protons, neutrons, and electrons.

3. What is the difference between atomic number and atomic weight?

4. How do ions differ from normal atoms?

5. In what ways do gamma rays differ from infrared rays?

6. What is the difference between absorption and emission lines? Draw diagrams illustrating each process.

7. Why is the spectrum of neutral helium different from the spectrum of hydrogen?

8. Compare the characteristics and purposes of the paraboloidal reflectors used for optical and radio telescopes.

9. Why do single-dish radio telescopes have much worse resolving power than do optical telescopes?

10. Distinguish between an ordinary radio interferometer and a very long baseline interferometer.

Numerical Problems

11. The wavelength of a photon is 1 mm. What is its frequency?

12. Compared with a yellow optical photon with a wavelength of 5,500 Å, what are the energies of photons with wavelengths of (a) 1 km; (b) 1 Å?

13. At what wavelengths will blackbodies with temperatures of (a) 125 K, (b) 3,300 K, (c) 50,000 K, and (d) 10^6 K produce their maximum radiation? In what wavelength domain does each maximum fall?

14. How much more radiant energy per square meter of surface will a blackbody with a temperature of 3,000 K produce than a blackbody with a temperature of 1,000 K?

15. Spherical blackbodies A and B have the same luminosity. Blackbody A is twice as hot as blackbody B. What is the radius of B compared with that of A?

16. How many times fainter can you see stars with a 20-cm telescope than with a 10-cm telescope?

17. What is the light-gathering power of the largest optical reflector relative to the largest refractor?

18. What is the visual resolving power of (a) a 20-cm telescope; (b) a 2-m telescope?

19. What are the focal lengths of the eyepieces that you would need to produce 100 and 200 power with a telescope of 240-cm focal length?

20. What is the approximate resolving power of a 10-m radio telescope working at a wavelength of 5 cm?

Thought and Discussion

21. What kind of particle defines a specific element?

22. What is the function of the neutrons in an atomic nucleus?

23. What part of the electromagnetic spectrum lies between (a) the visual and the X ray; (b) the visual and the radio?

24. How can a blackbody be bright in the visual spectrum?

25. What does the spectrum of ionized hydrogen look like?

26. Why will a penny dropped into a glass of water appear to float if looked at through the water's surface?

27. How does the degree of refraction change with wavelength?

28. What are the functions of the following spectrograph components: (a) collimator; (b) diffraction grating; (c) camera mirror?

29. For what two reasons are space telescopes necessary?

Research Problems

30. Find ways in which each domain of the electromagnetic spectrum is used for beneficial purposes.

31. Select a particular telescope: optical, radio, or one in space. Research the history of the device from inception through development to implementation. Find out why it was needed and why it has its particular design characteristics.

Activities

32. On a moonless night, record the different star colors you can see with the naked eye. Estimate the temperatures of the stars.

33. Measure the lens diameter of a pair of binoculars in millimeters. Cover one lens completely and look through the other at a group of stars like the Pleiades. Count the number of stars you can see. Then place a cover over the lens with a circular hole cut to half the size of the lens. Count the number of stars you see again. Reduce the opening even further and count again to see the effect of lens diameter.

34. Your radio is a form of spectroscope. As you tune across the dial, you find radio stations broadcasting at specific frequencies. These are actually emission lines in the radio spectrum. Use your AM radio to draw the spectrum as it might appear if you could photograph it. Estimate the relative strengths of these spectrum lines.

Scientific Writing

35. A children's encyclopedia has asked you to write an article on light and radio. Explain the similarities and differences between them in the simplest possible terms.

36. The U.S. space program is often criticized as too expensive. Write a justification of the program based on its value to astronomy.

PLANETARY ASTRONOMY

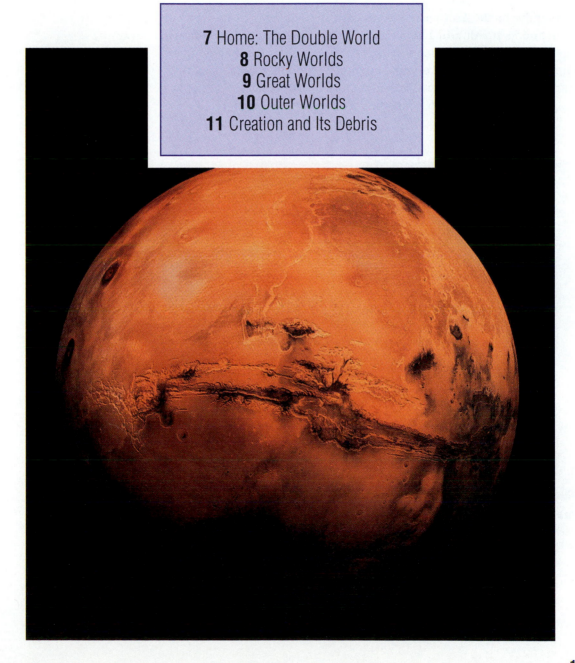

HOME: THE DOUBLE WORLD

The structure of the Earth and its Moon

7

Seven-tenths of the Earth is covered with liquid water in the form of **oceans,** and its surrounding blanket of gas, its **atmosphere,** is abundant in oxygen. Our planet's surface is mostly young, constantly changing, and teeming with life. In stark contrast, the nearby Moon is a cold, dead body (Figure 7.1). Each teaches us something about the other as we begin a broad study of planetary comparisons, of *comparative planetology*.

Figure 7.1
The crew of *Apollo 8* looks back at the blue, watery Earth over the bleak, cratered, rocky surface of the Moon.

7.1 THE DOUBLE PLANET

We grow up thinking of our Earth as a planet accompanied by a satellite, the Moon. In fact, we are part of a pair, a double world, the Moon in a sense serving as the fifth terrestrial (Earthlike) planet.

We explore them together. The Earth, always beneath our feet, is examined easily and directly. Our view of the Moon, however, is limited by our having to look through our turbulent atmosphere, and even then we can see only the lunar nearside. Detailed exploration requires spacecraft. Several have visited the Moon, culminating in the Apollo missions, six of which made successful soft landings, the first on July 20, 1969. The astronauts walked the landing sites, collected rocks and soil, and left scientific equipment behind. In all, the missions returned 380 kg of lunar material to Earth, a treasure that will be studied well into the next century.

The physical properties of the two bodies are summarized in Tables 7.1 and 7.2 on pages 154 and 155. The radius of the Earth is found easily from the apparent positions of stars (see Section 2.2) and its mass from the acceleration of gravity (see Section 5.1). From the Moon's distance and angular dimension (see Section 4.1.1), we derive a physical radius of 1,738 km, 0.272 that of Earth.

Your weight on the Moon would be only about one-sixth that at home.

The lunar mass can be determined from its gravitational effect on the Earth. As the Moon orbits the Earth, the Earth must orbit the Moon, the two swinging around a common point between them (Figure 7.2) once a

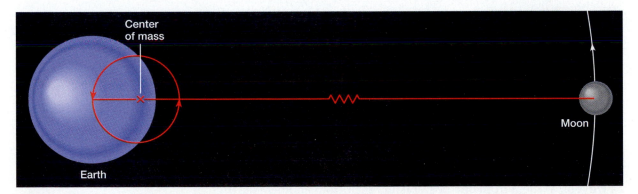

Figure 7.2
The Earth and Moon, like any two bodies, orbit a common center of mass placed on a line between them. The mass of the Earth is 81 times that of the Moon, so the Moon's distance from the center of mass is 81 times that of the Earth's. The Earth therefore has a small orbit about the Moon (indicated in red) whose center is actually inside the Earth. (The scale is distorted; the distance of the Moon from the Earth is actually 30 times the Earth's diameter, as seen in Figure 1.3).

TABLE 7.1
A Profile of the Earth

Planetary Data		Atmosphere (surface)	
Equatorial radius	6,378 km	Pressure	10,330 g/cm^2
Polar radius	6,358 km	Mean temperature	10°C
Mass	5.97×10^{24} kg	*Composition by Number*	
Mean density	5.5 g/cm^3	Nitrogen (N$_2$)	77%
Uncompressed density	4.5 g/cm^3	Oxygen (O$_2$)	21%
Escape velocity	11.2 km/s	Argon (Ar)	1%
Core temperature	7,000 K	Water (H$_2$O)	1%[a]
Core pressure	3.6 megabars	Carbon dioxide (CO$_2$)	0.03%
Age	4.6×10^9 years		

Layers	% Mass	Radii (km)	Layers[b]	Height (km)	Temperature (°C)
Solid inner core	1.7	0–1,220	Troposphere	0–12	20 to −60
Liquid outer core	30.8	1,220–3,480	Stratosphere	12–50	−60 to 0
Mantle	67.0	3,480–6,360			
Crust	0.7	6,360–6,370			

Planetary Features

Cloud cover: Considerable and variable water-vapor clouds.
Craters (meteor)[c]: A few obvious, most eroded and difficult to find.
Basins: Ocean basins produced by crustal spreading.
Mountains: Tectonic mountains caused by plate motion.
Life on the surface
Plate motion: 16 crustal plates in constant motion, producing rifts, mountains, and volcanoes; surface age varies from zero to 4.2 billion years.
Volcanoes: Considerable volcanic action caused by subduction of plates as well as plume-type volcanism.
Water: Considerable liquid water in ocean basins, solid (frozen) water at the poles, vaporized water in the atmosphere.

---

[a] Variable.

[b] The layering above the stratosphere is complex, temperature changing in different ways with height; the ionosphere lies above about 80 km.

[c] See Chapter 11.

lunar month. As a result of the Earth's motion around this *center of mass*, the nearby planets seem to shift back and forth, exhibiting a parallax effect as we look at them from one position and then from another (see Section 4.1.1). From the degree of shift, we find the Earth's orbital radius to be about 1/81 the radius of the lunar orbit. Since accelerations are inversely proportional to mass, the Moon's mass must be about 1/81 that of the Earth. From Kepler's generalized third law applied to spacecraft in orbit about the Moon we find a more precise modern value of 1.23% (1/81.3)

TABLE 7.2
A Profile of the Moon

Planetary Data		Layers	% Mass	Radii (km)
Mean distance from Earth	384,400 km[a]	Core	4[b]	400[b]
Mean radius	1,738 km	Mantle	84	400–1,670
Mass	7.3×10^{22} kg	Crust	12	1,670–1,740
Mean density	3.34 g/cm^3	Atmosphere	None	
Uncompressed density	3.3 g/cm^3			
Gravity	0.17 g_{Earth}			
Escape velocity	2.4 km/s			

Other

Magnetic field	Fossil only
Rotation period[c]	27.3 days
Axial inclination	5.7°

Planetary Features

Basins: Large number of giant impact basins from a few hundred km to over 3,000 km wide; those on nearside filled with lava to make maria.

Craters: Range from hundreds of km across to pits in rocks; vast number overlap in highlands; far fewer in maria; some young craters with large ejecta blankets and rays.

Highlands: 3.8 or more billion years old; heavily cratered by the late heavy bombardment.

Maria: Volcanic plains between 3 and 3.8 billion years old produced by lava that flooded impact basins after the end of the late heavy bombardment and that continued to about a billion years ago; relatively few craters; most maria are on nearside.

Mountains: Basin walls and ejecta blankets; no tectonic mountains.

Plates and plate motion: Single continuous plate; no motion.

Surface temperature: 383 K maximum daytime, 103 K minimum nighttime.

Volcanoes: Few ancient shield volcanoes and domes, none active.

Water: None.

[a] Varies from 406,000 km at mean apogee to 363,000 km at perigee.

[b] Upper limits; the core could be much smaller.

[c] Synchronously locked with revolution period about Earth; rotation period relative to the Sun is 29.5 days.

that of Earth. The lunar radius and mass show the acceleration of gravity (see Section 5.3) at the lunar surface to be 0.166 that of Earth. Thus your weight on the Moon would be only about one-sixth that at home.

The mutual gravity between the Earth and Moon has other effects. Everywhere along the seacoast, water periodically rises and falls with the **tides,** flowing first up and then down the slope of the beach between *high tide* and *low tide* (Figure 7.3). The interval between two successive high tides is 12^{h}25^m, equal to that between successive transits of the Moon

(a)

(b)

Figure 7.3
At Cutler, Maine, **(a)** the high tide is several meters above **(b)** the low tide that takes place six hours later.

across the meridian (first above the horizon, then below), demonstrating the lunar link. Tides are caused by a *differential* gravitational pull. Figure 7.4a shows that the lunar gravity is strongest on the side of the Earth closest to the Moon, less strong at the center, and weakest on the side away from the Moon. Subtract the arrow at the center (the average pull) from the other arrows. The result, shown in Figure 7.4b, is a stretching effect in the lunar direction. The oceans flow toward the line connecting the Earth to the Moon, producing two opposing *tidal bulges*, with shallower water in the perpendicular direction. Any coastal point on the rotating Earth will thus pass twice through high and low water between two successive meridian transits of the Moon above the horizon.

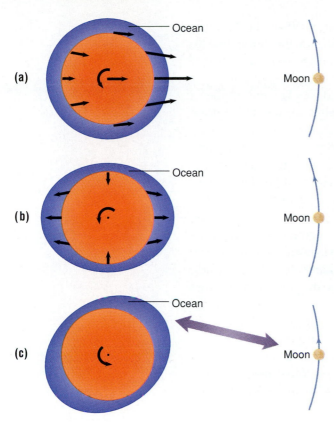

Figure 7.4
(a) The arrows show the relative strength of the lunar gravitational pull on Earth's oceans at different distances from the Moon. **(b)** The force at the Earth's center is subtracted from all the arrows. What remains is the differential force, which causes water to flow toward the line connecting the Earth and the Moon. **(c)** As a result of the Earth's rotation and the time it takes the water to flow, the tidal bulge leads the Moon. The depth of the ocean in these drawings is greatly exaggerated.

The ocean is rotating with the Earth, and the water takes time to adjust to the changing lunar pull. Therefore, the tidal bulges will be shifted forward in the direction of rotation (Figure 7.4c). A person on the beach will typically see the Moon pass across the meridian a few hours before high tide. Although the size of the tidal bulge in the open ocean is only about a meter, it can be greatly exaggerated in the shallower waters at the shore. The tide is not confined to the oceans: the Earth's solid body is also deformed in a tide that points toward the Moon with a bulge about 30 cm high.

The Sun produces tides with about half the effect of the Moon. Lunar and solar tides are independent. When the Sun and Moon are aligned, at new and full lunar phase, the two tides combine to produce the highest and lowest tides of the month, the *spring tides*. When the Moon is in its quarter phases, the solar tide partially fills in the lunar tide to produce *neap tides*, in which the range between high and low water is smaller.

The gravity of the Moon pulls back on the tidal bulge and applies a subtle but steady brake to the rotating Earth, causing our day to lengthen by 0.0015 second per century. The tidal effect of the more massive Earth on the Moon is much greater than the effect of the Moon on the Earth. Solid tides—the Moon has no oceans—have dissipated enough rotational energy to stop completely the rotation of the Moon relative to the Earth, forcing the Moon to keep one side pointing toward us (see Section 4.1.1). The Moon is therefore tidally locked onto the Earth in **synchronous rotation,** with equal rotation and revolution periods. The Moon *does* rotate on its axis relative to the Sun, however, turning once in a synodic month. As a result, someone on the Moon would see the Earth hang steadily in the sky, but the Sun would slowly rise, take half a synodic month to move across the lunar sky, then set.

The Earth and Moon are closely linked not just by gravity but also by age. Atomic nuclei are held together by the strong force produced by their protons and neutrons (see Section 6.1). For each element, there is a range in the number of neutrons the nucleus can contain. Most helium atoms have two neutrons and two protons, and because of these four nuclear particles are called ^{4}He. However, a small minority of helium atoms have only *one* neutron and are thus ^{3}He; such variations are called **isotopes.** Table 6.1 gives the atomic weights (protons plus neutrons) of the element's most common isotope. That of iron is ^{56}Fe (26 protons and 30 neutrons), but an iron ball also contains ^{57}Fe and ^{58}Fe. If an isotope has too many or too few neutrons, however, the nucleus is unstable and can change into a different kind of atom. Some of the decay processes involve the fourth, or *weak*, force. Though weaker than electromagnetism, it is far stronger than gravity, and—like the strong force—acts over only a short range. The change is accompanied by the release of energy in the form of particles and electromagnetic radiation, the latter including high-energy X rays and gamma rays. Some of these *radioactive isotopes* are exceedingly dangerous. Above atomic number 83 (bismuth), *all* isotopes are radioactive; this group includes such deadly elements as radium.

Radioactive isotopes provide us with invaluable natural clocks. Several kinds of rock contain radioactive elements that fall apart, or decay, into lighter daughter elements. The decay time of a radioactive isotope is indicated by its *half life*, the time it takes for half of it to change into something else. Usually, the shorter the half life, the more dangerous the isotope. ^{238}U (uranium), for example, decays into ^{206}Pb (lead), its half life 4.5

billion years. If you start with a kilogram of ^{238}U, after 4.5 billion years you will have 0.5 kg left, and after 9 billion only 0.25 kg. The surface rocks of the Earth were all once in a molten—that is, liquid—state, the result of great heat. When a rock solidifies, an initial ratio of ^{238}U to ^{206}Pb is sealed within. The ratio (^{238}U/^{206}Pb) decreases steadily at a rate given by the uranium's half life. The measured ratio of ^{206}Pb to ^{238}U thus gives the rock's age, the time since solidification. Several other isotope ratios are used as well.

The **4.6-billion-year age of the meteorites is identified with the age of the Earth-Moon system.**

The rocks of the Earth range in age from zero to 4.2 billion years. The Earth must be at least as old as its oldest rocks. The oldest rocks of the Moon date back even further, 4.3 to 4.5 billion years. The record is held by **meteorites,** pieces of rock and metal from interplanetary space that range from mere grains of sand to bodies many meters across that occasionally strike the Earth. (They will be addressed in detail in Chapter 11.) Almost all the meteorites are between 4.5 and 4.6 billion years old. The differences in all these ages are easily explained by the different times it took the bodies to cool after their creation. Since there is additional evidence (to be examined later) that everything in the Solar System was formed at about the same time, the age of the meteorites is identified with that of the Earth-Moon system.

7.2 THE EARTH'S SURFACE

The rocky surface of the Earth is broadly divided into two kinds of area. The higher **continents,** or dry land masses, occupy about 30% of the planet's surface. The remaining 70% are basins that are filled by ocean waters. The continents contain the oldest rocks, attesting to their relative permanency, whereas the seabed is nowhere more than about 200 million years old, revealing that it is continuously renewed and destroyed.

Rock consists of a variety of *minerals* that can be variously categorized. Chemically, most rocks are **silicates,** a broad class of mineral made of silicon and oxygen, often in combination with a variety of metals. Within this chemical class we find two basic kinds of crystal structure. The most common continental surface rocks are different forms of *granite*, a silicate characterized by coarse crystals. By contrast, the **basalts** have a fine-grained structure in which the crystals can scarcely be seen; they are made chiefly of calcium-bearing and heavy-metal silicates. Although they are found on land too, basalts dominate the ocean beds. The lighter rocks (including granites) that make the bulk of the continents sit atop this relatively heavy

(a)

Figure 7.5
(a) Mountains in the Chilean Andes typify those that ring the Pacific Ocean basin. **(b)** Earthquakes, which are caused by crustal slippage and can produce enormous damage, are commonly associated with mountain chains, as are **(c)** volcanoes. Here, Mt. St. Helens is seen in its 1982 eruption, which sent huge quantities of ash into the sky.

(b)

(c)

basaltic material rather like puffs of marshmallow in a cup of cocoa. Granites and basalts are made when molten rock from inside the Earth—**magma**—is extruded as liquid **lava** and solidifies. Various processes erode and crush rock; when mixed with plant and animal life, the particles become *soil*. Rock can also be transformed by pressures under the sea and within the Earth to produce a great variety of different kinds.

The Earth's surface has several characteristics that demonstrate a high level of activity. The most obvious continental features are *mountains* (Figure 7.5a). They are organized into groups and long chains (Figure 7.6), the greatest of which ring the Pacific Ocean basin and include the Rockies of North America and the Andes of South America. Within central Asia lies another huge group, of which Everest, 8.85 km above sea level, is the highest. Mountains, mostly confined to long sinuous ridges, also lie under the oceans. Continental mountains are constantly being worn away in part by rainfall, their eroded rocks and soil carried away to the oceans. Some process, therefore, must keep building new ones.

Figure 7.6
A map of the Earth made by satellite radar shows organized mountain ranges (red) ringing the Pacific Ocean and within southern Asia. More mountain chains lie on the ocean floor, the longest running down the middle of the Atlantic basin. The distribution of earthquakes and volcanoes is similar.

Mountain-building proceeds at an unnoticeable pace. Much more rapid and dramatic are **earthquakes** (Figure 7.5b), shocks produced when the Earth's upper layer, or **crust,** slips and suddenly moves, causing the ground to vibrate like a drumhead. Tiny, unfelt shocks occur every day; rare, huge rumbles can demolish cities. Although no place is immune, earthquakes are concentrated along mountain chains, both on land and at sea, showing the close relationship between mountain-building and the shaking of the crust.

Volcanoes (Figure 7.5c) are no less devastating than earthquakes. They usually appear as mountains that spew hot gases and solid materials that consist of rocks, finely divided ash, and lava. Most volcanoes do not erupt continuously: some have intervals of many years, even centuries, between active periods, while others are truly dead, remnants of activity long gone. A volcanic peak commonly contains a *volcanic crater*, a pit formed when mass was removed in an eruption, or a *caldera*, a depression

Volcanic activity is associated with mountain chains and earthquake zones.

Figure 7.7
The "Big Island" of Hawaii, topped by Mauna Loa, is a shield volcano 180 km across at sea level and half again as large at its base.

created when the mountain peak slumped into an underground chamber emptied of its magma. Volcanic activity, associated with mountain chains and earthquake zones, is concentrated around the Pacific basin (see Figure 7.5), the so-called "ring of fire."

There are several different kinds of volcanoes. Those with thicker, viscous lava like Mount St. Helens can explode violently. In contrast, the lava of a **shield volcano** is runny and flows easily. Over millennia, the outflow creates a broad mountain with a low slope. Shield volcanoes are *not* associated with mountain ranges. The "Big Island" of Hawaii (Figure 7.7), which contains inactive Mauna Kea with its observatories and the very active Mauna Loa and Kilauea, is the best example. Measured from base (the seabed) to top, it is 11 km high and the tallest mountain possible on Earth. If you placed more mass on top of it, the weight would make the mountain flow out at the bottom.

In spite of their prominence, volcanoes are far from the only source of magma. Huge amounts have periodically flooded from fissures in the crust to produce not mountains but vast, thick *lava plateaus*. Much of Washington, Oregon, and Idaho is covered by a lava plateau well over 2 km thick that was formed by multiple ejections over a period of 70 million years. Among the largest such areas is the Deccan plateau, which covers most of southern India. The ocean bed beneath the sediments is a vast basaltic lava plain that has issued from undersea crustal fissures associated with the mountainous ridges.

7.3 THE CONTRASTING LUNAR SURFACE

The Moon (Figure 7.8) is very different. It has neither water nor an atmosphere to produce clouds or haze. You can see with the naked eye that the lunar surface is divided into light and dark areas. With the telescope, we find that the bright areas, the **lunar highlands,** are covered with pits, or **craters,** of all sizes. The dark markings, or **maria** (Latin for "seas," singular *mare*), contain relatively few craters and consist of smooth dark rock set into basins in the lunar crust. Several maria are ringed by mountain ranges.

The lunar craters are depressions circled by elevated walls (Figure 7.9). They are *not* volcanic, but were produced instead by *impacts* (Figure 7.10). These impact craters are holes dug by collisions with meteorites. Interplanetary debris moves on orbits about the Sun with velocities of some 30 km/s. A 10-m-diameter body has a mass of 10^7 kg, and at 10 km/s—a typical collision speed—carries the energy-equivalent of a small nuclear bomb. When such a body hits the ground, some of the energy is transformed into heat that vaporizes it. Most of the energy, however, violently

Figure 7.8

(a) Photographs of the first (right) and last (left) quarters of the Moon show the rugged highlands and the smooth maria. (b) Names of features are given on the map, on which the maria are shown in blue and the highlands in orange; craters are indicated in violet, major ejecta blankets in green, rays in yellow, and mountain ranges in red. Linear structures are cross-hatched in black. The landing sites of the Apollo craft are shown by red dots and mission numbers. North is at the top. Turn the map upside-down when looking through a telescope.

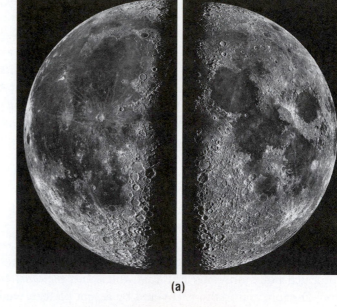

(a)

1. Langrenus	43. Albategnius
2. Geminus	44. Hipparchus
3. Macrobius	45. Horrocks
4. Stevinus	46. Rhaeticus
5. Rheita	47. Manilius
6. Metius	48. Autolycus
7. Fabricius	49. Aristillus
8. Franklin	50. Cassini
9. Cepheus	51. Archytas
10. Endymion	52. Bond
11. Atlas	53. Walter
12. Hercules	54. Regiomantanus
13. Fracastorius	55. Purbach
14. Piccolomini	56. Lexell
15. Posidonius	57. Arzachel
16. Theophilus	58. Alpatragius
17. Cyrillus	59. Alphonsus
18. Catharina	60. Ptolemaus
19. Julius Caesar	61. Herschel
20. Eudoxus	62. Pallas
21. Aristoteles	63. Archimedes
22. Vlacq	64. Plato
23. Zagut	65. Maginus
24. Pitiscus	66. Clavius
25. Nearchus	67. Blancanus
26. Barocius	68. Scheiner
27. Maurolycus	69. Tycho
28. Gemma Fricius	70. Longomontanus
29. Abulfeda	71. Wilhelm
30. Godin	72. Bullialdus
31. Agrippa	73. Guericke
32. Cuvier	74. Parry
33. Heraclitus	75. Eratosthenes
34. Licetus	76. Timocharis
35. Faraday	77. Copernicus
36. Stöfler	78. Pytheas
37. Fernelius	79. Lambert
38. Aliacensis	80. Euler
39. Werner	81. Gassendi
40. Apianus	82. Kepler
41. Bohenenberger	83. Aristarchus
42. Parrot	84. Mairan

A. Alpine Valley
B. Straight Wall
C. Straight Range

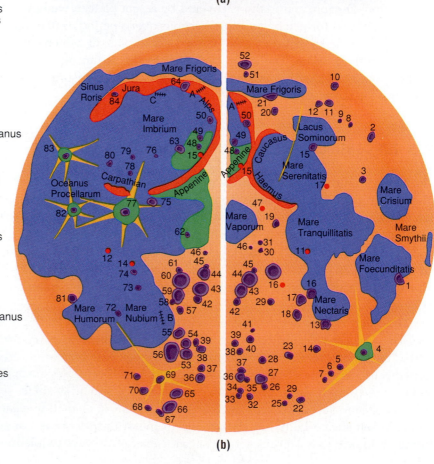

(b)

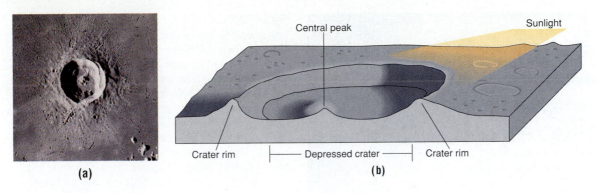

(a)   **(b)**

Figure 7.9

(a) Lunar crater Euler is 28 km in diameter. **(b)** A crater is a depression surrounded by a circular elevated wall. Large craters commonly have central mountain peaks. Shadows cast by slanting sunlight make the crater visible.

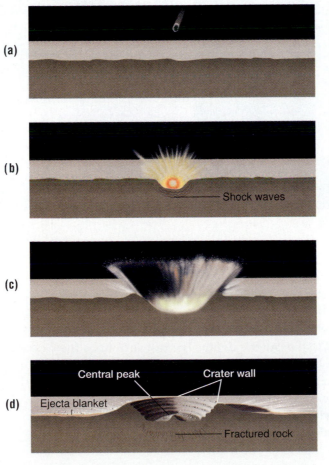

Figure 7.10

(a) A large body from space speeds toward the lunar surface. **(b)** It strikes and begins to vaporize as it sends powerful shocks into the rock. **(c)** Decompression of the compressed rock ejects vast amounts of broken rock to create the crater walls and the ejecta blanket. **(d)** The crater floor is now covered with rubble and solidified melted rock. If the impact is forceful enough, rebound of the crater floor pushes up a central peak and creates terraces in the walls.

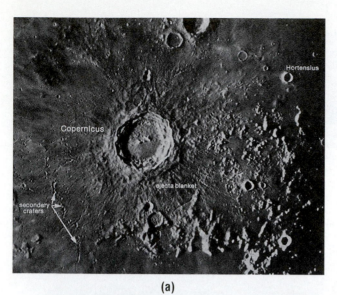

(a)

Figure 7.11

Three views of Copernicus (a 90-km-wide crater on the southern edge of Mare Imbrium) display all the properties of a massive impact. **(a)** An overhead photograph shows the elevated walls and the extensive ejecta blanket. Beyond the blanket are numerous secondary craters; inside the primary crater is a central peak. **(b)** We look over Mare Imbrium and the Carpathian Mountains at Copernicus on the horizon. Elongated secondary craters string away from the impact, as do the rays. The crater Pytheas, in the foreground, also displays a prominent ejecta blanket and some of its own secondary craters. The number of craters, however, is far lower than that seen in the highlands. **(c)** Looking directly into Copernicus we view the central peak, the rubble-filled interior, and terraced walls.

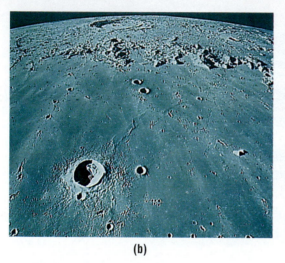

(b)

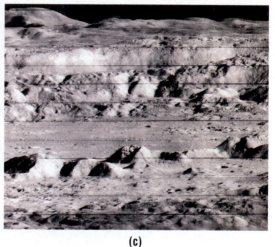

(c)

compresses the surface rock. Subsequent decompression, or rebound, expels a vast amount of **ejecta** consisting of fractured and melted debris, excavating a hole with a diameter some 10 times that of the impacting body and a depth about one-fifth the crater's width.

Some of the ejected material piles up to produce a circular crater wall with a height roughly equal to the crater's depth. The rest surrounds the new crater in a large **ejecta blanket** (Figures 7.10 and 7.11). Rubble and resolidified melted rock also fall back inside, partially filling the crater. If the force of the blow is sufficient to create a crater more than about 20 km in diameter, the force with which the crater floor rebounds from the shock is great enough to produce a central mountain peak (see Figures 7.9, 7.10, and 7.11), which then drags the crater walls down in a series of terraces. Big pieces blown out of a large crater will dig new pits of their own and

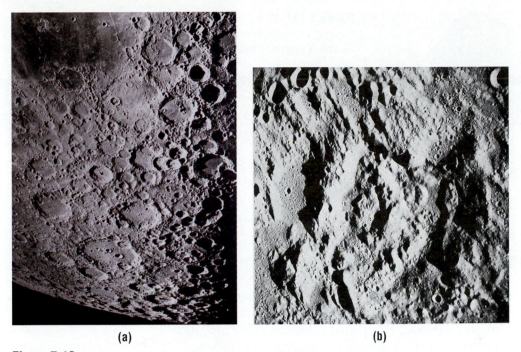

(a)   (b)

Figure 7.12

(a) The battered highlands of the Moon toward the south pole show an extraordinarily rugged terrain, with craters ranging down to only a kilometer across. By contrast, Mare Nubium, at upper left, is quite smooth. **(b)** This view of the lunar highlands near Sinus Medii was taken by an orbiting *Apollo 10* astronaut. It shows thousands of tiny craters superimposed on a crushed terrain. The oldest craters have been nearly destroyed by the younger ones; the smallest ones are only 50 m in diameter.

produce numerous secondary craters that surround the primary. Streamers of the rocky debris can arc across the lunar surface, creating strings of secondary craters and white *rays* (see Figure 7.11b). The rays from the large crater Tycho, deep in the southern lunar hemisphere, extend for a thousand kilometers and cover nearly an entire lunar hemisphere.

Tens of thousands of craters—some over 100 km across—are seen in the lunar highlands (Figure 7.12). The entire surface of the highlands is cratered, the devastation of the original surface complete. Craters pile on craters, showing that they must have been created over a period of time. The relative ages can be determined by the way in which various features lie on top of one another. Some craters are so badly beaten up (see Figure 7.12) that they are only barely recognizable: these must be the oldest. Even-earlier craters must have been entirely obliterated. Craters with few superimposed meteorite strikes are the youngest.

These relative ages can be calibrated with absolute ages determined by radioactive dating of samples from the Apollo missions. The majority of the highland rocks were fused from smaller smashed particles under

Figure 7.13
This lunar rock is a mixture of various kinds of older rocks that have fused under the force of a meteorite impact, making it look a little like broken concrete.

the force of meteorite impact (Figure 7.13). The highland rocks are all old, their ages mostly between 3.8 and 4 billion years. The heavy cratering must have taken place over a relatively short interval of less than 700 million years, from the time of the origin of the Solar System (4.6 billion years ago) to roughly 3.8 billion years ago. This period is known as the time of the *heavy bombardment*; the tail end of it, which produced the craters we see, is the **late heavy bombardment.** Chemically, the highland rocks consist of light silicates. Ancient intact crystals of varying composition can be assigned crude ages of 4.3 to 4.5 billion years, the large quantities showing that the Moon was at one time covered with a liquid magma ocean, one that may even have involved the entire interior.

By comparison with the densely cratered highlands, the maria look peaceful. Within them, the crater density (the number per unit area) is so low that undamaged surface can be seen (see Figures 7.11b, 7.12a, and 7.14a). The circular outlines of most maria indicate that they are contained by enormous impact craters or **impact basins**. The Imbrium basin is at least 1,100 km in diameter. The impacting bodies must have been tens or even a hundred or more kilometers across. The lunar mountain ranges are actually the basin walls, towering as much as 9 km (almost the height of Everest) above the basin floors.

The color and reflectivity of the dark mare rock suggests it is solidified lava that has at least partially filled the basins. Many craters in and around the maria look as if they have been flooded with the same material (see Figure 7.14a). Evidence for volcanism is also provided by long, sinuous valleys or *rilles* that look like riverbeds (Figure 7.14b) but are actually channels in which molten lava once flowed. Similar features are seen on Earth. Rock samples brought back by the Apollo astronauts show that the mare material consists of fine-grained volcanic basalts that are abundant

The circular outlines of most maria indicate that they are contained by enormous impact craters.

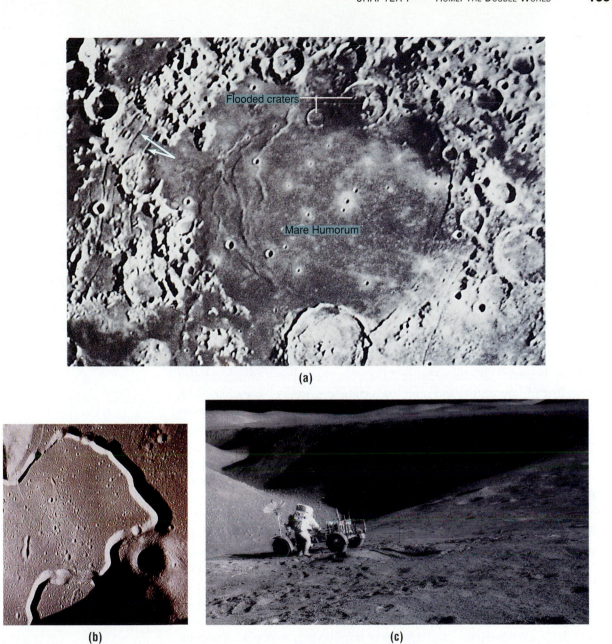

(a)

(b)   (c)

Figure 7.14

(a) The lava of the Mare Humorum's basin has buckled and has flooded nearby craters. Cracks parallel to the mare walls (arrows) are the result of slumping after solidification. **(b)** Ancient lava flows, erupting from cracks in basin floors, also ran in sinuous rilles. Here, Hadley Rille, photographed by the *Apollo 15* crew, snakes across the floor of eastern Mare Imbrium. **(c)** Astronaut James Irwin stands at Hadley Rille's edge. Note the bulky life-support system he had to carry, and the lunar rover, which enabled the astronauts to travel considerable distances.

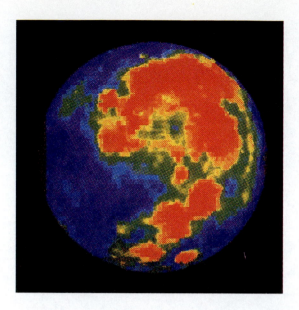

Figure 7.15
The spacecraft *Clementine*, which orbited the Moon in 1994, imaged the Moon at spectral absorption features produced by iron. High iron content, up to 14%, is indicated in red. The colored areas match almost perfectly with the lunar maria (compare with the map in Figure 7.8), showing the mare basalts to be iron-rich in agreement with laboratory measures made of Moon rocks. Titanium shows the same distribution.

in heavier iron-rich silicates (Figure 7.15). *Volatile elements*—which solidify only at low temperatures and melt (or turn to gas) easily—are greatly depleted relative to terrestrial basalts; there is no water at all. On the other hand, the rocks are relatively rich in *refractory elements* (those that melt at high temperature) such as iron and titanium.

There are no active volcanoes on the Moon. We see only some structures that look like ancient shield volcanoes and a few domes that apparently have been created by upward-pushing magma. The maria were not produced by lava from volcanoes, but are volcanic floods reminiscent of the kinds that made the thick plateaus of the American Northwest and southern India.

Among the oldest recognized basins (or combination of basins) is Procellarum, which formed before the heavy bombardment ended. Part of its original outline—an amazing 3,400 km in diameter—can be seen as Oceanus Procellarum and Mare Frigoris (see Figure 7.8). The other major nearside basins were created in a series of impacts that took place between about 3.8 and 3.9 billion years ago, beginning with Nectaris and ending with Imbrium and Orientale (just barely seen from Earth at the lunar edge). These impacts wiped out the highlands that once existed there and destroyed and hid much of the Procellarum basin.

The craters we see within the maria were created after the basins were formed and filled with lava. Some of these craters have flooded floors, and a few are only ghostly rings; these must have been formed after the creation of the basins but before the flooding lava filled them in. Therefore, there must have been a significant interval between the time of the great impacts and the eruption of the mare magma. The bulk of the mare lavas are between 3.1 and 3.8 billion years old, but the flows continued at an ever-decreasing rate until roughly a billion years ago. In spite of the apparent extent of the flows, they cover only 17% of the lunar surface and are typically only a few hundred meters thick. The total amount of magma extruded during the 700-million-year peak is really quite small, amounting to no more than that produced by Earth's Vesuvius.

Since even the oldest mare surfaces are lightly cratered, the heavy bombardment must have ended fairly quickly. Over the last 3.1 billion years, the cratering rate seems to have proceeded at an almost constant pace. The rays from the great craters Copernicus, Aristarchus, Tycho, and others lie over almost everything else, showing them to be relatively young. Tycho may be only 100 million years old. This lunar chronology (Figure 7.16) provides a key for studying other planets and their satellites. If the surface of another body is saturated with craters, we know it must have suffered through the late heavy bombardment and must be old. If the surface is not saturated, we can estimate its age from the crater density and the cratering rate calculated from the age and crater density of the lunar maria.

Almost all the maria are on the lunar nearside. The once-mysterious farside (Figure 7.17a) is covered with the usual craters and displays several large impact basins. The huge 2,300-km-wide Aitken basin (Figure 7.17b) near the lunar south pole is one of the largest on the Moon, as well as the deepest, descending 8 km below the surrounding terrain. However, the farside basins remain largely empty of mare lavas. Even when the

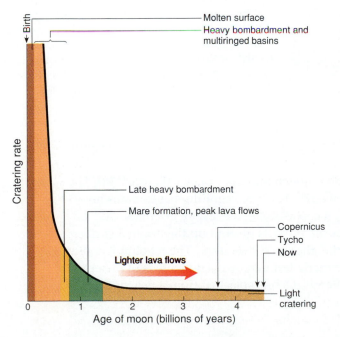

Figure 7.16

A rough chronology for the Moon shows that heavy cratering and basin formation took place very early in lunar history. From the time the Moon was 1.4 billion years old—3.1 billion years ago—the surface has been relatively quiet, with light cratering and diminishing lava flows.

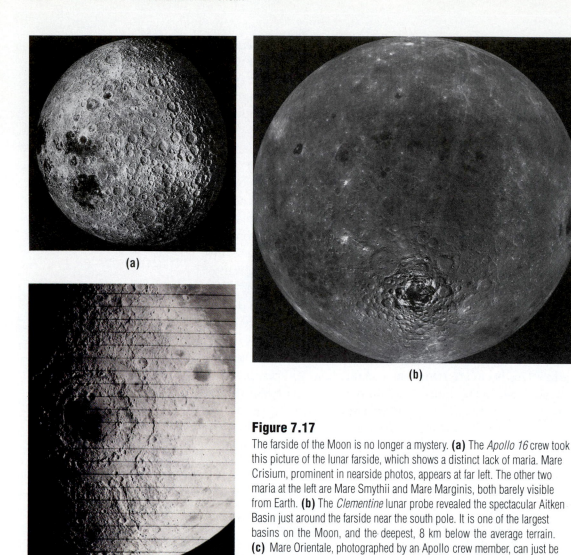

(a)

(b)

(c)

Figure 7.17

The farside of the Moon is no longer a mystery. **(a)** The *Apollo 16* crew took this picture of the lunar farside, which shows a distinct lack of maria. Mare Crisium, prominent in nearside photos, appears at far left. The other two maria at the left are Mare Smythii and Mare Marginis, both barely visible from Earth. **(b)** The *Clementine* lunar probe revealed the spectacular Aitken Basin just around the farside near the south pole. It is one of the largest basins on the Moon, and the deepest, 8 km below the average terrain. **(c)** Mare Orientale, photographed by an Apollo crew member, can just be glimpsed from Earth. From above, this striking feature is clearly a giant ringed crater that is only partially flooded with dark lava.

lavas do appear, they fill in only portions of the impact structures (Figure 7.17c).

On the maria, gritty plains stretch into the distance (Figure 7.18). The surface is covered with a pulverized soil, or **regolith,** that contains no organic or biological material. It consists of a fine, glassy grit made over billions of years as micrometeorite impacts ground up the original surface, the resulting heat turning the glassy crystals dark. The regolith is a few meters deep in the younger maria but may reach down tens of meters in the older, more beaten highlands. Debris splashed from a new crater can

Figure 7.18
Edwin Aldrin, Jr., of *Apollo 11*, deploys a scientific experiment package in Mare Tranquilitatis; the lunar module sits on its launch pad in the background. The sky is black; the "soil," or regolith, is a fine grit produced by micrometeorite bombardment.

dig through the regolith and expose lighter-colored material, forming the rays. But as the bright streaks are exposed to meteorite infall, they too turn dark, and the rays disappear; we therefore see rays only around young craters.

7.4 IMPACTS AND THE EARTH

The Earth ought to have been struck by meteorites at least as frequently as the neighboring Moon. Where then are the Earth's impact craters? Even though the oldest rocks on the Earth are 4.2 billion years old, the majority solidified after the late heavy bombardment and the creation of the great lunar basins. Most of the impact craters that formed later have been erased by mountain-building and by erosion by ice, wind, and water.

The Moon's numerous rayed craters, however, are geologically relatively young, so the Earth's surface should still show some identifiable impact craters. It does. The most famous is Meteor Crater near Winslow Arizona (Figure 7.19a), which was produced by an iron meteorite some 20 m in diameter that struck roughly 50,000 years ago. Other impact craters are harder to find because they have been severely eroded, filled in, or covered with vegetation, but about 150 are now confirmed (Figure 7.19b).

These collisions pale beside one that may have taken place 65 million years ago, at the time that the dinosaurs and some 95% of all other species disappeared. Meteorites are relatively rich in the element iridium. All over our planet, layers of clay deposited at the time of the dinosaurs' demise are also iridium-rich, suggesting that the Earth was struck by a massive projectile from space 5 km or more across. The impact would have raised

A huge impact feature in Mexico's Yucatán peninsula may be the scar of the great collision that killed the dinosaurs.

Figure 7.19
(a) Meteor Crater in Arizona displays characteristics associated with lunar craters, including an upraised rim, a deep depression more than a kilometer across, and an ejecta blanket of exploded debris. **(b)** Meteor Crater is dwarfed by 75-km-wide Lake Manicouagan, an ancient impact site in Canada.

(a)   (b)

huge clouds of dust and, if it hit the ocean, steam; it could also have triggered massive volcanism. The resulting change in climate caused by the blockage of sunlight could easily have killed large numbers of life forms. A huge impact feature that may be as much as 300 km across lies at the northern edge of Mexico's Yucatán peninsula and may be the remaining scar of the great collision.

7.5 INTERIORS

To understand more about the varied surface features of the Earth and Moon, we must probe into their interiors. The first clue to a planet's structure is its average density, found by dividing its mass by its volume (equal to $4\pi/3$ times the radius cubed). The Earth's average density is 5.5 g/cm³. The weight of the Earth's rock layers compresses the interior, raising the density; when this effect is taken into account, the uncompressed density drops to 4.5 g/cm³. However, the densities of surface rocks and volcanic basalts are typically between 2.6 and 3.0 g/cm³. Something very heavy, with a density well above 5 g/cm^3, must lie inside to compensate for the light stuff at the surface. Such a high density implies a heavy metal. Since metal is heavier than rock, and since the Earth's interior is hot (as we know from volcanoes), we might expect the Earth to have undergone **differentiation,** a process by which the heavy materials sink to the center under the force of gravity and the lighter float to the top. The most common metal is iron, followed by nickel. The Earth should therefore have an iron **core** that contains about a third of the total terrestrial mass.

A finer probe of the Earth is provided by earthquakes. A quake acts much like the clapper of a bell, causing the entire Earth to ring with *earthquake (seismic) waves* that penetrate the globe. There are two different kinds. The *primary wave (P wave)* is a compressional wave that vibrates the rock back and forth along the direction of motion. It is essentially a sound wave, but with a frequency too low for the human ear to hear. The *secondary wave (S wave)* is a shear, or transverse, wave that vibrates the rock perpendicular to the wave's motion, much like a wave in a pool of water. Both kinds are detected by *seismographs* placed in geological observatories around the world.

The two kinds of waves move at different speeds through the Earth. The delay in their arrival times at the seismographs tells how far they have traveled from their point of origin. Their speeds also depend on the materials, densities, and temperatures encountered along the paths and, like light waves, they are reflected and refracted at discontinuities. Analysis of the strengths and arrival times of the waves of large numbers of quakes have allowed geophysicists to build up a picture of the terrestrial interior. We now know where the various layers are and, with the aid of theory and laboratory measurement, even know how temperature and density change all the way to the center.

S waves will not pass through liquid and are never detected by seismographs placed in a broad zone centered on the side of the Earth opposite the quake (Figure 7.20a). We therefore know that the inside of our planet has a fluid core, and we can even measure its size. Refraction of the waves shows that the chemical composition also changes at the boundary of the fluid. The core extends 3,500 km from the center, somewhat over half the Earth's radius, and is almost exactly one-third of the terrestrial mass. It consists of a mixture that consists chiefly of iron, followed by nickel, a variety of lesser metals, oxygen, and sulfur.

Conditions within the core (Figure 7.20b) can be defined in terms of temperature, density, and **pressure.** Pressure represents the degree of compression, or the outward force that a gas, liquid, or solid would apply per square meter to a wall separating it from a vacuum. At the core's outer edge, the temperature is 5,000 K and the pressure is nearly 1.5 million *bars* (1.5 megabars), where a bar represents the pressure of the atmosphere at the Earth's surface. That pressure is high enough to crush iron from its normal density of 7.3 g/cm^3 to over 10 g/cm^3.

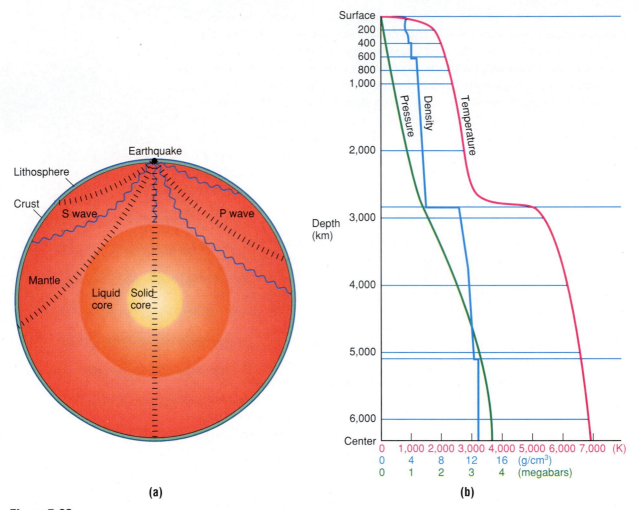

(a) (b)

Figure 7.20
(a) An earthquake sends S and P waves traveling into the rocky mantle toward the iron core. The S waves are stopped by liquid. Opposite the site of the quake is a huge shadow zone in which no S waves appear, allowing us to deduce both the existence and size of a liquid core. The waves are affected by density and temperature variations and are reflected and refracted by other discontinuities. At the surface of the Earth is the solid crust (narrow blue circle); it is firmly attached to the top part of the mantle, and with it makes the lithosphere (in green). **(b)** The graph shows how pressure (in megabars, millions of times Earth's atmospheric pressure), density, and temperature change as depth increases. Between the iron core and the mantle is a particularly sharp discontinuity in density and temperature.

The state of a substance depends on the balance between temperature and pressure. Increasing temperature makes atoms move faster and can turn a solid to a liquid and a solid or liquid to a gas. Increasing pressure acts oppositely. In the outer part of the core, the balance is such that the iron is molten and flows like water, a massive metal ocean. Temperature and pressure both rise toward the center under the crushing force of gravity. At a distance of 1,200 km from the center (5,200 km below the surface), pressure wins out over temperature, and the iron actually freezes or turns solid. At the very center, the temperature is nearly 7,000 K, hotter than the surface of the Sun! This solid, frozen nucleus contains 5% of the core's total mass and about 2% of the mass of the whole Earth.

Remarkably, the core makes its presence known to us directly. Its temperature decreases in the direction toward the surface, and it is therefore in a constant state of circulation through **convection.** If the bottom of a fluid is heated, it expands and becomes less dense. It can then rise, like a helium balloon in the air. But as it moves upward it loses heat, becomes denser, contracts, and falls. The Earth's rotation produces a more complex circulation pattern. Iron is an electrical conductor—that is, it allows the free movement of electrons among atoms—so the circulation of the iron generates a magnetic field. In its simplest form (Figure 7.21), the Earth's field is a *dipole* because it has two *magnetic poles*, one called north, the other south. The *magnetic axis*, the line that connects the poles, is tilted relative to the rotation axis by about 11°. A magnetized compass needle points toward the magnetic poles, giving the approximate direction to geographic north or south.

Surrounding the metallic netherworld of the core is a thick layer of rock that extends nearly to the surface; this **mantle** (see Figures 7.20 and 7.21) accounts for about two-thirds of the Earth's mass and is entirely dis-

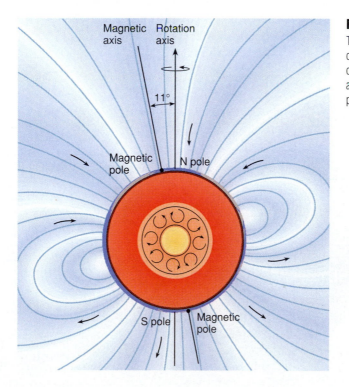

Figure 7.21

The Earth's dipole magnetic field, caused by terrestrial rotation and core convection, is tilted by about 11°. As a result, the northern hemisphere's pole is in extreme northern Canada.

tinct from the core. At the core-mantle boundary, the temperature and density suddenly drop. The mantle, made of relatively heavy iron- and magnesium-bearing silicates, is the source of the volcanic lavas.

At a depth of about 100 km, the mantle separates into two parts that can slide along each other. The upper one is the **lithosphere.** The topmost part of the lithosphere is the Earth's crust, which consists of light rock that has literally floated upward. On the very top is the lightest rock, the granitic continents. The thickness of the crust is highly variable. It is only 10 km or so thick beneath the seas, whereas it reaches downward 60 or more km beneath the continental land masses.

If the Earth is 5 billion years old, why is the interior hot; why has it not cooled off? It takes heat to melt ice. Conversely, when water freezes, it must give up heat. Likewise, the Earth's liquid core gives up heat as it slowly solidifies. Additional heat is produced by the decay of radioactive elements. The crust is an excellent insulator and dissipates heat only slowly, most of it escaping through volcanoes and other vents. The result is a planet with a very hot interior.

The Moon is very different. The low average lunar density of 3.3 g/cm^3 suggests that any iron core—if one exists at all—is very small, smaller than 400 km in radius and containing less than 2% to 4% of the total lunar mass (Figure 7.22). Unlike the Earth, the Moon has no dipole magnetic field, so the core must be solid throughout. Surrounding the core is a rocky mantle containing around 85% of the lunar mass; we know little about it. However, since the mantle is the obvious source of the mare lava basalts, it must have a relatively low inventory of volatile chemical elements and compounds and must also contain heavier minerals than the

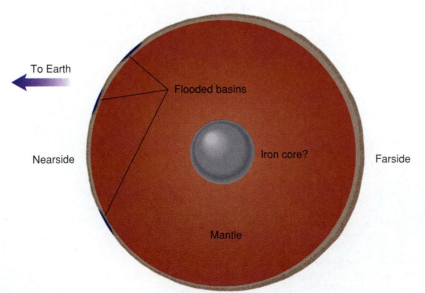

Figure 7.22

A cross section of the Moon shows a small iron core (which may be proportionally considerably smaller than represented here), a thick, rocky mantle, and a lower-density crust that is pulverized to about half its depth. The nearside crust is thinner than that on the farside.

highland crust. On top is the beaten, lighter crust, some 70 km thick. Geologists estimate that the force of the immense bombardment that created the jumble of craters has crushed and pulverized the top half.

Gravity measurements show that the lunar crust is thinner on the nearside of the Moon than on the farside (see Figure 7.22). It was easier for the lava to reach the surface on the nearside. As a result, we see the maria generally only on the nearside, the large basins of the farside remaining nearly empty.

7.6 CONTINENTAL DRIFT, PLATE TECTONICS, AND SURFACE FEATURES

A map of the Earth (Figure 7.23) shows that the eastern coastlines of North and South America fit the curves of western Europe and Africa, evidence that the land masses were once joined. More telling, you can fit the Americas with Europe and Africa in such a way that similar rock structures straddle the boundaries. Some 200 million years ago, these great continents were part of a single land mass called *Pangaea*; they are now separated by 5,000 km.

Some 200 million years ago, South America and Africa were part of a single land mass.

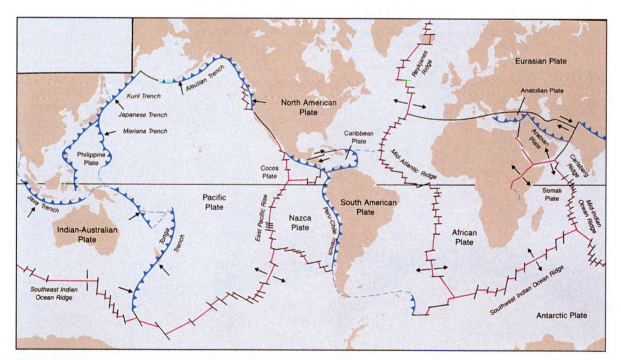

Figure 7.23

A map of the Earth shows that the outlines of the Americas fit against those of Europe and Africa, indicating that the land masses were once united. The continents float on 16 crustal plates that are in constant motion (arrows) relative to one another. Compare with Figure 7.6.

The Earth's lithosphere is divided into 16 separate **plates** (see Figure 7.23) that are in continuous motion relative to one another, resulting in global **continental drift.** This geologic restlessness is caused by convection within the upper part of the mantle. The part of the mantle below the lithosphere is so hot that it acts like a soft plastic that can flow, allowing the plates to move. Rising hot mantle rock forces its way to the surface and cracks the crust, generally beneath the oceans where the crust is thinnest. The result is the vast network of suboceanic ridges that girdle the globe and form one kind of plate boundary. All along these ridges we can see evidence of volcanic activity, magma pouring out from the mantle to make new crust. The entire island country of Iceland, at the northern end of the Mid-Atlantic Ridge, was made this way. As the lava from the vents in a ridge accumulates, it forces the opposing plates to move away from the plate boundary, carrying the seabeds and continents with them (Figure 7.24).

The evidence for such motion is overwhelming. The Earth's youngest rocks are those next to the suboceanic ridges. Toward the east and west the rocks become older, reaching a maximum of about 200 million years near the continents. We can actually *see* the drift taking place. Observations of celestial sources made with very long baseline interferometers (see Section 6.6) yield the distances between the individual radio telescopes to within a wavelength of the radiation observed. As a result, we can measure the separations between points in America and Europe to about a centimeter. The distance is observed to increase at a rate of about 5 cm per year.

Many of the major geologic features of the Earth are caused by **plate tectonics,** the set of deformations caused by plate motions that squeeze,

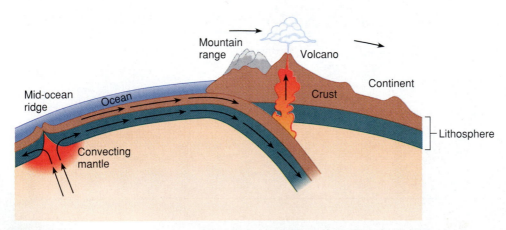

Figure 7.24
Upwelling mantle material forces its way through the surface, creating a mountain range or ridge and pushing the seabed—with its continents—aside. The oceanic crust and lithosphere are subducted beneath a continental plate, the collision raising mountains on the continent. As the subducted material heats, molten matter drifts toward the surface to produce a volcano.

pull, break, and stretch the Earth's crust (compare Figures 7.23 and 7.6). There are no spaces between the plates, so the boundaries at which they meet and push against each other can be the scene of awesome violence. If a plate holding a seabed runs into one holding a continent, the seafloor can dive downward in a process called *subduction*, and the continental edge is lifted into a mountain chain (see Figure 7.24). The Pacific Ocean, attacked from many sides, is shrinking, and mountains ring the basin. North America may some day collide with Asia. Where the ocean floor subducts we also find the deepest ocean trenches, like those off the east coast of Asia, and great island arcs like Japan and the Aleutians. When the subducted seabed heats and melts, some of the magma rises. Where it pops through the surface, it creates chains of volcanoes. The plates may also meet at a *fault* like the great San Andreas Fault in California (Figure 7.25), where the Pacific and North American plates slide past each other. When the plates suddenly slip, earthquakes rattle the ground. Mountains are also raised by direct collision between continental land masses. The towering Himalayas are buckled crust created when the subcontinent of India crashed ever so slowly into Asia.

Plate boundaries are not the only sites of rising mantle material, which also comes up in great **plumes** that probably begin at the core-mantle boundary. As the magma emerges from the crust at the top of the plume, it may produce a lava plateau or create an isolated shield volcano like Hawaii. The islands of the Hawaiian chain become smaller toward the northwest. The plate on which the islands sit is moving, but the plume is fixed. The island on top of the plume grows until plate motion carries it away, after which the island gradually erodes as it is replaced by another. The result is a scar, or *plume track*, in the crust.

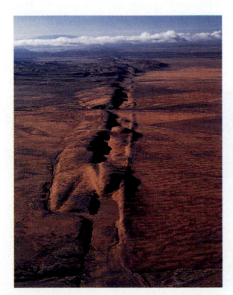

Figure 7.25
The San Andreas Fault of California is clearly visible as a seam in the Earth. The western (right-hand) side is moving northward, taking Los Angeles toward San Francisco.

The Moon, however, is now cold; it has but a single plate, no continents and no continental drift, neither volcanoes nor mountains of the sort we have on Earth. When the early Moon was hot and differentiating, the heavier silicates sank and the lighter rose to produce what eventually became the battered highlands. The lava plateaus of the maria were then created by the great basin-forming impacts. The force was apparently sufficient to fracture the lunar crust in a basin's interior. Hot heavy-silicate magma from the ancient lunar mantle subsequently worked its way up through the cracks and spread across part or even all of the basin floor in a series of flows. It pulled the rock at the edges of the maria apart, so more magma could escape through the weakened crust to run into the basin through the sinuous rilles. The lunar mass is so low that not even the decay of radioactive elements could keep the interior warm for long. As a result, the total magma output, while obvious even to the naked eye, is really quite small, and the Moon now presents us with no volcanic activity and few quakes.

7.7 ATMOSPHERES

The Earth's atmosphere is a mixture of gases consisting of 77% nitrogen (N_2) (counted according to the number of molecules or atoms per cubic meter), 21% oxygen (O_2), 1% atomic argon (Ar), and on average 1% water (H_2O) vapor (the proportion of water vapor is highly variable because of rainfall and evaporation from the oceans). Mixed in is 0.033% carbon dioxide (CO_2) and a vast number of trace gases that include helium, neon, krypton, and methane, all of which are measured in only a few parts per million. Nitrogen, which provides most of the atmospheric pressure, is a relatively inert gas (that is, it does not readily react chemically). Oxygen, however, is chemically active, allowing the chemical burning, or oxidation, of food to provide heat and energy for breathing animals. The water and the CO_2 provide an insulating blanket around the Earth through the **greenhouse effect** (Figure 7.26). Sunlight heats the ground. The Earth,

Water and carbon dioxide provide an insulating blanket around the Earth through the greenhouse effect.

Figure 7.26
The Earth is warmed partly by the greenhouse effect. Short-wave high-energy radiation from the Sun heats the ground, but long-wave infrared radiation is trapped by water and carbon dioxide in the atmosphere. Clouds and weather phenomena fill the troposphere, within which temperature drops with height. Above it is the stratosphere, in which temperature climbs as a result of the absorption of ultraviolet sunlight by ozone.

BACKGROUND 7.1   Changes in the Earth's Atmosphere

Our planet's thin atmosphere is affected by a variety of natural and artificial processes. Two gases are of great importance to human life, carbon dioxide (CO_2) and ozone (O_3). Carbon dioxide and water vapor are largely responsible for keeping the Earth warm, but too much would lead to thermal disaster. The amount of CO_2 in the atmosphere depends on the balance between the rates of its creation and destruction, which involve absorption by green plants and the oceans, the locking of carbon in rocks, and emission by animal life. There is great concern that this balance is being upset by the rapid burning of fossil fuels, which consist of plant life that long ago locked up vast amounts of carbon. Today's measurements show an increase in CO_2 from times past. If we do produce too much CO_2, the atmosphere will warm, perhaps to the point of endangering our own existence.

Yet there are many uncertainties. High CO_2 levels have been produced naturally between periods when the great glaciers covered the land and by extreme volcanic activity. We are still not sure how the Earth's various cycles work and to what degree the oceans can clean the atmosphere. But uncertainty should not mean complacency. If the atmospheric CO_2 content gets too high, the level could be irreversible.

Ozone in the high stratosphere prevents ultraviolet rays from reaching the ground. This gas is destroyed by chlorine and fluorine. Many chemicals involving these elements are or were used in industry and in the home (refrigerants like freon and propellant chemicals in aerosol cans, which have generally been banned, are examples). As these chemicals are released, they float upward and are broken down by sunlight. The free chlorine and fluorine atoms then attack the ozone.

There have been strong indications that the ozone layer is thinning. In particular, a hole in the layer that develops for natural reasons each summer over Antarctica has been getting larger. It is not entirely clear that manufactured chemicals are responsible, but it is certain that they have potential for doing serious damage. We depend on our thin atmosphere for life, and we must take care of it before it is too late to repair the damage.

---

however, behaves as a blackbody that radiates in the infrared (see Section 6.3.1), in which atmospheric water and CO_2 have myriad opaque spectrum lines. The heat has difficulty escaping, and the terrestrial temperature is some 35 K higher than it would be without these gases.

Air is surprisingly heavy. Over every square meter of Earth is a mass of 10,330 kg (14.7 pounds per square inch). The densest part lies at the surface (about 10^{19} molecules/cm^3), the pressure halving every 5 km in elevation and dropping by a factor of 10 for every 20 km. At the top of Mount Everest, the pressure is about a third that at sea level and there is barely enough oxygen to sustain a human being. At 100 km the pressure is down to 10^{-5} its value at sea level. There is no place where the atmosphere ends: it just thins until it blends with the gases of interplanetary space. Even at 200 km there is enough air to produce sufficient friction to degrade the orbit of a spacecraft.

In the lowest layer of our atmosphere, the *troposphere* (see Figure 7.26), which contains 85% of our air and most weather phenomena, the temperature drops with increasing altitude (look for snow on high mountain peaks). At the top of the troposphere, at an altitude of $10 - 15$ km (where the highest commercial jet aircraft fly), the temperature is down to $-60°C$. However, above it, in the *stratosphere*, temperature starts to climb as a result of absorption of ultraviolet sunlight by ozone (O_3), a process that both heats the stratosphere and keeps these harmful rays from penetrating to the ground.

At sufficiently high altitudes, above about 80 km, further absorption of sunlight causes molecules to be broken into atoms and atoms into ions, producing an extensive *ionosphere* that refracts broadcast radio waves back to the ground to make short-wave radio communication possible. Above 200 km altitude, the temperature is over 1,000 K (700°C). From the blackbody rules (see Section 6.3.1), you might expect the sky to be brilliant. But the air here is much too thin to be a blackbody. Its temperature refers only to the speeds of its atoms, not to the amount of radiation the gas gives off, and is therefore called a **kinetic temperature.**

The earliest atmosphere of the Earth was very different from what we breathe today. Formed after our planet had solidified from an early molten state, it likely consisted of water, carbon dioxide, and nitrogen. After the surface and the air had sufficiently cooled, the water would have precipitated into oceans, leaving CO_2 and N_2 behind. There the story might have ended, except for two extraordinary features of the Earth: liquid water and the development of life.

If all carbon absorbed by the Earth were released, our planet would have a dominantly CO_2 atmosphere.

Single-celled life-forms, which arose only a billion years after the planet was created, began the chemical conversion of carbon dioxide into oxygen. The evolution of life into multicelled organisms hastened the process. The deaths of living creatures places carbon into soil, then into rocks. When the rocks erode, some of the carbon flows into the sea and is used by small marine animals to make shells. Eventually they die and fall to the seafloor, where under pressure they become carbon-bearing rocks such as limestone. The seabeds are subducted under continents, and the carbon disappears into the Earth's mantle. Some of the carbon is liberated directly from continental rocks in the form of CO_2, while some finds its way back to the surface through volcanoes. The result of this cycle, in which much of the Earth's carbon is tied up in rock, is a drastic decrease in atmospheric CO_2. If all this absorbed carbon were released, our planet would have a dominantly CO_2 atmosphere with a pressure 60 times what it is now.

The effect of an atmosphere is easily seen by examining the Moon. Our companion effectively has no atmosphere at all, its gravity too weak to retain any gas for a significant amount of time. As a result, there is no air to cut down the heating rays of the Sun, and the surface rocks can reach a temperature of 110°C (383 K) during the long daylight hours. At night, the rocks cool to −170°C (103 K) in the absence of a heat-retaining atmospheric blanket.

7.8 THE MAGNETOSPHERE

The Earth's magnetic field, created in its liquid core, extends well into interplanetary space. The simple tilted dipole of Figure 7.21, however, is severely distorted by the action of the Sun. The Sun loses a tiny part of its mass each year through the **solar wind.** The outflowing gas consists of a mixture of ions (in this case mostly protons) and electrons. At the Earth, the solar wind has a density of about 10 to 100 particles per cubic centimeter (some 10^{18} times less than the density of the atmosphere at the Earth's surface) and is moving at some 400 km/s. The solar magnetic field, also a rough dipole, is carried away from the Sun by the charged particles of the wind. Solar rotation causes the field to wrap into a spiral, and at the distance of the Earth, the magnetic field lines are almost perpendicular to the line connecting the two bodies (Figure 7.27).

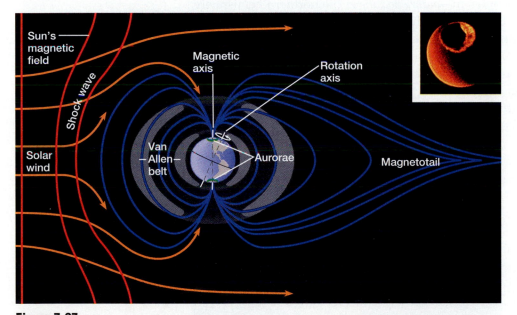

Figure 7.27
The simple dipole of the terrestrial magnetic field (see Figure 7.21) is compressed on the sunward side by the solar wind. The shock wave where the solar wind hits the Earth's field is actually about 15 Earth radii upstream (toward the Sun). Opposite the Sun, the magnetic field is stretched into a long tail that extends 1,000 Earth radii downstream. The solar magnetic field, coupled to the moving wind, is bent around the Earth. Some electrons and protons from the wind become trapped in the two Van Allen belts. The magnetic and electrical interactions form huge current rings around the magnetic poles, causing the upper atmosphere to glow (inset).

The Earth interacts with the solar wind somewhat like a boat plowing upstream. Its field is squashed inward, and the wind flows around the planet like a wave spreading out from the bow. The surface where the solar magnetic field and the wind first meet the Earth's field is actually a curved **shock wave.** In a shock wave, the particles cannot get out of the way fast enough, so they pile up, producing a sudden increase in pressure. The shock occurs about 15 Earth radii out. The wind and the magnetic field then spread out to encompass the planet and to push the terrestrial field into a huge tail that stretches 1,000 Earth radii downstream, in the direction away from the Sun. The entire structure is called the **magnetosphere.**

A fraction of the particles find their way inside the Earth's magnetic field and become trapped in two doughnut-shaped *Van Allen radiation belts.* The inner one is about 1.5 Earth radii from the terrestrial center, the outer one about 4. An astronaut orbiting close to the Earth, within about 1,000 km of the surface, will be inside the inner belt and protected from the energetic particles in the solar wind. The environment outside the belts is much more dangerous.

The magnetosphere is responsible for one of nature's loveliest sights, the *aurora.* The phenomenon—*aurora borealis* in the northern hemisphere, *aurora australis* in the southern hemisphere—is commonly called the northern and southern lights. A display (Figure 7.28) may begin quietly, with no more than a large patch of diffuse red or green light. From there it can grow to develop streamers that look like giant searchlight beams with pulses of radiation moving along them, or may take the form of huge draperies hanging in the sky. An hour or so later it all fades away, leaving only the perpetual stars.

Figure 7.28
A spectacular display of the aurora borealis, which is generated by huge electrical currents that ionize the upper atmosphere, hovers over the landscape.

BACKGROUND 7.2   A Day on the Moon

It seems likely that one day we will return to the Moon to set up permanent bases. Lacking air, our satellite is an impressive observatory site, and the farside is shielded from interfering radio radiation from Earth. What would a lunar day be like?

Pick a spot—for example, Mare Vaporum, near the center of the lunar disk as seen from Earth. The Sun would first appear when the Moon is in our first quarter. It would be night even minutes before sunrise, since there is no air to scatter sunlight and create twilight. The landscape would still be relatively bright, however, because of the reflected light from the third-quarter Earth. Your first hint of impending sunrise would be a glow from the solar corona, seen on Earth only during a solar eclipse. Then the tip of a mountain or crater wall to the west would suddenly light up as it caught the first direct solar rays. Since the lunar day is 29.5 Earth days long, the Sun would creep up at 1/29 the pace it does on Earth; at the Earth's equator it takes the Sun two minutes to vault the horizon, so on the Moon it would take about an hour.

Very slowly the temperature of the surface rocks would climb, an increase that would be of little consequence to any residents, since most would be living underground for protection from the effects of both temperature and the solar wind. It would actually be rather dangerous to spend too much time on the surface because of these high-speed particles. Only a meter or so down, however, the Moon's temperature remains quite stable and no particles can penetrate.

A week later, it is finally noon and the "new" Earth has disappeared. Another week, and the Sun begins to set. It turns dark as soon as the solar disk disappears, except for the now-brightening Earth and a few peaks that catch the last of the solar rays. Now you settle down for two weeks of long, cold lunar night. This pattern would be repeated day after lunar day, as it has for billions of years.

The flow of the solar wind across the Earth's magnetic field injects particles into the Earth's upper atmosphere and sets up gigantic electrical current rings about 20° across. This electrical energy ionizes the upper atmosphere and makes it glow as free electrons are recaptured by ions (Figure 7.27 inset). The rings' energy output of a trillion watts is greater than that used by the entire United States. The aurora is most commonly seen beneath the current rings, in Alaska and northern Canada. Solar activity, which can create great magnetic explosions on the Sun, can intensify the solar wind and produce disturbances within it that expand the current rings and extend the aurora toward the equator, allowing much of the population in lower latitudes to see it.

The Moon, with no magnetic field, has no magnetosphere. (Even if it did, there would be no aurora since there is no atmosphere.) The solar wind simply smashes into the ground.

7.9 WHY A DOUBLE PLANET?

Finally, we return to the Earth-Moon system as a unit and ponder its origin. Why is it double? There are several hypotheses for our duplicity. (A **hypothesis** is a theoretical idea presented for testing by experiment or observation.) Perhaps the Earth simply divided in two in the course of its formation. However, then we would expect the Moon to orbit in the Earth's rotation (equatorial) plane instead of close to the ecliptic plane and would expect more-similar chemical compositions. The same arguments might be presented against a co-accretion hypothesis, which holds that the Moon and Earth developed simultaneously from the same cloud of matter.

It is also unlikely that the Moon was captured from space by the Earth. A primitive, independent Moon would have approached the Earth on a hyperbolic orbit, gone around it once, and flown back out into space never to return. An external force, perhaps produced by a third body, would be needed to reduce the relative speeds, allowing capture into an elliptical path. No third body can be identified.

An independent body roughly the size of Mars is thought to have crashed into the young Earth to create the Moon.

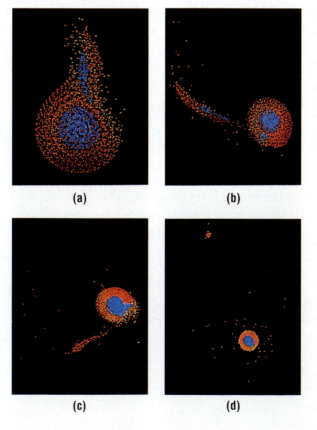

(a)   (b)

(c)   (d)

Figure 7.29
A computer model shows how the Moon could have been formed by collision with the early Earth. **(a)** An independent, differentiated body hits the Earth; **(b, c)** the cores have nearly merged and portions of the mantles fly away into orbit; **(d)** this gasified material consolidates into the current Moon, which has little core.

The hypothesis on which most scientists now tend to agree involves a massive collision (Figure 7.29). At a time near the formation of the Solar System, almost 4.6 billion years ago, a body roughly the size of Mars may have crashed into the young Earth. Both bodies were already differentiated and had large iron cores at their centers. The colliding body's core merged with that of the Earth upon impact. Part of the Earth's mantle was ripped away and, with that of the colliding body, was vaporized and sent into terrestrial orbit. The Moon subsequently accreted from this ring of gas and dust, then heated as a result, and differentiated. The hypothesis explains the low lunar iron content (though the regolith can be rich in iron, the Moon has little if any iron core), the Moon's deficiency of volatile elements (dissipated into space), and the orientation of its orbit. The concept is supported by violent events that involve the other bodies of the Solar System, to which we now turn.

▶ KEY CONCEPTS

Atmosphere: The gas that surrounds the Earth.

Basalts: Fine-grained metal-bearing silicates (as opposed to coarse-grained granites).

Continental drift: The motion of continents across the mantle.

Continents: Raised portions of the Earth made of light rock.

Convection: The up-and-down circulation of a heated fluid.

Core: The metallic (nickel-iron) central region of the Earth or Moon.

Craters: Pits caused by impacts (or, in the case of volcanoes, by explosions).

Crust: The thin, light, top layer of the Earth or Moon.

Differentiation: The separation of a planet's interior into layers of different composition.

Earthquakes: Vibrations caused by the slippage of the Earth's crust.

Ejecta: The debris thrown out of a crater by the impact that made it; this debris forms an **ejecta blanket** around the crater.

Greenhouse effect: The trapping of radiated heat by atmospheric carbon dioxide and water vapor.

Hypothesis: An idea tested by experiment or observation.

Impact basins: Large impact craters, sometimes filled with dark lava.

Isotopes: Variations of a chemical element caused by difference in neutron number.

Kinetic temperature: Temperature defined by atomic velocities.

Late heavy bombardment: The last period of meteorite fall before 3.8 billion years ago; it produced lunar highland cratering.

Lithosphere: A moving layer of rock that consists of the crust and the top part of the mantle; it is 10 km thick under the ocean, 60 km or more under the land.

Lunar highlands: The original lunar crust, crushed by heavy cratering during the late heavy bombardment.

Magma: Liquid rock inside the Earth; extruded onto the surface of the Earth as **lava.**

Magnetosphere: The structure around the Earth (or any other planet) filled with the Earth's magnetic field and particles trapped from the solar wind.

Mantle: The thick layer of hot rock that surrounds the Earth's or Moon's core.

Maria: Dark lava flows that fill many lunar impact basins.

Meteorites: Rocky or metallic debris from space that hits the ground.

Oceans: Basins filled with water.

Plates: The segments of the Earth's crust.

Plate tectonics: Deformations of the Earth's crust caused by continental drift.

Plumes: Rising columns of hot mantle material that break through the crust, producing shield volcanoes and volcanic floods.

Pressure: The outward force of compressed matter per unit area.

Regolith: Lunar (or planetary) soil produced by constant pulverization of rock by meteorites.

Shock wave: A wave of sudden increase in pressure.

Silicates: Rocks (including granites and basalts) made of compounds that contain silicon, oxygen, and a variety of metals.

Solar wind: A thin, ionized gas blowing from the Sun.

Synchronous rotation: Rotation with a period equal to the revolutionary period.

Tides: Periodic flows of water caused by the differential gravity of the Moon and Sun; generalized to mean any distortion caused by differential gravity.

Volcanoes: Vents in the Earth's crust through which magma, ash, and gas escape; **shield volcanoes** have runny lava that produces broad mountains with low slopes.

> **EXERCISES**

Comparisons

1. Compare the cores of the Moon and Earth.

2. Make a table to compare the properties of the Earth's mantle with those of the outer part of the core.

3. Compare the behaviors of S-type and P-type earthquake waves.

4. Compare the worldwide distributions of mountain ranges, earthquakes, and volcanoes.

5. Compare the ages and major features of the lunar highlands and the maria.

6. How does the Earth's troposphere differ from the stratosphere?

Numerical Problems

7. Show that the surface gravity of the Moon is 0.166 that of the Earth.

8. You have a kilogram of a radioactive substance that has a half life of 20,000 years. How much is left after 80,000 years?

9. If you could shrink the Earth to half its present diameter but maintain its mass, what would be its average density?

10. California and China are separated by about 5,000 km. At the present rate of continental drift, how long will it take before the two collide?

11. What altitude would you have to reach to arrive at a point where the pressure of the atmosphere is one-quarter its surface value?

12. Lunar craters A and B are respectively 2 and 0.5 km across. Roughly how deep are they relative to the surrounding plane? About how large were the meteorites that made them?

Thought and Discussion

13. Describe the gravitational action that produces a tide.

14. Name some cities that would *not* receive shear waves if there were an earthquake in Brazil.

15. What layer of the Earth produces volcanic lava?

16. On what basis do we surmise that the Earth's core is made of iron? Why do we surmise that the Moon has a small or nonexistent core?

17. What evidence suggests that part of the Earth's core is liquid?

18. What would happen if the Earth's liquid core suddenly turned solid?

19. Why is the center of the Earth so hot?

20. Why and how does a planet differentiate?

21. Why is part of the Earth's mantle in a flowable plastic state?

22. **(a)** What is the evidence for continental drift; **(b)** what process causes it; **(c)** what tectonic processes are associated with it?

23. How do carbon dioxide and water vapor keep heat trapped near the surface of the Earth?

24. How and why is the Earth's magnetic field distorted?

25. Where are the Van Allen radiation belts? With what are they filled?

26. Why is the aurora borealis less common over New York City than over Fairbanks, Alaska?

27. Where are the lunar mountain ranges? How were they formed?

28. **(a)** How are maria associated with impacts; **(b)** why do we believe the maria are solidified lava?

29. Explain why the lavas of the maria are heavier and denser than the rocks of the lunar highlands.

30. Why is the collision theory of lunar formation consistent with the small lunar core?

31. Why are there so few obvious meteorite craters on Earth?

32. Why are the lunar maria concentrated on the Moon's nearside?

Research Problems

33. The theory of plate tectonics met with considerable resistance during the twentieth century. When did this theory originate, and when and why was it finally accepted?

34. Examine a variety of astronomy texts that have appeared over the last century to see how the theory of the formation of lunar craters has evolved.

Activities

35. Take a compass outdoors at night and find north. How does the direction differ from that found from Polaris? Knowing that the magnetic pole is on Bathurst Island in far northern Canada, estimate from a globe the difference you would expect. Compare the two numbers. Where on Earth would the difference be greatest?

36. Over the course of the school term, keep track of earthquake and volcano reports that you see in the newspapers. Locate them on a world map and comment on the relation between them and other geographic features.

37. Draw a naked-eye sketch of the full Moon, showing the location of the maria. Compare your sketch with a lunar map. What was the smallest feature you could see?

Scientific Writing

38. A nonscientist claims that the Earth is unchanging. Refute this claim in an article for the scientifically literate.

39. NASA has decided to mount a major initiative to study the Moon, employing a variety of spacecraft and by making landings. You are enthusiastic about the project, but Congress is not. Write a letter to your representative that explains the importance of a study of the Moon with special emphasis on the Moon's significance for learning about the Earth.

ROCKY WORLDS

The contrasting natures of Mercury, Venus, and Mars

8

Despite the proximity of the Moon to the Earth, the two bodies are glaringly dissimilar. The other three terrestrial planets bear some close similarities to either the Earth or the Moon, but again, each is unique, each teaching us something about the others and about ourselves.

8.1 NATURE AND EXPLORATION

Although Mercury, Venus, and Mars are all close to Earth (Figure 8.1a), they were long shrouded in mystery. Venus (Figure 8.1b) is covered with clouds that perpetually hide its surface. Mercury (Figure 8.1c) has no significant atmosphere, so its rocky surface is accessible, but it is so close to the Sun that we can observe it effectively only in twilight, when our murky atmosphere provides a poor view. Mars (Figure 8.1d), however, with its surface markings and polar caps, appears intriguingly Earthlike.

Before any physical analysis of the planets is possible, we need distances. The distances in AU between the bodies of the Solar System are found from their orbital locations. The distance in kilometers from the Earth to *any* of these bodies then gives the number of kilometers per AU (that is, the distance in kilometers between the Earth and Sun) and thus the distances in kilometers between all the planets. In practice, we use Venus, measuring its distance from Earth in kilometers by **radar** (**ra**dio **d**irection **a**nd **r**ange).

Radar uses artificially produced radio signals sent to and reflected from the body being examined. An astronomical radar system is a radio telescope that can both transmit and receive. A radio signal at a known wavelength is sent toward a target planet (Figure 8.2). A small portion of the signal bounces off the body and is returned. The time it takes the radio beam to travel to the target and back at the speed of light gives a precise distance. The semimajor axes of the orbits of Mercury, Venus, and Mars, already known in AU, are thereby found to be 57.9, 108, and 228 million km respectively. From the planets' angular radii and distances we easily calculate respective physical radii of 2,400, 6,100, and 3,400 km. Mercury is

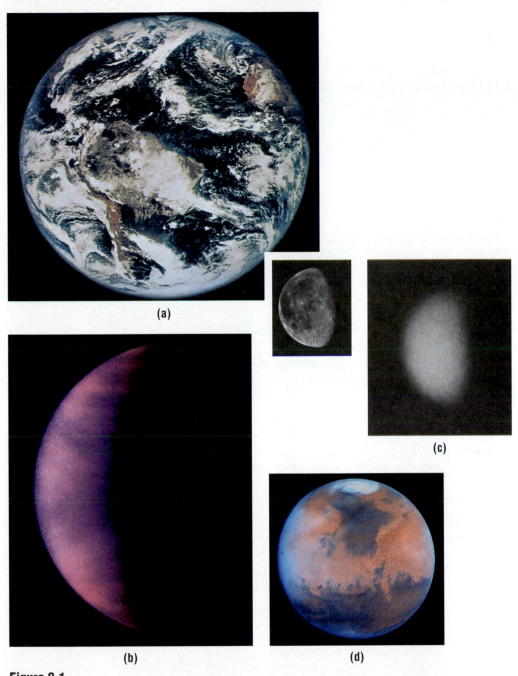

Figure 8.1
Even at best, the view of the terrestrial planets from Earth is limited. Here they are scaled according to size, with **(a)** the Earth and Moon as references. **(b)** To the Hubble Space Telescope, Venus displays wind-blown markings in a swirling cloud deck that obscures its surface. **(c)** Mercury's surface, viewed from the ground, presents only a featureless disk. **(d)** The Martian surface, however, reveals to Hubble a variety of permanent features and an Earthlike polar cap. The more prominent markings are easily seen even in a small telescope.

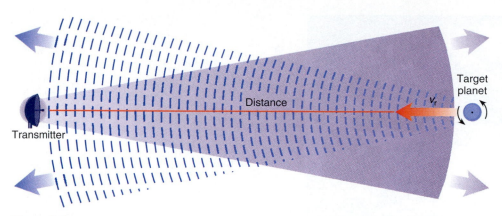

Figure 8.2

A radar telescope sends a radio beam (in violet) to bounce off a planet (dashed blue waves) that is rotating with its axis perpendicular to the page. The distance is found from the round-trip travel time of the signal, and the radial velocity of the planet (v_r) can be found from the Doppler shift. We can also determine the rotation period from the spread in the wavelengths of the returned signal.

40% larger than the Moon, Venus is nearly a twin of Earth, and the size of Mars falls roughly between that of the Moon and Earth.

Neither Mercury nor Venus has any natural satellites from which to find masses by applying Kepler's generalized third law. Instead, respective masses of 0.055 and 0.81 that of Earth were originally found from the gravitational perturbations that the two planets exert on each other and on the Earth. The two tiny satellites of Mars allow a determination of 0.11 Earth masses.

Earthbound telescopes, however, are no match for spacecraft, which can yield high-precision measures of radius and mass and can image planetary surfaces. All three planets have been visited by a variety of flybys and orbiters. The most effective have been *Magellan*, which orbited Venus between 1990 and 1993 and provided radar images with resolutions of up to 100 m, and the two Viking craft, both of which entered Martian orbit in 1976 and produced images of nearly the entire globe with resolutions that approached only 20 m. The Vikings also carried landers that descended to two different surface locations, examined the atmosphere and regolith, and searched for life.

8.2 ANALYSIS

To understand the planets, to learn of their natures from great distances, we apply laws that govern radiation.

8.2.1 Velocities and the Doppler Effect

We can determine not only the distances of nearby planets with radar but also their velocities and rotation periods. The apparent wavelength of an electromagnetic wave can be changed by relative motion between the ob-

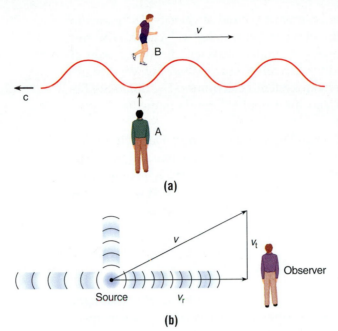

(a)

(b)

Figure 8.3
(a) A wave moves to the left at velocity c. The wave has a well-defined wavelength or frequency for observer A, who is watching it go by. Observer B, however, is moving to the right, into the wave, at some velocity v. This person will encounter the wave crests more frequently and will measure a higher frequency and shorter wavelength. If B moves to the left, with the wave, the frequency will appear lower and the wavelength longer. **(b)** A source of radiation moves at velocity v relative to the observer. This motion can be broken down into a motion in the line of sight, the radial velocity (v_r), and one across the line of sight, the tangential velocity (v_t). Only the radial motion produces a Doppler effect.

server and the source. In Figure 8.3a, person A stands next to the wave and counts the number of wave crests that go by every second to determine the wave's frequency. Person B, however, does not stand still, but moves to the right, into the waves. The waves hit this observer at a faster rate, the frequency appears higher, and the wavelength—the distance between crests— appears to be less. The faster the motion, the shorter the observed wavelength. Conversely, if B reverses direction and moves to the left, the frequency appears lower and the wavelength greater. This **Doppler effect** (named after Christian Doppler, who explained it in 1842) will also be observed if the source is moving instead of the observer. It is most commonly noted with sound waves: the pitch of a horn that approaches you is higher than its pitch as it recedes.

Any velocity relative to an observer has two components, one *along* the line of sight, the **radial velocity,** (v_r), and another *across* the line of sight, the **tangential velocity** (v_t) (Figure 8.3b). All that matters in the Doppler effect is the radial component. The magnitude of the Doppler effect—the **Doppler shift**—is given by the **Doppler formula:**

$$\lambda_{obs} - \lambda_{rest} = \lambda_{rest}(v_r/c),$$

or
$$v_r = c(\lambda_{obs} - \lambda_{rest})/\lambda_{rest},$$

where λ_{rest} is the wavelength that would be observed if the source and observer were at rest with respect to each other, λ_{obs} is that observed when the two are moving at the relative radial velocity v_r, and c is the speed of light. The radial velocity is negative for approach and positive for recession.

Radar allows the quick measurement of radial velocity. The planet in Figure 8.2 is approaching the observer with radial velocity v_r. The rest wavelength, λ_{rest}, is the known transmitted wavelength. After a wait of several minutes, the reflected wave returns, now with a new wavelength, λ_{obs}, that is measured by the radar system's electronics. The difference between the two wavelengths and the Doppler formula immediately give the planet's radial velocity, v_r.

We can do more. The planet in Figure 8.2 is rotating. Part of it is coming toward the observer relative to its center (that is, relative to v_r) and part is going away. Different portions of the planetary surface therefore produce slightly different Doppler shifts. The spread in returned wavelengths then gives the rotation speed, from which the rotation period can be found.

8.2.2 Spectra and Temperature

All planets shine in the optical spectrum by reflected sunlight. That light must pass through the planet's atmosphere, which superimposes its spectrum lines on the solar spectrum. The wavelengths of these spectrum lines will be shifted by the Doppler effect, allowing another way of finding the radial velocity. The strengths of the spectrum lines also yield information on the atmospheric composition.

In the radio spectrum and in the infrared, however, a planet *emits* energy because it is a blackbody warmed by the Sun. In Figure 8.4, the Earth is shown on the surface of a sphere of radius D with a planet radiating at its center. From Earth, we measure the energy (F, in watts/m^2) that passes through a square meter of this sphere every second. That same amount must pass through *each* square meter of the sphere, so the *total* energy

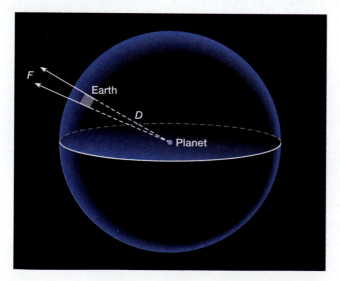

Figure 8.4

The Earth is on the surface of a sphere of radius D centered on a planet. The planet radiates energy with a luminosity of L watts. F is the amount of energy from the planet (in watts/m^2) that passes each second through 1 square meter of the surface of the sphere. L is therefore F times the sphere's surface area, or $4\pi D^2 F$.

passing through the sphere per second (in watts) is the sphere's surface area ($4\pi D^2$) times F, or $4\pi D^2 F$. The same amount of energy must leave the planet every second, and is therefore the planets luminosity (L). The effective temperature is then found from $T^4 = L/7.2 \times 10^{-7} r^2$ (see Section 6.3.1), where r is the planet's radius; that is, temperature is the fourth root of $L/7.2 \times 10^{-7} r^2$.

8.3 IRON MERCURY

We can now examine our planetary system. We cannot see features on Mercury's (Table 8.1) surface clearly enough to determine its rotation period. Radar, however, shows that it turns counterclockwise with a sidereal rotation period of 59 Earth days, two-thirds of its 88-day sidereal year, its

TABLE 8.1
A Profile of Mercury

Planetary Data		Atmosphere (surface)	
Distance from Sun	0.38 AU	Pressure	10^{-12} Earth
Radius	2,439 km = 0.38 R_{Earth}		
		Composition by Number	
Mass	3.3×10^{23} kg = 0.055 M_{earth}	Sodium	97%
Mean density	5.4 g/cm^3	Helium	3%
Uncompressed density	5.3 g/cm^3	Hydrogen and oxygen	Trace
Gravity	0.38 g_{Earth}		
Escape velocity	4.3 km/s		
Surface temperature	700 K (days)–88 K (nights)		

Layers	% Mass	Radii (km)	Other	
Iron core	60	0–1,750	Magnetic field	0.01 Earth
Mantle plus crust	40	1,750–2,440	Rotation period	58.65 Earth days
			Axial inclination	0°

Planetary Features

Cloud cover: None.

Craters: Numerous and much like the Moon's, except that craters are situated on intercrater plains; lower relief and smaller ejecta blankets caused by higher gravity; most of surface ancient, nearly 4 billion years old.

Maria: Several volcanically flooded basins.

Origin: Early collision may have blown away much of mantle.

Plates and plate motion: None.

Scarps: Younger than most craters; produced by shrinkage of planet.

Surface temperature: 700 K daytime maximum, 88 K night minimum, 348 K average; cold ice caps at rotation poles.

Water: Ice caps may be water; otherwise, very little.

Volcanoes: None known.

axis almost perfectly perpendicular to its orbital plane. This odd rotation is caused by the same kind of tidal locking that keeps the Moon pointing one face toward Earth. Mercury's high orbital eccentricity, however, produces a large variation in the tidal force. The rotation period is close to what the orbital period *would* be were the orbit circular with a radius equal to the perihelion distance.

Mercury's infrared and radio luminosity tell us that the effective temperature of the sunlit surface is 700 K (430°C), the result of the planet's proximity to the Sun. But because there is no insulating atmosphere, the temperature plunges to a chilly minimum of 88 K (−185°C) on the nighttime side. The surface rocks and regolith are excellent insulators, however, and only a meter below the surface the temperature is maintained at a more comfortable average of 348 K or 75°C.

Spectrographs aboard the *Mariner 10* flyby craft found evidence for atmospheric helium as well as for hydrogen and oxygen. The surface pressure, however, is a mere 10^{-12} bar, that is, 10^{-12} that of Earth's atmosphere. Some of this thin gas is likely captured from the solar wind and quickly escapes back into space as a result of the low gravity. Earth-based spectroscopy, however, shows the dominant atmospheric constituents to be sodium and potassium, which are heavy enough to be gravitationally trapped, at least for a while. Mercury has a magnetic field about 1% the strength of the Earth's. Ions from the solar wind ride the field lines down toward the surface. Because the atmosphere is so tenuous, the ions can slam into the ground and kick the abundant sodium and potassium from surface rocks.

Mariner 10 imaged nearly half of Mercury's surface (Figure 8.5). Impact craters are everywhere. The record of the late heavy bombardment has survived: the surface is roughly 3.8 to 4 billion years old. Mercury, somewhat like the Moon, is divided into highlands, lava-filled basins, and plains. The biggest feature is Caloris (see Figure 8.5b), an impact basin 1,300 km in diameter that looks like the Moon's Mare Orientale (see Figure 7.17c). The regolith of the planet, probably similar to that on the Moon, is made up of rock that has been battered and ground up by billions of years of bombardment.

For all the similarity to the Moon, however, there are some major differences. Mercury's volcanic plains are not dark like the lunar maria but are about as bright as the lunar highlands. They are also more beaten up, indicating that the lava flows are older than those on the Moon. In the highlands, craters do not fill the entire surface but are set onto extensive plains that show volcanic flooding older than that on the lunar surface.

Mercury's most unusual surface features are *scarps* (see Figure 8.5c), huge cliffs that can stretch for 500 km. They appear to have been produced when the planet shrank by a couple of kilometers as it cooled from the high temperatures generated during its formation, forcing its crust to buckle and slip. Because the scarps cross the craters, the contraction had to occur after the heavy bombardment ended.

Because of Mercury's proximity to the Sun, the surface of the sunlit hemisphere climbs to 700 K.

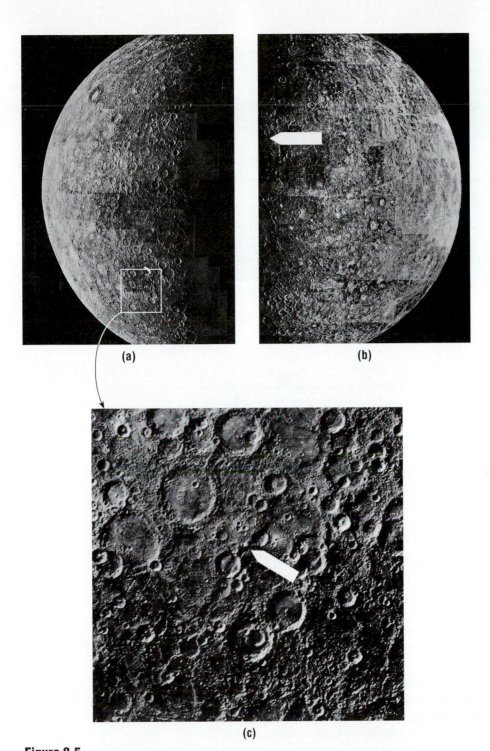

(a)   **(b)**

(c)

Figure 8.5

Two quadrants of Mercury observed by *Mariner 10* during **(a)** approach and **(b)** recession show a pro-liferation of craters that make the planet look much like the surface of the Moon. Caloris, one of the largest impact basins in the Solar System, is indicated by the arrow in (b). **(c)** A close-up of a section in (a) reveals a prominent scarp (arrow) 500 km long and up to 3 km high that has broken the large crater in the center.

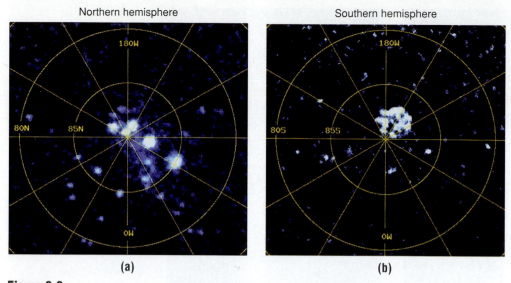

Figure 8.6
Radar maps of Mercury's **(a)** north and **(b)** south poles show bright patches of ice.

The cold poles of the Earth support massive icecaps. Mercury would seem too close to the Sun to have any such features. However, radar observations show bright reflective spots surrounding the poles that are almost certainly sheets of water ice at least 2 m thick (Figure 8.6). Since Mercury has no significant axial tilt, the Sun is at the horizon at these high latitudes, and the interiors of deep craters are permanently shadowed and cold. The ice may be protected from evaporation by a covering of dust.

Mercury's magnetic field, though weak, implies the existence of at least a partially molten iron core. Because of the planet's small size, the core should have cooled below the melting point of pure iron. There must be some impurities like sulfur present that lower the melting point and keep some of the core in the liquid state. An iron core is also indicated by Mercury's average density of 5.4 g/cm^3, since the interior has to be dense to compensate for the low density of the rock at and near the surface. Because Mercury is small, its uncompressed density (see Section 7.3) is still a hefty 5.3 g/cm^3, well above Earth's value of 4.4 g/cm^3, and the highest of any body in the Solar System. It must consequently have the largest iron core, occupying some 60% of its mass and over 70% of its radius (Figure 8.7). Surrounding the iron core is a rocky mantle only 700 km thick. The depth of the lighter surface crust is unknown.

Mercury's proximity to the Sun probably played a strong role in its strange construction: intense solar heat would have prevented lighter volatile elements from being incorporated into the planet during its formation. The result was a higher proportion of metals than that found in the Earth. Still, theory suggests there is too much iron. A violent collision between a primitive Mercury (which had already differentiated to an iron

Relative to its size, Mercury has the largest iron core of any planet in the Solar System.

Figure 8.7
Mercury's hot iron core is relatively the largest in the Solar System.

core and mantle) and another body may have blown away much of Mercury's original mantle, leaving it with a large quantity of iron. The newly transformed planet, perhaps remelted by the collision or some other process, then differentiated again, sending dirty iron to the center. Mercury thus provides additional evidence for the violent formation of Earth's Moon.

8.4 BLISTERING VENUS

Venus (Table 8.2) seems a near-twin of Earth. However, Venus's smaller distance to the Sun and slightly lower mass seem to have conspired to produce a planet that in many ways is profoundly different.

8.4.1 The Planet and Its Atmosphere

Radar, which can penetrate the Venusian clouds, shows that Venus rotates backward relative to Earth with a ponderous 243-day period. As a result, the planet's solar day is 117 Earth days long. The axial tilt is 177.3° (a value over 90° indicates reverse rotation), only 2.7° from perpendicular. It seems possible that a collision of the kind hypothesized for the creation of the Moon and for the removal of part of Mercury's mantle may have been responsible.

Because of Venus's position as the second terrestrial planet out from the Sun, we would expect it to be hotter than Earth but significantly cooler than Mercury. However, radio observations show the surface temperature to be 740 K or 470°C; a block of lead placed on Venusian ground would melt. Moreover, in spite of the long nights, the nighttime and daytime

TABLE 8.2
A Profile of Venus

Planetary Data			Atmosphere		
Distance from Sun	0.72 AU		Pressure	98 Earth	
Radius	6,051 km = 0.95 R_{Earth}				
Mass	4.87×10^{24} kg = 0.81 M_{earth}		*Composition by Number*		
Mean density	5.3 g/cm^3		Carbon dioxide		96%
Gravity	0.90 g$_{Earth}$		Nitrogen		4%
Escape velocity	10.4 km/s		Water, sulfur dioxide		
Surface temperature	730 K		Argon, carbon monoxide		Traces
			Layers		*Height (km)*
			Troposphere		0–110 km
Layers	*% Mass*	*Radii (km)*	*Other*		
Iron core	30?^{a}	0–3,000?	Magnetic field	None	
Mantle plus crust	70?	3,000–6,050?	Rotation period	243 Earth days backward	
			Axial inclination	177°	

Planetary Features

Craters: Randomly distributed across the planes, none older than about 500 million years.

Clouds: Solid cloud cover about 50 km high; sulfuric acid.

Mountains: All major mountains in Ishtar Terra; probably volcanic or produced by mantle downwelling.

Tectonics: Small compared to Earth; vertical motions rather than horizontal; most features volcanic; some faults; no plates or plate motion.

Volcanism: Most landforms are volcanic uplifts; huge number of shield volcanoes; coronae; surface heavily repaved with volcanic flows, most occurring in a global upheaval half a billion years ago.

Water: None on surface; very little in atmosphere.

_____
^{a}The interior structure is uncertain; core is probably solid.

temperatures are the same. The average temperature is therefore considerably *hotter* than Mercury's. The planet must be kept warm by a thick, insulating atmosphere (Figure 8.8).

Various probes dropped by spacecraft into the Venusian air have provided a superb measure of atmospheric composition and structure. The pressure at the surface is *98 bars* (that is, 98 times that at the surface of the Earth). The atmosphere is almost pure (96%) CO_2, which is responsible for a greenhouse effect that makes the planet over 500 K hotter than it would be without its atmospheric blanket. Nearly 4% of the atmosphere is nitrogen, and the remainder is a complex mixture of water vapor, sulfur dioxide, argon, carbon monoxide, neon, and other chemicals. The amount of water is only a ten-thousandth of that found in the Earth's atmosphere.

Remarkably, Earth and Venus have similar total inventories of carbon, oxygen, and nitrogen. On Earth, almost all the carbon is tied up in rock

The pressure at Venus's surface is 98 times that at the surface of the Earth.

Figure 8.8
Venus posed for *Mariner 10* in 1979. A banded cloud structure, revealing a thick atmosphere, converges toward the equator. Some of this structure is visible from Earth (see Figure 8.1b).

and in decayed vegetation in the form of oil and coal, but on Venus, it is all in the air. The reason probably lies in the planet's proximity to the Sun. There are neither oceans nor any chemical processes—including life—that can absorb carbon and lock it into the rocks. When Venus was young, its surface may have been relatively cool (although hotter than Earth's) and may have held a substantial amount of liquid water. The nearby Sun would have gradually evaporated the water to form an important greenhouse gas. The resulting greater atmospheric temperature in turn evaporated even more water and allowed less carbon to bind into rocks. With more water and CO_2 in the atmosphere, the planet's temperature kept climbing in a **runaway greenhouse effect.** The water vapor floated upward, where it was split by solar ultraviolet light into its constituent hydrogen and oxygen. The light hydrogen atoms escaped into space, and the water disappeared forever.

The clouds that surround Venus in a continuous sheet (see Figure 8.8) are sulfuric acid droplets, not water. The cloud cover is a result of the greenhouse effect and the intense surface heat. Since there is little water, sulfur dioxide in the atmosphere cannot become tied up in surface rocks, so it climbs upward where *photochemical reactions* (chemistry aided by sunlight) turn it into the acid. Spacecraft probes showed that the clouds lie about 50 km above the ground (Figure 8.9), much higher than the water clouds of Earth. Because the Venusian surface temperature is so high, Venus's troposphere is nearly 10 times thicker than Earth's, the temperature dropping steadily from 470°C at the surface to −70°C at an altitude of 110 km. Clouds will condense only when air is saturated with vapor. A gas

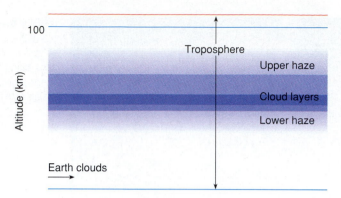

holds less vapor when it is cold than when it is warm, and sufficiently low condensation temperatures are achieved only at great altitude. Venus has no measurable magnetic field, so there is neither a magnetosphere nor an aurora.

Venus has no significant seasons, as the obliquity of its ecliptic is a mere 2.7°. The slow rotation produces a rather constant surface wind pattern that blows in the direction of planetary rotation at a speed of only 3 or 4 km/h. Near the cloud layer, however, the wind velocity increases to a fierce 300 km/hr because of solar heating. The winds blow the whole upper atmosphere around the planet in only four Earth days, giving optical observers a false impression of the rotation period.

8.4.2 The Surface

We can see nothing of Venus through the clouds. With the exception of lander images, everything we know about planetary surface structure comes from radar, which records radio reflectivity; in general, smooth areas are dark, rough areas bright. Figure 8.10 shows a map of the Venusian surface made by *Magellan*'s radar, the low areas colored blue and green and high ones yellow and red in order of elevation. The elevation reference level is set by Venus's average radius of 6,051 km. Though the variation in surface elevation is about 13 km, similar to that on Earth, the extremes are much less common. Earth's surface has distinct high and low areas, whereas Venus's is rather strongly concentrated at one level, two-thirds in broad, rolling plains. The highlands cover only about 10% of surface area (as opposed to the 40% occupied by Earth's continents). The lowlands make up the difference (about 25%), the deepest of them dropping nearly 3 km below the reference level.

Venus has two large high regions, one near the north pole called Ishtar Terra ("land of Ishtar"), about the size of the continental United States, and another, comparable to half of Africa, called Aphrodite Terra. Ishtar Terra contains the only mountain ranges, which ring a broad plateau known as Lakshmi Planum. To Lakshmi's east rise the towering Maxwell

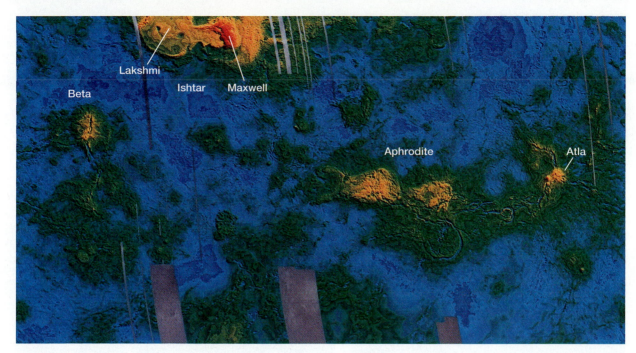

Figure 8.10

The surface of Venus between 70°N and 70°S latitude is seen here as mapped by *Magellan*'s radar. Altitudes are color coded: blue, 2 km below the mean planetary radius of 6,051 km; green, near the mean radius; yellow, 3 to 4 km high; and red, at a maximum of 11 km high. The blank strips are regions with no data. The flat map distorts the sizes of land masses toward the poles: Ishtar Terra is actually smaller than Aphrodite Terra.

Montes (*montes* is the plural of the Latin *mons*, "mountain"), which climb 10 km above the reference level, higher than Everest. Along the broad equatorial belt are several other high areas that include Beta Regio ("Beta region") and Atla Regio ("Atla region") at the eastern end of Aphrodite.

The planet displays a remarkable variety of volcanic landforms that are broadly distributed and do not concentrate along anything that might resemble a plate boundary. Most of the equatorial highlands, including much of Aphrodite Terra, appear to be large, volcanically induced rises and domes, pushed upward by upwelling mantle rock driven by convection. There may be as many as 100,000 volcanoes, most of them small shield volcanoes typically 10 km in diameter and a few hundred meters high, concentrated into extensive fields in the lowlands or rolling plains. Larger shield volcanoes (Figure 8.11a) up to 2 km high and 50 km across are concentrated more toward the highland rises. Some of the volcanoes are apparently active: bright material suggests relatively recent lava flows, and spectroscopy shows declining levels of sulfur dioxide indicative of a recent eruption.

Venus also has several volcanic forms that are not present on Earth. The most prominent are circular **coronae** (Figure 8.11b), broad, bowl-

(a)

Figure 8.11
(a) Radar data can be manipulated by computer to show how volcanoes such as Maat Mons might look as seen from near the planet's surface. The image of this 500-km-wide area is exaggerated vertically by a factor of 20 and is colored to match the effect of heavily filtered sunlight. Bright areas near the top may have been produced by recent volcanic flows. A 25-km-diameter meteor crater lies in front of the mountain. **(b)** A 200-km-diameter corona in the plains southwest of Beta Regio has been pushed upward by a flattened lava plume. The stretching of the rock has faulted and fractured the depressed center. **(c)** A lava channel 2 km wide runs for 300 km, from the top of the picture to the bottom, into a volcanic plain.

(b)

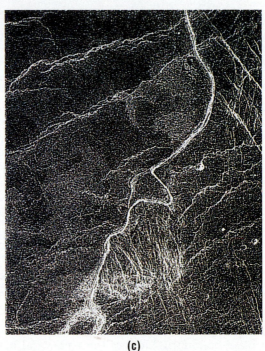

(c)

shaped, cracked rises 200 to 1,000 km across. They are commonly associated with volcanic flows. The volcanic rises and shield volcanoes attest to an enormous number of mantle plumes (see Section 7.6), some of them very large. The coronae seem to have been produced by plumes that had limited mass or that were active over only short periods of time and that flattened on reaching, and buckling, the crust. The majority of volcanic outpourings, however, have been great floods from fissures. Up to 80% of the planet seems to have been paved and repaved by volcanic outflows of one kind or another. We see rilles (channels) up to 2 km wide and an astounding 3,400 km long through which lava once flowed (Figure 8.11c).

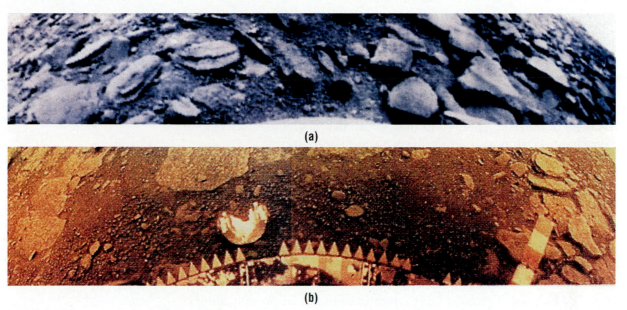

(a)

(b)

Figure 8.12

(a) *Venera 9* set down to the east of Beta Regio on a slope to a deep plain and imaged a panorama of boulders, some a meter across. In the upper corners we look to a distant horizon. **(b)** *Venera 13*, which landed to the south of Beta Regio in a rolling plain, revealed flat rocks and a basaltic regolith. The sawtooth ring is a part of the lander, and the odd-looking plate on the ground is an ejected lens cap.

Scientists of the former Soviet Union sent several Venera landers to Venus. They gave an eerie picture of the volcanic surface (Figure 8.12), revealing a dry, desolate, flat land that stretches toward the horizon. The surface rocks are basaltic, varying in detailed composition from one place to another, much as we find on Earth.

Venus is also pocked with almost 1,000 meteorite craters (Figure 8.13) that tell us something of its history. There is no trace of the late heavy bom-

Venus underwent a catastrophic global volcanic upheaval about half a billion years ago.

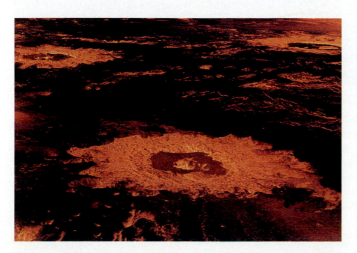

Figure 8.13

A view across the plains region of Lavinia Planitia in the southern hemisphere shows three meteor craters, each with a central peak. The crater in the foreground is 37 km across, that at upper right 63 km.

bardment. From the number of craters and the known cratering rate on the Moon, we find that they have been accumulating for only about half a billion years. In that time, they have not been much altered by volcanic action. They are distributed across the Venusian surface at random, and very few have been flooded with lava. Venus apparently underwent a catastrophic global volcanic upheaval about half a billion years ago that wiped away all of the older craters and then settled down. The presence of long lava channels supports the idea of such enormous outflows.

Venus has no long ridges that span the globe as they do beneath Earth's seas, nor are there volcano chains of the Hawaiian type that show plate motion over a plume. There is little evidence for sliding and colliding crustal plates. Venus's low areas are not ocean beds, and the highlands are not really continents. The planet seems to consist of one large plate. On Earth, tectonic (crust-deforming) activity is mostly horizontal, caused by plates sliding over the mantle. On Venus, the tectonic activity is vertical, caused almost entirely by large-scale up and down convection of mantle rock and by smaller plumes. Only a small amount of crustal spreading is

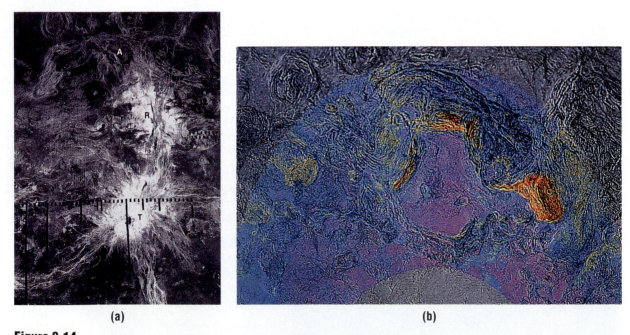

(a)   (b)

Figure 8.14
Venus exhibits limited tectonic activity. **(a)** Beta Regio is a large volcanic rise near the equator. It is split by Devana Chasma, a great rift valley. R and T mark two volcanoes, Rhea and Theia Montes.
(b) Ishtar Terra consists of a vast volcanic plateau, Lakshmi Planum, ringed to the west, north, and east by ridged mountains. Blue and then violet represent the lowest areas, yellow and then red the highest. The plateau contains a pair of giant calderas. The Maxwell Montes lie to the right.

revealed by cracks or faults of the kind that cut through Aphrodite Terra and Beta Regio (Figure 8.14a). These are similar to rift, or separation, faults in the Earth's continental blocks, exemplified by the Great Rift in eastern Africa. Smaller cracks can extend for hundreds of kilometers. Downwelling currents may produce highland crustal compression and ridges, the combination of the two over time creating complicated terrain.

The planet's small mountainous regions are oddly distributed about Lakshmi Planum (Figure 8.14b), a vast volcanic plateau 5 km high. It was flooded by a pair of vents now seen as immense calderas, one 3 km deep. Ishtar Terra was probably produced by upwelling and downwelling mantle currents, its towering mountains compressed and raised by a downwelling flow. The mountains, some of which have slopes greater than 30°, must be young: otherwise, the steep sides, which are softened by high atmospheric temperature, would by now have collapsed under their own weight.

8.4.3 The Interior

Venus's average density of 5.4 g/cm^3 indicates a metallic core similar in size to the Earth's, thereby constituting roughly a third of the planet's mass and half its radius. Given Venus's mass, the core ought to be molten, like Mercury's and part of the Earth's. The lack of a magnetic field, however, suggests that the interior is solid.

Although there are a variety of ideas, no one really knows why Venus, with a diameter and mass nearly equal to Earth's, should be so different from our own planet. Instead of driving plate motion, the convection currents of the mantle create volcanic rises and features associated with plumes. The dryness of a thick lithosphere may be important in suppressing plate development and motion. The extreme atmospheric temperature also heats, softens, and weakens Venusian crustal and lithospheric rock, making it subject to catastrophic flooding that periodically releases internal heat. The resulting rapid loss of heat could have cooled the core to the point of solidification.

Does Venus foretell the Earth's future? As the Earth's interior cools, plate tectonics may shut down and our planet may become much like the Venus of today, magma spilling from fissures to repave a single plate, mountain chains and basins vanishing forever. Conversely, when Venus was hotter inside, it may have had Earthlike plates and may have looked much like modern Earth, the difference between the two caused by a 30% difference in distance from the Sun and the resulting difference in solar heating, which changed the characteristics of the atmosphere and surface rocks.

8.5 INTRIGUING MARS

The ancient mythologies of the two planets flanking the Earth represent the extremes of human passion, love and war. The true contrast, however, is the reverse of that suggested by their names: Venus is a hellish place, whereas Mars (Table 8.3) is comparatively benign, even inviting to human exploration.

TABLE 8.3
A Profile of Mars

Planetary Data			*Atmosphere*	
Distance from Sun	1.52 AU		Pressure	0.07 Earth
Radius	3,394 km = 0.53 R_{Earth}			
Mass	6.4×10^{23} kg = 0.11 M_{Earth}		*Composition by Number*	
Mean density	3.9 g/cm^3		Carbon dioxide (CO_2)	95%
Uncompressed density	3.8 g/cm^3		Nitrogen (N_2)	2.8%
Escape velocity	5.0 km/s		Argon (Ar)	1.6%
Gravity	0.38 g_{Earth}		Oxygen (O_2)	0.13%
			Carbon monoxide (CO)	0.07%
			Water (H_2O)	0.03%
			Layers	*Height (km)*
			Troposphere	0–30 km
Layers	*% Mass*	*Radii (km)*	*Other*	
Iron core	15?	0–1,200	Magnetic field	$<10^{-4}$ Earth
Mantle	80?	1,200–3,300	Rotation period	24^h36^m
Crust	5?	3,300–3,394	Axial inclination	24°

Planetary Features

Basins: Two major impact basins in southern hemisphere.

Bulges: Two large volcanic bulges in northern hemisphere.

Cloud cover: Wisps of water clouds over high mountains; dry-ice clouds over poles.

Craters: Heavy cratering in southern hemisphere from late heavy bombardment; light cratering in the northern.

Faults: One great fault, Valles Marineris, caused by volcanic uplifting.

Mountains: Volcanic; none tectonic.

Plates and plate motion: One plate; no horizontal crustal movement; vertical movement produces bulges.

Polar caps: Permanent polar caps of dry and water ice in the south, water ice in the north; seasonal polar caps of dry ice.

Surface air temperature: Highly variable and dependent on latitude; 240 K (−33°C) maximum under full sunlight to 170 K (−103°C) at night; ground temperature can exceed 0°C.

Volcanoes: Several large shield volcanoes in northern hemisphere.

Water: Vapor in atmosphere, ice in polar caps and below surface; once flowed as liquid.

8.5.1 The View from Earth

Aside from its obvious orange-red color, Mars displays a variety of permanent surface features. From the apparent movement of dark markings (Figures 8.15a and 8.15b), we find an Earthlike rotation period of 24^h37^m and an axial inclination of 25°, only 1.5° greater than Earth's. Mars therefore has seasons similar to ours, though with two important differences. First, the Martian sidereal orbital period is 1.88 years, so each season is almost twice as long as on Earth. Second, because of Mars's high orbital eccentricity ($e = 0.09$), the planet is 20% closer to the Sun and receives 45% more sunlight at perihelion than at aphelion. Aphelion occurs during southern winter and perihelion during southern summer. As a consequence, the southern hemisphere has significantly hotter, shorter summers than the northern and colder, longer winters, quite unlike conditions on Earth.

Mars is enveloped by a thin, somewhat hazy atmosphere that occasionally contains small bright clouds. During the southern-hemisphere summer, a huge dust storm can obscure the planet's surface. Surrounding the rotation poles are caps of ice that seem similar to those of Earth (Figures 8.15c and 8.15d) and change dramatically with the seasons. The

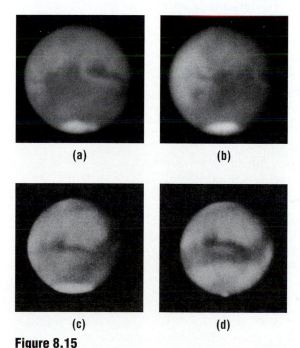

(a)   (b)

(c)   (d)

Figure 8.15
(a, b) Mars rotates on its axis, the dark markings moving to the right. The polar cap shrinks and the dark markings expand between the southern hemisphere's **(c)** late spring and **(d)** summer.

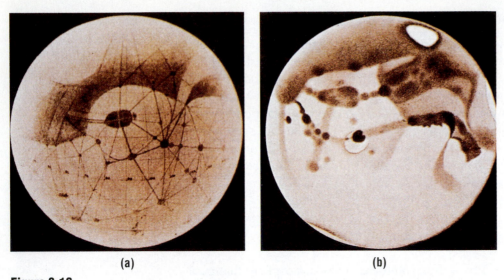

(a)   **(b)**

Figure 8.16
(a) Percival Lowell's 1894 drawing of Mars shows an interconnected set of canals. **(b)** The French astronomer E. M. Antoniadi's drawing, however, showed them to be spots. The dark curved line at left center is actually a great chasm that can also be seen at lower left in the Hubble image in Figure 8.1d.

southern cap can extend almost halfway to the equator. The dark markings also vary with the seasons.

These features pale beside another that brought Mars lasting public fame. In 1877, the Italian astronomer Giovanni Schiaparelli observed a network of fine lines that crisscrossed the planet (Figure 8.16a). The Italian word he used to describe the lines was translated into English as "canals," implying that they were artificial waterways. Martians, it was imagined, had constructed the "canals" to transport water from the poles into dry desert regions, allowing the dark "oases" to expand and blossom with thick vegetation during the summer. Some astronomers, however, resolved the canals into patterns of dots (Figure 8.16b). Under less than excellent conditions, the dots appeared as lines, producing the effect of canals. A far less glamorous explanation for the changes in the dark areas held that they were caused by wind patterns that shift dust around the globe. The resolution of the controversy and the discovery of the planet's real nature were finally revealed by closeup views made possible by the space age.

8.5.2 The Martian Atmosphere

Observation by spacecraft showed Mars's atmospheric pressure to be a mere 0.007 bar (0.7% that of Earth) and to consist of 95% carbon dioxide, 3% nitrogen, and 2% argon. The Martian atmosphere is remarkably like that of Venus but a ten-thousandth as dense. If we include the lithosphere

BACKGROUND 8.1   Martian Canals and the Scientific Method

The episode of the Martian canals teaches a lesson on how science should, and usually does, operate. The essential assumption of science is that nature is not capricious—experiments or observations must be *repeatable*. Only a few people could "see" the canals; those who could not were told that their equipment or their eyes were inferior. The early-twentieth-century French astronomer E. M. Antoniadi, who considered the canals false apparitions, used the best telescopes available and could not see them. Percival Lowell, who founded the Lowell Observatory in Arizona in the 1890s to study Mars, strained believability when he attempted to explain why smaller telescopes were better for the detection of delicate features than were larger telescopes. This kind of explanation is usually a clue that something is amiss.

The history of science is filled with similar instances, one of them a topic of more recent headlines. The Sun is powered by thermonuclear fusion, which creates heavy elements from light ones with the release of vast amounts of energy. On Earth, the process has been used to produce hydrogen bomb explosions. Scientists have tried to generate nonexplosive fusion for power production, but with limited success, because the process requires enormously high temperatures. In 1989, a pair of physicists apparently discovered a way of fusing elements at room temperature using a catalyst (a substance that aids a reaction without being changed by it). They immediately announced the results of their "cold fusion" experiments, and scientists everywhere took note. If the work was valid, here was a potential source of cheap, reliable energy. Unfortunately, few scientists could repeat the fusion reaction. So many failed, in fact, that it became clear that cold fusion does not exist.

The moral of these tales is that nature is reliable. The scientific method, which depends on experimental repeatability, operates to check phenomena that are observed only by particular fortunate or "sensitive" individuals. Scientists cannot become part of the science itself.

of the Earth and add up the total inventory of carbon and nitrogen, all three planets are similar, showing both unity of formation and development and the effects of liquid water and life in shaping the Earth's atmospheric evolution.

The Martian air also contains 0.03% water. That, however, is some 30 times less than we find in the Earth's atmosphere. If all the water in the Martian atmosphere could precipitate out, it would make a planetary ocean a mere millimeter deep. On Earth, liquid water is maintained by high air pressure, which prevents rapid evaporation. The air pressure on Mars, however, is too low to allow liquid water to exist: if formed, it would immediately return to the gaseous state.

Since Mars is farther from the Sun than is Earth and has a thinner atmosphere, it is colder. Though the surface air temperature under a summer noon Sun cannot reach much above 240 K (−33°C), the ground tem-

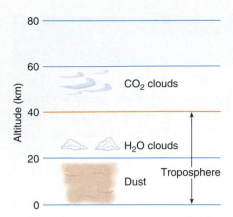

Figure 8.17

Water-ice clouds form in the Martian troposphere at about 30 km altitude over volcanic summits; dry-ice clouds develop twice as high over the poles.

perature can exceed 0°C. The CO_2 in the Martian atmosphere produces a greenhouse effect as it does on Earth, but it is much less effective as an insulator, raising the air temperature by only about 5 K. Consequently, temperatures drop sharply to 170 K (−103°C) or below at night: the water vapor in the air can then turn to fog and coat the rocks with frost.

Martian carbon dioxide precipitates as dry ice.

The troposphere (Figure 8.17) extends to a great height, its temperature steadily declining from the surface value of 240 K (−33°C) to about 150 K (−123°C) at 40 km altitude. Icy water clouds form at altitudes of about 20–30 km, over tall volcanic summits. The atmospheric pressure on Mars is not great enough to maintain liquid CO_2 (nor is it even on Earth), but at sufficiently low temperatures (−79° on Earth), it will freeze to become **dry ice.** Over the polar areas the Martian air is cold enough that dry-ice clouds form at altitudes of some 60 km. Carbon dioxide even precipitates as dry ice from the thin air to produce the seasonal polar caps vividly seen in Figure 8.15.

The Martian magnetic field is extremely weak, 10^{-4} that of the Earth or less, so there is no significant outer magnetosphere.

8.5.3 The Beaten Volcanic Surface

The major Martian features revealed by spacecraft are mapped in Figure 8.18. Like the Moon, Mars has two distinctly dissimilar hemispheres, though for different reasons. The line of division is roughly a great circle inclined about 30° to the equator. The southern hemisphere consists largely of heavily cratered highlands, typically 2 or 3 km above a reference level defined by the average atmospheric surface pressure. It contains two spectacular impact basins, Hellas (Figure 8.19) and Argyre, which are partially filled with lava and bright Martian dust.

The number of southern craters testifies that Mars, like Mercury and the Moon, preserved the record of the late heavy bombardment and that this portion of the surface must be ancient. For so many craters to persist

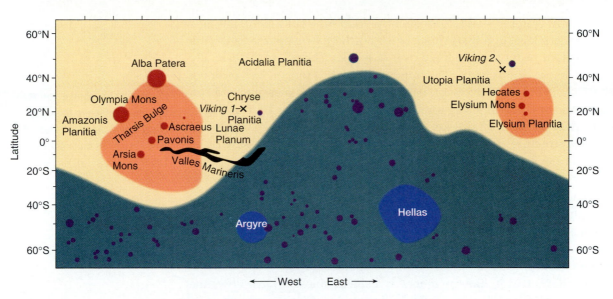

Figure 8.18

A color-coded map of Mars between 65° north and south latitude shows how the cratered uplands of the southern hemisphere (green) are separated from the volcanic plains of the north (yellow). Two volcanic bulges, Tharsis and Elysium, are orange, and volcanoes are red. A variety of the larger craters are indicated by violet, and the two prominent impact basins by blue. The *Viking 1* and *Viking 2* landing sites are also shown.

Figure 8.19

The Argyre basin, an impact crater 800 km across, partially filled with dust and lava, occupies the lower left quadrant of the picture. The upper part of the basin rim has been destroyed by another large impact.

so long, the average rate of erosion must be small, showing that Mars has had neither thick atmosphere nor liquid water over extended periods of time. Moreover, the land has not been destroyed by subduction caused by continental drift, implying that the crust is one solid plate.

Lightly cratered—and consequently youthful—volcanic plains near the reference level dominate much of the northern hemisphere. Set within the northern sector are two huge rises or bulges. The one to the east, Elysium Planitia, climbs to an elevation of 5 km and is topped by volcanoes. The Tharsis Bulge, to the west, is twice as high, more massive, and contains a trio of volcanoes at its top that tower 27 km into the thin air (Figure 8.20). They lie along a 1,000-km-long ridge and show all of Tharsis to be a volcanic structure, a huge mound pushed upward by the pressure of hot magma from below.

Far down the northwest slope of Tharsis is one of the most massive geologic structures in the Solar System, Olympus Mons (Figure 8.21). Though starting at a lower altitude than the three volcanoes at the top, it rises just as high, 25 km from base to peak, three times the equivalent height of the island of Hawaii. Nearly 600 km across, it would cover the state of New Mexico or most of New England. Its low slope (only 4°) clearly shows that it, like the others, is a shield volcano. Lava flows that have run down its side overlap, providing evidence for the growth of the structure. At its top is a group of calderas 90 km across, at its base a cliff that stands 6 km tall.

Shield volcanoes on Earth are limited in mass because plate motion eventually removes them from the upwelling mantle plumes that generate them (see Section 7.4). The enormous mass of Olympus Mons and the absence of a plume track reveal that there is no plate motion on Mars, a

Nearly 600 km across, the Martian shield volcano Olympus Mons would cover the state of New Mexico or most of New England.

Figure 8.20
Three volcanoes topping the Tharsis Bulge are seen at left. Valles Marineris, which consists of several individual canyons up to 9 km deep in the middle, stretches eastward from the bulge. Above the eastern end are the northern plains Lunae Planum and Chryse Planitia. Valles Marineris is the prominent linear feature seen from Earth in Figure 8.1d.

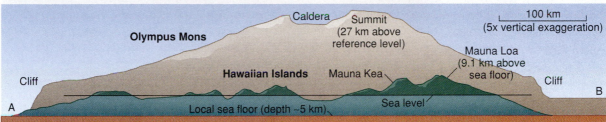

Figure 8.21
Olympus Mons, the most massive volcano in the Solar System, towers 27 km into the thin Martian air and sits on the northwest slope of Tharsis Bulge. At its top is a 90-km-wide caldera. The drawing compares the volcano with the much smaller Hawaiian chain.

finding consistent with the continued existence of craters from the late heavy bombardment. Olympus Mons has sat in that one spot, getting ever bigger, for much of Mars's history, its great height possible because of the planet's low gravity. The small number of craters on its slopes demonstrates that lava flowed perhaps only a half a billion years ago.

Mars may be one plate, but there is still spectacular evidence for ancient tectonic activity. No words can quite describe Valles Marineris (see Figure 8.20), a collective term for a set of huge interrelated canyons, named after the Mariner spacecraft, that stretch over 5,000 km of Martian terrain. These are fault canyons, caused by an expanded crack in the thick Martian crust created by the raising of the Tharsis Bulge. Its character changes considerably along its course. At the western end is a complex system of deep cracks that coalesce into parallel canyons that continually deepen toward the east. At the eastern end, the great valley opens out onto highly chaotic terrain.

The Viking landers take us right to the surface. *Viking 1* landed in the youthful northern plain Chryse Planitia, the "plains of gold." The camera

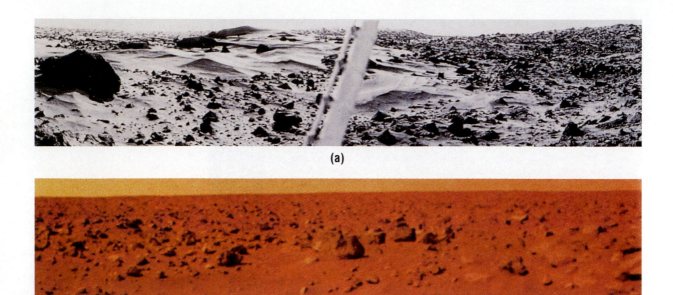

(a)

(b)

Figure 8.22
(a) *Viking 1* took this 100°-wide view of Chryse Planitia, revealing a bleak landscape filled with rocks and drifted sandy soil. **(b)** *Viking 2* showed us a rockier terrain than did *Viking 1*, with less drifted sand.

shows a nearly flat horizon and a surface spread with fine sand or dust that covers and surrounds thousands of rocks, some of them quite large (Figure 8.22a). *Viking 2* set down nearly half a planet away, at the edge of the bulge of Elysium Planitia. Here we see more rocks and less sand (Figure 8.22b). Both Vikings apparently landed in the ejecta fields of meteorite impacts that had fractured the rocks out of the solid volcanic surface. The rocks are heavily pitted, either as a result of exploding gas bubbles or wind abrasion.

Automated laboratories aboard the craft showed that the Martian regolith is rich in iron and is hydrated (that is, infused with water), rather like a terrestrial clay. The planet's red color is caused by the reaction of this iron with atmospheric oxygen—the god of war has rusty armor. The surface has, moreover, become crusted and compacted into a kind of hardpan.

The sand grains are very fine, only a thousandth of a millimeter or so in diameter. Slow winds of a few kilometers per hour cause the sand drifts seen near the large boulder at the left edge of the panorama in Figure 8.22a. Wind storms can easily raise the fine material and cover the entire planet in a dust storm. This dust—or rather the lack of it—produces the planet's dark markings as atmospheric circulation patterns blow the sandy soil free of the dark volcanic rocks.

8.5.4 Water

There is abundant evidence that the Martian surface once held vast amounts of water. The ancient cratered highlands of the south contain dry **runoff channels** that look like common terrestrial riverbeds, with branching tributaries flowing toward a main stream (Figure 8.23a). Because they have not produced much erosion, the runoff channels could not have been active for long, and since they are related to the oldest areas of Mars, conditions for running water could have been right for only a short period of time.

There is abundant evidence that the Martian surface once held vast amounts of water.

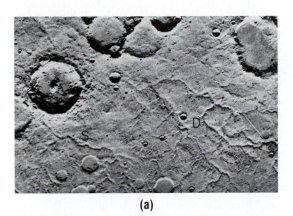

(a)

30 km

(b)

Figure 8.23

(a) Dry runoff channels in the Martian highlands show a branched structure quite similar to terrestrial river systems. **(b)** The outflow channel Ravi Vallis seems to erupt from the Martian surface. It once contained water that flowed into the northern plain Chryse Planitia.

Much more dramatic are great, dry riverbeds called **outflow channels** that are concentrated between Valles Marineris and Chryse Planitia. Unlike terrestrial rivers or the runoff channels, which accumulate from a series of tributaries into a broader stream, they arise seemingly from nowhere, directly from slumped chaotic terrain (Figure 8.23b). These were not quiet streams but massive outpourings of water that had thousands of times the flow rates of the greatest rivers on Earth. The running water massively eroded the land, the scars showing that the water ran from Tharsis into the plains.

Evidence for standing water is provided by the hydrated clays found by the Vikings and by a few terrestrial meteorites blasted off the planet by meteoric impact and identified because they contain gases similar to those in the Martian atmosphere. When sliced and analyzed, these rare meteorites reveal clays that appear to have been soaked in water for centuries. The floor of the northern canyon in the center of Marineris is dark with layers of sediment (Figure 8.24). To the east, the canyon walls look scrubbed, and where the canyon opens up at its eastern end there are clear effects of erosion. The canyon was apparently filled with a huge lake whose containing walls suddenly fell, allowing the water to flow out, scouring the land as it went.

Where did the water come from and where did it go? Meteorite craters provide a critical clue. The ejecta blankets of meteorite craters frequently have petallike flow patterns ending in low cliffs (Figure 8.25). They can be

Figure 8.24
You look onto the floor of what may once have been a massive lake that filled Candor Chasma. The dark surface consists of sediments.

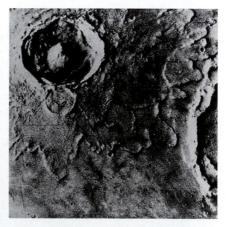

Figure 8.25
The petal pattern in the ejecta blanket surrounding this 18-km-wide crater suggests an impact in an icy surface.

explained by ice locked into the ground as permafrost. At impact, the ice melted and the ejecta flowed from the point of explosion in a quick-freezing mudflow. The abundance of underground ice nearer the poles also makes higher-latitude craters take on a softer, almost out-of-focus appearance; the ice-logged regolith is weaker, allowing the walls of the craters to slump with time.

Like Earth (and maybe Venus), Mars was initially endowed with a supply of water. Since the planet is farther from the Sun than is Earth, and colder, the water probably saturated the surface rocks as ice. Heat associated with the impact cratering of the heavy bombardment and ancient volcanism associated with the southern highlands could have melted the ground ice and caused the water to flow and cut the runoff channels. By the end of the late heavy bombardment, it had sunk back as permafrost.

Large-scale volcanism associated with the raising of the Tharsis Bulge and the Tharsis volcanoes again apparently liquefied the water, but this time on a much greater scale, causing it to gush from the ground to create the outflow channels. Carbon dioxide (as well as water vapor) released from the rocks in the process might then have generated a temporary thick greenhouse atmosphere that helped warm the planet and melt the rest of the ice. The events that took place well over a billion years ago are not at all clearly understood. But as the water poured into the northern plains, it may even have pooled into a large ocean. The result may have been a warm, wet climate. Precipitation could then have created glaciers in the southern highlands. Some of the water subsequently cycled back into the rock, freezing when the driving heat from volcanism terminated. The atmosphere dissipated into space along with the remaining water, leaving the planet's surface cold and arid.

More water lies at the poles. Beneath the seasonal dry-ice polar caps lie permanent polar caps that remain during the summers. The southern one (Figure 8.26) is about 500 km across, the northern twice that size. The

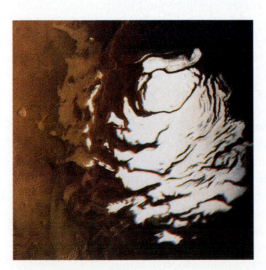

Figure 8.26
The southern permanent polar cap is made of frozen carbon dioxide and water ice.

BACKGROUND 8.2   Travels to the Terrestrials

A trip to Mercury would be dangerous at best, not just because of the searing heat, but also because of the solar wind, which is energetic enough to create the bulk of the sparse Mercurian atmosphere. The solar day, 176 Earth days long, would proceed with agonizing slowness. At that pace, the Sun would move along its daily path at an average rate of only 2° per Earth day, or a mere five minutes of arc per Earth hour. Watching a sunrise would be an exercise in extreme boredom, the event lasting on the average 18 Earth hours.

The day would proceed much as on the Moon, but with surface temperatures climbing to the lead-melting point. After an equally long sunset, the nighttime temperature would plunge to −185°C. Of course, you could live underground, where the temperature could be maintained fairly comfortably with a little air conditioning, but then you would see nothing at all. More likely you would choose to visit near one of the poles, where the Sun is shadowed by craters and you would have a supply of ice to replenish your water supply.

On Venus, the heat and atmospheric pressure are so great that it seems unlikely that anyone will ever land. Nevertheless, radar images and Venera pictures paint a fascinating landscape. What would it be like, for example, to stand in Lakshmi Planum and gaze at Maxwell Montes off in the distance, or to look into the rift valley of Beta Regio? No stars would be visible at night because of the clouds. As daylight approached, the sky would begin to brighten in the west because of Venus's backward spin. At noon, the sky would appear about as it does at home under a very heavy thunderstorm cloud deck, the kind that brews tornadoes. What sunlight does penetrate the clouds would be heavily absorbed and would have an orange cast to it, giving the landscape an eerie burnished glow.

Your protective suit would have to be strong and efficiently air conditioned, as without it you would be crushed by the extreme air pressure and cooked in a searing heat that is comparable to that inside an oven in a self-cleaning cycle. You would not have much weather to worry about, however; the forecast would be the same day after day, hot and dry, with a slight wind from the east. As the Sun sets in the east, it would gradually get darker but it would never cool off—the air is too good an insulator. You would probably want to leave as quickly as you could to go to the next planet, Mars.

Your spacecraft takes you to a location near the equator of the red planet during the southern-hemisphere summer. Though conditions are certainly more pleasant than on Venus, you could still not go outside without protective garments: the air is too thin to support human life, there is no oxygen to breathe, and lethal ultraviolet sunlight and solar wind particles both reach the ground. Nevertheless, you could get out of your spacecraft and easily walk around. Other than the lower gravity and the need for a spacesuit, a Martian day might seem rather like one spent in a high, very dry, cold desert rather like central Antarctica.

At sunrise, you see frost or maybe a little fog that would quickly dissipate. You might extract some ice from underground to bring back to the ship for drinking water. It is a cloudless day, though airborne dust tints the air a bit pink. The wind blows gently from the south, and by noon the ground temperature has risen nearly to the freezing point of water. As you walk, the regolith crunches under your feet, and you occasionally kick up little puffs of dust.

The Sun proceeds on its daily path much as it does at home, and near sunset you return to the ship to turn up the electric heat and prepare for a temperature drop of 100°C. While you anticipate the next day of exploration, Deimos shines faintly, like a slowly moving star. Perhaps Phobos will rise in the west, making you long for the more familiar Moon of home.

permanent south polar cap consists of dry ice overlying water ice, but the northern one contains only water ice. Because of the orbital eccentricity, the winds blow most strongly in southern summer (northern winter), depositing dust in the northern ice. Dust is an efficient absorber of sunlight, and it will cause the northern cap to become so warm in northern summer that all its dry ice evaporates away directly to a gas, leaving only frozen water.

8.5.5 Life

Life began on Earth easily and quickly, only a billion years after the planet's formation, after the end of the late heavy bombardment (see Section 7.5). Mars—like Earth—once had liquid water, a fundamental ingredient needed for life as we know it, and a significant atmosphere. Did life ever emerge on Mars?

The Viking cameras (see Figure 8.22) showed no forms of life nor evidence that any ever existed. The landscape looks desolate and forbidding. However, this sterile appearance does not preclude microscopic life, which still dominates on Earth and was the precursor of larger life-forms. Terrestrial life is based on organic molecules, those built from carbon atoms. The Viking landers carried a set of sophisticated experiments designed to search for signs of active microscopic life and for organic matter in the soil that might announce that life had once existed.

Shortly after a Viking craft settled to the Martian surface, an extendable arm (Figure 8.27) scooped up numerous soil samples. In one experiment, the dusty regolith was heated and the resulting gases examined with a *mass spectrometer*, a device that can deduce the chemical composition of a gas by the weights of the molecules that compose it. The Vikings found no organic molecules of any kind.

The regolith was also tested with various nutrient baths designed to feed any life-forms that might be present. A gas detector then searched for simple by-products of biological activity, like oxygen or carbon dioxide. In other experiments, the regolith was exposed to baths and gases containing

Figure 8.27

This 4.5 billion-year-old meteorite, knocked from Mars several million years ago in a giant collision with an asteroid, may contain evidence that life once existed on the planet.

radioactive ^{14}C. The regolith and the gas above it were then examined by a radioactivity detector to see whether any of the ^{14}C had been incorporated or emitted, which might have indicated its consumption by microbes.

For a brief and exciting moment it looked as if Viking had found signs of biological activity. On closer examination, however, it became clear that the source of the results was not life but a chemically active regolith oddly rich in peroxides (compounds containing oxygen atoms bonded together). The peroxides are created by the action of ultraviolet sunlight, a process not possible on Earth, where atmospheric ozone screens out these rays. The eventual conclusion was that neither Viking craft had found any evidence of life.

In 1996, however, a team of researchers announced the discovery of structures within a meteorite from Mars (Figure 8.27) that look suspiciously as if they had been created by primitive life forms. Close examination showed small tubules that are reminiscent of fossils of microscopic terrestrial life, and the chemical compositions of tiny globules are similar to those produced by ancient terrestrial bacteria. The evidence is circumstantial, but it shows the need to go back to the red planet.

8.5.6 The Interior

The interior of Mars remains mysterious. From the planet's average density of 3.9 g/cm^3 (and the low density near the surface), we predict an iron core that gravity measurements suggest is contaminated with some other material, probably sulfur. The size of the core is relatively similar to that of Earth. The absence of a magnetic field implies that the core has solidified, the result of relatively rapid cooling in a small planetary mass. The mantle is basaltic like our own, surface conditions suggesting a thick crust perhaps 100 km deep. Tectonic activity has been limited to the volcanic northern hemisphere. The activity has been mostly vertical, with no horizontal plate motion. Giant mantle plumes produced two massive bulges and a huge crustal crack (Valles Marineris) as well as several smaller ones.

The low Martian mass may explain the lack of life. The escape velocity is low, and a once thick atmosphere, heated by sunlight, can easily evaporate. Moreover, the low mass allowed the interior to cool, shutting down the magnetic field. With no magnetic protection, the solar wind slammed into the atmosphere, adding to its steady erosion. With low internal heat, Mars had insufficient tectonic activity to resupply much gas from the interior. As volcanic activity wound down, the carbon stayed in the rocks, the greenhouse effect diminished, most of the atmosphere escaped for good, and the planet turned into a giant refrigerator. Closer to the Sun, Venus became a runaway oven. Here we are in the middle, with our relatively high mass (which keeps plate motion going) and just the right amount of solar heating. Our neighbors are teaching us something about our own precious world.

Neither Viking craft found any evidence for life on Mars.

8.6 THE MARTIAN MOONS

Mars has a pair of odd little moons (Figure 8.28 and Table 8.4). Phobos and Deimos (from the Greek words for "fear" and "panic," the war god's attendants), discovered in 1877, orbit close to the planet, counterclockwise in its equatorial plane and in synchronous rotation. Deimos is 23,460 km from the planet's center and has a sidereal period of only 30^h18^m. Thus it would appear to move slowly on a daily path to the west, taking nearly 5.3 Martian days between successive moonrises. Phobos is considerably closer, 9,378 km from the Martian center and 6,000 km from the surface. Its sidereal period is 7^h39^m, less than the Martian rotation period, and it therefore rises in the west and sets in the east!

Phobos and Deimos were probably not formed with Mars but were most likely captured from the debris that pervades the Solar System. Such a capture requires an external force to slow the incoming body (see Sec-

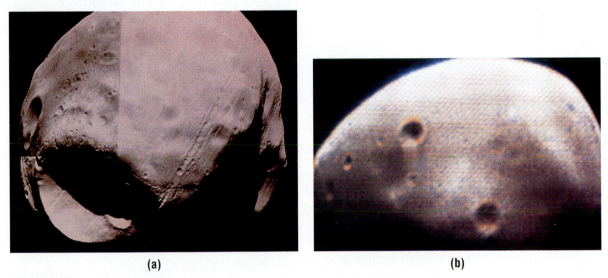

(a)   (b)

Figure 8.28
The Martian moons, **(a)** Phobos and **(b)** Deimos, are heavily cratered. The large crater on Phobos is 10 km across. The fractures radiating away from it were probably created by the impact.

TABLE 8.4
The Moons of Mars

Moon	Diameter (km)	Distance from planetary center (km)	Orbital Period
Deimos	12 × 16	23,460	30^h18^m
Phobos	20 × 28	9,378	7^h39^m

Satellite Properties

Small, rocky, and heavily cratered.

tion 7.7); friction with a once-thick atmosphere may have done the job. The satellites are too small for their gravity to shape them into spheres, and they remain irregular blocks. Phobos averages 25 km across, little Deimos 14 km. Both are heavily cratered. Phobos has a huge crater 10 km in diameter, about half the size of the satellite itself. A little larger, and the moon would probably have broken in half, further support for the hypotheses of planetary structure and properties that involve catastrophe. We will find more such support as we enter the realm of the Jovian planets.

▶ **KEY CONCEPTS**

Coronae: Volcanically uplifted, cracked, bowl-shaped regions on Venus 200 to 1,000 km across.

Doppler effect: The observed shift (the **Doppler shift**) in wavelength or frequency caused by relative radial motion.

Dry ice: Frozen carbon dioxide.

Outflow channels: On Mars, large channels created by eruptions or releases of vast amounts of water.

Radar: An active observational technique in which radio waves are reflected from a body to determine its distance, speed, and surface features.

Radial velocity (v_r): The relative speed of a body moving along the line of sight.

Runaway greenhouse effect: A spiraling effect brought about when high temperature produces more atmospheric carbon dioxide, which produces higher temperature, and so on.

Runoff channels: On Mars, ancient dry riverbeds, evidence of low-volume water flows.

Tangential velocity (v_t): The relative speed of a body moving perpendicular to the line of sight.

▶ **KEY RELATIONSHIP**

Doppler formula (for velocities much less than c):
$$\lambda_{obs} - \lambda_{rest} = \lambda_{rest}(v_r/c),$$
or
$$v_r = c(\lambda_{obs} - \lambda_{rest})/\lambda_{rest}.$$

▶ **EXERCISES**

Comparisons

1. Compare the following properties of the terrestrial planets and the Moon: **(a)** rotations; **(b)** craters; **(c)** magnetic fields; **(d)** cores.

2. Compare the basins of Mercury with those of the Moon.

3. What are the differences between the collision theories for the Moon and for Mercury?

4. Compare the atmospheres of Mercury and the Moon.

5. What do the atmospheres of the Earth, Venus, and Mars have in common? How are they different?

6. How and why do the permanent Martian polar caps of the north and south differ?

7. How do the volcanoes of Mars differ from those of the Earth?

8. Distinguish between the outflow and runoff channels on Mars. What do they tell us of early Martian conditions? How do they differ from the canals?

Numerical Problems

9. How long does it take a radar signal to reach and return from **(a)** an airplane 20 km away; **(b)** the Moon; **(c)** Venus, if the planet is at inferior conjunction?

10. If the radio (and infrared) luminosity of Venus were only a quarter of what is now observed, what would be the temperature of the surface?

11. What is the average heating rate of sunlight on Mars relative to that on Earth?

12. What is the latitude of the Martian arctic circle?

Thought and Discussion

13. How can we find the masses of Venus and Mercury?

14. How can Mercury have ice at its poles when its surface can get so hot?

15. What is the origin of Mercury's scarps?

16. Why do we believe that Mercury has the largest iron core of any planet in the Solar System?

17. How do we know that Venus's atmosphere is made of carbon dioxide?

18. Why are the clouds of Venus made of sulfuric acid and not of water?

19. How have upwelling and downwelling mantle material affected Venus's surface?

20. Why do we believe that Venus's surface was globally repaved with volcanic magma about half a billion years ago?

21. What effect has Venus's atmosphere had on the nature of the planet's surface and core?

22. How do we know there is water vapor in Mars's atmosphere?

23. Can it rain on Mars? Explain your answer.

24. Why are the largest Martian volcanoes on bulges?

25. What is the evidence for tectonic activity on Mars?

26. How do we know that water was once abundant on Mars?

27. How do we know that Mars's southern highlands are older than its northern plains?

28. Describe the different ways in which the terrestrial planets lose, or have lost, internal heat, and discuss how these different modes of heat flow have produced different planetary surfaces.

Research Problems

29. Before radio astronomy and radar revealed the nature of Venus, astronomers speculated about its surface conditions. Using library materials, examine the nature of some of the speculations made over the last century and comment on why the astronomers making them were misled.

30. Read a science-fiction novel about Mars. List the conditions the author adopts and evaluate them in the light of modern knowledge.

Activities

31. Compile a summary of spacecraft that have visited Mercury, Venus, and Mars, and give the functions of each.

32. Using photocopied library materials, such as *National Geographic* magazine, compile a booklet showing surface features of the Earth (volcanoes and the like) that have counterparts on Mercury, Venus, Mars, and the Moon. Under each, comment on how they differ from one planet to the other.

33. If Mars is near opposition, use a telescope to sketch its surface features. Particularly note the polar caps and their sizes.

Scientific Writing

34. For a feature article in your local newspaper, write about a hypothetical trip to Venus, taking the reader through the clouds, down to the surface, and then home. Convey the excitement of seeing a new world for the first time.

35. Write to a friend who has read the old ideas about the Martian canals and has become convinced that Martians exist. In your letter give counterevidence, from the past and present, that shows why Lowell was wrong.

GREAT WORLDS

The giants of the planetary system are more related to the Sun than to the Earth

9

Huge Jupiter and Saturn are entirely different from the little worlds near the Sun. Their characteristics are in some ways more solar than terrestrial, and with their collections of satellites, they behave something like miniature Solar Systems.

9.1 A PAIR OF GIANTS

Jupiter, with its four **Galilean satellites** (discovered by Galileo in 1609) and its banded clouds, and Saturn, with its magnificent ring system, are among the finest telescopic sights of the sky (Figure 9.1a). From its 47-second-of-arc diameter at opposition, we find Jupiter to have an astonishing equatorial radius of 71,500 km, 11.2 times that of Earth, making it the largest of the planets. Saturn (Figure 9.1b)—20 seconds of arc across—is number two, with a 10% smaller equatorial radius of 60,268 km, 9.45 times Earth's. Application of Kepler's generalized third law to the planets' satellite orbits yields respective masses of 318 and 95 Earth masses. Jupiter's mass is almost a thousandth that of the Sun and twice that of all the other planets put together.

A great deal has been learned about both planets from Earth-based observation. Real comprehension, however, requires that we move in closer. Four spacecraft have made flyby visits, the last two the famed Voyager craft. Launched in 1977, *Voyager 1* and *Voyager 2* skimmed by Jupiter's cloud tops in 1979, Jovian gravity propelling them past Saturn in 1980 and 1981 respectively. The Voyagers' 11 experiments included wide- and narrow-angle television cameras, which sent back extraordinary pictures. Equally important were devices that detected magnetic fields and charged particles, as well as various spectrometers (digital spectrographs). Though the two planets differ considerably in mass, they are close cousins, with similar external features and internal constructions. The latest spacecraft to visit Jupiter was *Galileo*. Launched in 1989, it went into orbit about the giant planet in 1995 and dropped a probe into its atmosphere. Because of the similarity between the two giant planets, some of what we learn from this experiment might be applied to Saturn as well.

Jupiter's mass is twice that of all the other planets put together.

228

(a)

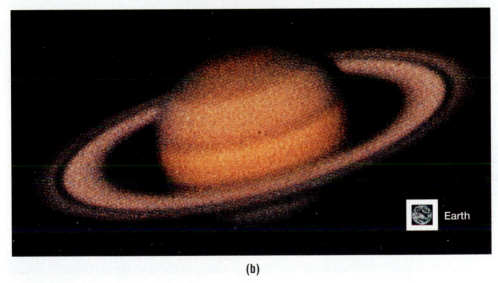

Earth

(b)

Figure 9.1
(a) Jupiter, with its alternating sets of bright zones and dark belts, and **(b)** Saturn are shown to scale, as is the inset Earth.

9.2 MAGNIFICENT JUPITER

The giant planet (profiled in Table 9.1) is an astonishing natural laboratory in which we can examine planetary atmospheres and even states of matter under extreme conditions.

TABLE 9.1
A Profile of Jupiter

Planetary Data			Atmosphere	
Distance from Sun	5.2 AU		**Composition by Number**[c]	
Equatorial radius	71,492 km = 11.2 R_{Earth}		Hydrogen	92%
Polar radius	66,854 km		Helium	8%
Mass	1.90×10^{27} kg = 318 M_{Earth}		Methane (CH_4)	0.09%
Mean density	1.33 g/cm^3		Ammonia (NH_3)	0.02%
Surface (cloud deck) gravity	2.4 g_{Earth}		Water	0.01%
Escape velocity (cloud deck)	59.6 km/s		Ethane	Trace
			Acetylene	Trace
Layers	**% Mass**	**Radii (km)**	Phosphine	Trace
Rocky core	4	0–13,000	Other gases	Trace
Central temperature 25,000 K				
Central density 50 megabars			**Other**	
Liquid metallic		13,000–60,000[a]	Magnetic field	14 times Earth
hydrogen mantle			Tilt of magnetic field axis	9.6°
Liquid molecular		60,000–68,000[a]	Rotation period (interior)	9^{h}55^{m}30^s
hydrogen mantle			Axial inclination	3.1°
Gaseous atmosphere		60,000–69,000[b]	Effective temperature	124 K

Planetary Features

Cloud cover: Apparent surface consists of deep clouds only; no solid visible surface.

Magnetosphere: 2-million-km radius, highly distorted by solar wind; contains extended magnetodisk.

Surface features: Dark belts and bright zones parallel to equator; many dark ovals; Great Red Spot in the southern hemisphere; highly turbulent.

Temperature: Effective temperature of 124 K is 15 K hotter than expected because of internal heat caused by contraction.

Winds: Direction correlates with belt-zone pattern; 350–400 km/s prograde wind in equatorial zone; winds reverse direction within equatorial belts, becoming retrograde.

[a]Layer radii approximate; liquid mantle blends gradually into gaseous atmosphere.

[b]Average radius.

[c]Abundances of atoms and molecules are relative to the number of hydrogen atoms, not hydrogen molecules.

9.2.1 The Cloudy Atmosphere

Jupiter's apparent surface is not solid. With the eye we see only an intricate network of colored clouds arranged in a series of stripes that run parallel to the equator, lighter **zones** separated by darker **belts** (see Figure 9.1). Though variable in brightness and width, they are permanent features.

As we watch, we can see individual cloud markings moving across the apparent disk of the rotating planet and can thus derive the rotation period, which is remarkably short and depends on latitude. Within the bright equatorial zone, Jupiter spins every 9^{h}50^{m}30^s. Well outside this region, the rotation period is 9^{h}55^{m}41^s. Jupiter is therefore in **differential rotation,** the atmospheric gases moving past one another at high speed. The planet's spin axis has a tilt of only 3° relative to the orbital perpendicular;

as a result, there are no significant seasonal changes during the 12-year sidereal orbital period.

This odd rotation can be understood through analysis of Jupiter's radio radiation. With the exception of the Sun, Jupiter is the strongest radio source in the sky. The radio data are divided into three broad domains. At the shortest wavelengths, those of a few centimeters and into the infrared, Jupiter behaves like a blackbody, producing **thermal radiation** because of its warmth. At longer radio wavelengths, however, it generates **nonthermal radiation** by processes that have nothing to do with blackbodies or heat. In the decimeter (tens-of-centimeters) wavelength range, the planet emits **synchrotron radiation,** which is caused by electrons moving near the speed of light and trapped in a magnetic field. Since an electron is a charged particle, a magnetic field will continuously deflect it (Figure 9.2a), accelerating it into a spiral path around a magnetic field line and causing it to radiate in the direction of its motion. The total energy radiated by the electron gas will be proportional to an inverse power of the frequency. The spectrum—a graph of brightness against frequency (Figure 9.2b)—will thus look very different from that of a blackbody, allowing us to distinguish immediately between the synchrotron and thermal mechanisms. The existence of synchrotron radiation shows that Jupiter has a powerful magnetic field.

At wavelengths of tens of meters, we observe **decameter bursts** of nonthermal radiation that vary regularly with a period of $9^h55^m30^s$ and are

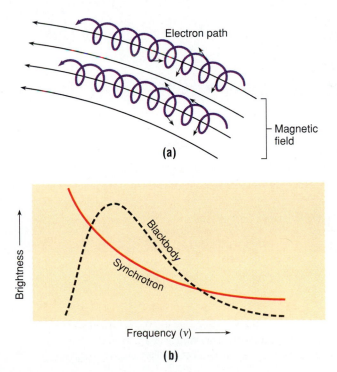

(a)

(b)

Figure 9.2

(a) Synchrotron radiation is produced by high-speed electrons caught in a magnetic field and deflected into continuous spiral paths. As they accelerate, they will radiate energy in the direction of their motion. **(b)** The sum of the radiation of all electrons, which are moving at different speeds, has a spectrum (red curve) that depends on a power of the frequency in a very different way from that of a blackbody (black curve).

caused by an interaction between Jupiter's magnetic field and one of its satellites. Jupiter's magnetic field is produced by and anchored in the planet's deep core, and therefore the burst period is the *true* rotation period of the interior and of the vast mass of the planet.

The differential rotation and the differences in rotation periods between the clouds and the interior are caused by atmospheric winds. The bright equatorial zone is a broad avenue of Jovian air moving from west to east, in the direction of rotation (*prograde*), with a fairly steady velocity of 350 km/h or more relative to that of the underlying planet (Figure 9.3). The equatorial clouds thus spin around the planet faster than does the interior, giving us the sense of a shorter rotation period. In the north and south equatorial belts, which flank the equatorial zone, the wind speeds decrease with increasing latitude until they are blowing east-to-west (*retrograde*). As the winds blow past one another, they produce beautiful turbulent patterns in which individual swirling eddies come and go, apparently clearing bands of clouds from the upper atmosphere, and allowing us to look to the lower clouds below. Farther from the equator the winds reverse again. Toward the poles the reversals become smaller, correlating with the less obvious distinctions between zones and belts.

Jupiter has its own powerful internal heat source.

Jupiter's cloud deck is cold, a result of great distance from the Sun. The effective temperature, derivable from the thermal radio radiation and

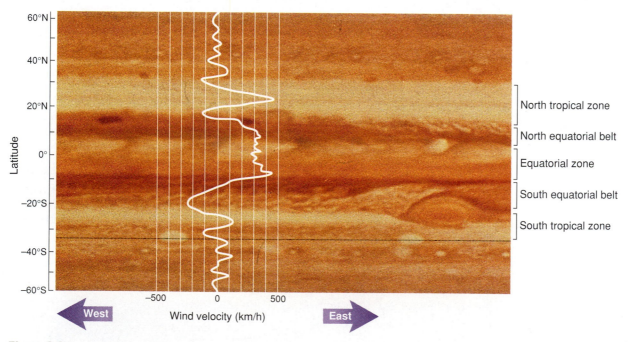

Figure 9.3

Computer processing allows a Voyager image of Jupiter to be spread into a flat map; the equator runs from left to right. The graph shows the speeds of the winds derived from Voyager data relative to the internal rotation period. Note the complexity of the clouds, the spinning eddies, and the numerous ovals.

the average radius, is a chilly 124 K (−146°C). However, sunlight should heat the planet to an effective temperature of only 109 K, 15 K lower. Because the luminosity of a blackbody is proportional to T^4, Jupiter radiates nearly $(124/109)^4$, or 1.7, times as much energy as it gets from the Sun. It has its own powerful internal heat source. *Galileo* showed that the winds maintain their speeds to great depths, suggesting that the internal heat, not solar heating, drives the wind patterns.

The visual appearance of the belts and zones does not depend on temperature but on reflective properties. In the infrared the belts are bright and the zones are dark, indicating that the belts are warmer than the zones and may be gaps in the upper cloud deck through which we can see the lower cloud layers. The minimum temperature of 100 K lies at the point where the air pressure is 100 millibars (10% that on the Earth's surface), which serves as an altitude reference level (Figure 9.4). Temperature rises with increasing depth and air pressure to over 300 K at a level 125 km farther down. Above the top cloud deck, the temperature climbs as the atmosphere thins into a great surrounding ionosphere.

Quite unlike the atmospheres of the terrestrial planets, Jupiter's atmosphere is dominated by hydrogen and helium. The *Galileo* probe shows that (by number of atoms) the atmosphere contains 92% hydrogen and 8% helium, very similar to the mixture found in the Sun, except that Jupiter's hydrogen is in its molecular form (H_2). The chemistry of minor compounds (constituting only 0.1%) is extraordinarily rich. Absorption lines of both methane (CH_4) and ammonia (NH_3) were found in the 1930s. Modern spectroscopy reveals traces of other **hydrocarbons** (hydrogen-carbon molecules) including ethane (C_2H_6), acetylene (C_2H_2), and propane (C_3H_8) made by sunlight acting on methane. We also see water (H_2O), phosphine (PH_3), and germane (GeH_4).

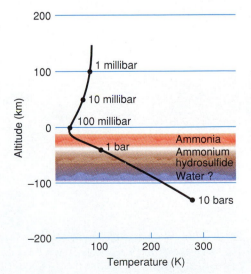

Figure 9.4

Jupiter has no solid surface, so we use the 100-millibar atmospheric pressure level (10% that at Earth's surface) as an altitude reference. Temperature and pressure both increase inward through cloud layers of different color, composition, and depth. Temperature also increases above the reference level.

These and other chemicals condense to produce clouds at different altitudes and temperatures. The highest are almost certainly made of ammonia crystals (see Figure 9.4). The next layer, detected by the *Galileo* probe, is likely to consist of ammonium hydrosulfide crystals (NH_4SH), made by the interaction of ammonia and hydrogen sulfide (H_2S). Additional support for such clouds was provided by a collision (to be examined in Chapter 11) between an interplanetary body and Jupiter in 1994, which dredged up large amounts of sulfur. Theoretical calculations suggest an even lower layer of frozen water clouds, but *Galileo* did not detect them, finding the planet to be unexpectedly dry. Unidentified trace materials produce the subtle colors of the clouds.

Everywhere within the Jovian atmosphere we find strange ovals, some white, others dark. The dark features (see Figure 9.3) appear to be holes through which we see lower layers. They are observed exclusively at a latitude of about 14° in the north equatorial belt. White ovals are seen only in the southern hemisphere. No one knows why there is such clear discrimination.

None of these features, however, compares with the **Great Red Spot** (GRS), a reddish whirlpool of spinning Jovian air and clouds that lies in the southern hemisphere at a latitude of about 25°S (Figure 9.5). Although highly variable in appearance, we know it has been present for at least 330 years. It is immense, 26,000 km by 14,000 km, big enough to hold two Earths side by side, and is apparently trapped between the oppositely di-

Jupiter's Great Red Spot is big enough to hold two Earths side by side.

Figure 9.5
Winds blow past Jupiter's Great Red Spot, a huge spinning oval of enormous turbulence.

rected winds of the south equatorial belt and the south tropical zone (see Figure 9.1). As a result, the GRS is forced to roll counterclockwise with a spin period of about six days. Its reddish color and dark appearance in the infrared shows it to be cold and elevated above its surroundings. Its origin and the way it is maintained are far from understood.

9.2.2 The Interior

From Jupiter's radius and mass, we derive a density of 1.3 g/cm³, only 30% greater than water and dramatically different from the densities of the terrestrial planets. Jupiter must be structured much differently from these smaller bodies and must be made of lighter materials.

One of the chief tools in the study of Jupiter's interior is the concept of **hydrostatic equilibrium.** Pressure increases toward the planet's center because of the increasing weight of the overlying layers. In the thin layer of matter in Figure 9.6, the pressure at the bottom is greater than at the top, creating an outward shove that exactly compensates the inward pull caused by the layer's weight. Consequently, the layer maintains a stable position. This principle helps us build a mathematical **model** that we can use to calculate the temperature, pressure, and state of matter at each point necessary for the layers to be self-supporting. For Jupiter to have its observed size and mass, it must be mostly hydrogen with about 8% helium (by number of atoms), consistent with the atmospheric composition. The model also shows that temperature and pressure climb rapidly within Jupiter, reaching 25,000 K and 50 megabars at the center, respectively nearly 4 and over 13 times the values found inside the Earth.

Like the Earth's, Jupiter's interior is divided into separate regions, although the details of its structure are very different (Figure 9.7). The planet is noticeably oblate, with a polar radius 7% smaller than the equatorial.

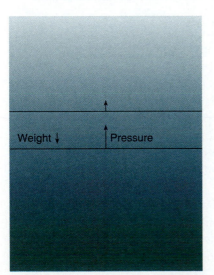

Figure 9.6

In hydrostatic equilibrium, the outward push caused by the pressure difference across a layer will support that layer's weight, keeping it from rising or falling.

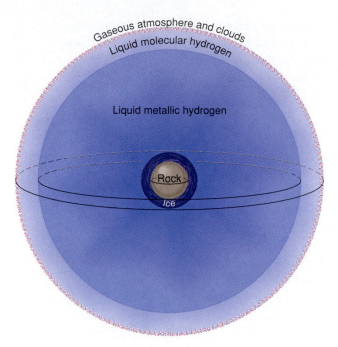

Figure 9.7

A model of Jupiter shows a hot core of heavy elements called "rock" (silicon, iron, and others) and "ice" (here a general term for water, methane, and ammonia). Surrounding the core is an inner mantle of liquid metallic hydrogen, encased by an outer mantle of liquid molecular hydrogen. (The sizes of the layers are uncertain.) About 8% of the atoms are helium. The liquid H_2 gradually merges with the gaseous atmosphere we see.

A body's oblateness depends on both its rotational speed and how its matter is concentrated. Detailed analysis of Jupiter's shape shows that its mass is strongly condensed to the center and that it has a core consisting of 12 or so Earth masses of **rock** and **ice** about 13,000 km in radius. In discussions of the Jovian planets, "rock" refers to common heavier elements like silicon, oxygen, and iron; "ice" is a general term for lighter compounds like water, methane, and ammonia. The terms are used even if these substances are liquid, as they may be in the intense 20,000–25,000 K temperature of the Jovian core. If we could remove the outer hydrogen and helium, the remaining core might resemble our own planet, though it would be vastly larger.

Surrounding the core is a mantle of hydrogen and helium in unusual states. At the low temperatures near the visible surface, hydrogen is in its gaseous molecular form; as we proceed inward, the pressure climbs so high that near a depth of about 1,000 km below the clouds, the molecular hydrogen should gradually liquefy. Recent laboratory data indicate that at a depth of about 10,000 km (about 15% of the way to the center), the pressure is so great that the liquid hydrogen should behave like a metal and conduct electricity. (The relative sizes of the different zones are not well known because of the uncertainty of the pressure at which hydrogen becomes metallic.)

Unlike the Earth, which is heated internally by core crystallization and radioactive decay, Jupiter is still hot from its formation 4.5 billion years ago. As it cools, it contracts and slowly releases its internal energy, the principal reason it is hotter than we would expect for a blackbody of the same size.

The pressure inside Jupiter is so great that its liquid molecular hydrogen can behave like a metal.

9.2.3 The Magnetosphere

Jupiter has an immense magnetosphere (Figure 9.8) over 4 million km across, caused by the powerful magnetic field generated by convection and rotation in its fluid, electrically conducting interior. To a loose approximation, the field is a dipole passing not quite through the center of the planet. Like the Earth's, Jupiter's field is tilted by about 10°. Electrons trapped in this field close to Jupiter produce the nonthermal decimeter synchrotron radiation.

The magnetic field is influenced by both the Sun and by the innermost Galilean satellite, Io. The satellite is covered with silicate-rock volcanoes that apparently contain enough sulfur to give the satellite its characteristic yellowish color. Some of the volcanic material escapes into space and is ionized. These ions are trapped in the magnetic field along with those that enter from the solar wind. Since the field rotates with the planet in less than 10 hours, the ions move much faster than the satellite, which orbits in 1.8 days. As the magnetic field whips past Io and slams ions into it, other atoms (including those of sulfur and sodium) are kicked away from the surface. The result is a doughnut-shaped ring, the *Io torus*, filled with an energetic mixture of ions and electrons that lies in the Jovian magnetic equator and is slightly tipped relative to Io's orbit (see Figures 9.8 and 9.9a).

In part as a result of Jupiter's fast rotation, the ionized gas spins itself out into a huge *magnetodisk* that lies roughly parallel to the Jovian magnetic equator. Like the Earth's magnetosphere, Jupiter's magnetic

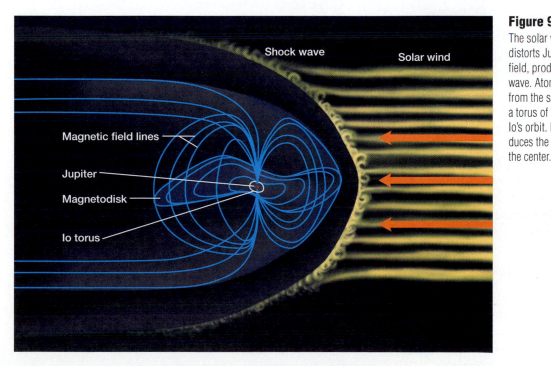

Figure 9.8

The solar wind hits and distorts Jupiter's magnetic field, producing a shock wave. Atoms knocked from the satellite Io create a torus of ionized gas in Io's orbit. Rotation produces the magnetodisk in the center.

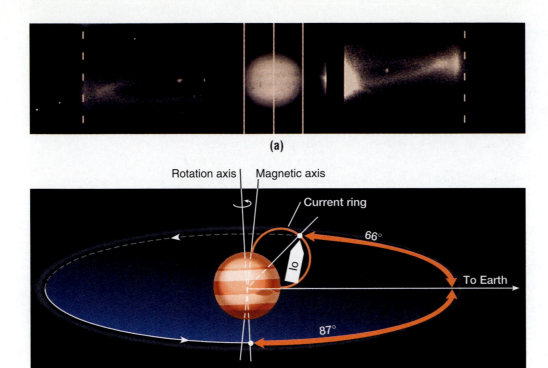

Figure 9.9

Io, less than six Jovian radii from the planet, is deep inside Jupiter's powerful magnetosphere. **(a)** A torus of ionized gas from Io caught in Jupiter's magnetic field wobbles around as the planet rotates. The bright spot to the right is an overexposed image of Europa. **(b)** A current ring associated with the decameter bursts connects Io to Jupiter. The bursts are strongest when Io is 87° ahead of the position in which it crosses (transits) Jupiter as viewed from Earth, and again 66° after transit.

environment is highly distorted by the impact of the solar wind, which compresses the magnetosphere into a shock wave (see Section 7.6) and blows the field into a long magnetic tail that stretches beyond Saturn's orbit 5 AU away!

As Io moves through, and disturbs, the Io torus, it generates a powerful ring of electrical current that flows between the satellite and Jupiter (see Figure 9.9b). The current in turn disturbs Jupiter's ionosphere and generates the long-wave decameter bursts (see Section 9.2.1), which are strongest when the satellite is 87° east (as viewed from Earth) and 66° west of the line that connects the Earth with Jupiter. The mechanism that generates the bursts is only dimly understood.

9.2.4 Jupiter's Satellites

Even binoculars will show you Jupiter's four bright moons, Io, Europa, Ganymede, and Callisto. Their positions change continuously as they orbit the planet, presenting an ever-entertaining sight. They make up a system (Table 9.2 and Figure 9.10) that stretches from Io, 400,000 km from

TABLE 9.2
The Jovian Satellites

Name	Distance (10⁶ km)	$(R_J)^a$	Period (days)	i^{oa}	e^a	Radius (km)	Mass (10²⁴ kg)	Density (g/cm³)
Metis	0.1280	1.79	0.295	0	0.00	20		
Adrastea	0.1290	1.80	0.298	0	0	12 × 8		
Amalthea	0.1813	2.54	0.498	0.4	0.00	135 × 75		
Thebe	0.2219	3.11	0.675	0.8	0.01	50		
Io	0.4216	5.89	1.769	0.04	0.00	1,815	89.4	3.57
Europa	0.6709	9.38	3.551	0.47	0.01	1,569	48	2.97
Ganymede	1.070	15.0	7.155	0.19	0.00	2,631	148	1.94
Callisto	1.883	26.3	16.689	0.28	0.01	2,400	108	1.86
Leda	11.09	155	239	27	0.15	8		
Himalia	11.48	161	251	28	0.16	90		
Lysithea	11.72	164	259	29	0.11	20		
Elara	11.74	164	259	28	0.21	40		
Ananke	21.2	297	631 r[b]	147	0.17	15		
Carme	22.6	316	692 r	163	0.21	22		
Pasiphae	23.5	329	735 r	147	0.38	35		
Sinope	23.7	332	758 r	153	0.28	20		

Satellite Features

Galileans: Outer two have silicate cores, deep icy mantles, surfaces of dirty ice; inner two have higher density, fewer volatiles, and smaller icy mantles because of proximity to Jupiter; Io is highly volcanic (sulfur) as a result of tidal heating; surface ages of satellites increase with distance as tidal heating declines.

Outer satellites: All small and rocky; two groups of four; outer group retrograde; are probably shattered captured bodies.

Inner satellites: Three are small, Amalthea intermediate size, all rocky.

---

[a]R_J, Jupiter radii; i, orbital tilt; e, orbital eccentricity.

[b]Retrograde orbit.

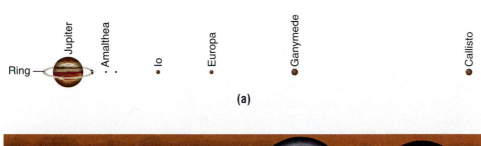

(a)

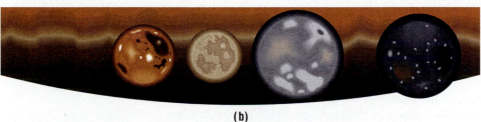

(b)

Figure 9.10
(a) The Galilean satellites (exaggerated in relative size by a factor of 6) are shown at their maximum distances from Jupiter. The four small inner satellites and the main ring are also indicated. On this scale, the two sets of outer satellites lie respectively about 0.6 and 1.2 m away from Jupiter. **(b)** The Galilean satellites are drawn to scale relative to the planet.

Jupiter (5.9 Jupiter radii, R_J), to Callisto at 1.9 million km (26.3 R_J). Io orbits in a mere 1.8 days, whereas more distant Callisto takes 16.9 days to make a circuit.

The Galilean satellites all move counterclockwise along nearly circular orbits in the plane of Jupiter's equator. Tides induced by Jupiter have made them synchronous rotators, so each keeps the same face pointed toward the planet at all times. They range in radius from 1,570 km (just smaller than that of the Moon) for Europa to 2,630 km (8% larger than that of Mercury) for Ganymede. Mutual perturbations and perturbations of passing spacecraft yield masses that range from 0.65 that of the Moon for Europa, the smallest, to double lunar for Ganymede.

The inner two satellites, Io and Europa, have densities of 3.57 and 2.97 g/cm^3 respectively, comparable to that of the Moon. They are probably made of similar material, chiefly silicate rock. However, the densities of the outer two, Ganymede and Callisto, are significantly lower, only 1.94 and 1.86 g/cm^3. They are too light to be made just of rock. About half their mass is probably frozen water—*real* ice.

Images taken by the Voyagers show that the surface ages of the four satellites correlate with distance from Jupiter. Callisto (Figure 9.11a), the farthest away, has the most ancient surface, one heavily covered with craters from the late heavy bombardment. Infrared spectra show absorption bands produced by ice. The craters are flattened, consistent with an icy exterior; at the low temperature of the surface, 150 K, ice has great strength but can still flow slowly, leading to crater deformation. The ice is far from pure. The satellite's surface is dark—a common characteristic among bodies of the outer Solar System. It seems to be coated or mixed with nonreflective carbon compounds. Callisto is probably differentiated, its ice surrounding a core of rock (Figure 9.11a inset).

Ganymede's surface (Figure 9.11b) is marginally younger. It has dark areas that, like Callisto's, are heavily cratered and are probably just as old. Ganymede has lighter regions that are less heavily cratered and younger, however, perhaps about the age of the lunar maria. The surfaces of the light areas are heavily covered with grooves (Figure 9.11b inset) that suggest the body has been slightly pulled apart. Again, infrared spectra indicate ice, consistent with the low density. Ganymede's internal structure is probably similar to Callisto's.

The two inner satellites are under the powerful influence of Jupiter. Europa (Figure 9.11c) is bright, glazed with relatively clean ice, and is the smoothest body in the Solar System, having features no more than 50 m high. Its high density, however, implies that the ice is only about 25 km deep and surrounds a large rocky core (Figure 9.11c inset). There are no large craters and the surface may be as young as 100 million years. The heat that has smoothed the surface may have liquefied the water beneath the icy skin into an ocean.

By all odds, however, Io reigns supreme in the annals of weirdness. It is covered with more than 200 volcanoes that are over 50 km across. A few

Jupiter's Galilean satellites range in radius from just smaller than that of the Moon to larger than that of Mercury.

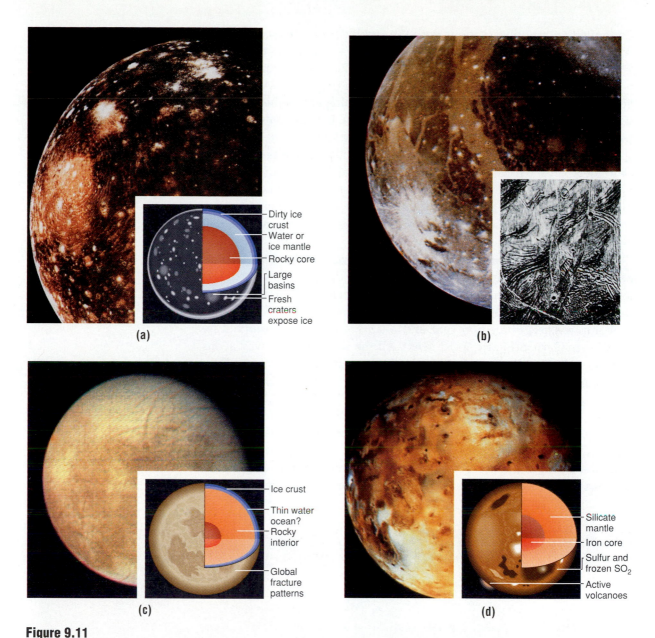

(a)

- Dirty ice crust
- Water or ice mantle
- Rocky core
- Large basins
- Fresh craters expose ice

(b)

(c)

- Ice crust
- Thin water ocean?
- Rocky interior
- Global fracture patterns

(d)

- Silicate mantle
- Iron core
- Sulfur and frozen SO_2
- Active volcanoes

Figure 9.11
(a) Callisto, an ancient body covered with impact craters, displays a huge multiringed basin called Valhalla. The satellite is differentiated (inset) into a rocky core and icy mantle. **(b)** Ganymede, the largest satellite in the Solar System, has old, heavily cratered dark regions and (inset) younger, lighter regions with grooved terrain. **(c)** Europa has a young, lightly cratered, smooth icy surface. It seems to have (inset) a rocky interior topped by a thin icy crust above a liquid ocean. **(d)** A global view of one hemisphere of Io shows many volcanic calderas. An iron core (inset) seems to be topped with a silicate mantle and a sulfurous crust.

are active and spew molten sulfur-loaded silicates that give the satellite its characteristic yellow-orange color (Figure 9.11d). Eight eruptions were seen during the *Voyager 1* flyby; six of them were still going four months later (and two more volcanoes had erupted) when *Voyager 2* made its pass.

Continued current observations with the Hubble Space Telescope reveal additional occasional eruptions. There are no impact craters, as the surface is continuously being repaved with the ejecta. The volcanic matter is expelled in fountains hundreds of kilometers high (Figure 9.12a); in addition, molten rock seems to gush from vents, running across the surface in rivers hundreds of kilometers long (Figure 9.12b). Io is the most volcanic body in the Solar System and like nothing else we know.

Io's period is almost exactly half that of Europa and a quarter that of Ganymede. Every time Ganymede makes a circuit, Europa and Io are in the same place. Io is therefore subject to a constant gravitational effect called a **resonance** that acts to pull the satellite about, and hence its orbital eccentricity and distance from Jupiter are continuously changing by small amounts.

Io is so close to Jupiter that it is subject to powerful tides. The strength of a tide depends on the inverse of the cube of the distance between the attracting bodies. As Io changes its semimajor axis, it is flexed and squeezed; the interior heats, melts, and pours through the surface. The interior probably consists of an iron core and a silicate mantle surrounded by a sulfur-rich crust (Figure 9.11d inset). All the water boiled away aeons ago. Even though Europa is farther from Jupiter, a resonance seems to have had something of the same effect and probably supplied the heat that smoothed the satellite's surface.

The progression of satellite characteristics with distance from Jupiter is reminiscent of the variation of planetary properties with distance from the Sun. The terrestrial planets, made of rock and iron, have relatively little in the way of light volatiles, substances that freeze or boil at low temperatures (see Section 7.3), because solar heat prevented their accumula-

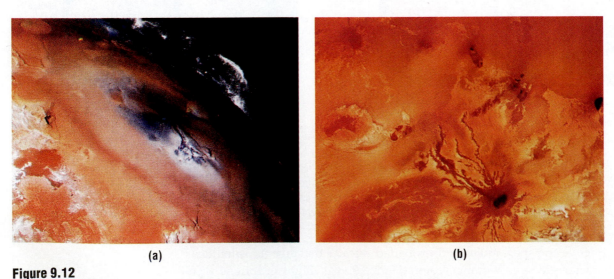

(a)   (b)

Figure 9.12
(a) Pele, one of Io's volcanoes, sends a giant plume far above the horizon of Io. **(b)** Rivers of molten rock seem to have run from the collapsed volcano Ra Patera.

tion during planetary formation. Distant Jupiter, on the other hand, is made *mostly* of volatile hydrogen. Jupiter had a similar effect on the Galilean satellites. When the planet was young its great heat prevented the incorporation of much water—hence ice—into Io and Europa, but distant Ganymede and Callisto were cold enough to accumulate and keep a great deal of it. Tidal heating subsequently caused the young-to-old progression of surface ages with distance from the planet.

A dozen other rocky satellites that belong to Jupiter are very different from the Galileans. Four lie 11.5 million km from Jupiter, and another set of four, which orbits backward, lies twice as far away. All were probably captured from interplanetary space and are approximately the same size, typically 20 km across. Four more small objects are found inside the orbit of Io. Except for Amalthea, which averages about 100 km across, they are about the same size as those in the outermost groups. Each set of four probably consists of fragments of larger bodies.

BACKGROUND 9.1   A Visit to Jupiter

A trip to Jupiter sounds interesting, but would you really want to go? The conditions make the worst of the terrestrial planets seem benign. The journey alone is arduous: like the Voyager craft, you would cruise across interplanetary space for almost two years. Still, that is survivable; after all, the crews of tiny sailing vessels regularly did much the same in the seventeenth and eighteenth centuries.

Arriving at Jupiter, where would you land? There is no accessible solid surface. Progressing inward, you would encounter the slushy beginning of a molecular-hydrogen ocean, but no person or machine could actually survive even part of the passage. There is a spot among the clouds where you could drift happily in a balloon at a comfortable temperature and pressure, although attempting to breathe the hydrogen/hydrocarbon atmosphere would quickly prove fatal. You had best go to the satellites.

Io would be a fascinating stop. You land on the side facing the giant planet, which appears 20° across in the sky, larger than the Big Dipper seen from Earth. You can easily watch the huge planet rotate. You orbit it in less than two days, the stars rising and setting much as they do on Earth, while immense Jupiter hovers stationary in your sky because of tidal locking. In the distance you may see a geyser spewing molten rock straight up, like the stream from a giant firehose, so high it disappears. To appreciate the eerie beauty of the place, however, you would have to be swathed in heavy shielding to protect you from Jupiter's energetic magnetosphere. It is doubtful, in fact, that you could take shielding heavy enough.

Proceed to the other satellites, and Jupiter retreats into the distance. From bright, shiny Europa it has shrunk to 12° across, and from grooved Ganymede to 8°. Even there, the magnetosphere's particle radiation field would probably be deadly. On ancient Callisto, the magnetosphere weakens to a point where you could survive, but for long-term safety you would need to withdraw to one of the tiny outer satellites, where the planet you want to study is about the size of the full Moon in our own heavens. Better to stay home and send a robot.

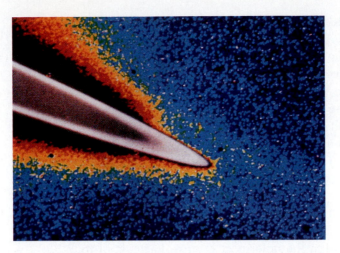

Figure 9.13
In this false-color image, Jupiter's main ring, only 30 km thick, is white (it is actually almost black). A fainter halo (here, red) begins at the inner edge of the main ring and spreads toward the planet to the north and south. Another faint ring (blue) stretches outward.

9.2.5 The Rings

The Voyagers imaged a faint set of **rings** in Jupiter's equatorial plane (Figure 9.13). The main ring is narrow and dark. At a distance of 123,000 to 129,000 km from the planet's center, it encompasses the two innermost small satellites, Metis and Adrastea, which inhabit a pair of lanes within it. Although over 250,000 km across, the ring is less than 30 km thick; it is made of countless dust grains a few thousandths of a millimeter across that orbit the planet. The individual grains are so small that electromagnetic and particle radiation either destroy them or kick them out of orbit. The result is two thicker components to the ring system, one that extends inward toward Jupiter and a broader one that stretches outward to a great distance. The dust must be replaced continuously, from the innermost satellites or possibly from as-yet-undiscovered bodies.

Jupiter opens the door to the wonders to come, as we will see upon departing for Saturn.

9.3 BEAUTIFUL SATURN

Although ringed Saturn (profiled in Table 9.3) bears a strong resemblance to its mighty relative Jupiter, there are notable distinctions caused by differences in mass, composition, and distance from the Sun.

9.3.1 The Atmosphere

Saturn's banded cloud pattern (Figure 9.14) is like Jupiter's but much fainter, and the bright equatorial zone is about twice as wide. The planet's lower gravity causes its atmosphere to be less compressed than Jupiter's, and the rate at which temperature climbs with increasing depth is consid-

TABLE 9.3
A Profile of Saturn

Planetary Data		
Distance from Sun	9.54 AU	
Equatorial radius	60,268 km = 9.45 R_{Earth}	
Polar radius	54,362 km	
Mass	5.69×10^{26} kg = 95.2 M_{Earth}	
Density	0.69 g/cm^3	
Surface (cloud deck) gravity	0.93 g_{Earth}	
Escape velocity (cloud deck)	35.5 km/s	

Layers	% Mass	Radii (km)
Rock and ice core	13%	0–16,000
Central temperature 15,000 K		
Liquid metallic hydrogen mantle		16,000–40,000?[a]
Liquid molecular hydrogen mantle		40,000?–59,000[a]
Gaseous atmosphere		59,000–60,000[a]

Atmosphere	
Composition by Number	
Hydrogen	98.5%
Helium	1.5%
Methane (CH_4)	0.01%
Ammonia (NH_3)	0.015%
Ethane	Trace
Acetylene	Trace
Phosphine	Trace
Carbon monoxide	Trace
Hydrogen cyanide	Trace
Other gases	Trace
Other	
Magnetic field	0.71 Earth
Tilt of magnetic field axis	0°
Rotation period (interior)	$10^h40^m30^s$
Axial inclination	26.7°
Effective temperature	95 K

Planetary Features

Cloud cover: Deep clouds only; no solid visible surface.

Magnetosphere: Much smaller and weaker than Jupiter's.

Surface features: Dark belts and bright zones parallel to equator (less prominent than Jupiter's); Great White Spot in northern hemisphere is seasonal storm.

Temperature: Effective temperature, 95 K, is 13 K hotter than expected because of contraction and liquid helium separation.

Winds: 1,700 km/s prograde wind in equatorial zone; otherwise, no particular correlation with belt-zone pattern.

[a]Based on the equatorial radius; size of liquid metallic hydrogen zone very uncertain.

Figure 9.14
A Voyager view of Saturn shows a weakly banded structure of zones and belts in addition to the famous rings.

erably lower (Figure 9.15a). As a result, the clouds (again, probably layers of ammonia and ammonium hydrosulfide) are formed farther down, and the contrasts are less visible. From infrared measurements we find an effective temperature of 95 K and a radiant flux 1.8 times greater than expected from solar heating alone (compared with 1.7 for Jupiter). Saturn's atmospheric composition—dominated by molecular hydrogen—is also similar to that of Jupiter's, except that it has much less helium (only 1.5%, counted by number of atoms).

The rotation period of the equatorial clouds is $10^h13^m59^s$, somewhat greater than Jupiter's. Voyager observations of Saturn's magnetic field showed the planetary interior to be rotating with a significantly longer period of $10^h40^m30^s$, the difference caused by violent equatorial winds that blow from west to east (prograde) at 1,700 km/h (Figure 9.15b). There are few east-to-west (retrograde) winds. This departure from Jupiter's pattern is a mystery.

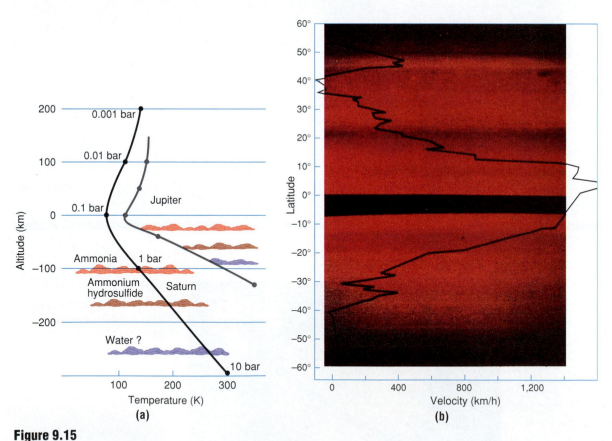

(a)   **(b)**

Figure 9.15
Saturn's banded atmosphere bears a strong similarity to Jupiter's, but with interesting differences.
(a) Saturn's visibly accessible atmosphere is much deeper than Jupiter's, the average temperature much colder, and the clouds are at lower levels. **(b)** Saturn's broad equatorial wind blows from west to east at 1,700 km/h. Another narrow, rapid jet blows to the east at about 45°N; otherwise there is little correlation with belts and zones.

Figure 9.16
The Hubble Space Telescope captured the detail of Saturn's Great White Spot in 1990 after it had already been blown a good way around the planet by winds. The upper edge is sculpted where the spot runs into a region of lower wind velocity. The image is in false color to enhance contrast.

There are a number of small, turbulent ovals set within the clouds, but no feature analogous to Jupiter's Great Red Spot. Saturn's claim to stormy fame is its *Great White Spot* (Figure 9.16), a reflective cloud feature last witnessed in 1990 that develops for a few months every thirty years during northern-hemisphere summer. Wrapped around the planet by winds, it appears to be caused by a bubble of ammonia gas that rises under the warmth of the summer Sun. Similar, though smaller, storm systems are also occasionally seen.

9.3.2 Inside Saturn

Saturn's low density of only 0.7 g/cm^3 is much smaller than Jupiter's. Gravitational compression causes a giant hydrogen-dominated planet's radius to change only slowly with mass. If you could add a large amount of mass to Saturn, you would also increase its gravity. Instead of a significant expansion in radius, there would be an increase in density. Triple the mass and the radius goes up by only about 20%, whereupon you have another Jupiter, which is nearly the largest size physically allowed. After a point, increasing the mass actually would cause such a planet to *shrink*.

Saturn's polar radius is 10% smaller than its equatorial radius, making it more oblate than any other planet in the Solar System. The principal cause is its rapid spin. Its lower density also helps it to bulge more at the equator than does Jupiter. For it to be *so* oblate, however, it must also be less centrally condensed than Jupiter, with a larger "rock" and "ice" core that contains the inner quarter of its radius (about 16,000 km) and 13% of its mass (Figure 9.17). Because of lower internal pressure, the inner mantle of liquid metallic molecular hydrogen is expected to be notably thinner

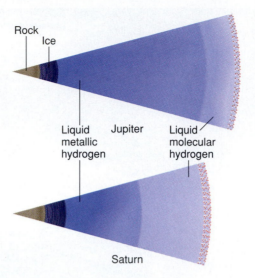

Figure 9.17

Saturn's rock and ice core is larger than Jupiter's, occupying roughly the inner 25% of the radius. The metallic hydrogen layer, on the other hand, is smaller, though its extent within either planet is uncertain.

than Jupiter's, stretching less than three-quarters of the way out from the center. As a result, Saturn's magnetic field is relatively weaker and its magnetosphere is only one-third the size of Jupiter's and more compressed by the solar wind. The upper part of the mantle, nonmetallic liquid H_2, extends most of the rest of the way to the surface. At the top is a gaseous molecular hydrogen atmosphere thicker than Jupiter's, the result of lower gravitational compression.

A mathematical model of Saturn reveals that its mass and radius are consistent with a gross composition of 92% hydrogen and 8% helium (by number of atoms), close to the Jovian and solar values. Saturn produces relatively more internal heat than does Jupiter, and must therefore have another interior energy source in addition to gravitational compression. Given the temperature and pressure of Saturn's deep interior, helium should condense into droplets. As the heavier helium falls inward, it releases gravitational energy: Saturn may glow in the infrared in part because of a deep helium rain, explaining the great depletion of atmospheric helium.

9.3.3 The Rings

Saturn's glory lies in its rings (Figure 9.18 and Table 9.4). Though first seen by Galileo in 1610, their true nature was not realized until observed by Christiaan Huygens in 1659. Three rings—A, B (the brightest), and delicate C—are easily visible from Earth, together stretching 274,000 km across, a bit more than twice the diameter of Saturn's planetary disk. The A and B rings are separated by a dark band called the **Cassini division** (discovered by Giovanni Cassini in 1675). The outer part of the A ring is also split by a narrow dark band called the *Encke division*. The D ring is barely visible from Earth, and the others were discovered by the Voyagers.

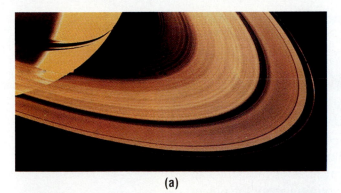

(a)

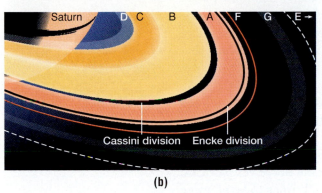

(b)

Figure 9.18
Saturn has seven major named rings. Rings A and B (separated by the Cassini division), C, and narrow F are seen in **(a)** this Voyager image and **(b)** colored in the map. Rings D, G, and E are indicated on the map but are not evident in the image. The Encke division is dark and clear in the A ring. Each ring divides into hundreds of ringlets. Note that Saturn can easily be seen through the rings.

TABLE 9.4
Saturn's Rings

Name	Distance 10^3 km	Distance R_S[a]	Width (km)	Thickness (km)	Mass M_S[b]
D	67–75	1.11–1.24	7,500		
C	74–92	1.24–1.52	17,500		2×10^{-9}
B	92–118	1.52–1.95	25,500	0.1	50×10^{-9}
Cassini division	118–122	1.95–2.02	4,700		1×10^{-9c}
A	122–137	2.02–2.27	14,600	0.1	10×10^{-9}
Encke division	137	2.26	35		
F	140	2.34	30–500		
G	165–174	2.75–2.88	8,000	100–1,000	10^{-20}
E	180–480	3–8	300,000	1,000	

Ring Features

Three main broad rings only a few tens of m thick; several fainter ones, both broad and narrow; narrow F ring shepherded by satellites; Cassini division between A and B rings caused by satellite resonance; Encke division in A ring swept clean by satellite; thousands of ringlets, rings composed of dust and ice-coated rocks a few cm to a few m across.

[a]Saturn radii.

[b]Saturn masses.

[c]Even though a gap, the Cassini division contains ring particles.

All have different characteristics. The outer E ring extends as far as eight Saturn radii from the planet's center, while the F ring is only a few kilometers wide.

The rings sit exactly above Saturn's rotational equator and cast a prominent shadow on the clouds below. The whole system is tilted through an angle of 26.7°. As the planet orbits the Sun we look at the rings alternately from the north and south. Twice during the 29.5-year sidereal orbital period, the rings are presented edge-on, whereupon they nearly disappear from view, revealing their extreme thinness (Figure 9.19).

In 1857, James Clerk Maxwell used the principles of mechanics to demonstrate that if the rings were solid, their revolution about the planet would tear them apart; they must, therefore, consist of finely divided material. Doppler shifts in fact reveal that the outer ring revolves more slowly than the inner one, in accord with Kepler's laws, establishing that they are made of billions of orbiting moonlets.

In spite of what we already knew, no one was prepared for the spectacular views from the Voyagers. Each main ring is divided into hundreds of small **ringlets** (see Figure 9.18), each only a few hundred kilometers across, and these are subdivided into still smaller ringlets (Figure 9.20a) a mere 20 or so kilometers wide. Ringlets even fill the Cassini division, which is not empty at all, just a region with substantially fewer particles. More astonishing, the Voyagers showed the A and B rings to be no more than a few tens of meters thick. The C ring and the Cassini division may be thinner yet. Only in the outer E and G rings does the thickness increase to the order of a thousand kilometers.

As a result of constant collisions, which disorganize their motions, the particles making up the A, B, and C rings must be several times smaller than the ring thicknesses. (The A, B, and C rings are those easily visible from Earth.) Ring transparencies determined from Voyager experiments

When Saturn's rings are edge-on, they nearly disappear from view.

Figure 9.19
Saturn's rings practically disappear, even to the Hubble Space Telescope, when they are edge-on to the line of sight. Only with great enhancement can a strip be seen, as in the box. The shadow they cast on the planet, however, is clearly evident. Bright satellites are seen to the left. When the rings are edge-on, very faint ones pop into view.

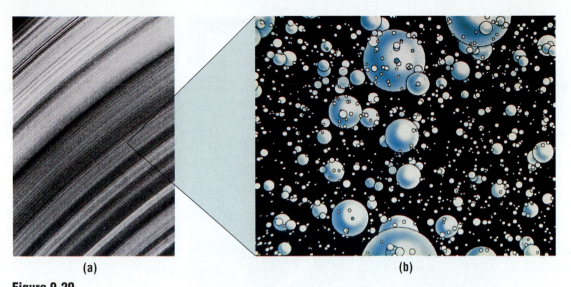

(a)   (b)

Figure 9.20
The rings are resolved into successively smaller units. **(a)** A high-resolution image of the B ring shows
that the ringlets are subdivided into yet more ringlets. **(b)** A computer simulation of a volume only 3 m
across taken from within a ringlet shows hundreds of dirty iceballs, most only a few centimeters wide.

reveal a distribution of particle diameters from approximately 1 cm to perhaps 10 m, into which fine dust must be mixed (Figure 9.20b). The particles are only a few centimeters apart; no spacecraft could survive a journey through the rings. Infrared observations show that the ring particles
are made partly of water ice or, at least, are coated with it, making the
rings bright and reflective. The much thicker E and G rings must be made
of fine dust to have spread out so far from the main ring plane.

Why are the main rings so thin? The particles are so close to one another that they collide constantly. A particle may therefore attain sufficient
energy to be launched out of the plane into an inclined orbit. When it re-
crosses the plane, it will invariably collide with another iceball, and the
subsequent collision lowers its inclination and returns it once again to the
plane. Individual moonlets cannot escape, and the result is a ring of extreme thinness.

The broad Cassini division is caused by a perturbing resonance (see
Section 9.2.4) with the satellite Mimas. The period of each particle at the
inner edge of the division is exactly half that of the satellite, so ring particles that wander into the Cassini division tend to be kicked back out of it.
By contrast, the Encke division is directly swept free of particles by the
tiny moon 1981 S13. However, the origins of the myriad ringlets are not
known. The breaks do not fit any resonances, and there are too many of
them. Small, unseen embedded satellites a kilometer or so across may be
involved. The strikingly narrow F ring (Figure 9.21) should diffuse outward with time, but two **shepherd satellites** flank it and keep the particles
in line.

**The rings may be relics of the
breakup of one or more larger
bodies.**

Figure 9.21
The narrow, braided F ring is kept together by the shepherd satellites shown in the inset.

The rings are inside a radius surrounding Saturn called the **Roche limit.** Within it, tides are strong enough to tear a fluid body apart. The effect on a solid body depends on its distance from the planet and on its size and strength. If it is large, weak, and sufficiently close to its parent planet, the stretching force induced by the tide can be greater than the mechanical and gravitational forces holding it together, and the body will be ripped to pieces. The rings are inside Saturn's Roche limit. They could be the remnants of an ancient dusty disk that developed around Saturn during the planet's formation. Satellites coalesced in the outer part of the disk, but within the Roche limit tides would have been too strong to allow a satellite to accumulate from the smaller particles. The icy ring particles, however, should darken with time under the impact of dust. Their brightness suggests a more recent origin. The rings may therefore be relics of the breakup of larger bodies, perhaps a moon or two or an icy interplanetary interloper that strayed too close and either shattered or struck and smashed a satellite. Continued impacts among the fragments would then have reduced them to small particles and dust.

9.3.4 Saturn's Satellites

Saturn has the largest number of known satellites (not counting ring particles), at least 18 (Table 9.5). Observations made with the Hubble Space Telescope during the 1995 ring-plane crossings revealed several other

TABLE 9.5
The Saturnian Satellites

Name	Distance (10^6 km)	(R_S)[a]	Period (days)	i^{oa}	e^a	Radius (km)	Mass (10^{24} kg)	Density (g/cm^{-3})
1981 S13[b,c]	0.137	2.26	0.60			20		
Atlas[c]	0.1376	2.28	0.60	0	0	20 × 15		
Prometheus[c]	0.1380	2.31	0.611	0	0.00	70 × 40		
Pandora[c]	0.1417	2.35	0.63	0	0.00	55 × 35		
Epimetheus	0.1514	2.51	0.69	0.34	0.01	10 × 50		
Janus	0.1515	2.51	0.70	0.14	0.01	110 × 80		
Mimas	0.1855	3.08	0.94	1.53	0.02	195	0.038	1.17
Enceladus	0.2380	3.95	1.37	0.02	0.00	250	0.084	1.24
Tethys	0.2947	4.89	1.89	1.09	0.00	525	0.755	1.26
Telesto[c]	0.2947	4.89	1.89	0	0	12		
Calypso[c]	0.2947	4.89	1.89	0	0	15 × 10		
Dione	0.3774	6.26	2.74	0.02	0.00	560	1.05	1.44
Helene	0.3774	6.26	2.74	0.2	0.01	18 × 15		
Rhea	0.5270	8.75	4.52	0.35	0.00	765	2.49	1.33
Titan	1.222	20.3	15.95	0.33	0.03	2,575	135	1.88
Hyperion	1.481	24.6	21.28	0.43	0.10	175 × 100		
Iapetus	3.561	59.1	79.33	14.7	0.03	720	1.88	1.21
Phoebe	12.952	215	550.5	175[d]	0.16	110		

Satellite Features

Titan: The one large, Galilean-type moon; thick, opaque nitrogen atmosphere denser than Earth's; hydrocarbons.

Smaller satellites: Varying properties; several satellites inside Titan's orbit are resurfaced, though Mimas is not; icy Enceladus is the brightest body in solar system; Dione is cratered on one side, shows evidence for activity on the other; Hyperion irregular; Iapetus bright on one side, dark on the other.

[a] R_S, Saturn radii; *i*, orbital tilt; *e*, orbital eccentricity.

[b] Temporary designation.

[c] Discovered by Voyager team members.

[d] Retrograde and captured.

small bodies near the rings, their number not yet sorted out. There is one Galilean-type satellite, the huge moon Titan. With a radius of 2,575 km, it is just smaller than Ganymede. The other confirmed satellites range in radius from Rhea (765 km) down to blocks of ice and rock only a few tens of kilometers across. Orbital radii (Figure 9.22) range from 2.26 Saturn radii (R_S) for the satellite that sweeps the Encke division (1981 S13) through Titan at 20 R_S to Phoebe at 215 R_S (which orbits backward and was probably captured).

Masses and densities have been determined for seven of the satellites. Since Titan's density is similar to those of the outer two Galilean satellites of Jupiter, it is probably about half water ice. The densities of the others are

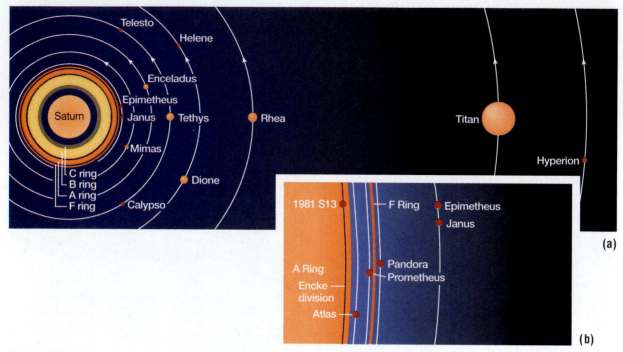

Figure 9.22

(a) The orbits of Saturn's satellites are shown out to Hyperion. Iapetus is more than twice as far as Hyperion, Phoebe (which orbits backward) almost nine times as far. The larger satellites are indicated in orange, their sizes magnified by a factor of 20 relative to Saturn. The smaller satellites are represented by red dots. **(b)** The diagram is magnified by a factor of 17 to show the inner satellites.

significantly lower. They may be as much as 65% water ice, the remainder likely some kind of silicate rock.

A sampling shows that each of the larger satellites has unique properties. Mimas (Figure 9.23a) is a battered body with an ancient surface much like that of Callisto. Cold, strong ice has preserved the craters, including one with a radius two-thirds that of Mimas itself, making the satellite look like George Lucas's "Death Star." Though next door, Enceladus (Figure 9.23b) is the most reflective body in the Solar System, and parts of it seem to have been paved in smooth, fresh ice by eruptions from watery volcanoes. The leading side of Dione (Figure 9.23c), which faces in the direction of its orbit, is heavily cratered, but the trailing side looks as though it has been resurfaced. The variations in the characteristics of these (and other) satellites are probably caused by ices with low melting points and by different levels of tidal heating, each related to differences in size and distance from Saturn.

The satellites outside Titan's orbit are strange. Hyperion (Figure 9.23d) is the largest irregular body known in the Solar System. It is distin-

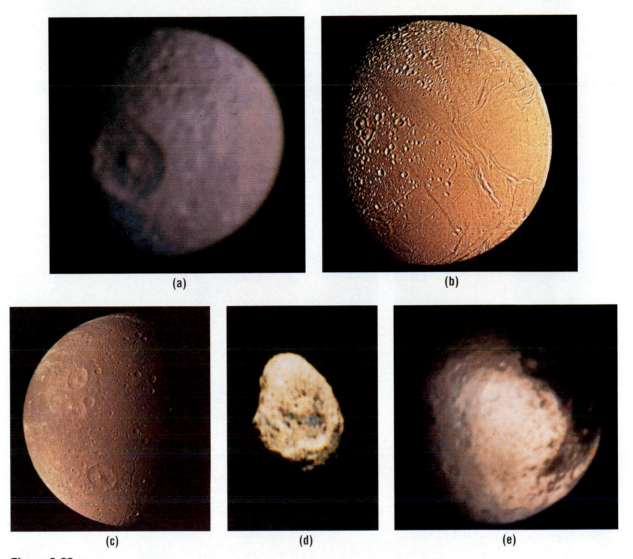

Figure 9.23
(a) Mimas is battered; **(b)** Enceladus is bright and icy. The huge plain at the lower right on Enceladus lacks visible craters and must be very young. **(c)** Dione is cratered but displays some signs of activity. **(d)** Hyperion is highly irregular. **(e)** Iapetus is dark on the leading side and bright on the trailing.

guished by chaotic rotation caused by gravitational perturbations and may actually have no rotation period. Not to be outdone, distant Iapetus (Figure 9.23e) has very two different faces, a fact noticed by its discoverer, Cassini, in the 1600s. The leading hemisphere is dark, the trailing one 10 times brighter. The dark material appears to be similar to a tar sand (sand laced with thick hydrocarbon tar). It is apparently being knocked off even more distant Phoebe by collisions, then swept up the leading side of synchronously rotating Iapetus.

Titan is the most intriguing of the satellites. Even in the 1940s, methane bands were seen in its spectrum, suggesting that it might have a fairly thick atmosphere. *Voyager 1* revealed a featureless yellowish haze so dense that its cameras could see nothing of Titan's surface (Figure 9.24a). Infrared observations from Earth, however, have dimly punched through the haze (Figure 9.24b), revealing surface features of some kind and establishing that the satellite is (as expected) in synchronous rotation, keeping one face always pointed toward Saturn.

Models constructed from Voyager data reveal that Titan's air is mostly—perhaps 80% or more—molecular nitrogen. The remainder is a mixture in unknown proportions of methane (CH_4) and maybe argon. The

(a)   (b)

(c)

Figure 9.24

(a) Titan displayed only a thick atmosphere to *Voyager 1*. **(b)** An infrared view taken with the Hubble Space telescope penetrates the haze to show surface features that have high reflectivities. **(c)** The artist depicts a cold nitrogen-hydrocarbon ocean that may wash Titan's mysterious shores.

surface pressure is actually somewhat greater than that of Earth's, but the temperature is a bone-chilling 95 K, and all the oxygen is trapped in ice. Chemical and photochemical reactions produce a remarkable array of hydrocarbons that include ethane, propane, ethylene, and methylacetylene (C_3H_4). The low spots of Titan may even be filled with liquid nitrogen/hydrocarbon seas (Figure 9.24c). The composition of Titan's atmosphere is largely a product of its distance from the Sun and the resulting low temperature. If we could warm the satellite and sublime the ice, oxygen freed by sunlight from the water vapor would convert the methane to carbon dioxide, making the satellite's atmosphere more like that of Mars.

The Voyagers are long gone, leaving a superb legacy of surprises and puzzles, mysteries that will be difficult to solve from Earth. Ground-based observations and those made with the Hubble Space Telescope are beginning to fill in some of the blanks associated with both planets, as are dedicated spacecraft. *Galileo* will spend two years in orbit about Jupiter studying the giant planet and its satellite system, and *Cassini*, now under construction, should arrive to orbit Saturn early in the next century, both adding more to our knowledge of the worlds we have directly explored.

▶ KEY CONCEPTS

Belts and **zones**: Jovian features, respectively, strips of darker, lower-altitude, higher-temperature clouds and strips of lighter, higher-altitude, lower-temperature clouds.

Cassini division: A resonance gap that divides Saturn's A and B rings.

Decameter bursts: Bursts of radio radiation from Jupiter, caused by the action of an electrical current ring that connects the planet and Io.

Differential rotation: Rotation in which different parts of a body rotate with different speeds and therefore with different periods.

Galilean satellites: Jupiter's four largest satellites; discovered by Galileo in 1609.

Great Red Spot (GRS): A huge reddish zone between Jupiter's south equatorial belt and south tropical zone.

Hydrocarbons: Chemical compounds based on hydrogen and carbon.

Hydrostatic equilibrium: A condition in which the upward push of pressure is balanced by the downward pull of gravity.

Model: A mathematical description of a physical system (for example, the changes of temperature, density, and pressure with depth) that allows predictions of observations.

Resonance: A gravitational phenomenon in which one orbiting body produces a large gravitational perturbation in another because one has a period that is a simple fraction of the other.

Ringlets: Small rings a few hundred km wide that make up Saturn's big rings.

Rings: In the context of the Jovian planets, encompassing belts of dust and debris.

Roche limit: A limit surrounding a planet within which a fluid body would be torn apart by tides.

Rock and **ice:** In the context of the Jovian planets, respectively a mixture of heavier materials like silicates and iron and a mixture of lighter volatiles like water, methane, and ammonia.

Shepherd satellites: Small satellites that organize and preserve narrow rings.

Synchrotron radiation: Radiation produced by fast electrons spiraling in a magnetic field.

Thermal radiation: Radiation produced as a result of heat, for example, blackbody radiation; **nonthermal radiation** is radiation produced by processes that are not the result of heat and cannot be related to temperature.

▶ **EXERCISES**

Comparisons

1. Compare the origins of centimeter, decimeter, and decameter radio radiation from Jupiter.

2. What are the periods and origins of Jupiter's different observed rotation periods?

3. How does synchrotron radiation differ from thermal radiation?

4. Compare the heights and temperatures of Jupiter's zones and belts.

5. Compare the different kinds of Jovian clouds.

6. How do Io and Europa contrast with Ganymede and Callisto?

7. Compare Saturn's winds with those of Jupiter.

8. How does Saturn's Great White Spot differ from Jupiter's Great Red Spot?

9. How is the origin of Saturn's internal heat different from that of Jupiter's?

10. Compare the natures and origins of the Cassini division and the Encke division.

Numerical Problems

11. Use the orbital periods and semimajor axes of Jupiter's satellites to show that Kepler's third law works for satellite systems as well as for the planetary system.

12. The Great Red Spot rotates around its center relative to the surrounding clouds. What is the speed of rotation at its edge? (Assume it is circular.)

13. Give the period of a hypothetical satellite that is in orbital resonance with Callisto.

14. If you were to lay Saturn's main rings on a field 100 m in diameter, what would their thickness be?

Thought and Discussion

15. How do Jupiter's belts and zones relate to its winds?

16. How does hydrostatic equilibrium keep Jupiter from expanding or contracting?

17. Why does Jupiter radiate more energy than it gets from the Sun?

18. What phenomenon produces synchrotron radiation?

19. Where do you expect Jupiter's magnetic field to be generated?

20. What is meant by an orbital resonance?

21. What features make Io unique?

22. Why do the internal and external characteristics of the Galilean satellites change with distance from Jupiter?

23. Why is Saturn more flattened at the poles than Jupiter?

24. Why is Saturn's density so low compared to Jupiter's?

25. How can you tell that Saturn's rings are not a solid sheet?

26. What keeps Saturn's F ring in place?

27. What is the significance of the Roche limit?

28. What hinders a clear view of Titan's surface?

Research Problem

29. Galileo discovered the four big satellites of Jupiter. Who discovered the others? When? With what equipment?

Activities

30. Use a telescope to make observations of Jupiter. Make a sketch of the cloud-belt pattern, including the location of the Great Red Spot. Chart the motions of the satellites until you can identify them. Estimate their periods and their distances from Jupiter.

31. Draw to scale the orbits of Jupiter's outer satellites (relative to those of the Galileans) and show their orbital inclinations.

32. Examine Saturn through a telescope and make a drawing. From the known angular and physical diameters of Saturn's disk, estimate the angular and physical diameters of the rings and of the Cassini division, and the distance of Titan from the planet.

Scientific Writing

33. You are an interviewer on a nationally syndicated television program, and a famous planetary astronomer is to be your guest. Write a script containing 20 questions that will illuminate the subjects of Jupiter and Saturn for your audience.

34. A high school science teacher writes you a letter asking how to make a scale model of the planet Saturn. Write a return letter in which you describe in some detail how to do it. Pick a reasonable scale for Saturn (say, a baseball), and tell where all the parts should be placed. The object is to provide a vivid demonstration for the students.

OUTER WORLDS

Uranus, Neptune, and Pluto float in the cold depths at the end of the planetary system

10

We now reach out to three cold worlds that further mystify and delight as they help illuminate the constitution and origin of the planetary system.

10.1 TWINS AND AN ODDBALL

Our Earthly view of these far worlds is poor indeed. Uranus (Figure 10.1a) presents a tiny disk just 4 seconds of arc across, and little surface detail is seen from Earth. Neptune (Figure 10.1b), with a disk half as large, is even harder to examine. Most of what we know comes from *Voyager 2*, which visited the planets in 1986 and 1989 respectively. Pluto (Figure 10.1c), a mere 0.1 second of arc across, is directly resolvable only with the Hubble Space Telescope.

Figure 10.1

(a) Uranus's small disk is surrounded by five satellites. Because of Uranus's 98° axial tilt, their orbital poles here point nearly at Earth. **(b)** Neptune is seen with its satellite Triton. **(c)** In Earthbound telescopes, Pluto is indistinguishable from a star.

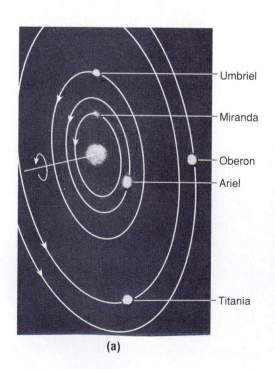

- Umbriel
- Miranda
- Oberon
- Ariel
- Titania

(a)

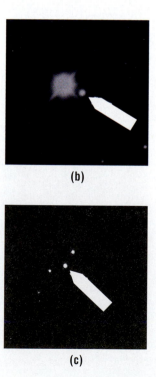

(b)

(c)

Uranus and Neptune are very nearly twins, with equatorial radii of 25,600 km and 24,800 km respectively, four times as big as Earth. Uranus has five medium-sized moons and Neptune has a big one that fits the class of Galileans (see Figure 10.1a). The satellites' orbits and gravitational effects on *Voyager 2* yield respective masses of 14.5 and 17.1 Earth masses—large, but between those of Earth and Jupiter. Observations of Pluto's single satellite, Charon (described below), allowed us to measure the planet's tiny radius of 1,190 km (0.19 Earth radii) and mass of only 0.002 Earth masses. It is by far the smallest planet in the Solar System and nothing like Uranus and Neptune. Indeed, it may be fair not to call it a planet at all.

10.2 CLOUDY ATMOSPHERES

In spite of superficial similarities, Uranus (Table 10.1) and Neptune (Table 10.2) are quite different from Jupiter and Saturn. (Pluto is examined by itself in Section 10.6.) Magnetic field measurements by *Voyager 2* give internal

We suspect that Uranus was knocked over in a giant collision.

TABLE 10.1
A Profile of Uranus

Planetary Data		*Atmosphere*[c]	
		Composition by Number	
Distance from Sun	19.19 AU	Hydrogen	90%
Equatorial radius	25,559 km = 4.01 R_{Earth}	Helium	9%
Polar radius	24,974 km	Methane (CH_4)	1%
Mass	8.68×10^{25} kg = 14.5 M_{Earth}	Other hydrocarbons	Trace
Mean density	1.29 g/cm³		
Surface (cloud deck) gravity	0.79 g_{Earth}	*Other*	
Escape velocity (cloud deck)	21.3 km/s	Magnetic field	0.74 Earth
		Tilt of magnetic field axis	58.6°[b]
Layers	*Radii (km)*	Rotation period (interior)	17^h 14^m
Rocky core	0–7,500	Axial inclination	97.9°[a]
Possible liquid water mantle	7,500–25,000	Effective temperature	59 K

Planetary Features

Cloud cover: No clouds; apparent surface consists of hydrocarbon haze.

Magnetosphere: Large but weak.

Surface features: Almost featureless; near-invisible bands, no belts or zones; a few barely visible clouds near the equator.

Temperature: Effective temperature of 59 K, same as expected from solar heating; little or no internal heat source.

Winds: Reverse pattern from Jupiter and Saturn; retrograde at least 350 km/h near equator; prograde at 600 km/h at higher latitude.

[a]Retrograde rotation.

[b]Offset by 30% from center.

[c]Relative to number of hydrogen atoms, not molecules.

TABLE 10.2
A Profile of Neptune

Planetary Data		Atmosphere[b]	
Distance from the Sun	30.06 AU	**Composition by Number**	
Equatorial radius	24,764 km = 3.88 R_{Earth}	Hydrogen	87%
Polar radius	24,343 km	Helium[c]	12%
Mass	1.02×10^{26} kg = 17.1 M_{Earth}	Methane (CH_4)	0.5%
Mean density	1.64 g/cm^3		
Surface (cloud deck) gravity	1.12 g_{Earth}	**Other**	
Escape velocity (cloud deck)	23.3 km/s	Magnetic field strength	0.43 Earth
		Tilt of magnetic field axis	46.8°[a]
Layers	**Radii (km)**	Rotation period (interior)	16^h 07^m
Rocky core	0–7,000	Axial inclination	29.6°
Possible liquid water mantle	7,000–24,000	Effective temperature	59 K

Planetary Features

Cloud cover: Several layers amidst a deep blue hydrocarbon haze.

Magnetosphere: Large but empty because of distance from the Sun.

Surface features: Occasional dark storm features and bright cirrus clouds.

Temperature: Effective temperature, 59 K, is 13 K above solar heating value, implying internal heat source.

Winds: Like Uranus, pattern the reverse of Jupiter's and Saturn's; retrograde at 1,600 km/h in equatorial region; prograde at higher latitudes.

[a]Offset 55% from center.

[b]Relative to number of hydrogen atoms, not molecules.

[c]Uncertain, but a high percentage, like that found in Uranus.

rotation periods of 17^{h}14^m for Uranus and 16^{h}07^m for Neptune, notably longer than those of the two giants. Neptune's axis is tilted through a normal 30°, but Uranus and its whole satellite system is oddly tipped on its side through 98° (see Figure 10.1a). Like Venus, Uranus rotates backward. We suspect that Uranus was knocked over in a giant collision with another body sometime in the early days of the Solar System.

Uranus at first showed *Voyager 2* merely an attractive blue-green disk (Figure 10.2a). Only when the color is enormously enriched by computer (Figure 10.2b) can we discern a pattern of stripes running parallel to the equator. Immense contrast enhancement eventually also showed some whitish clouds near the equator that moved along with rotation (Figure 10.2c). Neptune, on the other hand, displays more obvious belts and bright elongated clouds (Figure 10.3). Voyager observations revealed the equatorial winds of Uranus to be blowing at 350 km/h (similar to Jupiter's wind speed), Neptune's at a powerful 1,700 km/h (comparable to Saturn's). However, the equatorial wind directions are retrograde, *opposite* those of the giant planets.

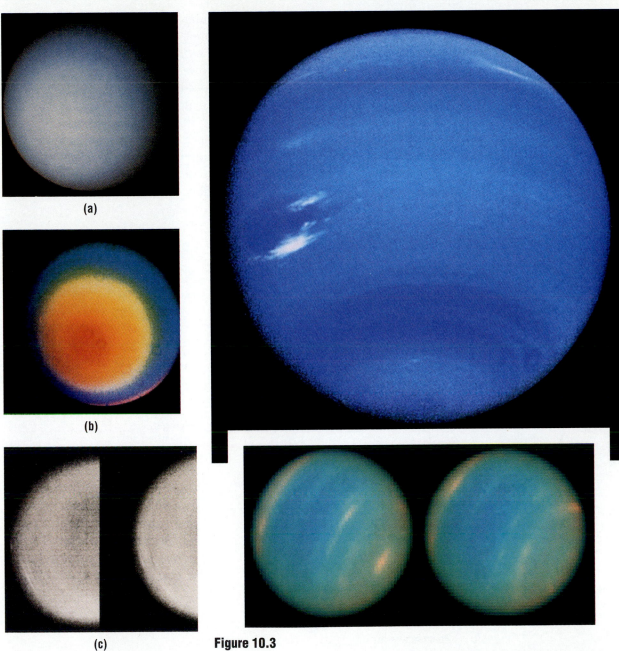

Figure 10.2

(a) A hypothetical astronaut aboard *Voyager 2* would have seen Uranus as a blue-green sphere with no features to mar its surface. **(b)** Great color enhancement shows bands of clouds encircling the south pole and parallel to the equator, which lies near the edge of the disk. **(c)** Contrast enhancement shows white clouds near the equator that rotate with the planet.

Figure 10.3

Neptune has familiar cloud belts and whirling cloudy ovals. At left center is the Great Dark Spot. Surrounding its lower edge is a zone of high, wispy clouds that streak out to the right. Toward bottom right is another, smaller, dark spot. More recent images (inset) made by the Hubble Space Telescope of opposite hemispheres of Neptune show that the dark spots have gone away.

The atmospheres of Uranus and Neptune are dominated by molecular hydrogen, and like Jupiter contain a solar mix of helium (8% for Uranus and 10% for Neptune). No helium condensation has taken place (as it has for Saturn), as expected for planets this size. Methane abundances, however, are some five and ten times Jovian for Neptune and Uranus respectively. Methane is chiefly responsible for the planets' colors, since it strongly absorbs red sunlight, leaving only blue to be reflected.

Atmospheric temperatures are very low as a result of great distance from the Sun. Neptune's minimum temperature (Figure 10.4) is only about 50 K, as is Uranus's. As a result, the highest clouds are made of frozen methane (CH_4) rather than ammonia. Below them are probably

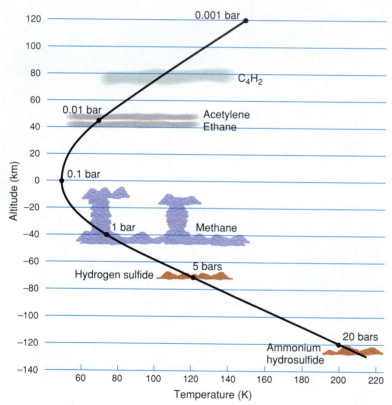

Figure 10.4

Neptune's atmospheric temperature profile looks much like those of the other Jovian planets. (Uranus's minimum temperature is similar.) The dominant clouds (40 km below the reference level at a temperature of 80 K) are made of methane. Rising columns may produce clouds and ovals. Below the methane clouds may be clouds of hydrogen sulfide (70 km down at 120 K) and ammonium hydrosulfide (120 km down at 220 K). Above them are high, smoggy hazes of various hydrocarbons.

layers of frozen hydrogen sulfide (H_2S) and ammonium hydrosulfide. Uranus's clouds are smoothly layered, whereas those of Neptune appear to form rising plumes (see Figure 10.4). Above the clouds are hydrocarbon hazes. Kinetic temperatures in thin outer atmospheres climb to 800 K.

Like Saturn, Uranus has no ovals akin to Jupiter's Great Red Spot (GRS). *Voyager 2* showed Neptune to be dominated by a huge dark oval near latitude 20°S, the *Great Dark Spot* (GDS) (see Figure 10.3). It was about half the diameter of the GRS and also spun counterclockwise, but with a 16-day period. The GRS, however, was trapped and rolled between winds that blow in opposite directions; the GDS was in a band of steady retrograde winds that flowed to the west. It is thought to have been the top of a plume of upwardly convecting Neptunian air that had a deep but unknown origin. The southern edge of the GDS was bordered by a line of high bright clouds. Farther down, at 55°S, was another dark oval. Observations taken with the Hubble Space Telescope in 1994 (Figure 10.3 inset) show that both dark spots disappeared sometime within the last five years, confirming that the GDS was really quite different from the GRS.

10.3 INTERIORS

The planetary interiors also separate Uranus and Neptune from Jupiter and Saturn as different kinds of worlds. Their respective densities of 1.29 and 1.64 g/cm^3 indicate that they are made of heavier materials. Because of gravitational compression, the radius of a hydrogen/helium planet like Jupiter or Saturn changes only slowly with mass (see Section 9.3.2). Although Saturn has a mass less than a third that of Jupiter, its radius is almost as large. If Uranus and Neptune were made mostly of hydrogen and helium like the two big Jovians, we would expect them to be only slightly smaller than Saturn, yet each is less than half Saturn's size. They must therefore be made largely of denser matter, most likely water and rock, that causes them to contract gravitationally, (see Section 9.2.3): in spite of the atmospheric compositions, only a sixth of their total masses is hydrogen. The equatorial bulges also indicate lower central concentrations of mass. The planets may have rocky cores some 7,000 to 8,000 km in radius surrounded by deep mantles of liquid water (plus methane and ammonia), topped by cloudy atmospheres of light hydrogen and helium.

Uranus has no measurable flow of internal heat.

Like Jupiter and Saturn, Neptune (with an effective temperature of 59 K) radiates more energy—2.6 times as much—than it would if it were heated only by the Sun. Remarkably, however, even though Uranus is closer to the Sun and receives more solar heat, its effective temperature is the same. In spite of the close similarity between the two planets, Uranus has *no* measurable flow of internal heat, a unique condition among the giant planets.

Since the masses of these planets are too low to allow the internal structure needed to separate liquid helium from hydrogen (see Section 9.2.3), any internal heat must be that left over from the planets' formations. Either Uranus has already dissipated its internal heat, or the heat flow is being blocked by outer layers. Its smooth cloud deck is consistent with no heating from below and with a tightly layered structure that may block convection. In turn, Neptune's cloud belts, bright cloud strips, and

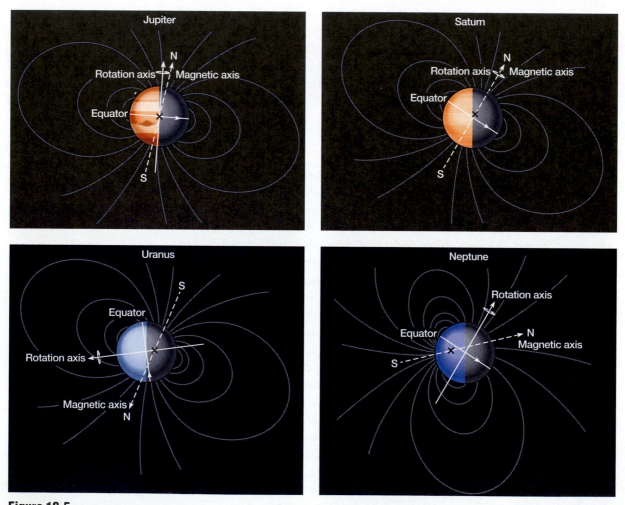

Figure 10.5
The magnetic fields of Uranus and Neptune are oriented and centered (at the X) quite differently from those of Jupiter and Saturn.

plumes are consistent with rising warmed columns of air related to the outpouring of its internal heat. Uranus's high axial tilt and the resulting extreme seasons may cause its different atmospheric structure. The south pole is now well heated. As Uranus orbits and sunlight again heats the equator, the structure may change and the internal heat—if there is any—may be released. We will have to wait years to see.

Voyager 2 found magnetic field strengths of Uranus and Neptune to be respectively about 74% and 43% Earth's field strength, roughly similar to Saturn's. However, Uranus's dipole is steeply tilted by 59° relative to its rotation axis; it is not even approximately centered, but is a third of the way to the planetary surface (Figure 10.5). Neptune's magnetic dipole is tilted by 47°, not quite so much; but is centered even farther out, over half way out toward the surface. The strange fields enhance the similarity of Uranus and Neptune. They are probably related to the planets' internal constructions, but no one knows how or why. The fields produce the usual magnetospheres, but they are weak and relatively empty because of the great distance from the Sun.

10.4 RINGS

In March 1977, astronomers watched Uranus pass in front of a star in order to determine the planet's angular diameter and to study the Uranian atmosphere. However, the star winked in and out before the planet reached it: something was getting in the way. After Uranus moved off the star, the events were repeated in reverse. The only reasonable conclusion was that Uranus has rings (Figure 10.6a). Unlike Saturn's, however, they are very narrow. Further investigation and *Voyager 2* observations revealed a total of 11 (Figure 10.6b). All are within two Uranian radii of the planet's center and within the Roche limit. Except for the ε ring, which is up to 100 km wide, and a broad sheet close to Uranus, they are only a few kilometers wide (Figure 10.6c), reminiscent of Saturn's F ring.

As *Voyager 2* passed behind the rings, the dimming and flickering of its radio signal allowed an assessment of ring constitution and structure. The ring particles range in size from a few tens of centimeters to a few meters, and the big ε ring exhibits the same kind of ringlet structure possessed by the Saturnian system (Figure 10.6c inset). The particles are almost black and are believed to be covered with carbon compounds. Like Saturn's F ring, the ε ring is shepherded by a pair of tiny satellites, Cordelia and Ophelia, each only 15 km in radius. We suspect the other rings to have shepherd satellites as well to keep them from diffusing outward. *Voyager 2* also revealed a hundred bands of fine dust that pervade the ring plane.

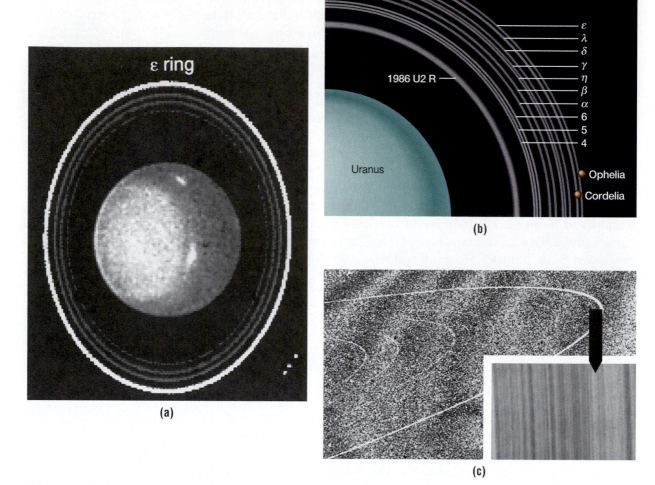

Figure 10.6
(a) An image of Uranus taken with the Hubble Space Telescope reveals rings that were first detected when they passed in front of a distant star. The image also shows a light cap around the southern pole and a pair of white cloud features. **(b)** A map made from more detailed observations from Earth and by *Voyager 2* shows 11 rings. Except for 1986 U2 R, the thicknesses of Uranus's rings and the diameters of the two inner satellites (which shepherd the ε ring) are 50 times smaller than shown. **(c)** *Voyager 2* viewed the narrow rings with high resolution. The detail in the ε ring is shown in the inset.

Because of Uranus's low gravity, its hydrogen atmosphere is distended and extends past the rings. The particles are subject to atmospheric drag and spiral into the planet. The major ring particles should last no more than 10 or 100 million years, and the fine dust should disappear in far less time. Therefore, the rings need to be resupplied. We think they are a way station for the smashed-up debris of collisions, particularly between comets and Uranus's satellites. The little rocks become trapped by shepherding satellites, collide, and produce dust; small grains swept from the rings by the atmosphere produce the dust sheet. All the particles inevitably wind up colliding with Uranus.

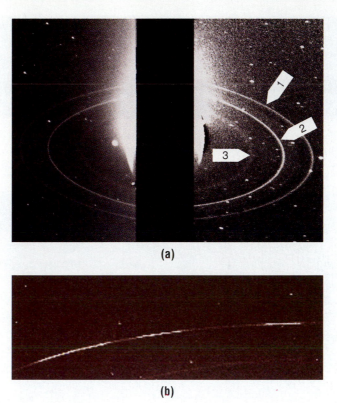

(a)

(b)

Figure 10.7
(a) A composite of two images made by *Voyager 2* shows three of Neptune's rings encircling the planet (which is behind the vertical bar). Ring 4, which is between rings 1 and 2, is not seen here. **(b)** Ring 1 shows arcs in which particles are consolidated.

Observation from Earth and *Voyager 2* also revealed four separate rings around Neptune (Figure 10.7a) within 2.5 Neptune radii of the planet's center, all enmeshed in a dust sheet. The brightest two rings are only a few tens of kilometers wide, the inner of them shepherded by the satellites Despina and Galatea. A ring interior to these is bright but broader, about 2,500 km wide. The particles that make the outer ring are not distributed evenly but clumped into short arcs (Figure 10.7b) by a gravitational resonance with the satellite Galatea.

10.5 SATELLITES

Like Jupiter and Saturn, Uranus is rich in satellites (see Figure 10.1a), though there are none of Galilean size. The five largest (Figure 10.8 and Table 10.3) range between only 235 km and 760 km in radius. The others are the usual collection of big rocks. All 15 rotate synchronously with their orbital periods within Uranus's magnetosphere.

The 10 little moons are irregular and very dark. They are probably covered with organic carbon compounds and are likely to be the sources of the ring particles. The bigger five—Oberon, Titania, Umbriel, Ariel, and Miranda—have densities averaging 20% greater than Saturn's intermedi-

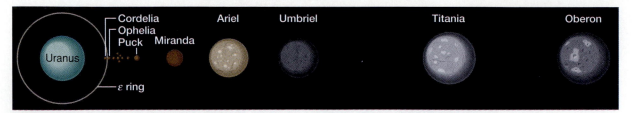

Figure 10.8

The 15 moons of Uranus are magnified by a factor of 40 relative to their orbital sizes and to Uranus.

	Distance		**Period**			**Radius**	**Mass**	**Density**
Name	$(10^6 km)$	$(R_U)^b$	*(days)*	i^{ob}	e^b	*(km)*	$(10^{24} kg)$	(g/cm^3)
Cordelia	49.75	1.95	0.35	0.1	0	15		
Ophelia	53.76	2.10	0.38	0.2	0.0	15		
Bianca	59.16	2.31	0.44	0.2	0	20		
Cressida	61.77	2.42	0.46	0.0	0	35		
Desdemona	62.66	2.45	0.47	0.2	0	30		
Juliet	64.36	2.52	0.49	0.1	0	40		
Portia	66.1	2.59	0.51	0.1	0	55		
Rosalind	69.93	2.74	0.56	0.3	0	30		
Belinda	75.26	2.94	0.62	0.0	0	35		
Puck	86.01	3.37	0.76	0.3	0	75		
Miranda	129.8	5.08	1.41	3.4	0.0	235	0.069	1.35
Ariel	191.2	7.48	2.52	0.0	0.0	580	1.26	1.66
Umbriel	266.0	10.4	4.14	0.0	0.0	585	1.33	1.51
Titania	435.8	17.1	8.71	0.0	0.0	790	3.48	1.68
Oberon	582.6	22.8	13.46	0.0	0.0	760	3.03	1.58

TABLE 10.3
The Satellites of Uranus[a]

Satellite Features

Sizes generally increase outward from Uranus; inner two shepherd the ε ring; small satellites very dark and may provide material for dark rings; larger five are brighter; among outer three, Oberon and Umbriel have old, cratered surfaces, but Titania is younger and icy; inner two have been reworked; rockier than Saturn's intermediate satellites.

[a] All but last five discovered by Voyager 2.
[b] R_U, Uranus radii; i, orbital tilt; e, orbital eccentricity.

ate satellites, indicating constitutions of more rock and less ice, about 50% of each, and are brighter than the inner satellites. The larger bodies are differentiated, rock on the inside, bright ice on the outside.

The larger satellites exhibit a variety of surfaces. Oberon, the most distant, is saturated with impact craters from the heavy bombardment. Though nearby and of nearly the same size, Titania displays crustal cracks and evidence that it has been extensively resurfaced by the extrusion of water ice. Umbriel (Figure 10.9a), the darkest, also has an old surface. The inner two satellites have been subject to considerable activity. Bright Ariel

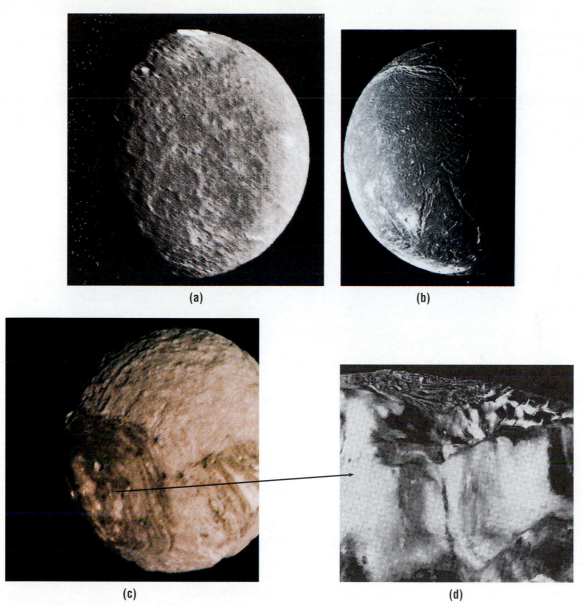

(a)

(b)

(c)

(d)

Figure 10.9

(a) Umbriel's surface is old, with rayless craters; **(b)** Ariel is partially covered with bright ice and displays great deep valleys; **(c)** Miranda has a trio of heavily grooved ovals associated with cracks close to 10 km deep (the images are not on the same scale). **(d)** *Voyager 2*'s data allowed computer reconstruction of views across Miranda's grooved terrain. The cliffs (exaggerated here) are the greatest straight drop in the Solar System.

(Figure 10.9b) has far fewer craters than the others and gigantic valleys tens of kilometers deep. Its surface has been repaved by icy volcanism, the probable result of ancient tidal resonances (see Section 9.2.4). Miranda's cratered and cracked surface (Figure 10.9c) features three oval structures some 200 km across. These are outlined by a series of deep concentric grooves and associated with crustal cracks up to 10 km deep (Figure

TABLE 10.4
The Satellites of Neptune[a]

Name	Distance (10^6 km)	$(R_N)^b$	Period (days)	i^{ob}	e^b	Radius (km)	Mass (10^25 kg)	Density (g/cm^3)
Naiad	48.0	1.9	0.29	0	0	30		
Thalassa	50.0	2.0	0.31	4.5	0	40		
Despina	52.5	2.1	0.33	0	0	90		
Galatea	62.0	2.5	0.43	0	0	75		
1989 N2[c]	73.6	2.9	0.55	0	0	95		
Proteus	117.6	4.7	1.21	0	0	200		
Triton	354.8	14.3	5.88	157[d]	0.0	1,350	2.14	2.07
Nereid	5,513	223	360.2	29	0.75	170		

Satellite Features

Five small, two intermediate, and one big satellite; two of inner six shepherd a ring; outer two including Triton probably captured; Triton has nitrogen/methane atmosphere, effective temperature of 38 K, large frozen nitrogen/methane polar caps, and terrain that features crustal cracks, basins, and nitrogen geysers; Nereid in highly elliptical orbit.

[a]All but last two discovered by *Voyager 2*.

[b]R_N, Neptune radii; i, orbital tilt; e, orbital eccentricity.

[c]Temporary name.

[d]Retrograde.

10.9d). The moon may have been shattered into pieces, its odd features created as the orbiting parts reassembled. More likely, the features were produced by upwelling convective plumes that stretched and broke the crust.

Neptune has a lesser family of moons (Table 10.4). Five small ones under 100 km in radius are bunched near the planet, within or near the rings. About double the ring distance out is larger Proteus, some 200 km in radius. These six satellites are dark, like the small Uranian moons, and are probably surfaced with some kind of carbon-based material. Nereid, also small, is 5.5 million km from the planet and has an orbital period of nearly an Earth year. Its high orbital eccentricity of 0.75 suggests that it is a captured body.

Finally, Triton (Figure 10.10), 1,350 km in radius, is in a league with the Jovian Galileans and Titan. Unlike the other large satellites, however, its orbit is *not* in the plane of the parent planet's equator but is tilted by 157°. It revolves *backward*, suggesting that it was *also* once independent, captured from interplanetary space perhaps by collision with another body or by encounter with a once-thicker Neptunian atmosphere. This conclusion is consistent with the lack of large Galilean-type satellites orbiting Uranus: neither planet formed any on its own, as did Jupiter and Saturn.

Triton is truly distinctive. Its density is high for an outer satellite, 2.07 g/cm^3 (exceeded only by Io and Europa), indicating that a large frac-

Neptune's Triton is in a league with the Jovian Galileans and Titan.

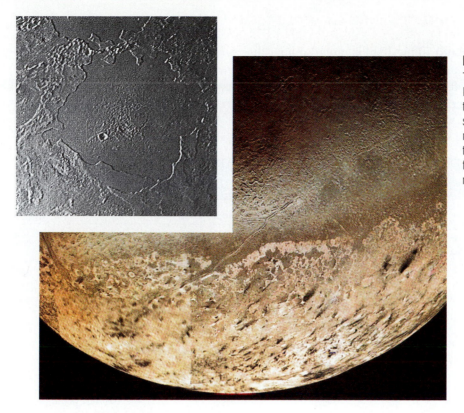

Figure 10.10
Triton displays a remarkable variety of landforms reminiscent of the terrestrial planets. At the bottom is the huge south polar cap. Toward the top (to the north) are vast areas of odd, cracked terrain (inset) that contain basins flooded with water/methane/ammonia "magma" from below.

tion—some 75%—is rock. Thus we see a continuance of the reversal of the rock-to-ice ratio with distance from the Sun first indicated by the Uranian satellites. The maximum water content occurs at Saturn. Triton also has an atmosphere. Although its surface pressure is only 2×10^{-5} bar, the atmosphere can produce high hazes. It is composed almost entirely of nitrogen mixed with a small amount of methane. Triton is so cold, a mere 38 K, that the nitrogen and methane freeze and fall to the surface as frost and snow, producing a vast polar cap that extends to a latitude of 25°. Within the polar cap are geysers spewing nitrogen into the thin air and forming elongated clouds as they are blown downwind.

Triton's equatorial region displays cratered areas and a strange mottled surface. The long streaks seen in Figure 10.10 are ditch-like cracks that reveal geologic activity. Closeups show basins (Figure 10.10 inset) that, like the cracks, have been flooded by volcanism, the "lava" an extruded mush of water, methane, and ammonia. Triton is probably highly structured, with a water-ice crust atop a liquid mixture of water, methane, and ammonia, which in turn rides on a rocky core. Its activity indicates internal heating, possibly caused by tides that changed and flexed the satellite during capture. This body is like nothing we have seen before. Remarkably, as we leave Neptune behind, we encounter another odd individual.

10.6 PLUTO

We have long known that Pluto (Table 10.5) is much smaller than Uranus and Neptune and that it rotates much more slowly, but its nature was not at all clear. Some of the mystery was dispelled in 1978 with the discovery that Pluto has a satellite (Figure 10.11). Charon was immediately seen to be

TABLE 10.5
A Profile of Pluto

Planetary Data		Atmosphere	
Mean distance from Sun	39.53 AU	Pressure 10^{-6} bar	
Equatorial radius = polar radius	1,190 km = 0.0018 R_{Earth}		
Mass	1.3×10^{22} kg = 0.0022 M_{Earth}	*Other*	
Mean density	1.8 g/cm^3	Magnetic field strength	Not known
Surface gravity	0.04 g_{Earth}	Tilt of magnetic field axis	Not known
Escape velocity	1.1 km/s	Rotation period (interior)	6.39 days
		Axial inclination	122.5°a
		Effective temperature	40 K

Layers	% Mass	Radii (km)	Satellite Data	
Rocky core	75	0−900?	*Charon*	
Icy mantle	25	900−1,190	Distance from Pluto	19,640 km
			Radius	590 km
			Mass	1.1×10^{21} kg
			Density	1.2 g/cm^3

Planetary Features

Not a Jovian, but similar to Triton; nitrogen/carbon monoxide/methane atmosphere, large frozen N/CO/CH$_4$ polar caps; wide temperature range as a result of high eccentricity; atmosphere probably all freezes out at aphelion; Charon, Pluto's large moon, has water ice.

aRetrograde rotation.

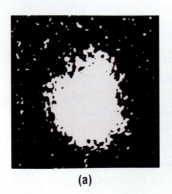 

(a)   (b)

Figure 10.11
(a) Pluto's satellite, Charon, was discovered as a blip on the planet's upper right edge. Turbulence in the Earth's atmosphere makes the images fuzzy and does not allow their easy resolution. **(b)** Pluto and Charon are nicely resolved by the Hubble Space Telescope.

large relative to its planet. Its orbital period is 6.4 days, the same as Pluto's rotation period; each is tidally locked on to the other. The orbit has a radius of 19,400 km and is highly tilted, Charon moving retrograde with a high orbital inclination of 122°. Because of the tidal locking, Pluto is also turning retrograde.

Kepler's generalized third law applied to the mutual orbits of Pluto and Charon, as determined both from ground-based telescopes and the Hubble Space telescope, shows that Pluto has a mass only 0.002 times that of the Earth, a sixth that of Earth's Moon. Charon's mass is still uncertain, but is roughly 12% that of Pluto.

Between 1985 and 1991, Pluto's equatorial plane and the plane of Charon's orbit were directed at the Earth, so the bodies crossed in front of each other in a long series of mutual eclipses. These events provided a wealth of information. If the orbit is known, so are the orbital velocities. The durations of the eclipses allow the computation of the radii of the two bodies. Pluto's radius is 1,190 km, less than half that of Mercury and only two-thirds that of our Moon. Charon has a radius of just 590 km. With a density of about 1.8 g/cm^3 (and possibly higher), Pluto is at least 60% rock. Charon's uncertain density of 1.7 g/cm^3 tentatively indicates a similar structure.

The spectrum of Pluto shows absorptions of gaseous methane and of nitrogen and carbon monoxide ices. The little planet has an atmosphere with a surface pressure about a millionth that of Earth's in which N_2 dominates. At Pluto's temperature of around 40 K, some of the atmosphere freezes to the surface as a nitrogen/carbon monoxide/methane snow that surrounds the poles and lies in a complex pattern across the rest of the planet (Figure 10.12). The brightness variations probably indicate the presence of basins and relatively new impact craters. At aphelion, the temperature drops so low that all the atmosphere falls to the ground and disappears. Charon's spectrum shows features from water ice instead of methane absorptions. Brightness variations observed during eclipses of Charon by Pluto reveal frost patterns on the satellite vaguely similar to those seen on Pluto.

Pluto bears an uncanny resemblance to Triton.

Pluto bears an uncanny resemblance to Triton. The two are nearly the same size, mass, and density, with similar atmospheres and polar caps. Pluto and Triton represent a kind of body different from the other planets and their satellites. We seem to be staring out at a pair of the few remaining large **planetesimals** (the word means "infinitesimal planets"), primitive, solid bodies that gave birth to the planets themselves. Pluto and once-free Triton could have been perturbed by the large planets of the Solar System into orbits that brought them under Neptune's influence. Neptune captured Triton directly, whereas Pluto was trapped in a gravitational resonance in which its orbital period (averaged over a long interval) is exactly 3/2 that of Neptune. As part of the process, primitive Pluto

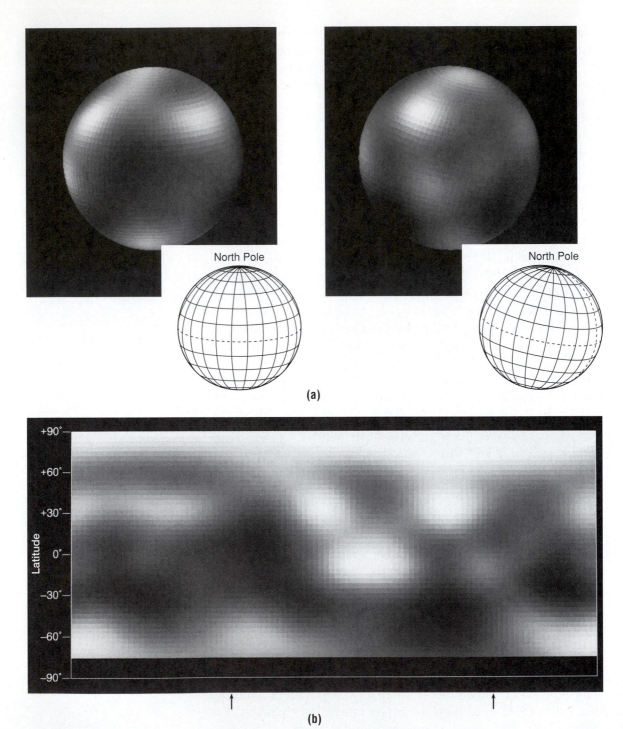

Figure 10.12

(a) Opposite hemispheres of Pluto observed with the Hubble Space Telescope (their orientations shown in the insets) reveal bright snow or frost blanketing much of the northern polar regions; some appears around the south pole as well. **(b)** An equatorial map shows reflection features that may indicate impact basins. The arrows show the regions centered in the hemispherical pictures above.

collided with another body to create Charon, much as the Moon was created violently from the primitive Earth (see Section 7.9).

There may be many more such bodies beyond the orbit of Neptune. In the next chapter, we examine more evidence for the existence of planetesimals, and finally put the whole Solar System together.

BACKGROUND 10.1   Life at the End of the Planetary System

Earlier, we spent a day on blistering Mercury, which is tucked so close to the Sun that lead would melt on its surface. What is the other extreme like?

Travel now to an outpost on Pluto—the end of our planetary system, where we stare out into the vast depths of interstellar space. The trip alone is daunting. It took *Voyager 2*, with accelerating gravitational assists by Jupiter, Saturn, and Uranus, a dozen years to make the journey to Neptune.

Finally, you arrive on this dim, cold world at a time when it is at its average distance from the Sun, 40 AU. Put on your pressure suit and climb out of your craft. A strange Sun shines overhead in a black sky. There is no apparent disk! To your eye, the Sun appears only as a blinding star. Look away, and a huge, dim, quarter moon more than 3° across hangs over the horizon. Crunching across a nitrogen/carbon dioxide/methane snow frozen to only a few tens of degrees above absolute zero, you arrive at your cabin, where you can finally breathe oxygen without your spacesuit.

Your day is a long one. Very slowly the Sun moves across the sky. It was noon when you arrived; 1.5 Earth days later, that distant star has finally set. Although Charon looms nearly overhead, steadily going through its phases, it stays motionless in the sky, the captive of mutual synchronous rotation. Night sets in, and the temperature drops even further. At least your view of the stars is good.

Time to call Earth. Pick up the phone and start talking. No one answers. You talk for about an hour and a half, telling the mission commanders or perhaps your family what life is like 6 billion kilometers from home. You hang up. Traveling at the speed of light, your "Hello, I'm fine" *has just reached the orbit of Neptune*. It will arrive at Earth five hours after you began speaking. Six or seven hours after that, your phone rings.

Gaze outward into space, away from the Sun. There are no planets going through their familiar oppositions. Look toward the Sun: all planets are inferior. With luck you might glimpse Jupiter at greatest elongation a mere 7° from the daylight star. Earth is almost lost. In your loneliness, you pull out your telescope for a glimpse of home. There it is, a fragile blue dot against the frigid sky. It's going to be a long year.

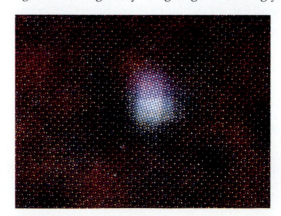

Figure 10.13
Earth is seen from Neptune in one of *Voyager 2*'s last images as it left the planetary system for deep space.

KEY CONCEPT

Planetesimals: Primitive bodies of the Solar System that preceded the formation of the planets.

EXERCISES

Comparisons

1. Graphically compare the relative angular diameters of Jupiter, Saturn, Uranus, Neptune, and Pluto at opposition to show the relative ease of seeing surface features.

2. Compare the winds of Uranus and Neptune with those of Jupiter and Saturn.

3. Why do the colors of Uranus and Neptune differ from those of Jupiter and Saturn?

4. Contrast the Great Dark Spot, the Great Red Spot, and the Great White Spot. How are they similar? How do they differ?

5. How does Uranus's internal heat differ from Neptune's?

6. Compare the internal structures of Uranus and Jupiter.

7. Compare the magnetic field structures of Uranus and Neptune with those of Jupiter and Saturn.

8. Contrast the rings of Uranus and Neptune, and compare these with the rings of Saturn.

9. List similarities and differences between Triton and Pluto.

Numerical Problems

10. What are the latitudes of Uranus's tropics and arctic circle?

11. About what year will Pluto and Charon again mutually eclipse each other?

Thought and Discussion

12. How might the surface appearances of Uranus and Neptune relate to the flow of internal heat stored within each planet's interiors?

13. Why do the particles of Uranus's rings have short lifetimes?

14. How can the rings of Neptune and Uranus be so narrow?

15. Why are Uranus's smaller inner satellites dark and its outer ones brighter?

16. Why do we think Triton was once in free orbit around the Sun?

17. Name the geologically most active bodies of the outer Solar System. What could be responsible for their activity?

18. Why does Pluto spin so much more slowly than do the Jovian planets?

19. How did Triton and Pluto fall under Neptune's influence?

Research Problem

20. Explore **(a)** pre-Voyager observations of surface features on Uranus and Neptune; **(b)** early false reports on rings of Neptune; **(c)** early observations of the angular diameter of Pluto that gave a false idea of its size and mass.

Activities

21. Compare the Jovian planets. Number them 1 through 4, and against these numbers plot the planets' radii, masses, densities, effective temperatures, ratios of internal to external heating, rotation periods, number of satellites over 1,000-km radius, number of satellites between 100- and 1,000-km radius, and number of small satellites.

22. At opposition, Uranus is visible to the naked eye and is an easy sight in binoculars. *Sky & Telescope* and *Astronomy* magazines annually publish maps showing its path. Use them to find the planet and sketch its position among the stars.

Scientific Writing

23. Write an essay for a textbook (like Background Essay 10.1) on what life would be like on Triton.

24. You are a public relations officer with NASA, whose leadership plans to send a spacecraft to explore Pluto. Write a press release to convince the public that such exploration is a good idea.

CREATION AND ITS DEBRIS
Asteroids, comets, meteors, and the creation of the Solar System

11

Ancient heavy cratering shows that there was once an astonishing number of small bodies—planetesimals—in orbit about the Sun, so many that we believe the planets were assembled from them. If so, the scraps of planetary creation ought to be all around us.

11.1 METEORITES

From over the nighttime horizon comes a brilliant ball of light, leaving a long trail behind it (Figure 11.1) like a torch thrown across the sky. You have just seen a **meteor** (from the Greek *meteoron*, "a thing in the air") of a special kind called a *fireball*. It was caused by a **meteoroid** (a piece of interplanetary debris) speeding along on a solar orbit at 20 to 40 km/s and colliding with the Earth. At an altitude of roughly 75 km, the violent shock wave set up in the atmosphere by the meteoroid traveling at supersonic speed (that is, faster than sound) heated and ionized the surrounding air. The recombination of ions and electrons then released the energy to pro-

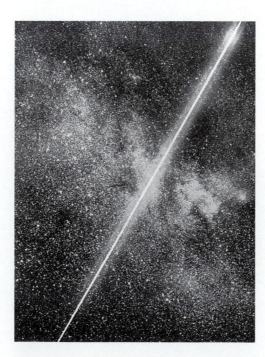

Figure 11.1
A bright fireball flashes past the distant stars.

duce the visible event. Small meteoroids vaporize in the air or explosively disintegrate. If sufficiently large, however, perhaps the size of a marble, one may survive to strike the ground to become a **meteorite** (Figure 11.2).

Meteorites are surprisingly common: some 800 with masses of at least 10 g fall on the Earth every day. They range in size from tiny scraps to bodies a few meters across with masses of several metric tons and, in rare instances, to far beyond. We find three basic kinds (Table 11.1). The **irons**

(a)

(b)

(c)

(d)

Figure 11.2
(a) This iron meteorite from Canyon Diablo in Arizona is about 40 cm across and weighs a quarter-ton. **(b)** Large crystals are seen when an iron meteorite is cut, polished, and etched with acid. **(c)** This carbonaceous chondrite has been broken to show its interior. **(d)** Millimeter-sized chondrules are set in a dark, grainy, silicate matrix.

TABLE 11.1
Types of Meteorites

Type	Subtype	Frequency	Composition	Formation
Stones	Carbonaceous Chondrites	5%	Water, carbon, silicates, metals	Primitive
	Chondrites	81%	Silicates	Heated under pressure
	Achondrites	8%	Silicates	Heated
Stony irons		1%	50% silicates, 50% free metal	Differentiated
Irons		5%	90% iron, 10% nickel	Differentiated

(Figure 11.2a) are the most easily identified. They are combined with at least 5% to 10% nickel. If sliced, polished, and etched with acid, they show beautiful crystal structures (Figure 11.2b), indicating that they have undergone a long, slow period of cooling. The irons are similar in composition to what we might find at the Earth's core and thus seem to have come from inside large differentiated bodies. Meteor Crater in Arizona (see Figure 7.19) was made by a huge iron some 100 m across.

Stones are far more common than irons but seem rare because they are so hard to distinguish from ordinary rocks. Eighty-six percent of the stones are **chondrites,** which consist of a dark, fine-grained silicate rock matrix with small spherical glassy inclusions or *chondrules*. The chondrites look as if they have been melted and resolidified, and many appear to have been smashed apart and welded back together. However, 5% of the stones, the **carbonaceous chondrites** (Figures 11.2c and 11.2d), show no such effects and are abundant in carbon and water. They have not been altered and appear to be among the true primitives of the Solar System. The remainder of the stones are **achondrites** that contain no chondrules and have been altered by heat. Some resemble the Earth's mantle in that they are made of basalt and look as if they have undergone volcanic processing. The third general kind of meteorite, the **stony irons,** constitutes only about 1% of all meteorites. These are made of silicates intermixed with roughly 50% free iron.

The different kinds of meteorites are nearly all about 4.5 billion years old. The oldest object we know is a carbonaceous chondrite dated at 4.56 billion years. Since it is primitive and unaltered, we identify its age with that of the Solar System. The meteorites thus seem to be remnants of the Solar System's formation. Certain isotope ratios in meteorites are similar to those found in exploding stars and in the atmospheres of stars that are losing matter back into space. Some meteorites contain ^{26}Mg, the daughter product of highly radioactive ^{26}Al, which has a half life of only 720,000 years; others have been found to contain tiny bits of carbon crystallized by high temperature and pressure—diamonds—a trillion per gram. Their presence can be explained by shock waves from exploding stars acting on interstellar carbon. These discoveries allow us to go back in time *before* the origin of the Solar System and explore the source of the raw material. Most remarkably, chondrites contain complex organic compounds, including amino acids, the building blocks of proteins and of life. The acids are not terrestrial contaminants but belong to the meteoroids or their parent bodies. The meteorites thus may help illuminate the origin of life itself.

The meteorites seem to be remnants of the Solar System's formation.

11.2 ASTEROIDS

On January 1, 1801, the Italian astronomer Giuseppe Piazzi discovered a new member of the Solar System and named it Ceres after the ancient deity of Sicily. Its distance was later calculated at 2.8 AU. At seventh magnitude, vastly fainter than Mars or Jupiter, it had to be very small; we now know its radius is only 457 km. A year later, Heinrich Olbers found Pallas, also at 2.8 AU; discoveries of Juno (2.7 AU) and Vesta (2.4 AU, and visible to the naked eye) followed. It was beginning to look as if these small bodies or **asteroids** might be pieces of what had once been a planet. We now know that is not the case. What they are is much more interesting.

11.2.1 Number and Orbits

By 1872 over 100 asteroids were known. The 15 largest are listed in Table 11.2. An asteroid's name is preceded by its discovery number: 1 Ceres, 2 Pallas, 334 Chicago. Asteroids are easy to discover. Ceres, for example, moves an average 0.2° per day or 30 seconds of arc per hour. On a long photographic exposure or CCD image, an asteroid will therefore leave a short streak (Figure 11.3). Over 20,000 have been found, and more than 6,000 have known orbits. In general, asteroids orbit the Sun on roughly circular paths that lie between the orbits of Mars and Jupiter (Figure 11.4a); they are especially concentrated into a **main belt** between 2 and 3 AU from the Sun.

Meteorites are just small asteroids.

Many asteroids group into families with closely similar orbits; other families consist of broken parts of a once-larger body. Three families are of highly practical interest: the Amors approach the orbit of the Earth, the

TABLE 11.2
The Fifteen Largest Asteroids

Name	Radius (km)	a^a (AU)	P^a (years)	e^a	$i^{\circ a}$	Reflectivity (%)	Rotation (days)	Kind[b]
1 Ceres	457	2.77	4.61	0.10	11	0.10	9.1	C
2 Pallas	261	2.77	4.61	0.18	35	0.14	7.8	C-like
4 Vesta	250	2.36	3.63	0.10	7	0.38	5.3	S-like[c]
10 Hygiea	215	3.14	5.59	0.14	4	0.08	18	C
511 Davida	168	3.18	5.67	0.17	16	0.05	5.1	C
704 Interamnia	167	3.06	5.36	0.08	17	0.06	8.7	—
52 Europa	156	3.10	5.46	0.12	7	0.06	5.6	C
15 Eunomia	136	2.66	4.30	0.14	12	0.19	6.1	S
87 Sylvia	136	3.49	6.52	0.05	11	0.04	5.2	C[d]
16 Psyche	132	2.92	5.00	0.10	3	0.10	4.2	M
31 Euphrosyne	124	3.16	5.58	0.10	26	0.07	5.5	C
65 Cybele	123	3.43	6.37	0.13	4	0.06	6.1	C[d]
3 Juno	122	2.67	4.36	0.21	13	0.22	7.2	S
324 Bamberga	121	2.68	4.41	0.29	11	0.06	29.4	C?
107 Camilla	118	3.49	6.50	0.08	10	0.06	4.8	C

[a]a, semimajor axis; P, orbital period; e, orbital eccentricity; $i°$, tilt of the orbit to the ecliptic plane.

[b]C asteroids are probably carbonaceous chondrites, S are ordinary chondrites or stony irons, M are probably irons.

[c]Basaltic achondrite.

[d]Possibly like M.

Figure 11.3
Eros moved while its picture is being taken, instantly identifying itself as an asteroid.

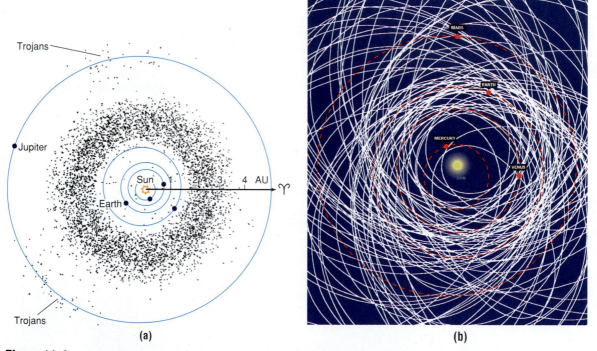

(a)   (b)

Figure 11.4
(a) The positions of 5,011 numbered asteroids, plus those of the terrestrial planets and Jupiter, are plotted for May 1, 1992. They are concentrated in the main belt between 2 and 3 AU. The *Trojan asteroids*, however, orbit at stable positions 60° ahead and behind Jupiter, and a few AAAOs are near or have crossed the Earth's orbit. (♈ indicates the direction of the vernal equinox.) **(b)** An expanded view of the inner Solar System shows the orbits of AAAOs, asteroids that have the potential for colliding with the terrestrial planets.

Apollos cross it, and the Atens have semimajor axes less than that of the Earth. Asteroids in these families are known collectively as AAA Objects, or **AAAOs** (Figure 11.4b). About 1,000 are known. None of the known AAAOs has an orbit that intersects that of the Earth, but constant perturbations produced by the planets give all AAAOs the capability of colliding

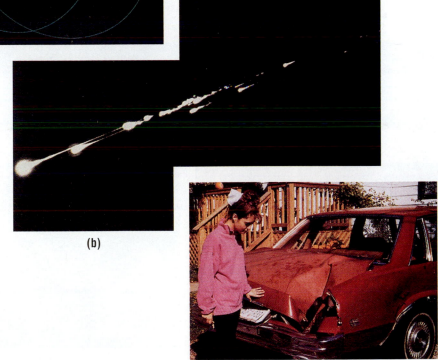

Figure 11.5
(a) The elliptical orbits of four meteoroids (in fact, small AAAO asteroids) once carried them out well past the orbit of Mars. **(b)** The Peekskill (New York) meteor broke into countless pieces upon entering the Earth's atmosphere in 1992. **(c)** One of them, a 12-kg meteorite, bashed a woman's car.

with us. In four instances, we have been able to reconstruct the meteoroid orbits of recovered meteorites from the paths of their meteors (Figure 11.5a). All were highly elliptical, with perihelia near the Earth's orbit (hence the collision) and aphelia between the orbits of Mars and Jupiter. Meteorites—like those in Figure 11.2—are just small asteroids that have been gravitationally perturbed by Jupiter into elliptical orbits to become unrecognized AAAOs. Further perturbations by the planets eventually aim them at us, and they scream through the sky as meteors before finally coming to rest on the ground (Figures 11.5b and 11.5c). A big one would produce an impact crater and perhaps disaster.

11.2.2 Physical Properties

Most asteroids are so small and far away that they appear through Earth-bound telescopes only as points. Radii and structures have been observed for a few, however, with the Hubble Space Telescope (Figure 11.6a), flyby spacecraft (Figure 11.6b), and radar (Figure 11.6c). In addition, Vikings and other craft have imaged the moons of Mars (see Figure 8.29), which are most likely captured asteroids. Fortunately, there are a variety of other methods, the most general of which involves a combination of optical and infrared brightness measurements that simultaneously provide radii and average percentage reflectivities.

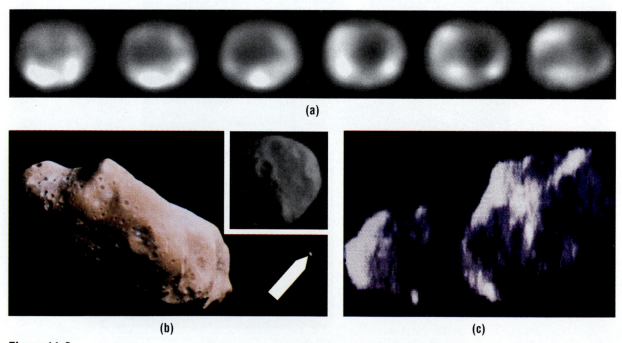

(a)

(b)   (c)

Figure 11.6

(a) The Hubble Space Telescope made multiple images of the asteroid 4 Vesta. The motion of the deep, dark crater at the center shows the asteroid rotating through a quarter of its 5.3-hour "day." The reflectivity of the asteroid at different wavelengths shows the inside of the crater to be heavy mantle basalt, whereas the bright areas are more like the Earth's lighter crust. **(b)** The *Galileo* spacecraft reveals the beaten surface of 243 Ida, 52 km long. The different colors indicate a variety of materials. Ida has a tiny satellite 1.5 km across (arrow), seen in closeup in the inset. **(c)** Radar imaging shows that 4179 Toutatis is a double asteroid, two rocks each 3 or 4 km across in contact (or near contact) with each other.

Large asteroids are rare, the number increasing rapidly with decreasing size down to billions of boulders, rocks, and pebbles, consistent with the observed sizes of meteorites and just what we would expect if asteroids are products of collisions. They have been grinding one another down from larger bodies nearly since the dawn of the Solar System. The collisional activity is vividly illustrated by the irregular shapes and battered surfaces of 4 Vesta, 243 Ida (see Figure 11.6b), and the Martian moons.

The masses of a few asteroids can be evaluated by the perturbations they exert on Mars and on one another. Ceres, the largest, has a mass of only 2×10^{-4} that of the Earth; Pallas and Vesta have masses about 20% of this value. The densities of Ceres and Pallas, derived from their masses and volumes, are about 2.7 g/cm^3, typical of rock. But there is considerable variety: 10 Hygeia's density is only 2.1 g/cm^3, while 4 Vesta's is about 3.6 g/cm^3, showing that more than rock is involved in Vesta's structure. Hubble observations of Vesta in fact show a huge crater (see Figure 11.6a) that has apparently dug through a light silicate crust into a heavier silicate mantle, the structure strongly suggesting a differentiated body.

From the average density and the number of asteroids with specific dimensions, we find that most of the mass in the asteroid belt is concentrated in the larger objects. As numerous as they are, the little asteroids do not count for much. Ceres is estimated to possess about a quarter of the total asteroidal mass. All the asteroids combined would make a mass of about 10^{-3} that of the Earth: together they would not constitute anything close to a respectable planet.

Taken together, the asteroids would not constitute anything close to a respectable planet.

11.2.3 Origins

We can relate different kinds of meteorites to different kinds of asteroids. Asteroid properties are in turn related to location within the asteroid belt. Those near the inner edge of the main belt tend to be bright and to have spectra (which include absorption features that give clues to the nature of the surface minerals) that fit those of ordinary chondrites and stony irons. Asteroids with properties similar to iron meteorites are a bit farther away. Dark asteroids near the outer edge of the main belt at 3 AU are almost certainly the progenitors of carbonaceous chondrites. Even darker asteroids orbit yet farther out (between 4 and 5 AU from the Sun): these bodies produce few if any meteorites and seem to be covered with some kind of carbon residue like so many of the satellites and ring particles of the outer Solar System (see Sections 9.2.4, 9.3.4, 10.4, and 10.5), the low temperatures promoting an unusual, but uncertain, chemistry.

Like the carbonaceous chondrites, the outer asteroids seem to be primitive, unchanged over the aeons. Those nearer the inner edge of the main belt are the remnants of differentiated bodies that have been smashed to pieces. The iron meteorites are the cores of differentiated asteroids and the achondrites are the mantles; the stony irons may come from the boundary layers.

Following a collision between asteroids, some of the debris will be scattered to form new, smaller asteroids, but much will fall back to the parent body, leaving its surface a pile of fusing rubble. When another collision knocks this outer crust away, the result is a stony asteroid, which if it hits the Earth will be an ordinary stony meteorite.

11.3 COMETS

The sight of a great **comet** (Figure 11.7), its gossamer tail streaming across the sky, brings everyone out to look. Over the centuries, comets have been regarded as messengers of doom and destruction; to the contrary, life on Earth may have been impossible without them.

11.3.1 Appearance and Orbits

A typical comet has three basic parts. At the core is a dark, tiny, **nucleus,** the solid interplanetary body itself. Surrounding the nucleus is a large, bright, **coma** that can be a degree or more across and over 100,000 km in diameter. Streaming out of the coma are two kinds of **tail** that may be sev-

Figure 11.7
The gaseous coma of Halley's Comet surrounds a tiny nucleus (only a thousandth the size of the black dot). The gas tail streams out from the coma away from the Sun in nearly a straight line; the dust tail fans away to the left.

eral degrees and millions of kilometers long. One tail is narrow and points nearly straight back from the Sun. Spectroscopy shows that it radiates emission lines of ionized gas, so it is called the **gas tail.** It tends to be blue because of emissions of carbon molecules in the blue part of the spectrum. The other tail is generally curved and fan-shaped. Its spectrum is that of sunlight scattered from small grains; this **dust tail** is yellowish, the color of sunlight. Perspective effects can make it seem to point in any direction. Which tail is the brighter depends on the particular comet.

Comets orbit the Sun on highly elliptical paths and are easily seen only when they are within the confines of the terrestrial planets. **Short-period comets** (Figure 11.8a) have orbital periods less than about 200 years (Table 11.3). The shortest known is that of Encke's Comet, which goes about the Sun in only 3.3 years. The short-period comets generally move counterclockwise about the Sun and have low tilts to the ecliptic, under

(a)

(b)

Figure 11.8
(a) The short-period comet Giacobini-Zinner (named, as is traditional, after its discoverers) has a period of 6.6 years and sports a short but bright gas tail. The camera tracked the moving comet during the exposure, so the stars appear as streaks. **(b)** Comet West of 1976, a long-period comet, had a spectacular dust tail. Upon rounding the Sun, its nucleus broke into five pieces.

TABLE 11.3
Selected Comets

Name	Year of Discovery	q(AU)[a]	e[a]	i[°a]	P (years)	Characteristic
Short-Period						
Biela	1772	0.86	0.76	13	6.6	Vanished after 1851
Encke	1786	0.34	0.85	12	3.3	Shortest period
Giacobini-Zinner	1900	1.03	0.71	32	6.6	Visited by spacecraft
Halley's	——	0.59	0.97	162	76	Brightest
Long-Period						
De Chéseaux's	1744				——[b]	Six dust tails
Donati's	1858	0.58	0.996	117	——	
Morehouse	1908	0.95	1.0	140	——	Very dusty
Great January	1910				——	Rivaled Halley's
Arend-Roland	1956	0.32	1.0	120	——	Sunward spike
Ikeya-Seki	1965	0.008	1.0	142	800	Came very close to Sun
Bennett	1970	0.54	0.996	90	——	
West	1976	0.20	1.0	43	——	Nucleus broke into four pieces
Hyakutake	1996	0.22	0.9999	125	100,000	Came very close to Earth

[a]q, perihelion distance; e, orbital eccentricity; i, orbital tilt.
[b]No listing means the period is very long.

35°. Aphelion points range across the planetary system, but are typically near the orbit of Jupiter. Most of these comets are dim and have modest tails.

The orbits of the short-period comets are well known and—as first shown by Edmund Halley—their returns into the inner Solar System can be predicted. He recognized that the bright comets of 1531, 1607, and 1682 had similar orbits and were actually the same comet making reappearances. Halley then predicted that the comet would return in 1758–59, which it did—16 years after his death. His successful prediction provided solid support for Newton's laws. The comet, which has a 76-year period and returned in 1846, 1910, and 1985 (see Figure 11.7), was named in his honor.

The **long-period comets** (Figure 11.8b and Table 11.3) have periods over 200 years, most far longer. Some orbits are so eccentric that in the vicinity of the Earth they cannot be discriminated from parabolas (see Section 5.4), so their periods cannot be measured and are estimated to be in excess of a million years. Long-period comets have no respect for the ecliptic. As many proceed retrograde as prograde, and their orbital tilts are random. The majority of the known comets are of the long-period variety.

Perhaps once a generation we will witness a great comet, one that dwarfs the dozens of others that pass by the Earth every year. Great comets can provide spectacular interludes in our perpetual searches of the heavens. Several have already been depicted here. Comet West of 1976 (Figure 11.8b) is among the brighter comets ever seen. However, even this beauty is surpassed by some that appeared in the nineteenth century (which were certainly enhanced by the darker skies of the time). The great comet of 1843 had a tail 300 million km long, and the central coma of the great comet of 1881 (Figure 11.9a) was visible in broad daylight. Certainly one of the more spectacular was De Chéseaux's Comet, which had six separate dust tails (Figure 11.9b). We never know when another great comet will appear, as the events are truly random. Comet Hyakutake (Figure 11.10), for example, took us by complete surprise. Discovered in late January

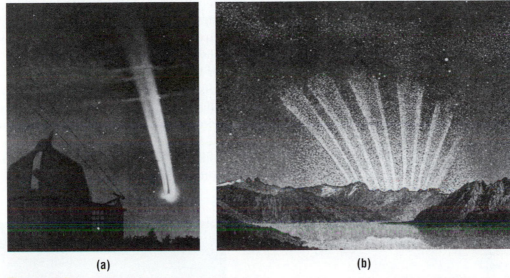

(a)   (b)

Figure 11.9
(a) The great comet of 1881, Tebbutt's Comet, was so bright that its central coma was seen in daylight.
(b) The six tails of De Chéseaux's Comet rose above the morning horizon in 1744.

Figure 11.10
Comet Hyakutake of 1996 took the world by storm, passing only 40 times farther from the Earth than the Moon, its bright tail easily seen even from town.

of 1996, it brightly passed 15 million km from Earth only two months later, its tail stretching 40° from the Little Dipper to Coma Berenices.

Halley's Comet is in a class by itself. Although it has a short period, it is so bright that it clearly deserves to be called a great comet. Its period is roughly the length of a human lifetime, which gives most of us an opportunity to see it once and a favored few the chance to view it twice. It has been recorded throughout much of written history and appears in the Bayeux Tapestry (Figure 11.11a), which chronicles the Norman invasion of Britain in 1066, and in Giotto's *Adoration of the Magi*, painted in 1303 (Figure 11.11b).

In our own time, Halley's has produced a brand of popular mania. Its 1910 return was one of the showiest ever, as the comet came into perihelion and was brightest when it was close to the Earth. The comet was used widely in advertising, and apparently inspired a thwarted blood sacrifice of a woman in Oklahoma by a band of fanatics. More intriguing, astronomers predicted that on the night of May 18–19, 1910, the Earth would actually pass through the tail. Shortly before, the deadly gas cyanogen had been discovered in the comet's spectrum, so there was widespread fear that life on Earth might be destroyed. On that night some people packed their windows and doorways with wet rags to keep out the fumes, while sophisticates in New York held rooftop parties to celebrate the end of the world. Entrepreneurs became rich selling comet hats and comet pills to ward off the evil effects. Nothing happened. The tail of a comet is so vacuous that no effects were seen at all.

The 1985 encounter with Halley's Comet, one of the most disappointing in its 2,000-year recorded history, was seen reasonably well only from the southern hemisphere. Yet the hype was enormous and sold a great many telescopes to people who had little idea of how to use them. No doubt we can expect more of the same when this comet returns in 2061.

(a)

(b)

Figure 11.11
(a) Halley's Comet is seen embroidered as a heavenly sign into the Bayeux Tapestry, a pictorial history of the Norman conquest. The (inaccurate) Latin inscription reads, "They wonder at the star." **(b)** In a painting by the Florentine master Giotto the comet probably represents the star of Bethlehem.

11.3.2 Physical Nature

In 1950, Fred Whipple of Harvard proposed the basic model of a comet, the *dirty snowball*. Comet nuclei are made of a variety of ices, mostly water, into which are mixed large quantities of small solids ranging in size from a few millimeters down through grains of dust less than 10^{-3} mm across. As a comet approaches the Sun, it warms and its ice starts to evaporate to a gas. The comet's gravity is low, and the gas expands into space under the force of its own pressure. Dust particles are carried away with the gas; together they form the coma. The force of the ejection is strong enough to produce a rocket effect that slightly changes the comet's orbit, sometimes making it difficult to chart its future course.

The basic model of a comet is the dirty snowball.

Ultraviolet sunlight breaks down the water, causes chemical reactions that make new compounds, partly ionizes the gas, and excites the electrons of atoms and molecules into outer orbits. When the electrons cascade back downward, they produce an emission spectrum. Simple molecules like CH, CN, C_2, and CO dominate the mixture, but we also find formaldehyde (H_2CO), methyl alcohol (CH_3OH), methyl cyanide (CH_3CN), and evidence from spacecraft for much heavier, though unidentified, molecules. Mixed in are also a number of free elements. Hydrogen freed in the reactions spreads out around the coma in a cloud that can be a million or more kilometers across.

A comet's gas tail is a creature of the solar wind (see Section 7.8) and the Sun's magnetic field. Since the solar wind is ionized, it interacts with the field, dragging it outward from the Sun. The Sun is rotating, however, with a period of about a month. The result is that the outbound magnetic field lines wrap into a spiral, much like sprays of water from a rotating water sprinkler (Figure 11.12a). As the comet plows into the magnetic field lines, they are captured by the comet's electrically charged gases. The field wraps around the comet like a wind sock and stretches the ionized gases into a long tail pointed away from the Sun (Figure 11.12b). The interaction between a comet's gases with either energetic sunlight or with the solar wind can be so strong that it even produces X rays (Figure 11.12c).

As the comet's ices evaporate, dust is released and pushed away by the pressure of the intense solar electromagnetic radiation. Since the particles have considerably more mass than the molecules of the comet's ejected gas, they lag behind the comet in its orbit, and the dust tail thereby spreads outward into a great fan (see Figure 11.8b).

A comet's tiny nucleus is so small it can be studied directly only by spacecraft. In 1985 and 1986, an international armada of spaceships equipped with magnetic sensors, dust detectors, spectrometers, and cameras was launched to rendezvous with the returning Halley's Comet. The European Space Agency's *Giotto* craft, which carried an imaging camera,

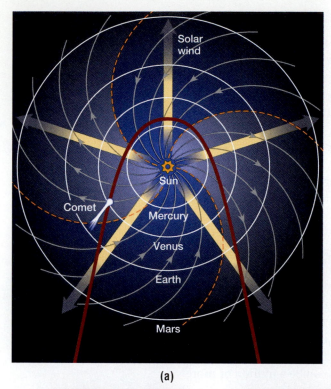

(a)

Figure 11.12
(a) As the solar magnetic field is dragged into the planetary system by the solar wind, the Sun's rotation twists it into a spiral. **(b)** The comet's gas tail is formed by the interaction of the Sun's wind and magnetic field with the comet's ionized coma. The comet captures the magnetic field, bending it (orange lines) into a shock wave. **(c)** The *Rosat* satellite discovered powerful X rays coming from the sunward side of the coma of Comet Hyakutake.

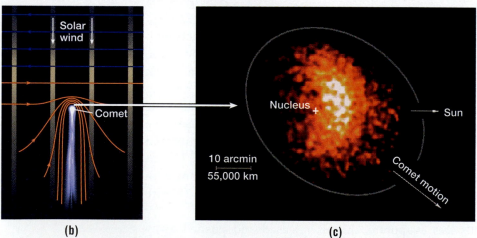

(b)   **(c)**

passed only 600 km from Halley's nucleus (Figure 11.13). The nucleus has the shape of a giant peanut, 8 × 16 km across, and is one of the blackest bodies known in the Solar System, continuing the progression of darkness (that is, nonreflectivity) with distance from the Sun. On the daylight side is a dark area thought to be a crater of some sort, and a small "hill" on the nighttime side seems to catch sunlight. The most prominent features are

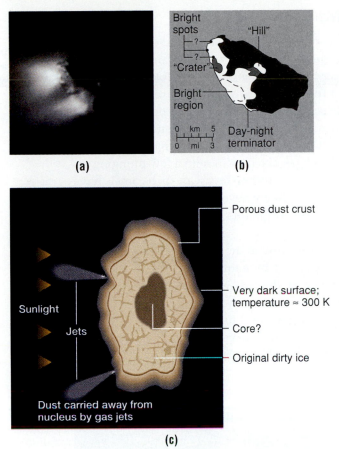

(a)

(b)

Figure 11.13
(a) The nucleus of Halley's Comet is revealed by ESA's *Giotto* spacecraft; the bright sunward side is to the left. The natures of the features identified in **(b)** as a "hill" and "crater" are uncertain. Gas spews from jets or geysers under the action of sunlight. **(c)** A model of Halley's Comet suggests a rocky core surrounded by dirty ice topped by a dust crust. Gas and dust stream out of the crust in jets.

(c)

geysers of gas filled with dust that spray from bright spots on the sunlit half.

From the orbital changes resulting from the rocket effect and the rate at which mass was observed to be lost, Halley's mass could be estimated at about 10^{14} kg (a mere 10^{-11} that of Earth), indicating that its density may be as low as 0.25 g/cm^3. The surface seems to be extremely porous and fragile. There may be a rocky core at the center (Figure 11.13c) surrounded by undisturbed dirty ice.

Comet nuclei typically lose 1% of their mass at each solar passage. The long-period comets are the brighter because their nuclei have lost the least matter. With the exception of Halley's, the short-period comets are faint. After enough perihelion passages, a comet's nucleus will lose all its volatiles and either disintegrate or reveal itself. Biela's Comet, discovered in 1772, broke in two at its 1846 perihelion passage, was seen at its next return in 1852, and then vanished. If a comet's nucleus does not completely disintegrate, it may look like an Apollo asteroid: up to half of the AAAOs may be dead comets.

11.3.3 The Sources of Comets

If comets are constantly disintegrating, there must be a huge pool of them. Several trillion long-period comets are thought to be contained within the **Oort comet cloud** (named after the Dutch astronomer Jan Oort), which extends 100,000 AU outward, a third of the way to the nearest star (Figure 11.14). The vast majority of these comets are on orbits that never take them close to the planetary system, and they therefore have neither comas nor tails. Over the ages, however, their orbits are disturbed by passing stars and massive clouds of interstellar gas, which will send a few plunging inward toward the Sun. Since they come from so far away, such comets have orbital periods so long that they will likely never be seen again, and since the Oort cloud is so spread out, the orbital tilts to the ecliptic will be randomly distributed.

Because they tend to stick fairly close to the ecliptic plane, most of the short-period comets must be supplied by another reservoir, the **Kuiper belt** (after Gerard Kuiper), which is aligned with the ecliptic plane and may be only about 200 AU in radius (as shown at the top of Figure 11.14). We cannot see comets within the Oort cloud itself: they are too far away and too faint. However, the Kuiper belt is accessible. As of 1996 we have found and plotted the orbits of over 20 such bodies, each a couple hun-

Several trillion long-period comets are thought to be contained within the Oort comet cloud.

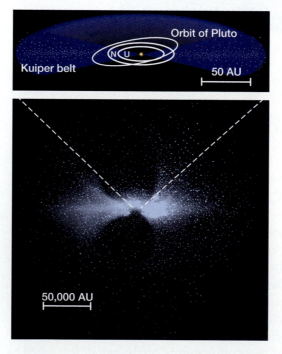

Figure 11.14
Comets are distributed in the great Oort cloud. The diagram above shows the Kuiper belt outside the orbit of Neptune.

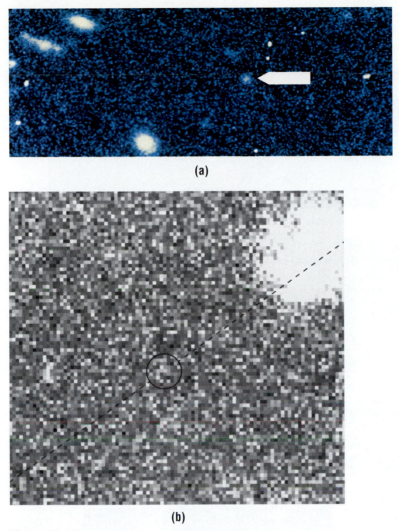

(a)

(b)

Figure 11.15
(a) The first Kuiper belt body found, 100-km-wide 1992 QB1, orbits just beyond Pluto.
(b) The Hubble Space Telescope seems to have revealed large numbers of tiny bodies, no more than a few kilometers across, within the belt.

dred kilometers across. Some are inside Neptune's orbit, but a few have semimajor axes as large as 45 AU, notably larger than that of Pluto (Figure 11.15a). Other observations with the Hubble Space Telescope possibly reveal tiny bodies no more than a few tens of kilometers wide, their number suggesting a Kuiper belt population of 200 million of them (Figure 11.15b). Perturbations by the giant planets eject some of the Kuiper belt comets from the Solar System and move others inward toward the Sun, to even smaller orbits, where with periods of only a few years, they are quickly destroyed or turned into bodies that look like asteroids. Pluto and Triton may be no more than giant examples, both trapped by Neptune.

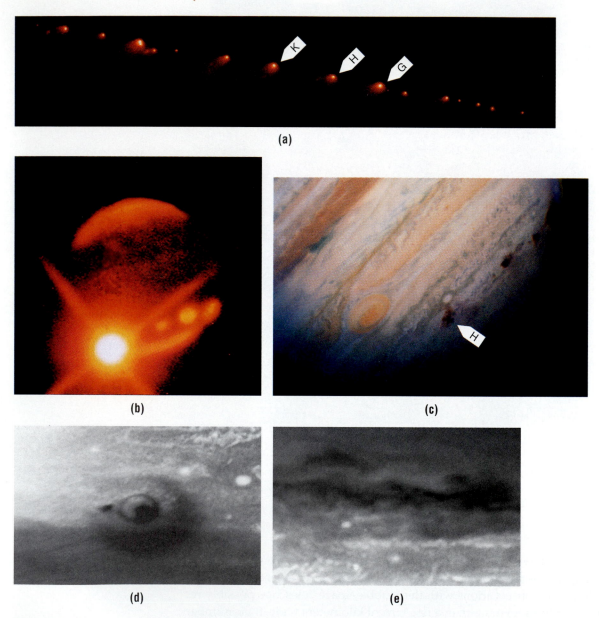

(a)

(b)   **(c)**

(d)   **(e)**

Figure 11.16
(a) The Hubble Space Telescope captured the fragments of Comet Shoemaker-Levy 9 strung out in a 600,000-km-long line headed toward Jupiter. The largest fragments are well under a kilometer across. **(b)** An infrared view of Jupiter shows the fireball produced by the impact of the large fragment. **(c)** Another Hubble image shows the dusty scars left in Jupiter's upper atmosphere by several impacts. **(d)** The dark impact scar of one fragment, seen on July 18, 1994, is **(e)** only five weeks later stretched out by winds above the cloud deck.

The number of comets is huge, but their combined mass is not. From estimated masses, all the comets rolled together would perhaps make a hundred Earths.

11.3.4 The Great Comet Crash

Sometime within the last century or two, a comet—now known as Shoemaker-Levy 9—passed through the satellite system of Jupiter and was slowed down and captured into orbit around the giant planet. Solar gravity gradually perturbed the orbit until the comet dipped well inside Jupiter's Roche limit, and it was torn into nearly two dozen fragments each at most only a few hundred meters across (Figure 11.16a). The orbit eventually lengthened until Jupiter simply got in the way, and over a six-day period one fragment after the other plunged into the Jovian atmosphere, heated from the shock of entry, and exploded. Moving at 60 km/s, the amount of energy liberated was staggering, hundreds of times greater than the world's entire nuclear arsenal. The strikes took place just around Jupiter's nighttime side, so they could not be viewed directly from Earth, but the planet's rotation quickly moved the impact sites into view, allowing us to see the resulting hot fireballs (Figure 11.16b). Condensing dust from the exploded debris subsequently produced dark impact scars in the Jovian atmosphere (Figure 11.16c).

The impacts dredged up sulfur from below, helping to confirm the existence of Jupiter's ammonium hydrosulfide cloud layer. A small amount of water was detected, but given the dryness of Jupiter, as found by the *Galileo* probe, it must have come from the comet. More than anything else, however, the event vividly showed that giant impacts between planets and interplanetary debris are not things of the past (see Section 7.4).

11.4 METEORS, METEOR SHOWERS, AND THE ZODIACAL LIGHT

Standing outdoors at three in the morning, facing in the direction of the Earth's orbital motion, you will typically see at least three meteors an hour. Most will be small and faint, quickly extinguishing themselves in the upper atmosphere. Several times a year you can see a **meteor shower** (Figure 11.17 and Table 11.4) that might provide more than a meteor a minute. If you trace the paths of these shower meteors backward, you will see that they all appear to emanate from a particular point in the sky, the **radiant.** Meteor showers are named after the constellation that contains their radiant: the Perseids of August 12 come out of Perseus.

Meteors and meteor showers are intimately linked to comets. As cometary nuclei heat and disintegrate, they leave behind a trail of debris, probably surface dust compacted into fragile, rocklike structures. The orbits of the particles gradually diverge into a band. When the Earth crosses

Figure 11.17
Several meteors radiating from the constellation Perseus flash across the sky.

TABLE 11.4
Prominent Meteor Showers[a]

Name	Date[b]	Hourly Rate[c]	Comet
Quadrantids[d]	January 4	100	?
Lyrids	April 20	10	1861 I
Eta Aquarids	May 5	30	Halley
N. Delta Aquarids	July 29	15	?
Alpha Capricornids	August 1	10	?
Perseids	August 12	80	Swift-Tuttle[e]
S. Delta Aquarids	August 13	10	?
Orionids	October 21	20	Halley
Leonids[f]	November 17	10	1866 I
Geminids	December 13	80	3200 Phaethon[g]

[a]Selected from about 100 known annual showers.

[b]Peak; meteors can usually be seen on many days to either side.

[c]For a dark sky; number can vary considerably.

[d]Named after the obsolete constellation Quadrans near the Big Dipper.

[e]1862 III.

[f]Source of spectacular meteor storms every 33 years.

[g]Classified originally as an asteroid; probably a dead comet.

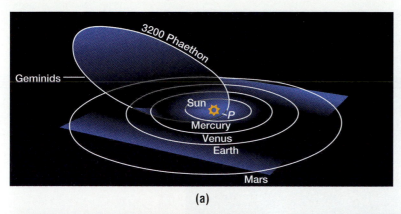

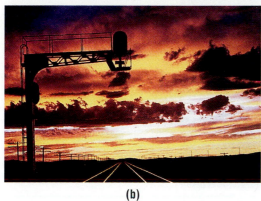

(a)

Figure 11.18
(a) A stream of meteoroids follows the path of the nearly dead comet 3200 Phaethon (once thought to be an asteroid). When the Earth crosses the orbit, we see the Geminid meteor shower. **(b)** Parallel railroad tracks appear to come from a point in the distance. **(c)** Meteoroids on parallel tracks (solid arrows) hit the atmosphere and as we look at them (dashed arrows), appear to be coming from a point or radiant in the sky.

(b)

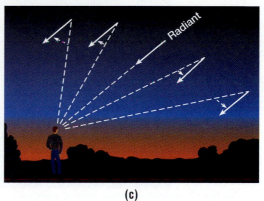

(c)

the orbital path of the stream (Figure 11.18a), we see a shower. The railroad tracks in Figure 11.18b are actually parallel to one another, but they look as if they diverge from a point on the horizon. To a good approximation, the little meteoroids are moving along parallel tracks (Figure 11.18c). When they hit the Earth's atmosphere and become visible, they seem to be coming from the shower's radiant. The radiant for a particular shower depends on the relative movement of the Earth and the parent comet.

The particles may be unevenly distributed along the comet's orbit, resulting in a large assembly of them moving around the Sun. If the Earth collides with the dense cloud, a **meteor storm** (Figure 11.19) will ensue. The Leonids of November 17 ordinarily do not produce many meteors, but every 33 years the Earth and the dense collection of debris collide and a lucky viewer may see thousands of meteors per hour falling from the sky. The Leonid storm is due again in 1999.

However they are distributed, the tiny particles gradually spread out from the original orbital path and lose sight of their siblings. These are the sources of the sporadic meteors, the ones that do not belong to particular

Figure 11.19
An old woodcut gives a spectacular impression of the Leonid storm of November 13, 1833.

Figure 11.20
The zodiacal light, sunlight scattered from interplanetary dust, is seen along the ecliptic after evening twilight or before dawn.

showers. The fine debris of a comet that is ejected to form the dust tail also spreads out into the Solar System, where it combines with dust produced by asteroid collisions. The asteroids and the short-period comets, those ripe for destruction, concentrate toward the ecliptic. The dust scatters sunlight and produces a faint conical glow, the **zodiacal light,** which rises from the horizon (Figure 11.20) along the ecliptic after dark or before dawn. The scattering of radiation produces a drag effect on the particles and makes them spiral into the Sun. They must then constantly be replaced by comet disintegration and asteroid collisions.

11.5 THE CREATION OF THE SOLAR SYSTEM

Strong evidence suggests that asteroids, comets, and perhaps bodies like Pluto and Triton are the remains of the planetary building blocks, the planetesimals. The asteroids lie in the ecliptic plane and go around the Sun counterclockwise, as does (more or less) one family of comets. If not for Jupiter's gravitational influence, the asteroids would probably have assembled into a small terrestrial planet. The compositions of comet nuclei and asteroids correlate with distance from the Sun, as they do for the planets and their satellites. The inner asteroids are relatively bright and rocky, the more distant asteroids darker and more primitive, the distant comets dark and icy, the two probably overlapping in characteristics.

The Sun appears to have been made from a contracting cloud of interstellar gas that was rich in molecules and infused with fine dust (this event is the subject of Chapter 15). The cloud, which was rotating and producing the Sun at its center, was subject to the law of conservation of angular momentum. As it squeezed down, it spun faster, flattening into a disk called the **solar nebula** (Figure 11.21a). This remnant of solar formation was the birthplace of the planets. The disk became the ecliptic plane. All the planetary motions, the counterclockwise orbits and spins, represent the rotation direction of the pre-planetary disk.

Within this disk, the original interstellar dust was subject to a variety of forces and influences. Among the most important controlling factors

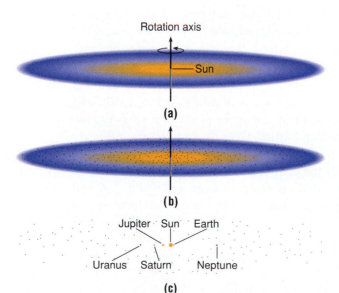

Figure 11.21
In this model, the solar nebula **(a)** began with a disk around the newly forming Sun. The upward pointing arrow shows the rotation axis. **(b)** Planetesimals formed in the disk as a result of dust accretion and gas condensation. **(c)** The planetesimals eventually accreted into planets and planetary cores as the solar nebula, dissipating under the action of the new Sun, was accreted into the outer layers of the Jovian planets. Some planetesimals remained in the disk outside the planetary system; those inside that were not swept up were ejected or thrown into the Oort cloud.

was temperature. At the outermost fringes it was very low, in the tens of degrees K. Toward the inner Solar System it climbed sharply, near 2,000 K close to the forming Sun, where the dust would have been vaporized. As the solar nebula cooled, the gas could condense onto the original dust grains to build larger particles of matter. Refractory molecules, like metallic oxides and silicates, enter and leave the gaseous state at high temperature and could condense into and onto solid grains everywhere in the nebula, even near the Sun. Others—volatiles like water, methane, and ammonia—could condense only at great distances from the Sun, beyond several AU. The composition of the solid raw material of the planets therefore depended on how far from the Sun it was formed and on its local environment.

Within the disk (see Figure 11.21a) everything was moving in the same direction at velocities governed by Kepler's laws of motion. At a specific radius from the Sun, all the grains had nearly the same speed and moved slowly relative to one another. When two bumped together they could fuse. As the grains were growing fatter by gas condensation, they also enlarged by accretion of other particles. Heat generated on impact (or by gigantic electrical discharges—lightning bolts—within the nebula) could have melted some of the solid material to produce the chondrules seen in primitive meteorites. The effect of the collisions snowballed as increasing particle sizes caused the capture of still more particles. When the particles became big enough, gravity began to play a role, and the larger ones grew at the expense of the smaller. Within perhaps 100,000 years, the dust had congealed into trillions of small bodies, the planetesimals (Figure 11.21b). The warmer ones near the center of the budding Solar System were rocky, the colder ones farther away icy. During all this activity, the new, hot Sun was driving the gas out of the inner Solar System, ending gas condensation.

In the inner Solar System, four bodies finally won out, becoming rocky Mercury, Venus, Earth, and Mars.

The planetesimals continued to collide and grow. The larger ones dominated because of their gravitational pulls and began to consume the others. The result was hundreds, perhaps thousands, of larger orbiting pre-planetary bodies. Gravitational perturbations caused their orbits to cross, resulting in violent merging collisions. In the inner Solar System, four bodies finally won out, becoming rocky Mercury, Venus, Earth, and Mars. In the outer Solar System, the accretion of planetesimals made the cores of the great planets Jupiter, Saturn, Uranus, and Neptune (Figure 11.21c).

The gravitational pulls of the growing planets were so high that the impacting planetesimals imparted enormous energy upon collision and the new planets melted and differentiated. The last of the collisions ripped away some of Mercury's mantle, created the Earth-Moon system, and tipped Uranus on its side. The rotations of the terrestrial planets, including Venus's backward spin, were set by the collisions as well.

In the outer Solar System, where the solar nebula was still more or less intact, the great gravity of the cores began to sweep the hydrogen and helium into vast envelopes, completing the formation of the Jovian planets. Their rapid spins threw some of the gas outward into a disk, from which their satellites condensed in much the same way as the planets.

The sizes of the planets depended on the distribution of mass within the solar nebula while they were being created. The new Sun probably swept much of the gas from its immediate neighborhood, and as a result, the planets get bigger out to Jupiter. Mars turned out small because Jupiter's gravity ejected planetesimals from the region in which the planet was being formed. Beyond Jupiter, the density of matter in the original solar nebula dropped with distance, so the outer planets grew more slowly. By the time Uranus and Neptune grew to the size where they could attract significant hydrogen and helium atmospheres, the solar nebula had largely dissipated, leaving them with lower masses and minimal light outer layers. After a hundred million years, there was a Sun and a retinue of eight cooling planets, the terrestrials developing solid crusts, then mantles.

It is likely that the comets supplied all the water needed to fill the Earth's oceans.

Trillions of planetesimals were left to a variety of fates. A large fraction was swept up by the planets and their satellites. By that time the planetary surfaces had turned solid and the colliding meteoroids produced permanent craters. We see the end product of this massive violence in the craters of the late heavy bombardment. The giant planets agitated the icy planetesimals—the comets—and huge numbers of them made it into the inner Solar System. They crashed into the Earth and other terrestrial planets, bringing supplies of volatiles that had previously been driven away, particularly water and perhaps even carbon. It is likely that the comets supplied all the water needed to fill the Earth's oceans, providing the necessary conditions for life.

As planetesimals developed between Mars and Jupiter, those near the giant planet were gravitationally kicked into eccentric orbits toward the Sun and in turn gravitationally stirred other planetesimals, prompting them to collide. Each collision produced more objects, causing further collisions; instead of accumulating, the bodies of the asteroid belt ground themselves down. The few remaining large bodies like Ceres were lucky to avoid great hits. The parent bodies of the present asteroids could never have been very large, and how they became hot enough to differentiate is still a mystery. Heat generated in the radioactive decay of ^{26}Al (whose daughter product is seen in meteorites) may have been in part responsible. Intense solar magnetic fields generated when the Sun was very young and just forming may also have heated the closer asteroids. As a result, the most primitive asteroids are the farthest away.

Jupiter hurled some asteroids and vast numbers of distant, icy planetesimals out of the Solar System altogether. Uranus and Neptune, with smaller gravities, threw cometary planetesimals into the Oort cloud. Planetesimals beyond the planetary system stayed more or less in place as comets in the Kuiper belt, their number not great enough ever to have formed a planet. A few really large planetesimals, such as Pluto and Triton, lurk at the outer fringes of the Solar System and may be representative of many more, similar bodies.

Though the pace of planetary and Solar System evolution slowed dramatically, it has never ceased. The planets continued to cool; the terrestrial planets developed atmospheres from their interiors and—along with the satellites of the Jovian planets—established unique volcanic and tectonic characteristics that on Earth and Venus are still highly active. And when we stand outdoors on a dark night, we can still watch a few remaining planetesimals orbit the Sun or even slice through the atmosphere as meteors.

We seem to have explained everything neatly: mathematical models calculate terrestrial planets forming in about the right positions. Yet a great many mysteries and uncertainties remain, and much of this satisfying picture may be wrong. We have not explored our amazing system in complete detail and do not even have a full census of its contents. Whatever the final picture assembled, however, one result was a great blue ball, the third planet from the Sun, the birthplace of life. From it we look out on the rest of our home, the Solar System.

▶ **KEY CONCEPTS**

AAAOs: Asteroids in the Amor, Apollo, or Aten families, all of which come close to the Earth's orbit.

Asteroids: Small bodies that orbit the Sun, mostly between Mars and Jupiter; concentrated in a **main belt** between 2 and 3 AU from the Sun; different kinds relate to meteorite classes.

Comet: A fragile interplanetary body with a **nucleus** of dirty ice that is heated by the Sun to produce a surrounding **coma** and **tails** of gas and dust; **short-period comets** have periods less than 200 years and come from the Kuiper belt; **long-period comets** have random orbits longer than 200 years and come from the Oort cloud.

Dust tail: The diffuse, fan-shaped tail of a comet, caused by sunlight scattered from dust released by the nucleus.

Gas tail: A comet's tail of ionized gas, which points away from the Sun and is structured by the Sun's wind and magnetic field.

Kuiper belt: A disk-shaped reservoir outside the orbit of Neptune that produces the short-period comets.

Meteor: A bright tube of ionized atmosphere caused by the passage of a meteoroid.

Meteorites: Meteoroids that land on the Earth; classified as **stones, irons,** and **stony irons;** stones are subclassified as **chondrites** (those with chondrules), **carbonaceous chondrites** (primitive chondrites with a high carbon content), and **achondrites** (those with no chondrules).

Meteoroid: A piece of interplanetary debris that produces a meteor when it enters the Earth's atmosphere.

Meteor shower: A display of meteors that appears to emanate from a specific part of the sky, the **radiant;** seen when the Earth passes near a comet's orbit.

Meteor storm: An intense meteor shower caused by a concentration of cometary debris.

Oort comet cloud: A reservoir roughly 100,000 AU in radius that contains trillions of comet nuclei; the origin of the long-period comets.

Solar nebula: The disk of gas and dust around the forming Sun, out of which the planetesimals and the planets grew.

Zodiacal light: A band of light in the zodiac caused by sunlight scattered from cometary and asteroidal dust.

▶ EXERCISES

Comparisons

1. Distinguish among meteors, meteoroids, and meteorites.
2. Distinguish between ordinary and carbonaceous chondrites.
3. What is the difference in formation between the iron and the stony-iron meteorites?
4. What are the differences between short- and long-period comets?
5. What are the differences in the structures of the Kuiper belt and the Oort comet cloud?
6. How do a meteor shower and a meteor storm differ?

Numerical Problems

7. Assume that Jupiter is at the vernal equinox. In what constellations of the zodiac would you expect to find the Trojan asteroids (see Figure 11.4a)? What is the angular extent of one set of the Trojan asteroids as viewed from Earth?
8. Estimate the total mass of the comets, making clear all your assumptions.

Thought and Discussion

9. What are chondrules? How might they relate to the origin of our Solar System?
10. Why do we think that iron meteorites were once part of the interiors of larger bodies?

11. What are the relationships between meteorites and asteroids?
12. What evidence is there for interstellar material in asteroids?
13. What observations confuse the distinction between asteroids and comets?
14. Why do comets lack tails when they are at Saturn's distance from the Sun?
15. What roles do the solar wind and the solar magnetic field play in the creation of comet tails?
16. Why do the meteors in a shower appear to come from a radiant?
17. Why are the distant planetary satellites icy, whereas the terrestrial planets are rocky?
18. How were dust particles able to accumulate into planets in the early days of the Solar System?
19. What heating mechanisms could have differentiated some of the asteroids? Which of the mechanisms might explain why the more distant asteroids are the more primitive?
20. How was the Oort cloud filled with comets?
21. How was the Kuiper belt filled with comets?

Research Problem

22. The *Circulars* of the International Astronomical Union give information on new astronomical discoveries. If they are available in your library, make a list of the comets that have passed perihelion in the last three years. What is the ratio of the numbers of short- to long-period comets? How does the number of comets vary with brightness? What does that tell you about the number of large comets relative to the number of small comets?

Activities

23. Plot the percentage reflectivities of the 15 brightest asteroids against their distances from the Sun. What is the significance of your results?

24. Average the known densities of the satellites of each of the Jovian planets. Graph these average densities against distance from the Sun. Comment on what your graph may mean and how the differences may have come to be.

Scientific Writing

25. Someone sends you samples of rock and iron, claiming that they are meteorites. Write a letter to your correspondent. Take either side of the issue, that they are or are not meteorites, and explain the reasons for your opinion.

26. You have been invited to give a lunch talk to your local Rotary Club, and you choose as your topic the origin of the Earth. Write the script for a 20-minute talk explaining how the Earth formed. Your audience has no background in astronomy and will be unfamiliar with technical terms.

PART

III

STELLAR ASTRONOMY

THE SUN
The Solar System's central engine and a fine example of a star

12

The planets are by-products of the formation of the central body of the Solar System, the Sun (Figure 12.1). One of 200 billion stars in our Galaxy, its characteristics fall near the middle of the ranges of stellar properties (*stellar* means "relating to stars"), and it is the only star on which we can see detailed activity, thus providing a model for understanding general stellar behavior.

12.1 IMPRESSIVE ATTRIBUTES

The Sun is like nothing we have yet encountered. Through a protective filter, we can see it is about half a degree across, the same angular size as the Moon. But the Sun is almost 400 times more distant than the Moon. Its radius is thus 400 times larger, an astonishing 6.96×10^5 km (Table 12.1); 109 Earths could be placed across it. From the Sun's volume and its mass of 2×10^{30} kg (333,000 Earths, calculated from the Earth's orbit and Kepler's generalized third law) we find an average density of 1.4 g/cm³, quite close to that of Jupiter. Yet the two are nothing alike. If the Sun were liquid (like

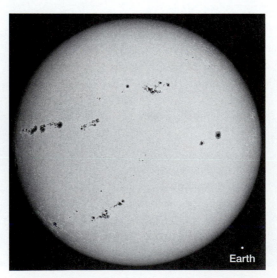

Figure 12.1
The Sun's diameter is 109 times that of Earth (which is represented here to scale by the dot at lower right). The solar photosphere, covered with spots, darkens toward the limb.

TABLE 12.1
A Profile of the Sun

Radius	6.96×10^5 km
Mass	1.99×10^{30} kg
Mean density	1.4 gm/cm^3
Gravity	27.9 g$_{Earth}$
Escape velocity	618 km/s
Magnetic field strength	About 5 times Earth
Rotation	
Equatorial period	25.4 days
Axial tilt	7°15′
Effective temperature	5,780 K
Luminosity	3.83×10^{26} watts

Layer[a]	% Mass	Radii (km)	Temperature (K)	Pressure (bar)	Density (g/cm^3)
Core	40	$0-2 \times 10^5$	1.5×10^7	1.5×10^{11}	160
Radiative envelope	59	$2 \times 10^5 - 5 \times 10^5$	4×10^6	6×10^8	2
Convective envelope	1	$5 \times 10^5 - 6.96 \times 10^5$	1×10^6	1×10^4	0.01
Photosphere	——	6.96×10^5	5,780	0.1	10^{-9}
Chromosphere	——	——	10,000	low	low

[a]Temperature, pressure, and density given here refer to the center of the core and to a point midway through the other layers. The photosphere is only a few hundred km thick and the chromosphere a few thousand km thick. The masses of the outer three layers are insignificant. The radius of the corona is ill defined and extends many solar radii from the photosphere, eventually merging with the solar wind.

Jupiter) or solid (like the Earth), gravitational compression caused by its huge mass would make the density *much* higher than we observe. The Sun must therefore be a gas. Unlike any of the planets, the Sun is gaseous throughout.

Also unlike the planets, the Sun radiates entirely by itself, from energy generated within, with an astounding luminosity of 3.85×10^{26} watts. To place this huge number in perspective, allow your local power company to run the Sun for a mere second. The bill would come to 7×10^{18}—the gross national product of the United States for the next 7 million years. To be so bright at visible wavelengths, the Sun must also be hot. The radius, luminosity, and the Stefan-Boltzmann law (see Section 6.3.1) yield an effective surface temperature of 5,780 K. From the Wien law, the peak of the blackbody curve falls at 5,020 Å. The combined radiation at all wavelengths makes a shaft of sunlight appear a soft yellow-white to the eye.

It would cost the gross national product of the United States for 7 million years for your local electric power company to run the Sun for a second.

12.2 THE QUIET SUN

Astronomers have long divided solar phenomena into two parts: the **quiet Sun,** a collection of features that are always present, and the **active Sun,** which includes features that come and go and are associated with intense magnetic activity.

12.2.1 The Thin Grainy Photosphere

The quiet Sun is divided into three principal layers, by far the most prominent of which is the brilliant apparent solar surface, the **photosphere.** Ions within the photosphere strongly absorb outflowing radiation, making the gas highly opaque. As a result, the edge, or **limb,** of the Sun is razor-sharp as seen from distant Earth, giving the gaseous Sun the false appearance of being solid.

The Sun is not uniformly bright across its surface (see Figure 12.1) but becomes steadily darker away from the center toward the limb. **Limb darkening** shows both that the gas is not *perfectly* opaque and that temperature increases with depth, as expected. At the center of the apparent solar disk (Figure 12.2) we see to a certain depth within the gaseous photosphere. Toward the limb, we see through about the same thickness of gas, but since we look at an angle, we do not look as deep. The solar limb therefore appears cooler to us than the center, and from the Stefan-Boltzmann law, darker. Measurement of limb darkening is used to find the rate at which temperature changes with depth, allowing us to check theories of photospheric structure.

The photosphere is not smooth but is covered by millions of tiny **granules** (Figure 12.3a), bright cells about 700 km (at the solar distance about a second of arc) across, separated by their own widths and set into a darker background. Each granule typically lasts for only a few minutes before it disappears and is replaced by another (Figure 12.3b). The spectra of the granules and of the darker background are Doppler-shifted to shorter and longer wavelengths respectively, showing that the granules are rising and that the darker matter is falling (Figure 12.3c). The granules are thus seen to be the tops of giant convection cells that bring heat to the surface from below. They radiate their energy, cool, darken, and fall.

The solar spectrum (Figure 12.4) displays a great number of absorption lines in a classic example of Kirchhoff's third law (see Section 6.3.2).

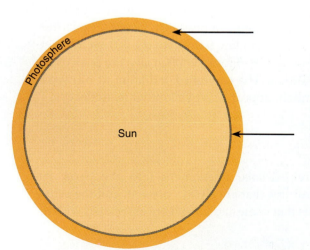

Figure 12.2

When we look in the direction of the arrows from Earth (which is far to the right, off the page) at the center of the photosphere, our vision penetrates to deeper, hotter, and brighter gases than when we look near the limb. The thickness of the photosphere is greatly exaggerated here for clarity; on this scale, it would be only 0.01 mm thick.

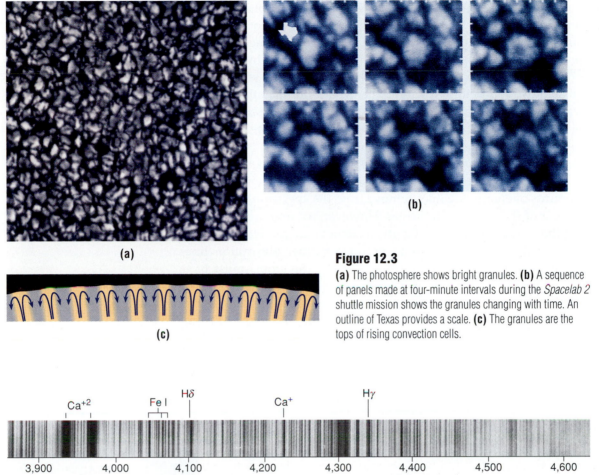

(a)

(b)

(c)

Figure 12.3

(a) The photosphere shows bright granules. **(b)** A sequence of panels made at four-minute intervals during the *Spacelab 2* shuttle mission shows the granules changing with time. An outline of Texas provides a scale. **(c)** The granules are the tops of rising convection cells.

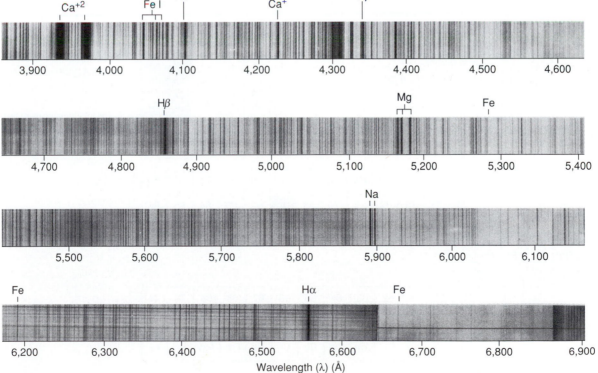

Figure 12.4

The optical solar spectrum is dominated by hundreds of absorption lines, mostly of common metals.

In the simplest sense, the lower, hotter layers of the photosphere generate a continuous spectrum. The upper layers constitute a gas of lower temperature and density and thus superimpose absorption lines. The most prominent lines are those of ionized calcium, Ca^+, followed by the Balmer series of hydrogen (the collection of lines that arise from the second orbit), neutral sodium, and neutral calcium. The majority of the other lines are produced by common metals like iron, chromium, and titanium. Some 70% of the natural elements (Figure 12.5) are observed in the solar gases, and we are confident that they are in fact all present.

We can use the spectrum to derive the chemical composition of the Sun. The *strengths* of a particular atom's—or ion's—absorption lines (the amounts of energy the lines remove from the spectrum) depend on the number of those atoms in the gas. However, we first must take into account that not all the atoms of a given element are *capable* of creating a particular line. For hydrogen to produce Balmer absorption lines (those commonly seen in the visual spectrum), the atoms must have their electrons in the second orbit (see Section 6.3.2 and Figure 6.13). Within the gas, electrons are knocked from one orbit to another by atomic collisions. The higher the temperature, the more energetic the collisions, and the greater the percentage of hydrogen atoms with electrons

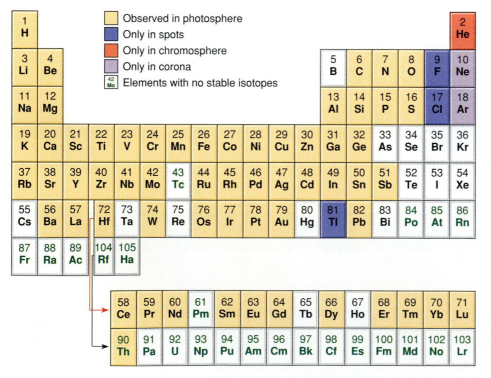

Figure 12.5

The periodic table arranges the chemical elements by increasing atomic number according to their chemical properties. The two rows at the bottom are inserted as indicated. Elements identified in the Sun are shown in color according to the solar layer or region in which they are found.

in outer orbits. But at the temperature of the solar photosphere, only one hydrogen atom out of every 10^{10} will have an electron in orbit number two. As a result, very few hydrogen atoms can absorb the Balmer lines. The lines of ionized calcium, however, arise from the ground state, so nearly *every* ion is capable of producing the lines. When we take these different absorption efficiencies into account, we can calculate the ratio of the number of ionized calcium atoms to the number of hydrogen atoms. In spite of the fact that the Ca^+ lines are stronger than the hydrogen lines, there is only one Ca^+ ion for every half-million hydrogen atoms! Similar analysis for other calcium ions gives us the total ratio of calcium to hydrogen.

Analysis of the entire solar spectrum reveals that the photosphere is 92% hydrogen (by number of atoms) and 8% helium (the helium abundance derived from observation of the outer solar layers and the solar wind). *All the other atoms constitute no more than about 0.1%.* Generally, the greater the atomic number of the element, the less there is of it (Figure 12.6), though lithium, beryllium, and boron are extremely scarce. There is also a rise around iron, the *iron peak,* that includes cobalt and nickel. These and other patterns provide clues to the ways in which the elements were created in the birth of the Universe and in stars. Above atomic number 2, the relative abundances of the atoms are similar to what we find in the Earth and in stony meteorites. Our Earth, the other terrestrial planets, and the planetesimals, are effectively a distillate of the solar gases.

Our Earth is effectively a distillate of the solar gases.

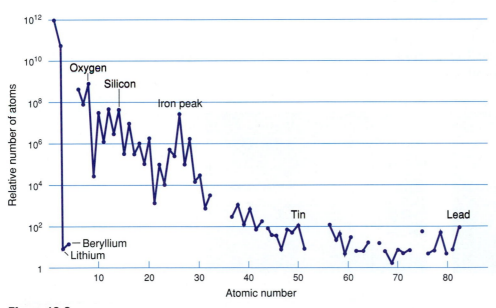

Figure 12.6
The abundances of the elements in the Sun are plotted according to powers of 10 against atomic number. The gaps indicate elements that have yet to be observed.

12.2.2 Outer Layers

Lying above the photosphere is a thin layer a few thousand kilometers thick, the **chromosphere** (Figure 12.7a), that produces so little radiation that it can be seen only during an eclipse or with special instruments. Because it is a gas under low pressure, its spectrum (Figure 12.7b) consists of emission lines, prominent among which are emissions of helium. This element was in fact discovered in the Sun before it was found on Earth, hence its name, from the Greek word for the Sun, "helios."

If we photograph the Sun using just the light radiated by one of these emission lines, we can take the chromosphere's picture (Figure 12.8a). Projecting from the chromosphere's base are millions of tiny needles of rising gas, *spicules*. Like the granules, the spicules last only a few minutes before they vanish and are replaced by others. Temperature, found from the strengths of spectrum lines, steadily declines upward in the photosphere to a minimum of about 4,500 K at the top (Figure 12.8b). Then the spectrum shows that as we enter the chromosphere, temperature increases to over 10,000 K. But the chromosphere's low density prevents it from being a blackbody, so it cannot emit much radiation. The temperature is a kinetic temperature (see Section 7.7) that describes only the speeds of the atoms in the gas.

As the photospheric gases churn over during convection, they create sound waves that move outward into the lower-density gas; there the speed of sound is lower, and the waves become shock waves (see Section 7.8). The shocks dissipate their energy into the surrounding gas, heating it and producing the spicules. Convection also generates waves and turbulence in the solar magnetic fields; the resulting energy is then deposited into the chromosphere. However, a complete theory of the formation and heating of the chromosphere still eludes us.

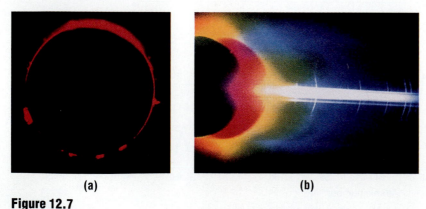

(a)   (b)

Figure 12.7
(a) The red rim of the chromosphere surrounds the eclipsed Sun. **(b)** The narrow chromosphere spreads out into an emission spectrum.

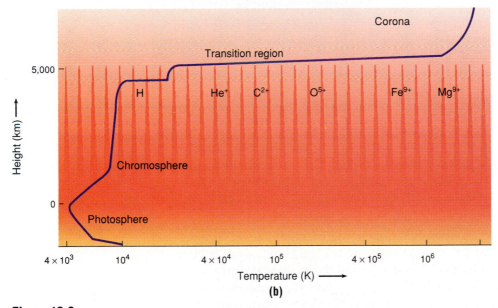

(a)

(b)

Figure 12.8

(a) A photograph of the Sun taken with only the light radiated by the Hα emission line shows the chromosphere and its upward-projecting spicules. **(b)** Temperature drops within the photosphere to about 4,500 K, then rises in the chromosphere. At the top of the chromosphere, a few thousand kilometers above the photosphere, temperature rapidly rises in a transition region to over 2 million K in the corona. Observed ions are shown along the curve.

Figure 12.9
Shining from a thin, magnetically confined gas heated to 2 million K, the solar corona is seen in full glory during a total solar eclipse.

Above the chromosphere lies the outer halo of the Sun, the **corona** (Figure 12.9, see also Figure 4.10). Superimposed upon a continuous spectrum (produced by photospheric light scattered from electrons) are emission lines of metals like iron ionized up to 16 times. For atomic collisions to strip so many electrons from an atom requires temperatures over 10^6 K. The coronal average of 2 million K is hot enough to produce X rays, but the density is so low that the corona is far from being a blackbody: it is visually dim and hidden by the blue sky. To see the corona properly, we must either wait for an eclipse or go into space (where the X rays may also be observed). The corona is too energetic to be heated by shock waves. It is a creature of solar magnetism, a principal feature of the *active* Sun.

12.3 THE ACTIVE SUN

We pay the Sun little heed except to note its warmth. Only when we see an aurora or our radio or television broadcasting goes awry might we have direct contact with the active Sun. In spite of the hidden nature of solar activity, we are slowly beginning to see that it is critical to life on Earth.

12.3.1 Sunspots and the Magnetic Activity Cycle

In 1610, Galileo found dark spots on the Sun, **sunspots** (see Figures 12.1 and 12.10a). They tend strongly to appear in pairs, and the pairs in groups called **centers of activity.** The spots range to 20,000 km or more in diameter, much larger than the Earth. Sunspots have dark centers surrounded by lighter rings. With central temperatures of about 4,500 K, sunspots radiate 35% as much light as the photosphere and appear dark only by contrast. They are flat depressions with no granulation a few hundred kilometers below the photosphere, the lighter outer portions sloping upward to the unspotted surface. The spots are born as small dark spaces between the granules and then grow. They can live from a few hours to months, depending on the sizes they attain.

The spots appear to move slowly across the solar disk as a result of solar rotation. The Sun is a differential rotator (see Section 9.2.1), near the equator taking 25 days to turn and near the poles closer to 30 days. The rotation axis is tilted only about 7° from the perpendicular to the ecliptic. The rotation is in part responsible for a dipole magnetic field that aligns with the rotation axis and has a strength about five times Earth's.

A sunspot's magnetic field is some 5,000 times that of Earth's.

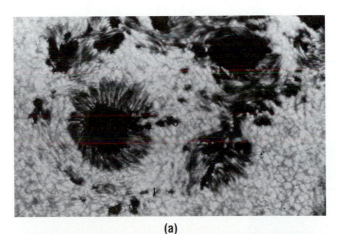

(a)

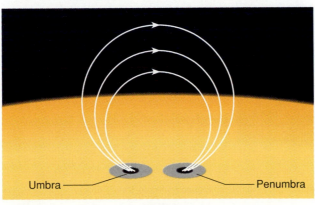

Umbra ——

—— Penumbra

(b)

Figure 12.10

(a) A huge sunspot group shows individual spots with dark centers and lighter surrounding rings. **(b)** A typical sunspot pair is created by a magnetic field that loops out of the Sun and cools the photospheric gases.

This magnetic field is enormously magnified within the spots to a strength some 5,000 times that of Earth's. The field loops out of the photosphere from one spot of a pair to the other (Figure 12.10b). The magnetic field somehow prevents convection and the flow of energy, thus chilling the spots. The chromosphere above the spots is bright (Figure 12.11) rather than dark, a manifestation of magnetic energy. As the magnetic loops grow and then disintegrate, so do the spots.

Spectrum lines produced by atoms that are within a magnetic field split into multiple components (Figure 12.12a). The stronger the field, the greater the splitting. The effect allows us to map the strengths and directions of magnetic fields all over the Sun (Figure 12.12b). We see that the leading spots of one hemisphere—those ahead in the direction of rotation—have identical magnetic directions (north or south), and the following spots the opposite magnetic directions. In the other hemisphere, the directions are reversed.

In 1851, the German astronomer Heinrich Schwabe discovered that the number of sunspots on the solar surface varies with an approximate 11-year period originally called the **sunspot cycle** (Figure 12.13a). As a new cycle begins, a few spots appear in both hemispheres at solar latitudes near 45°. The number of spots slowly grows with time, and their average latitudes creep toward the solar equator (Figure 12.13b). When the average latitude reaches about 10° to 15° north and south, the count rapidly diminishes to near zero. As one cycle begins to run out, a new one starts at high latitudes. During one cycle the magnetic directions stay the same, the leading spots having the same magnetic direction as the pole of

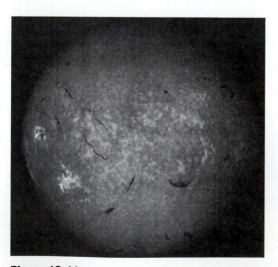

Figure 12.11
A photograph of the solar chromosphere shows that the active regions are bright. Sunspots in the photosphere below the chromosphere are not visible. The long dark streaks are filaments, prominences (see Figure 12.17) seen against the surface.

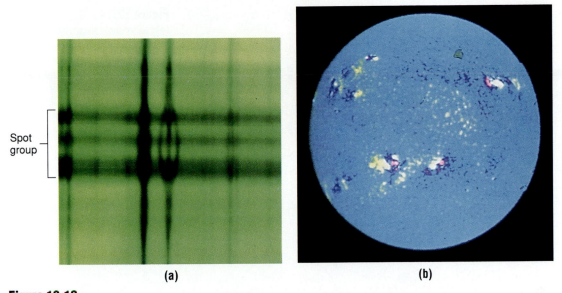

(a)   **(b)**

Figure 12.12

(a) The dark horizontal strips are the spectra of sunspots. Absorption lines from the unspotted photosphere are single but are split by the intense magnetic field within the center of activity. **(b)** Such spectra allow us to produce a magnetic map. Dark blue areas show a north magnetic direction, yellow south. Magnetic directions of spot pairs are the same in one hemisphere, the opposite in the other.

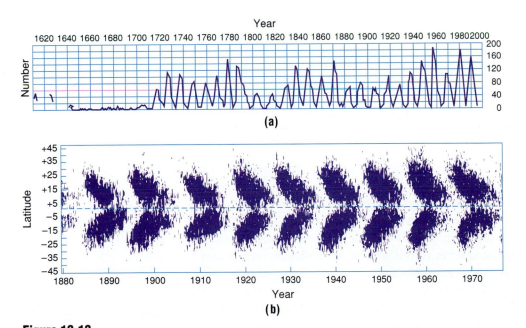

Figure 12.13

(a) The sunspot cycle is shown for 1615–1996; the average monthly number of spots is given on the vertical axis. The variation from one maximum to the next is quite large, and the spots were nearly absent between 1645 and 1710. **(b)** A more detailed diagram shows how the positions and numbers of spots change with solar latitude. At any given time, the spots occupy a large range in latitude.

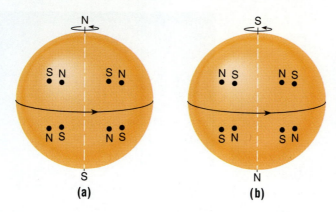

Figure 12.14
(a) Leader spots always have the same magnetic direction as the pole of the hemisphere that contains them; the followers have the opposite magnetic direction. During one 11-year cycle, the magnetic directions stay the same. **(b)** During the next cycle, all the magnetic directions are reversed.

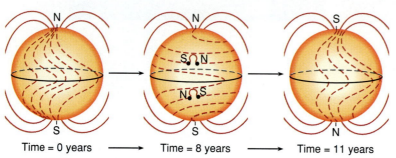

Time = 0 years ⟶ Time = 8 years ⟶ Time = 11 years

Figure 12.15
In the principal theory of the solar cycle, at the start of an 11-year cycle the Sun's magnetic field is a simple dipole. The field lines are frozen into the solar gases and differential rotation causes them to be wrapped around the Sun. They become concentrated and pop through the surface to produce centers of activity and magnetically opposite spot pairs. The greater the wrapping, the more spots we see. When the field lines become thoroughly tangled, however, they break down and reorder with reversed direction to start a new cycle.

their resident hemisphere (Figure 12.14a). In the next cycle, however, they are reversed (Figure 12.14b), as are those of the poles. Eleven or so years later, all the magnetic directions reverse again. The 11-year sunspot cycle is really a 22-year **magnetic activity cycle.**

The principal hypothesis for the origin of the solar cycle involves solar rotation. The gases in the interior of the Sun are ionized. Convection and motion stirred up by rotation cause this electrified matter to circulate, generating the Sun's dipole magnetic field. A new cycle begins with a simple dipole whose field lines are locked into the ionized gases of the surface (Figure 12.15). However, because the Sun rotates faster at low latitudes than at high, the field lines become stretched out along the equator and eventually spiral around the Sun. This effect, combined with convection, twists and concentrates the field lines into dense ropes. The intense magnetism produces a pressure that makes the magnetic ropes float upward with the convection cells. When a magnetic rope is squeezed through the surface, it produces a pair of spots. We therefore see why the magnetic directions of spot pairs must be the same in one hemisphere and the opposite in the other. As the magnetic field becomes more wound up, activity increases. Finally, after about 11 years, the tightly wrapped field breaks apart and the number of spots decreases. The field then reorders itself, but with reversed directions.

This hypothesis, however, does not account for other puzzling observations, for example for streams of solar gas that flow in alternating directions parallel to the solar equator and that drift from the pole to the equator over the entire 22-year cycle. A complete theory of the solar cycle is still out of our grasp.

12.3.2 Prominences, Flares, and the Solar Wind

The magnetic loops can extend far into space where they concentrate coronal gases to regions above centers of activity and give the corona great structure (Figure 12.16a). The hot gas of the corona is best seen in X-ray images (Figure 12.16b) that dramatically outline *coronal holes*, regions where the coronal gas is faint or absent. The X-ray-bright areas are further connected by fainter large-scale loops. The solar wind (see Section 7.8), which causes the Sun to lose matter at a rate of some 10^{-13} of a solar mass

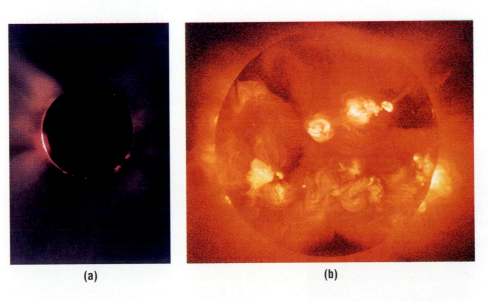

(a)   (b)

Figure 12.16
(a) The optical corona is broken into loops and streamers extending far into space.
(b) X-ray images show coronal holes and bright spots above active regions on the Sun.

per year, is somehow accelerated from the solar surface by these same magnetic fields, much of the wind pouring from the coronal holes, where it meets little resistance.

The solar limb (see Figure 12.7a) displays several *prominences*, sheets and ropes of relatively cool gas confined between oppositely directed magnetic regions. They can extend tens to hundreds of thousands of kilometers into the corona. When seen against the chromosphere (see Figure 12.11) they appear as dark *filaments*. Some prominences sit quietly in the corona (Figure 12.17), others condense from the corona and rain matter down onto the photosphere, and still others suddenly erupt into space.

Much more violent phenomena, sudden explosive brightenings, or **flares,** appear in the chromosphere in the spaces between spot pairs (Figure 12.18a). Flares range from minute flickerings to events that can encompass nearly a whole activity center. With temperatures up to 20 million K and perhaps much greater, most of a flare's energy is emitted in the X-ray spectrum (Figure 12.18b). A large flare can liberate 1% of the amount of radiation that the entire Sun releases in a second.

Flares originate in the corona and are caused by instabilities in the magnetic loops that arch from the photosphere between spot pairs (Figure 12.19). The mechanism is still not well understood, but if magnetic lines become too twisted and tangled, energy may be liberated in a giant spark that causes electrons and protons to accelerate upward and downward along field lines. When the downward electrons hit the chromosphere, they heat it, producing energetic X rays as well as the optically visible chromospheric flare. The event drives evaporated gas from the chromosphere, pushing it into the corona. This intense magnetic activity strongly suggests that the corona is heated with energy deposited by the powerful magnetism.

The activity has a powerful effect on Earth. Huge bubbles of confined corona can be buoyed upward by magnetic forces. Reorganization of the magnetic fields can release a bubble into the solar wind in a *coronal mass*

A large flare can liberate 1% of the amount of radiation that the entire Sun releases in a second.

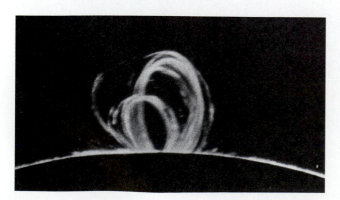

Figure 12.17
A quiescent cool prominence can hang in the corona for hours or even days as it traces out a magnetic field loop.

(a)

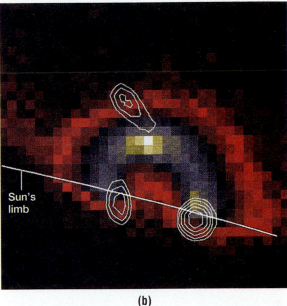

(b)

Figure 12.18

(a) A huge solar flare seen in visual light illuminates the chromosphere above a center of activity. **(b)** An Earth-orbiting X-ray satellite imaged a flare in profile at the solar limb. The colors represent emission by lower-energy X rays from energetic gas confined within a magnetic loop. The lower pair of white contours show the locations of high-energy X rays produced by the flare as particles slam into the chromosphere (producing a visual flare). The origin of the upper set of white contours is not understood; they show X rays that suggest temperatures as high as 200 million K.

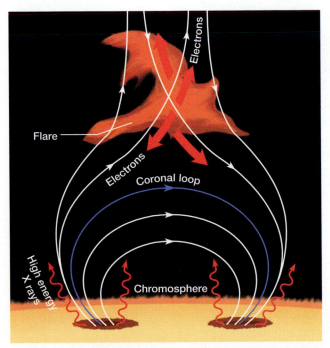

Figure 12.19

If the field lines from a magnetic loop that create a spot pair become too twisted and tangled, they short-circuit and release their energy in a solar flare. Electrons and protons move along field lines. The liberated high-speed electrons generate powerful X rays when they are stopped.

Figure 12.20
A spectacular composite image shows a coronal mass ejection surrounding an eruptive prominence.

ejection (Figure 12.20), enhancing the wind by 10% or 20% and creating a shock wave. The ejection may or may not be accompanied by a flare. The gas and the shock travel through the Solar System, and when they encounter the Earth produce terrestrial magnetic disturbances and auroras. Displays of the northern lights, and all their accompanying effects, therefore correlate with the solar magnetic cycle.

Much more important are the climatic effects of the cycle and the energetic radiation associated with it. The shorter the interval between sunspot maxima, the warmer the Earth; the longer the interval, the colder. During the period between about 1645 and 1715, when sunspots and magnetic activity effectively disappeared (see Figure 12.13a), Europe was plunged into a "little ice age." Indirect indicators of historic solar activity yield similar results. Recent studies show that two-thirds of neighboring sunlike stars display similar magnetic behavior. Astronomers thus speculate that the solar cycle operates only about two-thirds of the time. It has been "on" now for over 200 years; it could well go "off" again in the next century. A deep understanding of the active Sun will help us predict and prepare for such an event.

12.4 THE SOLAR INTERIOR

No kind of ordinary combustion could keep the Sun luminous for long: even if it were made of gasoline, it would burn for only a few thousand years. In the 1800s, William Thomson (Lord Kelvin) and Hermann von Helmholtz thought they knew how the immense solar luminosity is produced. The Sun must contract under the force of its own gravity, the squeeze raising the interior temperature. The hot gas then radiates. The Sun would need to contract about 20 m per year to produce the observed luminosity of 4×10^{26} watts; at that rate it could live for 10^8 years. However, geologists showed that the Earth's rocks are far older than that. Although gravity is the force that raises the interior temperature, it cannot have kept the Sun illuminated for the 4.6-billion-year age of the Solar System.

The key to solar energy was provided by Albert Einstein in his theory of relativity. If you accelerate a body toward the speed of light, its mass increases (see Section 5.5). Application of the necessary force requires energy. Einstein showed that energy is readily converted into mass—and mass back into energy—through the famous relation

$$E = Mc^2.$$

Because of the large value of the speed of light (c), a tiny amount of mass can produce an enormous quantity of energy. If fully converted, a gram of matter could light a 100 watt light bulb for 30,000 years and provide enough energy to serve your household needs for a million years.

The solar luminosity requires that 2×10^9 kg of mass be converted into energy every second.

The common isotope (see Section 7.1) of helium (^{4}He) has four particles in its nucleus (two protons and two neutrons), while ordinary hydrogen (^{1}H) has one (a proton). Neutrons and protons have very nearly the same mass, yet the helium atom was known to be 0.7% lighter than four hydrogen atoms. In 1920, the British astrophysicist Sir Arthur Eddington surmised that if four hydrogen nuclei could be converted, or fused, into a helium atom, the missing mass might appear as energy. The solar luminosity requires that 2×10^9 kg of mass be converted into energy every second. By the end of the decade, Cecilia Payne-Gaposchkin of Harvard showed that stars are made mostly of hydrogen. Therefore, nearly 0.7% of the solar mass might become available for the fusion of hydrogen into helium, enough to enable the Sun to live for 10^{11} years.

By 1938, Cornell's Hans Bethe and Charles Critchfield had consolidated a variety of additional findings and discovered the **proton-proton (p-p) chain** (Figure 12.21). The first step in the chain is to bring two protons so close together that the strong (nuclear) force can overcome the electrostatic repulsion created by like positive charges (see Section 6.1), which can be accomplished by throwing the particles together at great speeds. Such speeds can be reached by heating the gas to extremely high temperatures, as gravity does in the solar center. The process is therefore known as **thermonuclear fusion,** or in the jargon of astronomy as **nuclear burning,** and requires a minimum temperature of about 7 million K.

However, even the strong force cannot keep two protons together. For two protons to fuse, one must lose its positive charge and decay into a neutron. The decay is accomplished with the ejection of a **positron,** a positively charged electron. This particle is an example of **antimatter,** which consists of particles whose charges (and/or other properties) are reversed relative to normal matter. The proton thus becomes a neutron, and the

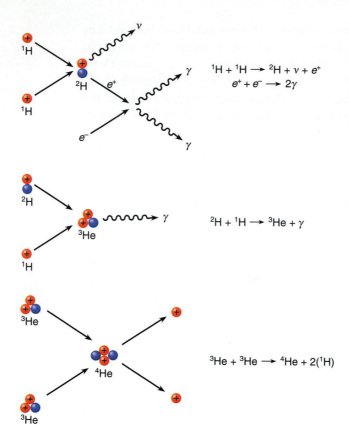

Figure 12.21
The proton-proton chain begins with the merger of two hydrogen nuclei (^{1}H) or protons (red), one converting to a neutron (blue), to form a deuterium atom (^{2}H), a positron (e^+), and a neutrino (ν). The ejected positron hits an electron to create a pair of gamma rays (γ). The deuterium immediately encounters another proton, which makes light helium (^{3}He) and another gamma ray. Two ^{3}He atoms then collide to produce ^{4}He with two protons left over.

$$^1\text{H} + {}^1\text{H} \longrightarrow {}^2\text{H} + \nu + e^+$$
$$e^+ + e^- \longrightarrow 2\gamma$$

$$^2\text{H} + {}^1\text{H} \longrightarrow {}^3\text{He} + \gamma$$

$$^3\text{He} + {}^3\text{He} \longrightarrow {}^4\text{He} + 2({}^1\text{H})$$

nucleus turns into a heavy isotope of hydrogen, ^{2}H (called *deuterium*). The positron quickly encounters a negative electron. The collision annihilates them both, their mass appearing as energy in the form of a pair of gamma rays. The fusion reaction also produces another particle, a **neutrino,** a massless (or nearly massless) neutral particle that carries energy at or near the speed of light. The deuterium nucleus subsequently absorbs another proton, which makes light helium (^{3}He) and another gamma ray. Two ^{3}He nuclei finally collide and make normal ^{4}He. The two protons left over are ejected into the surrounding gas. The result is that four protons are turned into one helium atom and energy is released in the form of gamma rays and neutrinos.

The solar gases are highly opaque, and a newly created gamma ray flies only a fraction of a centimeter before it is absorbed and re-emitted by an atom. The energy works its way outward randomly, moving first forward then backward, but on the average each re-emission takes place at a lower-temperature position that is farther from the center. From the blackbody laws, the wavelengths of the emitted radiation lengthen, but since the total energy stays the same, more photons must be emitted to compensate. What starts off as a single deadly gamma ray in the Sun's center is emitted from the photosphere 200,000 years later as hundreds or thousands of visual photons.

There was originally enough hydrogen fuel in the solar core to last 10 billion years.

Once we know how the Sun's energy is generated, we can calculate how long the Sun will live. To do that, we use the principle of hydrostatic equilibrium (see Section 9.2.2) to construct a *solar model* that gives temperatures, densities, and pressures at all points within the interior. We find that the central density is 160 g/cm³, 10 times denser than lead, but the temperature is so high—15 million K—that matter is still in the gaseous state. The temperature is great enough to sustain thermonuclear fusion within the inner 30% of the solar radius (3% of the volume), a region known as the **solar core** (Figure 12.22). The compression is so great that the core includes *40%* of the Sun's mass. At the calculated fusion rate (allowing for changes in structure as the fuel supply diminishes), there was originally enough hydrogen fuel in the core to last 10 billion years. Because the Solar System is about 5 billion years old, about half the core's hydrogen has been converted to helium and the Sun is roughly halfway through its allotted lifetime.

Surrounding the solar core is the **solar envelope,** where the temperature is too low to allow nuclear burning. The envelope provides the insulating blanket needed to keep the core hot and transports energy from the interior to the outside. From the model, we find that the envelope is divided into two parts. In the inner portion, the gas is quiet and energy moves by radiation (as it does in the core). The quiet inner envelope keeps the helium created in the core by thermonuclear burning from mixing with the rest of the Sun. The outer envelope, however, moves energy by convection, the result of which can be seen in the various aspects of granulation.

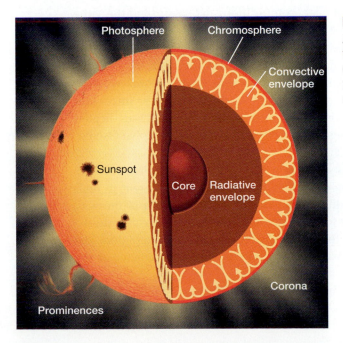

Figure 12.22

A cutaway model of the Sun shows the core, the radiative and convective envelopes, and the surrounding photosphere, chromosphere, and corona.

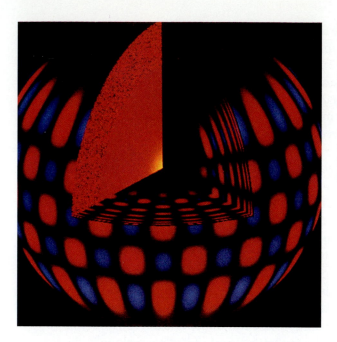

Figure 12.23
The Sun pulsates with a principal period of about 5 minutes, some parts of the surface moving inward (red) and others moving outward (blue) at the same time. After 2.5 minutes the directions reverse. A huge number of other oscillation periods have been observed. Oscillations such as these tell us a great deal about the solar interior.

Remarkably, we can check the theory of the solar interior by direct observation. Superimposed on the granulation is a complex up-and-down pattern of **solar oscillations** (Figure 12.23) that have a range of motion of only a few tens of kilometers. The entire Sun is ringing in space like a huge bell, its vibrations driven by the mass motions of convection. The principal period of oscillation is 5 minutes, but there are hundreds of lesser oscillations with periods between 4.5 and 6.7 minutes. A bell's natural vibration frequencies depend on its size, composition, and shape, and the same is true of the Sun. From the observed solar vibrations we find that convection extends about 30% of the way into the Sun (see Figure 12.22), that the helium content agrees with that derived from other methods, and that the convective envelope maintains the same pattern of differential rotation found at the surface. Observations of solar oscillations are so important that an entire network of solar observatories around the globe is dedicated to them, the Sun never setting on the aptly named GONG (Global Oscillation Network Group) project.

We can reach all the way to the Sun's core by observing the neutrinos created in the proton-proton chain. Matter is almost transparent to neutrinos, and typically it would take a block of lead an eighth of the way from here to the nearest star to stop one. As a result, the neutrinos, unlike electromagnetic radiation, fly directly from the core unimpeded. An amazing 66 billion should pass through every square centimeter of the Earth (and of you) every second.

Although neutrinos are difficult to capture, they can still initiate a variety of nuclear reactions. The little particles can therefore be detected and counted. One neutrino detector in Japan spots flashes of light when electrons in water molecules are struck by high-energy neutrinos. The detector is directional and clearly shows that the neutrinos are coming from the Sun, confirming that nuclear burning indeed takes place in the core. Yet various experiments consistently show between 30% and 60% of the number of expected neutrinos. The deficiency is confounding and confusing. Numerous explanations have been put forward, including a lower solar core temperature, but none is fully satisfactory and consistent with the solar model.

The leading hypothesis involves not the Sun but the neutrinos themselves. Matter is complex. The proton and electron actually represent only one of three known, successively heavier, *families of matter*. The heavier version of the electron is called the muon and the heaviest the tau particle. Each is associated with its own kind of neutrino, and each neutrino is paired with an antineutrino (which spins in the opposite direction from the "normal" neutrino). Of the six kinds, neutrino detectors can catch only normal electron neutrinos. Theory tells us that if neutrinos have a small amount of mass, they can switch back and forth from one kind to another through interactions with electrons in the solar interior and cannot change back after they leave the Sun.

Whatever the outcome of the neutrino problem, the Sun may be telling us something valuable about the nature of matter, as well as helping us to understand our terrestrial environment and the natures of the billions of other stars in the sky, to which we now lift our sight.

▶ KEY CONCEPTS

Active Sun: The collective phenomena of the solar magnetic cycle.

Antimatter: Matter that consists of particles with reversed charges; a **positron** is a positive electron.

Centers of activity: Magnetic regions of the solar surface that contain sunspots and associated phenomena.

Chromosphere: The hot, transparent layer just above the photosphere.

Corona: The hot (2 million K), low-density layer above the chromosphere.

Flares: Sudden releases of magnetic energy in the corona that brighten the chromosphere.

Granules: Bright cells in the photosphere, about 700 km across, caused by convection.

Limb darkening: The darkening of the surface of the photosphere toward its edge, or **limb**; caused by the transparency of the photosphere and the inwardly increasing temperature.

Magnetic activity cycle: The 22-year cycle in the amount of solar magnetic activity, which produces the 11-year **sunspot cycle.**

Neutrino: A massless (or nearly massless) particle that carries energy at or near the speed of light.

Photosphere: The bright apparent surface of the Sun.

Proton-proton (p-p) **chain:** A fusion reaction that turns four atoms of hydrogen into one atom of helium.

Quiet Sun: The constant phenomena of the Sun, which do not take large part in the solar magnetic cycle.

Solar core: The inner 30% of the solar radius, in which nuclear reactions are produced.

Solar envelope: The thick layer of the Sun that blankets the core and transports solar energy to the surface.

Solar oscillations: Multiple movements or vibrations of the solar surface.

Sunspots: Cool, dark solar regions caused by intense magnetic fields that inhibit the flow of energy.

Thermonuclear fusion (nuclear burning): The process of energy generation by the combination of lighter atoms into heavier atoms.

▶ EXERCISES

Comparisons

1. What is the distinction between the quiet Sun and the active Sun?

2. Compare the temperatures of the photosphere, chromosphere, and corona.

3. How do prominences differ from filaments?

4. Compare the heating mechanisms of the corona, chromosphere, and photosphere.

5. What are the similarities and differences between (a) electrons and positrons; (b) neutrinos and photons?

6. What is the difference between the solar core and the solar envelope?

Numerical Problems

7. What is the solar rotational velocity at the equator? What is the maximum Doppler shift for the sodium lines at 5,893 Å?

8. When do you expect the next solar maximum?

9. The element mercury is not found in solar spectra. On the basis of the abundances of other elements, estimate the ratio of the number of mercury atoms to the number of hydrogen atoms.

10. What mass of antimatter per day would you need to run a power plant that produces energy at the rate of 100 million watts?

11. What mass of hydrogen would the power plant in Question 10 use if it ran on hydrogen fusion?

Thought and Discussion

12. Why does the limb of the Sun appear so sharp?

13. Why is the Sun redder at the limb than at the center?

14. If energy passed through the outer solar envelope by radiation instead of by convection, how might the photosphere look?

15. What can we learn about the Sun from solar oscillations?

16. Why are the ionized calcium lines so strong in the Sun relative to those produced by more abundant hydrogen?

17. How do we know the temperature of the corona?

18. Why does the chromosphere have emission lines and the photosphere absorption lines?

19. How do the magnetic fields of sunspots alternate from one cycle to the next?

20. How are magnetic fields on the Sun detected and measured?

21. How do flares in the corona produce the observed chromospheric flares?

22. How do the corona and solar activity relate to the solar wind and to the Earth?

23. What are the steps of the proton-proton chain?

24. What characterizes antimatter?

25. Why are we not killed by the gamma rays emitted in the solar core?

26. You reside on a world orbiting another star. How might you sense the existence of an activity cycle on our Sun even though that body appears to you only as a star and you cannot see its surface?

Research Problem

27. Using data found in popular astronomy magazines, extend the sunspot cycle given in Figure 12.13a to the present.

Activity

28. Use a telescope to observe the Sun (by projection of the image only) over a period of two weeks. Sketch the positions and structures of sunspots and how their appearance changes from limb to center. Why are spots elongated near the limb?

Scientific Writing

29. For a nature magazine read by people who are not scientifically trained, describe what we do not understand about the Sun. Examine particularly the neutrino problem and give a sense of the depth of our ignorance.

30. For a children's magazine, take a hypothetical ride to the Sun straight into its core. Describe the sights along the way from the corona into the center and convey the awesome nature of the journey.

THE STARS
The observed properties of the stars

Reasoning from our study of the Sun, we define *stars* as bodies that either derive, have derived, or will derive their energy principally from thermonuclear fusion. The study of the stars in turn helps us learn more about our own Sun; it also allows us to probe outward, helping us to understand the nature of the Universe.

13.1 DISTANCE AND MOTION

The first step in building our knowledge of the stars (Figure 13.1a) is to measure their distances from Earth. The fundamental method of determining a star's distance involves measurement of its *parallax*, a concept used in Section 4.1.1 to find the distance to the Moon. Nearby stars A and B in Figure 13.1b are in the ecliptic plane. As the Earth orbits the Sun, both will appear to move back and forth against the background of more-distant stars (Figure 13.1c). The angular shift observed for real stars is very small, the largest only 1.54 seconds of arc for Proxima Centauri, a dim companion to α Centauri; this tiny angle is that subtended by a nickel seen face-on at a distance of 3 km. A star's **parallax** (p), always measured in seconds of arc, is actually defined as half the total angular shift. The distance of a star (d) is measured in **parsecs** (pc), where

$$d(\text{pc}) = 1/p''.$$

As p goes down, d goes up. Star B has half the parallax of star A, and its distance is double that of A (Figure 13.1d). The parallax of Proxima Centauri is 0.77"; therefore its distance is $1/(0.77) = 1.30$ pc.

Place star A in Figure 13.1 at a distance of 1 parsec. Its parallax is 1", which is also the angular radius of the Earth's orbit as seen from the star; that is, at a distance of 1 pc, the Astronomical Unit subtends an angle of 1 second of arc. If we know the angle subtended by a body (here not an actual body but the AU) of known size, we can find its distance (see MathHelp 2.1). At 1 pc a star would be 206,265 AU away. The AU is 1.50×10^8 km long, so 1 parsec also equals 3.09×10^{13} (31 trillion) km. Proxima Centauri is therefore $1.3 \times 206,265 = 268,000$ AU, or 4×10^{13} km distant.

The light we see from the nearest star left it more than four years ago.

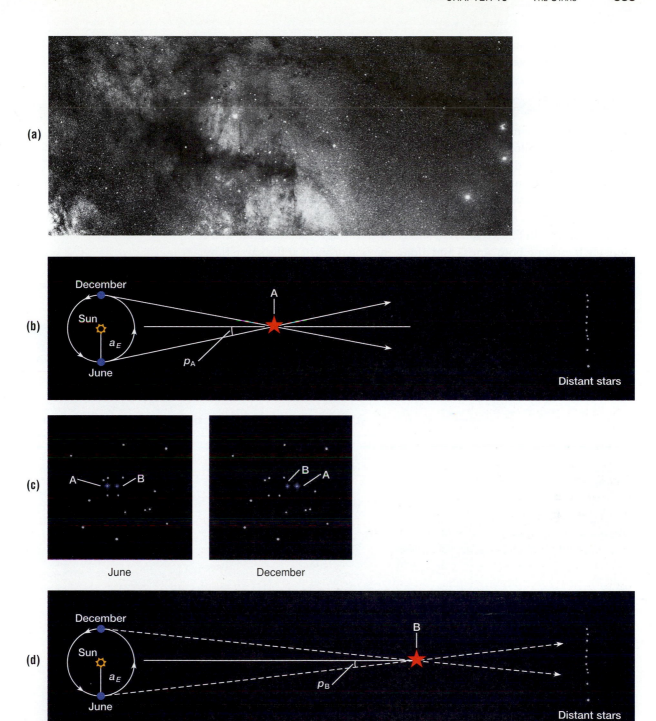

Figure 13.1
(a) The stars, seemingly scattered against the celestial sphere, are all at different distances from Earth.
(b) As the Earth orbits the Sun, star A displays a parallax shift. The astronomical parallax (p) is half the total angular shift. It is also the angle subtended by the semimajor axis of the Earth's orbit. **(c)** Simulated photographs of stars A and B made six months apart show how they change their positions against the background stars, which are assumed to be infinitely far away. **(d)** Doubling the distance from star A to star B halves the parallax ($p_B = 1/2\ p_A$).

A more familiar unit places stellar distances in perspective. The commonly used **light-year** (ly) is the distance a beam of light (or a photon) will travel in a year at 299,793 km/s. Multiplying by the number of seconds in a year (3.16×10^7), we find the length of the light-year to be 9.47×10^{12} km. Thus the parsec is also equal to 3.26 ly, and Proxima Centauri is 4.3 ly away: the light we see from this star left it more than four years ago! Current technology allows us to measure parallaxes as far as 300 pc—almost 1,000 ly—away, and parallaxes are known for thousands of stars. Data on stars within 3.5 pc are given in Table 13.1.

To the naked eye, the stars stay in fixed positions over our lifetimes. In fact, however, they are all in motion, appearing fixed only because of their great distances. In Figure 13.2, the stars travel through space relative to the Sun, each moving both across and along the line of sight at the same time. The motion across the line of sight, the *tangential velocity* (see Section

TABLE 13.1
The Closest Stars[a]

Name	α^b h	m	δ^b °	′	Distance (pc)	μ^c ("/yr)	v_t (km/s)	v_r (km/s)	v_s (km/s)	Apparent Visual Magnitude (V)	Absolute Visual Magnitude (M_V)	Spectral Class
Proxima Cen	14	30	−62	41	1.29	3.68	23	−23	33	11.00	15.40	M5 V
α Cen A	14	33	−60	50	1.33	3.68	23	−25	34	−0.01	4.87	G2 V
α Cen B								−21	31	1.33	6.21	K1 V
Barnard's Star	17	57	+04	33	1.83	10.31	88	−108	139	9.54	13.25	M5 V
Wolf 359	10	56	+07	03	2.39	4.71	52	13	54	13.53	16.68	M8 V
Lalande 21185	11	04	+36	02	2.53	4.78	56	−84	101	7.50	10.49	M2 V
Sirius A	06	45	−16	43	2.63	1.33	17	−8	19	−1.46	1.42	A1 V
Sirius B	06	45	−16	43	2.63	1.33	17	−8	19	8.68	11.56	D
Luyten 726–8 A	01	38	−17	58	2.68	3.36	43	29	52	12.45	15.27	M5 V
Luyten 726–8 B								32	54	12.95	15.77	M6 V
Ross 154	18	50	−23	49	2.92	0.72	10	−4	11	10.6	13.3	M4 V
Ross 248	23	42	+44	12	3.16	1.59	24	−81	84	12.29	14.80	M6 V
ε Eri	03	33	−09	27	3.27	0.97	15	16	22	3.73	6.14	K2 V
Ross 128	11	48	+00	49	3.34	1.37	22	−13	26	11.10	13.50	M5 V
Luyten 789–6	22	39	−15	20	3.41	3.26	51	−60	79	12.18	14.60	M7 V
ε Ind	22	03	−56	47	3.46	4.70	77	−40	87	4.69	7.00	K4 V
Cin 2456 A	18	43	+59	37	3.47	2.30	38	0	38	8.90	11.15	M4 V
Cin 2456 B								10	39	9.69	11.94	M5 V
61 Cyg A	21	07	+38	45	3.48	5.25	85	−64	106	5.21	7.55	K5 V
61 Cyg B						5.17	83	−64	105	6.03	8.37	K7 V
Procyon A	07	39	+05	13	3.50	1.25	20	−3	21	0.38	2.71	F5 IV
Procyon B										11.7	14.0	D

[a]Several stars are double; the components (indicated by A and B) are listed separately. The A component is the brighter, B the fainter.

[b]α and δ indicate the star's *right ascension* and *declination*, celestial measurements analogous to terrestrial longitude and latitude. See the introduction to the star maps in Appendix 1.

[c]μ indicates proper motion.

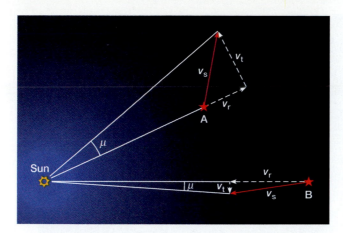

Figure 13.2
Star A is moving along the red arrow with speed v_s and is moving both across the line of sight with tangential velocity v_t and away from the observer with a positive radial velocity v_r. The motion across the line of sight makes the star move through (greatly exaggerated) angle μ each year. Star B, farther away, is moving in a different direction at a different speed, and v_s, v_r, v_t, and μ are all different. The observer sees the two stars moving apart. In each case, the true velocity $v_s^2 = v_t^2 + v_r^2$.

8.2.1), produces an angular displacement (per unit time) called the **proper motion** (μ). Point your finger at a person walking along a sidewalk, and you will see your hand trace through an angle of so many degrees per minute of time, the rate of angular motion (the person's "proper motion") depending on his or her distance and walking speed (a second person farther away walking at the same speed will appear to shift position more slowly). A star 100 pc away moving at 100 km/s across the line of sight relative to the Sun will appear to shift steadily relative to very distant stars at a rate of 0.05″ per year. If the star is 50 pc away, it will appear to move through twice that yearly angle, 0.10″ per year. The largest proper motion is 10″/yr for dim Barnard's Star (Figure 13.3), only 1.8 pc away. Though the angles are small, the shifts are cumulative, and in time will doom the familiar constellations to extinction (Figure 13.4).

The velocity *along* the line of sight, the star's *radial velocity*, the speed toward or away from us, can be found from the Doppler effect (again, see Section 8.2.1). Almost all stars have absorption lines similar to those found in the Sun (Figure 13.5a). If the star is moving away from us, like star A in Figure 13.2, the lines are shifted to longer wavelengths (Figure 13.5b), "to the red." If the star (star B in Figure 13.2) is moving toward us, the lines are shifted to shorter wavelengths, "to the blue." Measurement of the wavelengths of the lines (see Section 6.5) allows us to determine the radial velocity through the Doppler formula. Speeds are typically a few tens of km/s (see Table 13.1), though we find a small number with values greater than 100 km/s.

If we know the proper motion of a star as well as its distance, we can calculate the tangential velocity. That combined with the radial velocity gives us the actual motion of the star in three-dimensional space (see Figure 13.2). Tangential velocity (v_t) and radial velocity (v_r) are perpendicular to each other. The true velocity of the star, the **space velocity** (v_s), is given through the Pythagorean theorem, written as the **velocity equation,**

$$v_s^2 = v_t^2 + v_r^2.$$

The Sun is moving at a speed of 20 km/s toward a point between Hercules and Lyra.

(a)

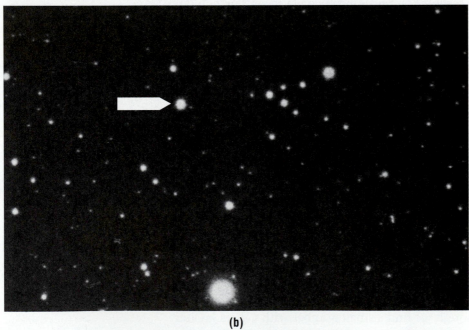

(b)

Figure 13.3
Barnard's Star (visible only telescopically), caught **(a)** in 1937 and **(b)** in 1960, sails against the distant background stars of Ophiuchus at a rate of 10 seconds of arc per year.

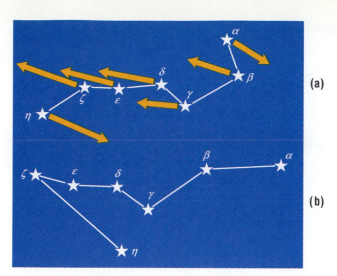

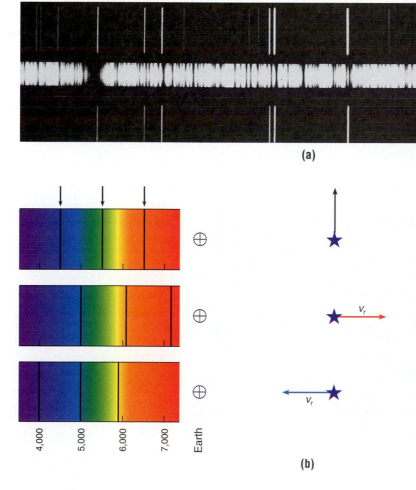

Iron spectrum

Stellar spectrum

Iron spectrum

(a)

4,000   5,000   6,000   7,000   Earth

(b)

Figure 13.5
(a) Several iron absorption lines in the spectrum
of Arcturus can be identified from their counterparts in the comparison emission spectra. The
stellar lines exhibit small but measurable Doppler
shifts toward the red caused by a recession velocity of a few km/s. (Most of the shift in this case is
produced by the orbital motion of the Earth.)
(b) The arrows indicate the laboratory wavelengths of three absorption lines produced by
imaginary astronomical bodies. In the top spectrum, the body moves only across the line of
sight; there is no radial velocity and no Doppler
shift. In the middle spectrum, the body is receding at a velocity a tenth that of light (greatly exaggerated to show the shift) and the lines are shifted
to the red by a few hundred Ångstroms; in the
bottom spectrum, a shift to the blue shows that
the body is approaching.

Measurements of motions are made relative to the Sun—but the Sun must be moving too. Think of walking through a crowd at a party. The guests are milling about in random directions. However, the people in the direction in which you are walking are *on the average* coming at you, those in the other direction going away. Your brain automatically processes this information to tell you how *you* are moving. Astronomers do the same thing with the local starry neighborhood, the crowd at our local party. From the average motions and velocities of stars all over the sky, we find the Sun is moving at a speed of 20 km/s toward a point in the northern hemisphere midway between the classical outlines of Hercules and Lyra.

13.2 BRIGHTNESSES

In 130 B.C., Hipparchus divided the stars into six brightness categories or **magnitudes** that extended from first magnitude for the brightest stars to sixth for the faintest he could see (see Section 3.3). These magnitudes are now called **apparent magnitudes** (*m*) because they represent the brightnesses of stars as they appear in the sky (Figure 13.6). Astronomers of the nineteenth century found that the first-magnitude stars were roughly 100 times brighter than the sixth; the term "brightness" here describes the amount of energy from the star passing through a unit area per second at the Earth. Hipparchus' classification was then placed on an exact scale in which five magnitude divisions correspond precisely to a factor of 100 in brightness. One magnitude thus corresponds to a brightness ratio of the *fifth root* of a hundred, or 2.512 . . . . Stars of the first magnitude are 2.512

Figure 13.6

The zodiacal constellation Gemini illustrates stars with a range of apparent magnitudes. Pollux is first magnitude, Castor and γ Geminorum second, ε and μ third, δ and ζ fourth. The fainter ones run through fifth and sixth. Note the different colors: Pollux, ε, μ, and others are distinctly orange.

TABLE 13.2
Magnitudes and Brightness Ratios[a]

Magnitude Difference	Brightness Ratio
1	2.51
2	6.31
3	15.85
4	39.81
5	100.00
6	251
7	631
8	1,585
9	3,981
10	10,000
15	1,000,000
20	100,000,000

[a]If star A is one magnitude brighter than B, it is 2.51 times brighter; if A is two magnitudes brighter, it is 6.31 times brighter, etc.

times brighter than those of the second, stars of the second magnitude 2.512 brighter than those of the third; a difference of two magnitudes thus corresponds to a ratio of 2.512^2, of three to 2.512^3, and so on (Table 13.2). To go beyond five divisions, combine the ratios: six magnitude divisions are a factor of 100×2.512 in brightness, ten magnitudes $100 \times 100 = 10,000$, and so on.

The scale was set by establishing the average apparent magnitude of a group of faint stars around the north celestial pole as 6.0. The very brightest stars then fall into apparent magnitude 0 and even -1. Venus at its brightest is -4, and the full Moon and Sun are respectively -12 and -27. Through the telescope, we see stars fainter than 6th magnitude: a small telescope allows you to see easily to 10th and a big one able to detect nearly to 30th.

We no longer use the human eye to measure magnitudes, but instead employ electronic devices matched to the eye's color response, which is greatest in the yellow-green part of the spectrum. Such magnitudes are now called **apparent visual magnitudes** (V). We can also view the sky in different wavelength bands, allowing us to measure infrared, ultraviolet, and a variety of other magnitudes. With modern techniques, we can measure the apparent visual magnitudes of stars to better than a hundredth of a division. The properties of the 30 brightest stars in the sky (listed according to decreasing V) are given in Table 13.3.

TABLE 13.3
The Thirty Brightest Stars[a,b]

Proper Name	Greek-Letter Name	α^c h m	δ^c ° ′	Apparent Visual Magnitude (V)	Distance[d] (pc)	Absolute Visual Magnitude (M_V)	μ (″/yr)	v_t (km/s)	v_r (km/s)	v_s (km/s)	Spectral Class
Sirius	α CMa	06 45	−16 42	−1.46	2.65	1.42	1.320	13	−8	15	A0 V
Canopus	α Car	06 24	−52 41	−0.72	70	−5	0.034	11	21	24	F0 II
Rigel Kentaurus[e]	α Cen A	14 40	−60 50	−0.01	1.33	4.37	3.68	23	−25	34	G2 V
	α Cen B			1.33	1.33	5.71			−23	31	K1 V
Arcturus	α Boo	14 16	+19 11	−0.04	10.3	−0.10	2.28	49	−5	49	K1 III
Vega	α Lyr	18 37	+38 47	0.03	7.5	0.65	0.35	12	−14	19	A0 V
Capella[e]	α Aur A	05 17	+46 00	0.93	13.3	−0.31	0.43	25	30	39	G8 III
	α Aur B			0.76		0.14					+G1 III
Rigel	β Ori	05 15	−08 12	0.12	265	−7	0.004	5	21	21	B8 Ia
Procyon	α CMi	07 39	+05 13	0.38	3.4	2.71	1.25	20	−3	21	F5 IV
Achernar	α Eri	01 38	−57 14	0.46	27	−1.7	0.14	18	16	24	B3 V
Betelgeuse	α Ori	05 55	07 24	0.50	320	−7	0.027	41	21	46	M2 Ia
——	β Cen	14 04	−60 22	0.61	95	0.61	0.031	14	6	15	B1 III
Altair	α Aql	19 51	+08 53	0.77	5	2.30	0.66	16	−26	30	A7 V
Aldebaran	α Tau	04 36	+16 31	0.85	19	−0.49	0.20	18	54	57	K5 III
Antares	α Sco	16 29	−26 26	0.96	190	−5.4	0.024	22	−3	22	M1.5 Ib
Spica	α Vir	13 25	−11 10	0.98	67	−3.2	0.054	17	1	17	B2 V
——	α Cru A[e]	12 27	−63 06	1.58	120	−3.8	0.030	17	−11	20	B0.5 IV
	α Cru B[e]			2.09	120	−3.3	0.034	18	−1	18	B1 V
Pollux	β Gem	07 45	+28 02	1.14	10	1.00	0.63	32	3	32	K0 III
Fomalhaut	α PsA	22 58	−29 37	1.16	6	2.02	0.37	12	7	14	A3 V
Deneb	α Cyg	20 41	+45 17	1.25	500	−7.2	0.005	12	−5	13	A2 Ia
——	β Cru	12 48	−59 41	1.25	150	−4.6	0.042	30	16	34	B0.5 III
Regulus	α Leo	10 08	+11 58	1.35	22	−0.38	0.25	26	6	27	B7 V
Adhara	ϵ CMa	06 59	−28 58	1.50	190	−4.9	0.002	2	27	27	B2 II
Castor	α Gem	07 36	+31 53	1.58	15	0.72	0.20	14	6	15	A1 V
——	γ Cru	12 31	−57 07	1.63	14	−0.5	0.27	18	21	28	M3 III
——	λ Sco	17 34	−37 06	1.63	67	−2.5	0.029	9	−3	9	B2 IV
Bellatrix	γ Ori	05 25	+06 21	1.64	35	−1.08	0.018	3	18	18	B2 III
Elnath	β Tau	05 26	+28 36	1.65	36	−1.13	0.18	30	9	31	B7 IV
Miaplacidus	β Car	09 13	−69 43	1.68	48	−1.73	0.18	41	−5	42	A2 IV
Alnilam	ϵ Ori	05 36	−01 12	1.70	460	−6.6	0.004	8	26	27	B0 Ia
Al Nair	α Gru	22 08	−46 58	1.74	37	−1.10	0.20	35	12	37	B7 IV

[a]Compiled principally from the *Bright Star Catalogue*, D. Hoffleit and C. Jaschek, Yale University Observatory, New Haven, 1982.

[b]Several stars in the list are double or multiple. If two components of a double star can be resolved through the telescope (α Cen, α Cru), they are listed separately; if the stars are similar (Spica) or multiple (Castor), the components' characteristics are combined.

[c]See the introduction to the star maps in Appendix 1.

[d]Distances above about 50 pc are derived from spectral classes.

[e]The combined apparent visual magnitudes of the double stars α Centauri, Capella, and α Crucis are −0.28, 0.08, and 1.05, respectively.

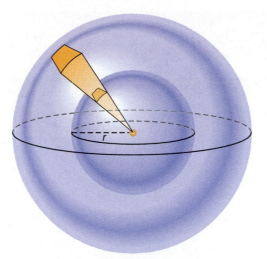

Figure 13.7
The energy radiated by a star per second (its luminosity, L) spreads out over a sphere of radius r. The apparent brightness of the star to an observer on the sphere is proportional to the energy passing through 1 square meter of the sphere (the shaded area) per second, or the star's luminosity divided by the sphere's surface area. The outer sphere has twice the diameter of the inner one but four times the surface area, so the energy must spread itself out four times as thinly. The apparent brightness of the star is thus inversely proportional to the square of the distance from it.

The apparent visual magnitude of a star depends both on its luminosity and on the inverse square of its distance. At any distance r in Figure 13.7, the photons and the energy radiated by the star at the center must spread over a sphere of that radius. The surface area of a sphere is $4\pi r^2$. The apparent brightness of the star at distance r then depends on its luminosity (L) divided by $4\pi r^2$. As the size of the sphere increases, the apparent brightness therefore decreases according to $1/r^2$. If you double your distance from a star, it will appear only a quarter as bright.

To compare the brightnesses of stars independently of distance, astronomers use **absolute visual magnitudes** (M_V), defined as the apparent magnitude a star would have if it were placed at a standard distance of 10 pc. If a star is distance d away and you move it to 10 pc, it will be $(d/10)^2$ times brighter (or as bright). As an example, Deneb (α Cygni) has an apparent visual magnitude of $+1.3$ and is 450 pc away. If moved to 10 pc, it becomes $(450/10)^2 = 2{,}025$ times brighter, which corresponds to 8.3 magnitudes (see Table 13.2). Deneb's absolute visual magnitude is therefore $1.3 - 8.3 = -7$ (three magnitudes brighter than Venus's maximum apparent magnitude). Procyon has $V = 0.4$ and a distance of 3.5 pc. If moved to 10 pc it becomes $(3.5/10)^2 = 0.12$ times as bright, or 2.3 magnitudes fainter, so M_V is 2.7. The Sun, with an apparent visual magnitude of -27, has a modest absolute visual magnitude of $+4.83$, and at 10 pc would appear about as bright as the faintest star in the Little Dipper (see Figure 3.3).

Apparent and absolute magnitudes and distance are conveniently related through the **magnitude equation,**

$$M = m + 5 - 5 \log d,$$

where d is distance in parsecs (see MathHelp 13.1 for a discussion of logarithms). The range of absolute magnitudes is staggering. There are a few rare stars with $M_V = -10$. Others are as faint as $M_V = +20$. We find stars

We find stars 15 magnitudes—a million times—both fainter and brighter than the Sun.

MathHelp 13.1 Logarithms

Logarithms (logs) are in constant use in all branches of physics and mathematics. The logarithm of a number is the power to which 10 must be raised to yield that number. For example, $10^2 = 100$, so the log of 100 is 2; $10^6 = 1,000,000$, so the log of 10^6 (or "log 10^6") is 6. Exponents need not be integers: log 10 is 1, and log 20 is 1.3. Tables of logarithms are available in libraries, and many hand calculators are designed to supply them readily.

Logarithms provided a means for simple calculation in the days before computers. If you multiply 10 by 100 you get 1,000, or $10^1 \times 10^2 = 10^3$. The numbers 10 and 100 are multiplied, but the exponents, or logarithms, are added. To multiply two numbers, all you need do is to look up the logs, add them, and look up the number in the table that corresponds to the sum. Even though they are no longer used much for calculation, logarithms are important parts of many equations, among them the magnitude equation.

15 magnitudes—a million times—both fainter and brighter than the Sun. The brightest stars are 30 magnitudes more luminous than the faintest—a factor of a trillion.

The visual magnitude samples only a portion of the spectrum: radiation at other wavelengths is not included. We can calculate ratio of the total luminosity to that in the visual wavelength band from a star's temperature and the blackbody laws. We can therefore relate M_V to actual luminosity.

13.3 STELLAR SPECTRA AND THE SPECTRAL SEQUENCE

Distances, motions, and magnitudes are important in understanding our stellar surroundings. But to understand the *physical* natures of stars we need to look at their spectra. The first such examinations puzzled the spectroscopic pioneers of the nineteenth century: only a few stars had spectra matching that of the Sun. Some had very obvious hydrogen absorption lines, while others had none at all; it looked as if stars might be chemically different from one another. Several classification schemes were developed to organize the observations, culminating in one created by Edward C. Pickering at Harvard in 1891. He ordered stellar spectra by letters A through O according to the strengths (darknesses and widths) of their hydrogen lines, A representing the strongest. Sirius and Vega (Figure 13.8a) went to the top of the list, while Betelgeuse and 30 Her (Figure 13.8b), with molecular absorption bands of titanium oxide (TiO) and insignificant hydrogen lines, fell into class M.

Pickering and his assistants, who included Williamina P. Fleming, Antonia Maury, and Annie Jump Cannon, found that some of the original

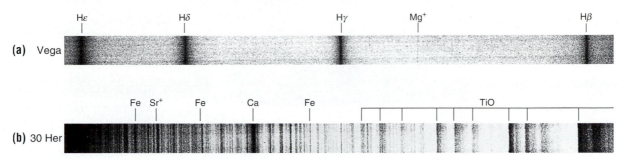

Figure 13.8

(a) The blue-violet spectrum (between about 3,900 Å at the left and 4,900 Å at the right) of Vega has overwhelmingly strong hydrogen lines (H$_\beta$, H$_\gamma$, H$_\delta$, and H$_\epsilon$ of the Balmer series). **(b)** The spectrum of 30 Herculis has no significant hydrogen lines, a strong line of neutral calcium, and absorption bands of titanium oxide.

classes were redundant or unneeded and that better continuity among other absorption lines was afforded by reordering the classes. By 1901 the system was in place, the vast majority of stars falling within the seven major classes of the **spectral sequence,** OBAFGKM (Figure 13.9). The Sun, with strong Ca$^+$ lines (see Section 12.2.1) and relatively weak hydrogen, is class G. Cannon (Figure 13.10a) also decimalized the system: A0 through A9 is followed by F0 through F9, and so on. In this expanded scheme, the Sun is class G2. By the time Annie Cannon died in 1941, she had classified more than 350,000 stars. Her name is one of the most revered in all astronomy.

The spectral sequence was known almost from the outset to be related to temperature. The classes correlate nicely with blackbody colors. Class O stars are bluish with temperatures of about 40,000 K, A stars are white and about 9,000 K, the G2 Sun is yellow-white at 5,800 K, and class M stars are reddish and about 3,000 K. The basic properties of the classes are summarized in Table 13.4; the way in which stellar temperatures change with temperature is shown graphically in Figure 13.11.

Analysis of stellar spectra in the 1920s, particularly by Harvard's Cecilia Payne-Gaposchkin (Figure 13.10b), demonstrated that in spite of their spectral differences, the photospheres of all the stars of the standard sequence OBAFGKM have similar solar-type chemical compositions (see Figure 12.6). *The origin of the sequence has to do not with composition but with temperature-dependent ionization and excitation* (Figure 13.12). In class M, the temperature is sufficiently low to allow molecules to form in abundance. The hydrogen lines are absent because they arise from the second orbit, and the temperature is too low to bump a significant number of electrons outward. However, neutral metals like calcium and sodium have powerful lines because they arise from the ground state and no collisional pre-excitation is needed (see Section 12.2). In the photospheres of warmer stars, in class K, energetic collisions between atoms and molecules cause most of the molecules to break up, and calcium starts to ionize: TiO

All the stars of the standard sequence OBAFGKM have similar solar-type chemical compositions.

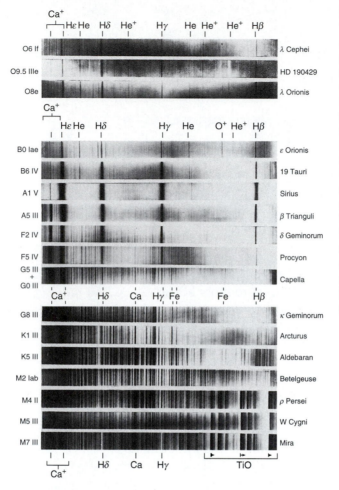

Figure 13.9

The spectral sequence (OBAFGKM) descends from class O at the top to M at the bottom. The blue-violet spectra run from about 3,800 Å on the left to about 4,800 Å on the right. The Balmer lines of hydrogen are strongest in the A stars and then decline both upward and downward. Helium lines become prominent in class B and ionized helium in O. Metal lines, particularly ionized calcium (Ca^+), increase in strength downward from class A, and in class G the neutral calcium (Ca) line starts to become prominent. The coolest M star is dominated by bands of titanium oxide (TiO) and has no hydrogen absorptions. The Roman numerals attached to the stars' spectral classes are luminosity classes.

(a)

(b)

Figure 13.10

Our understanding of the spectra of stars owes a huge debt to **(a)** Annie Jump Cannon (1863–1941), who provided the observational foundation of the spectral classification system in use today, and **(b)** Cecilia Payne-Gaposchkin (1900–1979), an observer and theoretician who helped explain the physical basis of the system.

TABLE 13.4
Properties of the Spectral Classes

Class	Characteristic	Color	Effective Temperature (K)	Examples
O	He⁺, He	Blue	28,000–50,000	χ Per, ε Ori
B	He, H	Blue-white	9,900–28,000	Rigel, Spica
A	H	White	7,400–9,900	Vega, Sirius
F	Metals; H	Yellow-white	6,000–7,400	Procyon
G	Ca⁺; metals	Yellow	4,900–6,000	Sun, α Cen A
K	Ca⁺; Ca; molecules	Orange	3,500–4,900	Arcturus
M	TiO; other molecules; Ca	Orange-red	2,000–3,500	Betelgeuse

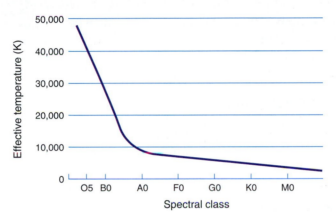

Figure 13.11
Temperature declines steadily along the spectral sequence from hot O to cool M.

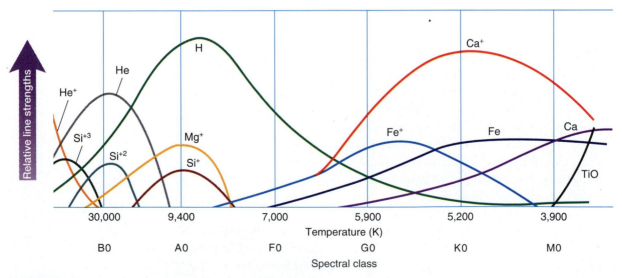

Figure 13.12
Different atoms, ions, and molecules produce their spectra at different temperatures. Molecules are populous only in class M. Neutral metals are present at low temperatures. Toward higher temperatures, more highly ionized species are created.

disappears, and the lines of neutral calcium and sodium weaken and are replaced by those of Ca$^+$. The calcium begins to be doubly ionized in class G and the Ca$^+$ lines weaken. During the progression, more electrons are collisionally excited into hydrogen's second orbit. Hydrogen lines become visible in class K and continue to increase in strength all the way to A. The metals become more highly ionized and their lines disappear from optical view (they can be seen in the ultraviolet), making the A spectra appear supremely simple.

As we proceed to stars hotter than class A, the hydrogen lines weaken. At 9,000 K the atomic collisions become vigorous enough to begin to produce significant ionization. The number of neutral hydrogen atoms then declines, and hydrogen begins to lose its ability to produce absorption lines. The temperature now becomes so great that high-speed collisions can elevate electrons into the second orbit of neutral helium, which has double the energy of the second orbit of hydrogen. The He$^+$ lines then become strong near 20,000 K in the B stars. At higher temperatures, helium ionizes and the He^{+2} lines become strong within class O.

13.4 THE HERTZSPRUNG-RUSSELL DIAGRAM

From the Stefan-Boltzmann law, the energy radiated by a star per unit area per second increases along with the fourth power of its temperature. We might therefore expect hotter stars, those of classes O and B, to be more luminous than the cooler stars of class M. Correlations between spectral class and luminosity were made independently by the Danish astronomer Ejnar Hertzsprung in 1908 and by the American astronomer Henry Norris Russell in 1913. The final result, the **Hertzsprung-Russell** (HR) **diagram** (Figure 13.13), is the single most important tool of stellar astronomy.

Stars placed on the HR diagram do indeed lie nicely along a main band from lower right (low temperature and luminosity) toward upper left (high temperature and luminosity). The great surprise was the discovery of some cool (and therefore orange or red) stars that, instead of being dim, are very luminous. These can be seen in a secondary band that extends up and to the right. The only way a star can be both bright and cool is to have a large radius, which produces a large surface area. Hertzsprung discriminated between the two broad classes of cooler, redder, stars by calling the dimmer ones (those on the main band) **dwarfs** and the brighter ones **giants**. Though the main band is now called the **main sequence**, all its stars are still also called dwarfs. The Sun is therefore a G2 dwarf or G2 main-sequence star.

It is easy to compare stellar diameters with that of the Sun by means of the equation $L = 7.2 \times 10^{-7} R^2 T^4$, where L is luminosity and R is radius (see Section 6.3.1). Write it first for the star ($\star$), then for the Sun ($\odot$) and divide one by the other. The constant cancels out and we are left with

$$L_\star/L_\odot = (R_\star/R_\odot)^2 (T_\star/T_\odot)^4.$$

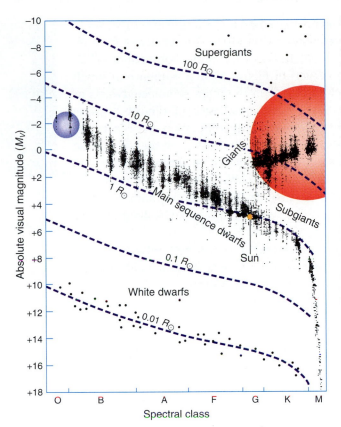

Figure 13.13
An adaptation of a famous version of the HR diagram compiled over 60 years ago by W. Gyllenberg of the Lund Observatory in Sweden shows the giant branch to be very distinct from the main (dwarf) sequence. A few supergiants are sprinkled across the top, and there is a sequence of white dwarfs running across the bottom. The relative sizes of stars in solar units are indicated by the colored circles and by dashed lines. The sequence of cool, dim red dwarfs (which, compared to other stellar sizes shown, are about as big as the dots used to represent them) seems to plunge downward because the red dwarfs radiate much of their energy in the invisible infrared and thus appear especially faint at visual wavelengths.

Here, the stellar luminosity, radius, and temperature ($L_\star$, $R_\star$, and $T_\star$) are expressed in terms of the solar values. The above equation can be solved for ($R_\star/R_\odot$), whence

$$R_\star/R_\odot = \sqrt{(L_\star/L_\odot)}/(T_\star/T_\odot)^2.$$

A red M giant has a temperature of 3,500 K, 0.61 that of the Sun. The giant's absolute visual magnitude is around 0, five magnitudes (100 times) brighter than the Sun. Infrared radiation not accounted for by its visual magnitude actually makes the giant about 600 times more luminous than the Sun. The radius of the star is then $\sqrt{600}/0.6^2 = 68$ times larger than the Sun. If placed at the Sun, the star would extend to 0.34 AU, almost as large as the orbit of the planet Mercury.

Sprinkled across the top of the HR diagram in Figure 13.13 are the **supergiants,** stars even more luminous and larger than the giants. A red M supergiant with an absolute visual magnitude near −4 is 4 magnitudes (40 times) brighter than the red-giant example discussed above. If the two stars have the same temperature, the M supergiant is $\sqrt{40}$ times, or over 6 times, larger than the M giant. Such a star has a radius of 2 AU, 1.4 times the size of the orbit of Mars. A few rare red supergiants near the top of the HR diagram have diameters comparable to the orbit of Saturn. Stellar radii, indicated in Figure 13.13 by dashed lines, also change along the

A few red supergiants have diameters comparable to the orbit of Saturn; the white dwarf Sirius B is only three-quarters the size of Earth.

main sequence, becoming larger as we climb toward the upper left. An O dwarf at the top of the main sequence has a radius 20 times that of the Sun. The radius of Proxima Centauri, an M5 dwarf, is only about a third solar.

Finally, we see at the lower left of Figure 13.13 a sequence of much dimmer stars. To be so much fainter than the main sequence dwarfs of the same temperature, they must be very small. Sirius B (Figure 13.14) is 10,000 times dimmer than its famous brilliant companion, yet is actually hotter. It is only three-quarters the size of Earth. Because of the color of the first ones found, these tiny stars became known as **white dwarfs**. The white dwarfs have peculiar chemical compositions, usually photospheres with all helium or all hydrogen. As a result, their spectra cannot be classified on the standard system, and they are placed on the HR diagram here by temperature.

Spectral classification has been extended to include luminosity. Class A supergiants (Figure 13.15a) have much narrower hydrogen absorption lines than the dimmer class A dwarfs (Figure 13.15b). Collisions among atoms in a gas minutely jiggle and smear atomic energy levels. As a result, the absorption lines are also smeared and broadened. The denser the gas, the closer the atoms are to one another, the more frequent the collisions, and the broader the lines. The masses of giants and supergiants are spread within an immense volume, and their photospheric densities are vastly lower than those of the dwarfs. As a consequence, the collision rate is lower and the absorption lines are narrow. The broadening effect reaches a maximum among the white dwarfs (Figure 13.15c). These tiny stars have amazingly high densities and the lines are correspondingly broad. Other spectral classes have similar density-dependent features. We can therefore determine the part of the HR diagram occupied by a star—dwarf, giant, supergiant, or white dwarf—from its spectrum alone.

In 1943, W. W. Morgan, P. C. Keenan, and E. Kellman divided the HR diagram into **luminosity (MKK) classes** (Table 13.5 and Figure 13.16) that range from Roman numeral I for supergiants, through III for giants, to V

Figure 13.14
Sirius B, smaller than Earth, shines dimly next to its brilliant companion.

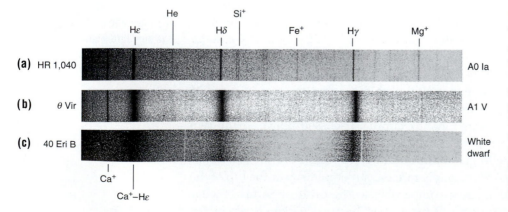

Figure 13.15
Three stars, all about the same temperature, have very different absorption line widths. **(a)** Those of the supergiant HR 1040 are very narrow. **(b)** The lines of θ Virginis, an ordinary main-sequence dwarf, are broader. **(c)** The spectrum of the white dwarf 40 Eridani B shows immensely broad lines.

TABLE 13.5
The Luminosity Classes

Class	Types of Stars	Examples
Ia	Luminous supergiants	Betelgeuse, Deneb
Ib	Less luminous supergiants	Antares, Canopus
II	Bright giants	Polaris, θ Lyr
III	Giants	Aldebaran, Arcturus, Capella
IV	Subgiants	Procyon
V	Main sequence (dwarfs)	Sun, α Cen, Sirius, Vega, 61 Cyg
D	White dwarfs	Sirius B, Procyon B, 40 Eri B

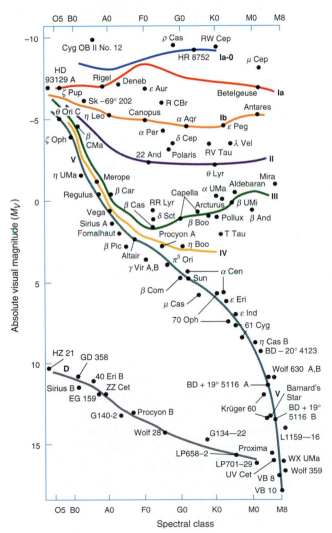

Figure 13.16
This modern HR diagram shows the average positions of the luminosity classes and the locations of about 100 representative stars. Because of their odd spectra, the white dwarfs are placed by temperature.

for main-sequence dwarfs. Classes II and IV respectively represent intermediate *bright giants* and *subgiants;* the white dwarfs are referred to as D. Class I is divided into Ia for bright supergiants, Ib for fainter ones. The MKK luminosity classes are appended to the Harvard spectral types, as seen in Tables 13.1 and 13.3. Betelgeuse is an M2 Ia star, Arcturus is K1 III, and the Sun is G2 V.

The MKK system allows us to find a **spectroscopic distance** to a star. We take a spectrogram, compare the spectrum to standards in the MKK atlas, find the class, and place the star on the HR diagram to determine the approximate absolute visual magnitude. Since $M = m + 5 - 5 \log d$, and m can be measured,

$$\log d = (m - M + 5)/5,$$

telling us how far away the star is.

The bright stars of the upper main sequence and the giants are easy to see, and therefore we count large numbers of them. Well over half the 30 brightest stars in Table 13.3 are of class A or B. There is only one K dwarf and no M dwarfs, so we might assume that these dim stars are rare. The truth is just the opposite. If we isolate a large volume of space and count the stars of each spectral class within it, what we find is astonishing. The number of stars climbs steeply as we proceed down the main sequence. The vast majority of stars, some 70%, are dim M dwarfs. The seemingly common B stars constitute only 0.1% of the total, and rare class O dwarfs a mere 0.00004%! Fewer than one star in a hundred is a giant, and the supergiants are about as common as O stars. Because they are so remarkably faint, we simply do not see with the naked eye the enormous numbers of red M dwarfs that flock around us.

> The vast majority of stars, some 70%, are dim M dwarfs.

13.5 STELLAR MASSES

In the years since the nature of the HR diagram was first established, we have discovered that the main sequence is a *mass* sequence: mass is the most important factor controlling the luminosity, temperature, and aging processes of a star.

> Mass is the most important factor controlling the luminosity, temperature, and aging processes of a star.

13.5.1 Binary Stars

Stellar masses are derived from observation of **binary** (double) **stars** (Figure 13.17a). Of the closest stars (listed in Table 13.1), 60% are members of double, triple, or more complex systems. The partners of a binary are bound by gravitational attraction and therefore each must orbit the other. (See Appendix 2 for a list of easily observable double and multiple stars.) In many instances we can watch the stars move around each other and plot the orbit of the fainter (B) component of the pair relative to the brighter (A) component (Figure 13.17b). From the orbit, we know the period (P) in years and the semimajor axis (a) in seconds of arc. If we know the distance, we can find a in AU (Figure 13.17c). From Kepler's third law,

$$P^2 = a^3/(M_A + M_B),$$

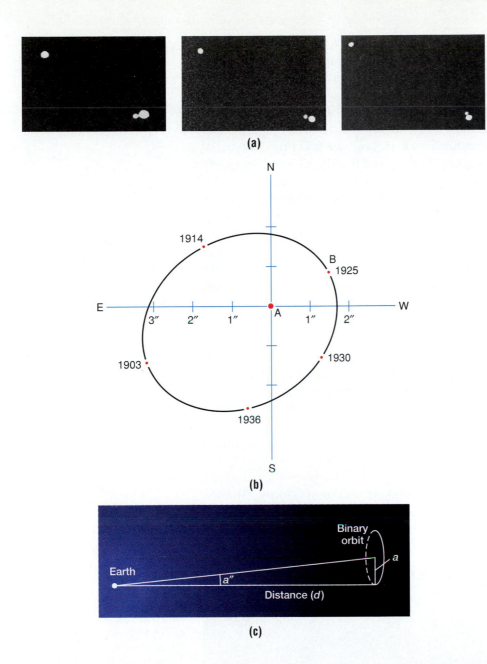

Figure 13.17
(a) Krüger 60, one of the nearest stars to Earth, is a double whose components show obvious orbital motion over only a dozen years.
(b) The 45-year orbit of Krüger 60 B about Krüger 60 A was constructed from years of observations like those shown in (a). Compare the orbit with the observations.
(c) The semimajor axis of a binary star orbit a distance d away subtends a tiny angle of a'', from which the semimajor axis a in AU can be found.

where $(M_A + M_B)$ is the sum of the masses of the two stars in solar units $(M_\odot)$ since a and P are also in solar-system units (years and AU). Therefore,

$$(M_A + M_B) = a^3/P^2.$$

The sum of the masses gives us limited information. If we can locate the system's center of mass, we can find the ratio of the masses, just as we did for the Earth and Moon in Section 7.1. If star B is three times the distance from the center of mass as star B, then star A is three times as massive as star B ($a_B/a_A = M_A/M_B$, where a_A and a_B are the distances at any time of A and B from the center of mass). If we know both the sum and the ratio of the masses, we can determine the *individual* masses. As an example, say the sum of the masses $M_A + M_B = 2\ M_\odot$ and $M_A/M_B = 2$. Then $M_A = 2\ M_B$. Therefore, $2\ M_B + M_B = 3\ M_B = 2\ M_\odot$, $M_B = 2/3\ M$, and $M_A = 4/3\ M_\odot$. From observations such as these, we find that Centauri A, a G2 star like the Sun, has a mass of $1.1\ M_\odot$ and that the K1 V B-component, which orbits in 84 years, measures $0.89\ M_\odot$.

Visual binaries can be resolved by telescope. If the members of a binary star are too close together to be separated visually, we might observe the system as a **spectroscopic binary.** If the two stars are of comparable brightness, we can see absorption lines from both. As each star swings about the other—actually about the system's center of mass—we see the lines from each move back and forth in the spectrum as a result of the Doppler effect (Figure 13.18b). If the orbit lies in the line of sight (as it does in Figure 13.18a), we can measure the actual orbital velocities, which with the period, gives us the orbital circumferences and radii. The sum of the individual orbital semimajor axes equals the semimajor axis of the orbit of one star about the other (see Figure 13.18a), so we can use Kepler's third law to find the sum of the masses. The individual semimajor axes give us the mass ratio, and therefore we can find the individual stellar masses as before.

Normally, however, the plane of the orbit will not lie along the line of sight, but will be tilted at some angle to it. We therefore can find only lower limits to the actual velocities and individual masses. (For example, if the orbit is perpendicular to the line of sight, the radial velocity variation

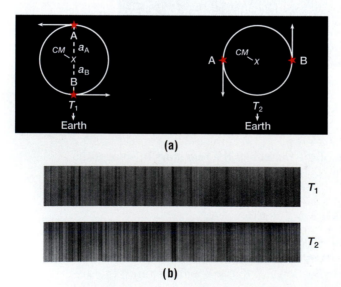

(a)

(b)

Figure 13.18
(a) A pair of identical stars, A and B, orbit a common center of mass (*CM*) with identical semimajor axes, a_A and a_B. They are too close to be individually resolved at the telescope, but the spectrum reveals the pair. **(b)** At time T_1 both stars are moving across the line of sight and there are no Doppler shifts in the spectrum. At time T_2 the stars have orbited 90° and now A is moving toward us and B moving away. A's lines are Doppler-shifted to the blue, B's to the red.

is zero no matter what the true velocities.) Nevertheless, we can still measure the ratio of orbital velocities, V_A/V_B, which equals M_B/M_A, allowing us to find the ratio of masses.

We have no way of knowing the orbital tilt unless it is near zero, that is, unless the orbit is aligned close to the line of sight. Then the stars may pass in front of each other, each cutting off part or all of the light of the other, and we see an **eclipsing binary** (Figure 13.19). If the system is also a

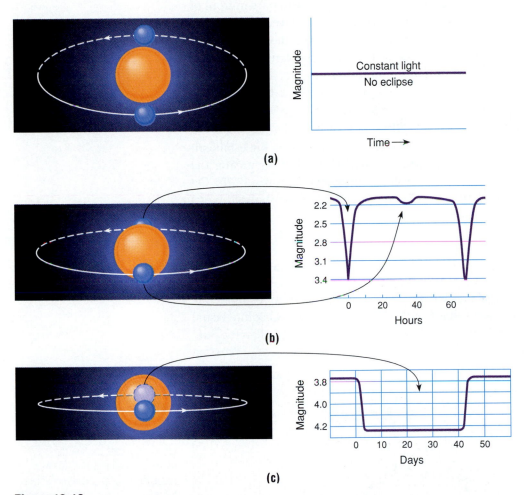

Figure 13.19

You look at an orbiting binary. The smaller blue star orbits the larger orange star on the left; the resulting light curve is on the right. **(a)** The tilt of the orbit to the line of sight is high, the stars do not get in each other's way, there is no eclipse, and the magnitude is constant. **(b)** The tilt is less, and once per orbit each star cuts off part of the light of the other. When the smaller star is partially hidden, the combined light of the pair drops. The sharpness of the deep minimum reveals a partial eclipse, as in the example of the star Algol. We also see a shallow secondary eclipse when the small star cuts off part of the light from the big one. **(c)** If the orbital tilt is low enough, the smaller star can be completely hidden behind the larger, the eclipse is total, and the minimum has a flat bottom, as it does for ζ Aurigae. Half a period later, the small star cuts off part of the light of the big one, and the eclipse is annular (not shown in the light curve).

spectroscopic binary, we can use its **light curve**—a graph of magnitude plotted against time—to get a remarkably complete picture of the stars' properties, making eclipsing binaries crucial in the construction of theories of how stars work.

Because of the stars' curved limbs, the detailed shape of the light curve will depend on the tilt of the orbit to the line of sight. The derived tilt and the velocities from the spectroscopic data then allow us to find the semimajor axes of each orbit and thus the stellar masses. Since we know the orbital speeds, the time required for the smaller star to pass in front of the larger gives the larger star's diameter, and the time required for the smaller star to disappear behind the larger one gives the smaller's diameter. If we know the luminosities of the stars, we can then use the Stefan-Boltzmann law to find effective temperatures.

13.5.2 The Mass-Luminosity Relation

The most important result of the study of binary stars was the establishment of the **mass-luminosity relation,** which shows that the greater the mass of a main-sequence star, the greater its luminosity (Figure 13.20). On the average, the luminosity is roughly proportional to (symbolized by $\propto$) the 3.5 power of the mass, or

$$L \propto M^{3.5}.$$

One star with twice the mass of another will be *10 times* as bright (between $2^3 = 8$ and $2^4 = 16$). If the mass is increased by a factor of 10, luminosity climbs by a factor of 3,000. If we reverse the relation, we can estimate the mass of any main-sequence star from its luminosity. Vega's absolute visual magnitude is 0.65, making it some 40 times more luminous than the Sun. From the graph in Figure 13.20 we find a mass close to 2.5 $M_\odot$. Proxima

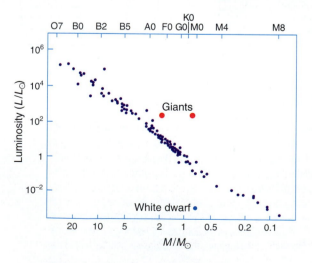

Figure 13.20
Stellar luminosities are plotted against main-sequence masses derived from binary star orbits to show the mass-luminosity relation. Main-sequence spectral classes are across the top. Giants, supergiants, and white dwarfs do not fit the relation.

Centauri, on the other hand, with a luminosity only 1/30 solar, has a mass of only 0.45 $M_\odot$.

The lowest mass measured for a star is about 0.08 $M_\odot$, which represents the end of the main sequence. Below this critical mass, bodies are not hot enough inside to fuse their hydrogen into helium by the proton-proton chain. The mass of the upper end of the main sequence is harder to ascertain because the observed curve in Figure 13.20 stops at about 25 solar masses. From the luminosities of the brightest stars, we estimate that the highest mass is in the neighborhood of 100 to 120 $M_\odot$.

Stars off the main sequence deviate considerably from the mass-luminosity relation and do not define any particular relation of their own. However, supergiants clearly have higher masses than the less luminous giants. They range well into the tens of solar masses, whereas the ordinary giants have masses typically in the neighborhood of 1 to 5 $M_\odot$. The class of white dwarfs serves up the greatest surprise. Analysis of the orbit of Sirius B (see Section 13.4) about brilliant Sirius A shows the dimmer companion, a star smaller than Earth, to have a mass of 1.05 $M_\odot$. The Earth has only 1/100 the radius of the Sun and only a millionth the volume. Compression of a solar mass into a volume the size of our planet would produce an average density of *1 million g/cm³*. A sample the size of a golf ball taken from the interior would have a mass of some 35 metric tons, the equivalent of 15 full-sized Cadillacs. The resulting high density of the photosphere produces the amazingly broad absorption lines seen in Figure 13.15c. How is such density possible? Remember that the atom is mostly empty space, so there is plenty of room for compression.

The different regions on the HR diagram reflect different states of **stellar evolution,** the stages and changes related to the stellar aging process. All main-sequence stars produce their energy by the fusion of hydrogen into helium. When the hydrogen fuel runs out, main-sequence stars must die. As we will see in Chapter 15, high mass O stars do not live long. They turn into supergiants, then explode. Lower-mass stars live longer, becoming giants and then white dwarfs.

> A sample from the interior of a white dwarf the size of a golf ball would have a mass of some 35 metric tons.

13.6 DISTRIBUTION

We cannot treat stars simply as individuals. We must also deal with their collective behavior, which ultimately tells us a great deal about them.

13.6.1 The Galaxy

All the stars you see at night, plus 200 billion others and great amounts of interstellar matter, constitute our **Galaxy** (Figure 13.21a). The great majority of its stars are found within a flat **galactic disk** about 25,000 pc (25 kiloparsecs) across, the structure that makes the Milky Way (Figure 13.22). The

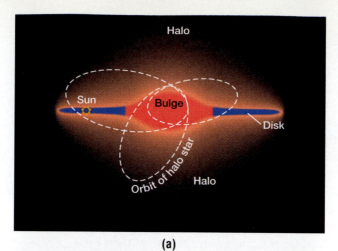

(a)

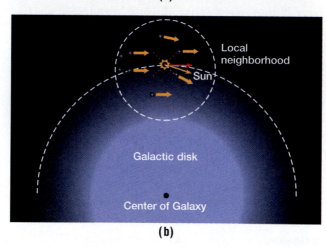

(b)

Figure 13.21
(a) An edge-on view of the Galaxy reveals a disk densely populated with stars encompassed by a thinly populated halo; both the disk and the halo condense into a thick central bulge. The orbits of halo stars are inclined to the disk and are elliptical. **(b)** In the disk, the local neighborhood of stars has a circular orbit about the galactic center. The stellar orbits are not identical, however, so the stars slowly drift past one another, resulting in the observed proper motions and radial velocities.

Figure 13.22
Our Sun is in the galactic disk. We therefore see the disk, crowded with millions of stars, as a band of light around us, the Milky Way.

disk is surrounded by a sparsely populated spherical **galactic halo**. The disk condenses inward toward a denser **galactic bulge** that surrounds the galactic center.

The different spectral classes are intimately related to galactic location. The hot, blue, massive stars of the main sequence lie only within the Galaxy's disk and therefore along the plane of the Milky Way. Even with the naked eye, we see that the blue stars of Orion, Perseus, and other constellations are near the Milky Way's plane. Even though the Galaxy's disk contains stars all the way down the main sequence (including our Sun), it, as well as the disks of other galaxies similar to our own, takes on a bluish color (Figure 13.23).

The stars of the disk go around the center of the Galaxy on fairly circular orbits at speeds of about 200 km/s (Figure 13.21b). The small radial and space velocities seen for most of the stars in Table 13.3 are caused by small orbital differences. (If you walk among a crowd of people, slight differences in walking speed make some people drift past or fall behind you.) The stars of the halo, however, are on elliptical orbits. Some local ones are plunging through the disk, and as the Sun whips past them, they appear as the high-velocity stars of Table 13.1. These high-velocity stars are principally red dwarfs of the lower main sequence. Direct examination of the halo also shows numerous red giants, but no massive stars, neither supergiants nor blue O and B stars. The halo is populated mostly with lower mass red stars.

Spectroscopic analysis shows that the halo stars are different in other ways. Relative to the stars of the disk, they are deficient in elements heavier than helium (all loosely called "metals" in the jargon of astronomy).

Figure 13.23
M 31, the great Andromeda Galaxy, is a relatively nearby galaxy that closely resembles our own. The bluish disk of the system, colored by the light of brilliant blue O and B stars, is tipped at an angle to the line of sight, making it look elliptical. The reddish bulge contains bright red giants but no blue stars. The fuzzy objects to the upper right and at the lower left edge of the large galaxy are small companions.

Halo metal abundances are typically about a hundredth of the solar metal abundance, though the range is quite large—from only a ten-thousandth or so for a few rare stars up to about half solar. The stellar content of the bright central bulge of our Galaxy is similar, red stars giving it the appearance more of the halo than of the disk. Metal abundances, however, range from those found in the halo to even greater than those found in the Sun.

13.6.2 Clusters

Within the Galaxy, stars commonly congregate into **clusters.** There are two kinds (a short list of each is given in Appendix 2). **Open clusters** are loose, ragged-looking systems of a few dozen to a few thousand stars contained in volumes ranging from under 1 pc to roughly 10 pc in radius. They are confined to the galactic disk, which contains thousands of them (Figure 13.24).

Stars not allied with clusters are found all over the HR diagram. The HR diagrams of open clusters, however, are different, though varied. The Hyades (Figure 13.24a) has a main sequence that proceeds upward from faint stars, then stops abruptly in class A. There are four K giants, but no upper main sequence and no supergiants. However, the main sequence of the Pleiades (Figure 13.24b) goes into the hot end of class B, and the combined HR diagram of the Double Cluster in Perseus (Figure 13.24c) extends into the O stars. The Double Cluster has no giants, but instead several supergiants.

The reason for the differences among clusters is age. As we will see, the stars of a cluster are all born at nearly the same time, and as the cluster ages, its high-mass stars die first. A young cluster will have an intact main sequence, but an older cluster will be missing its more-massive stars. The giants in the Hyades developed from the dwarfs of class A, and the M supergiants in the Double Cluster are the progeny of O stars.

Clusters are crucial in the establishment of the stellar distance scale. Because the 45-pc distance to the Hyades is accurately known, the cluster's HR diagram can be constructed according to absolute visual magnitude. To this diagram, we add all the nearby stars for which we have parallaxes. As a result, we know the absolute magnitudes for lower-main-sequence stars up through about class A and those of several giants. The other clusters' HR diagrams, however, are constructed with *apparent* visual magnitude, which is possible because all the stars within a cluster are at the same distance from us. Now in your mind lay the HR diagram of the Pleiades (Figure 13.24b) over that of the Hyades (Figure 13.24a) and line up the lower axes. Magnitudes are logarithmic: if you add them together you multiply brightness. As a result, you can slide the diagram of the Pleiades up and down over that of the Hyades until the main sequences fit

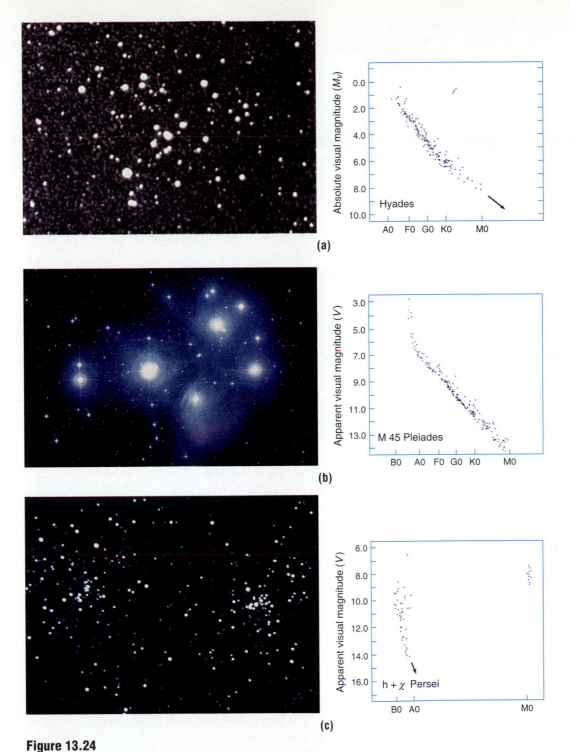

Figure 13.24

Photographs of three clusters are shown on the left; on the right are their HR diagrams. **(a)** The nearby Hyades, which makes the head of Taurus, is missing much of its upper main sequence but contains some giant stars. **(b)** The Pleiades, also obvious to the naked eye in Taurus, contains a host of B main-sequence stars. The wisps are clouds of interstellar dust that scatter starlight. **(c)** The Double Cluster in Perseus, which can just be seen with the naked eye as a small fuzzy enhancement of the Milky Way, has a nearly intact main sequence and several supergiants. The Hyades' HR diagram is plotted with absolute visual magnitude, the others with apparent visual magnitude.

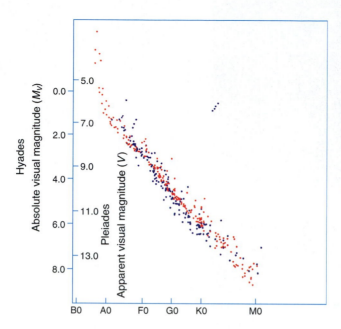

Figure 13.25
The HR diagram of the Pleiades (red dots) from Figure 13.24b is superimposed over that of the Hyades (blue dots) from Figure 13.24a. The lower axes are aligned and the diagram for the Pleiades has been placed so that its main sequence fits on that of the Hyades. The magnitude difference, $V - M_V = 5.5$, is read from the vertical axis and converted to distance.

together (Figure 13.25). On the left-hand axis, you note the apparent magnitudes for the Pleiades that correspond with the absolute magnitudes for the Hyades and find the difference, $m - M = 5.5$. The magnitude equation (see Section 13.4) then yields a distance of 125 pc, almost three times that of the Hyades. You can consequently replot the HR diagram of the Pleiades with absolute magnitude and know the luminosities of its B stars. Parallax measurements give 150 pc, so we adopt the average of 138 pc.

This procedure of *main-sequence fitting* can now be applied to the Double Cluster using the extended main sequence of Figure 13.25 to find not only the Double Cluster's distance but also the absolute magnitudes of even brighter main-sequence stars and supergiants. The establishment of the HR diagram is completed with a full sampling of open clusters. We can therefore know the luminosities of all the various kinds of stars, allowing us to determine through the method of spectroscopic distances (see Section 13.4) the distance of any star whose spectrum we can obtain and classify. We can now wander our Galaxy and know where we are, and can even begin to look into other galaxies.

Though they make fine sights through the telescope, open clusters pale beside the **globular clusters** (Figure 13.26a). A typical globular contains perhaps 100,000 stars, its spherical shape the natural result of the stars' great combined gravity. The stars are usually heavily concentrated toward the center, then thin outward into space; commonly some 90% will be found within about 25 pc of the center. The view from inside a dense globular must be spectacular. The heart of the globular cluster M 15 in Pegasus (Figure 13.26b) has 7,000 stars in a spherical volume a mere 0.2 pc

The heart of the globular cluster M 15 has 7,000 stars in a spherical volume a mere 0.2 pc across.

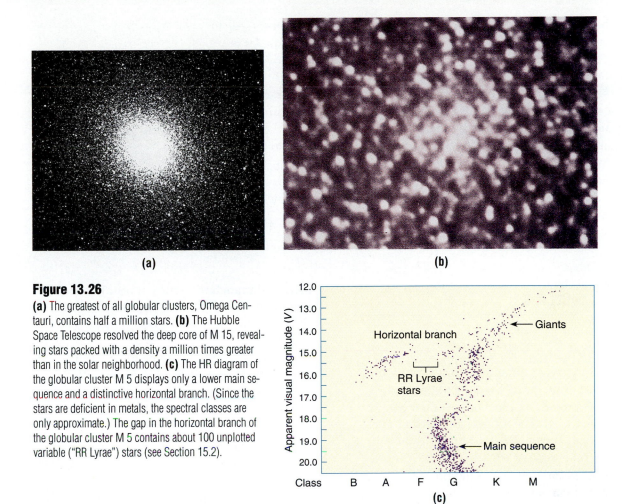

(a)   **(b)**

Figure 13.26

(a) The greatest of all globular clusters, Omega Centauri, contains half a million stars. **(b)** The Hubble Space Telescope resolved the deep core of M 15, revealing stars packed with a density a million times greater than in the solar neighborhood. **(c)** The HR diagram of the globular cluster M 5 displays only a lower main sequence and a distinctive horizontal branch. (Since the stars are deficient in metals, the spectral classes are only approximate.) The gap in the horizontal branch of the globular cluster M 5 contains about 100 unplotted variable ("RR Lyrae") stars (see Section 15.2).

across, a seventh the distance between the Sun and α Centauri! (The "M" in M 15 stands for Charles Messier, who compiled an early catalogue of clusters and other objects in the late 1700s. The complete list is given in Appendix 2.) Other globular clusters, however, are looser organizations with far fewer stars and less central condensation. The smallest globulars have radii of only 10 pc or so and just a few thousand stellar members.

Only 146 globular clusters are known in our Galaxy, and the total probably does not exceed 200. As viewed from Earth, the clusters concentrate heavily toward the galactic center, with some 60% found within Sagittarius, Scorpius, and the neighboring constellations. Except for this region, however, the globulars do not particularly concentrate toward the Milky Way but are spread out all over the sky. Moreover, the radial velocities tend to be high, a few topping 200 km/s. We are looking at members of the Galaxy's halo. Consistently, globular clusters are commonly highly deficient in metals compared with the Sun.

The profound difference between open and globular clusters is seen in their HR diagrams (Figure 13.26c). The main sequence of a globular is severely shortened; there are no stars in the main sequence above class G. Globulars have giant branches that curve up and to the right from the main sequence, and, most distinctively, display a **horizontal branch** that is absent in open clusters. The distances of the globular clusters can be found by main-sequence fitting and by the apparent magnitudes of their brightest giants, allowing us to probe deeply, and with accuracy, into the galactic halo.

The origin of the differences among stars and their clusters involves both stellar evolution and the evolution of the Galaxy as a whole, to which we now turn as we begin to explore the cycles of stellar lives.

KEY CONCEPTS

Binary stars: Double stars with gravitationally bound components: the components of **visual binaries** are detected directly through the telescope; **spectroscopic binaries** are detected by the Doppler shifts of their components; the stars of **eclipsing binaries** have orbital planes close to the line of sight and thus eclipse each other.

Clusters: Gravitationally bound groups of stars; **open clusters** are loose clusters found in the Milky Way, **globular clusters** are rich in stars and reside in the galactic halo.

Dwarfs: Stars of the **main sequence**, the main band of stars that climbs from lower right to upper left (luminosity increasing with temperature) on the HR diagram; MKK class V.

Galaxy (the): The collection of 200 billion stars and interstellar matter in which we live; it consists of a **galactic disk** surrounded by a sparsely populated spherical **galactic halo** that merges with the disk in a central **galactic bulge**.

Giants: Large stars that are brighter than the main sequence for a given temperature; MKK classes II and III.

Hertzsprung-Russell (HR) diagram: A plot of stellar magnitudes against spectral types.

Horizontal branch: A branch of roughly constant magnitude extending left from the giant branch in HR diagrams of globular clusters.

Light curve: A graph, in which the magnitude of a star that varies in brightness is plotted against the time.

Light-year (ly): The distance light travels in a year.

Luminosity (MKK) classes: The categorization of stars by luminosity and size (dwarf, giant, subgiant, supergiant).

Magnitude: A measure of stellar brightness, in which 5 magnitudes correspond to a brightness factor of 100. **Apparent magnitudes** (m) are measures of the brightnesses of stars as they appear in the sky; **apparent visual magnitudes** (V) are measured by a yellow-green detector that responds like the human eye; **absolute visual magnitudes** (M_V) are the apparent visual magnitudes that would be seen at a distance of 10 pc.

Parallax: In stellar astronomy, half the total angular shift of a star as the Earth goes around the Sun.

Parsec (pc): The distance at which the Earth's semimajor axis subtends 1 second of arc; 1 pc = 3.26 light-years = 206,265 AU.

Proper motion (μ): The angular motion of a star across the line of sight.

Space velocity (v_s): The velocity of a star relative to the Sun.

Spectral sequence: The categories of stars organized by temperature (OBAFGKM).

Spectroscopic distance: Distance estimated from absolute magnitude as determined by spectral class.

Stellar evolution: The process of change in a star brought about by its aging.

Supergiants: Huge luminous stars, plotted across the top of the HR diagram; MKK class I.

White dwarfs: Small dim stars about the size of Earth, plotted across the bottom of the HR diagram; class D.

 KEY RELATIONSHIPS

Distance:
$$d(\text{pc}) = 1/p''.$$

Magnitude equation:
$$M = m + 5 - 5 \log d \ (d \text{ is distance in pc}).$$

Mass-luminosity relation:
$$L \propto M^{3.5} \text{ (average) for main-sequence stars.}$$

Stellar luminosity:
$$L_\star/L_\odot = (R_\star/R_\odot)^2/(T_\star/T_\odot)^2.$$

Stellar radius:
$$R_\star/R_\odot = \sqrt{(L_\star/L_\odot)}/(T_\star/T_\odot)^2.$$

Velocity equation:
$$v_t{}^2 + v_r{}^2 = v_s{}^2.$$

 EXERCISES

Comparisons

1. Distinguish among tangential velocity, radial velocity, and proper motion.

2. What is the difference between apparent and absolute magnitude?

3. Compare the dimensions of dwarfs, giants, supergiants, and white dwarfs.

4. Compare the strengths of **(a)** the neutral and ionized helium lines and **(b)** the singly and doubly ionized silicon lines in the spectrum of an O9 star.

5. What is the difference between **(a)** an M6 V and an M6 III star; **(b)** an F0 III star and an M0 Ia star?

6. How do eclipsing binaries differ from spectroscopic binaries?

7. How do globular clusters and open clusters differ?

8. Give the fundamental differences between the Galaxy's halo and its disk.

Numerical Problems

9. A star appears to shift back and forth over the year through an angle of 0.05 second of arc. How far away is it in parsecs and light-years?

10. If you lived on Jupiter, what would be the parallax of Proxima Centauri?

11. The Hγ line of a star (at a wavelength of 4,340 Å) is shifted by 0.05 Å to the red. What is the star's radial velocity?

12. A star has a proper motion of 0.05"/yr, a parallax of 0.03", and a radial velocity of 10 km/s. What is its space velocity?

13. A 2nd-magnitude star is how many times brighter than a 14th-magnitude star?

14. A star has an apparent visual magnitude of 11.15 and is a kiloparsec away. What is its absolute visual magnitude?

15. An M2 giant and dwarf have $M_V = 0$ and +10 respectively. How many times bigger than the dwarf is the giant?

16. What is the diameter of a K5 II star relative to the Sun, assuming that the absolute visual magnitudes approximate the total stellar luminosities?

17. The spectrum of a star shows it is an M giant. Its apparent visual magnitude is 10.0. How far away is it?

18. The stars of ξ Boötis orbit one another with a period of 150 years. The semimajor axis of star B about star A is 33 AU. What is the combined mass of the system?

19. The center of mass of the system in Question 18 is found to be 45% of the way from the brighter component to the fainter one. What are the individual masses?

Thought and Discussion

20. The star 61 Cygni was chosen by Friedrich Bessel for parallax study because of its high proper motion. What was his reasoning?

21. To measure a parallax, the background stars are used as a reference. They must have parallaxes too. What is the effect of their parallax on the distance measurement of a nearby star?

22. How could you determine the direction of the motion of the Sun among the local stars by proper motions alone?

23. What is the spectral class of a star whose spectrum contains: TiO bands; strong hydrogen lines; ionized helium lines?

24. How do we know that the spectral sequence is a temperature sequence?

25. How can you tell an A dwarf from an A supergiant spectroscopically?

26. Where on the HR diagram do we find the most stars? Where do we find the fewest?

27. Show that if a is in AU, P is in years, and masses are in solar units, $(M_A + M_B) = a^3/P^2$ for binary stars.

28. Show approximately where on its orbit Krüger 60 B is today.

29. The primary component of ξ Boötis is spectral class G8 V and the secondary is K4 V. Use the mass-luminosity relation to estimate their masses. How do they compare with those derived in Exercise 19? How might you account for any differences you find?

30. What data are used to derive the radii of the components of an eclipsing binary star?

Research Problem

31. Cecilia Payne-Gaposchkin was instrumental in the discoveries of how stars shine. Using library materials, examine her career, list some of her contributions to astronomy, and comment on how they were made.

Activities

32. Make your own HR diagram from the stars listed in Tables 13.1. and 13.3. Color those from each table differently. What profound differences are there between the two sets of stars? Why do these differences exist?

33. From Tables 13.1 and 13.3 make a graph of space velocity against spectral class. Use colored symbols to discriminate among dwarfs, giants, and supergiants. Explain the meanings of the correlations.

34. Using binoculars, make drawings of the Pleiades, the Hyades, and Coma Berenices (see Appendix 2). Estimate the angular diameters of the clusters.

Scientific Writing

35. A planetarium has a newsletter for its season pass-holders. Write an article for the interested public on the HR diagram that introduces the concepts of spectral classes, dwarfs, giants, and supergiants (without going into detail about the mechanism of the formation of the spectrum itself).

36. You are planning to write a science fiction novel that describes life in a Solar System in which the Sun is binary and Jupiter is replaced by a KV dwarf. Write a proposal to a prospective publisher explaining what the system would be like, how it would appear from Earth, and how it would appear from α Centauri.

STAR FORMATION
How stars are born out of the gas and dust of the Milky Way

14

The Sun and Solar System were created 4.6 billion years ago from a spinning disk of gas and dust that had its origin in the dark cold clouds that litter interstellar space.

14.1 THE INTERSTELLAR MEDIUM

Vast amounts of matter are found in the spaces between the stars. This complex **interstellar medium** consists of atomic and molecular gas as well as tiny solid grains of dust that produce much of the Milky Way's structure.

14.1.1 Interstellar Gas

The most obvious features of the interstellar medium are bright clouds of illuminated gas called **diffuse nebulae** (*nebula* is Latin for "cloud"). Several, like the magnificent Orion Nebula (Figure 14.1), are visible through small telescopes (see Appendix 2). Diffuse nebulae are always associated with O (or hot B) stars and therefore, like most of the interstellar medium, are confined to the galactic disk. The stars in the middle of the Orion Nebula yield a spectroscopic distance of 420 pc. From the nebula's angular diameter of about a degree, the physical diameter is 8 pc. The diameters of other diffuse nebulae range from only a fraction of a parsec to over 100 pc.

Diffuse nebulae radiate emission-line spectra (Figure 14.2) and are prime examples of Kirchhoff's third law (see Section 6.3.2). The Balmer lines of hydrogen are prominent. If an electron in the ground state of hydrogen (Figure 14.3) absorbs energetic radiation with a wavelength less than 912 Å, it will be ripped away and the atom ionized to a bare proton. Only stars hotter than spectral class B2, which have temperatures higher than 25,000 K, radiate enough ultraviolet light to ionize the hydrogen. The free electron eventually recombines with another proton. It can land in any orbit, and as it jumps downward from one orbit to another toward the ground state, it creates the emission lines. Similar emission lines are also seen from helium, oxygen, and carbon. The free electrons also slam into atoms in various states of ionization and kick their electrons into

Figure 14.1
The Orion Nebula, seen here in a mosaic of images made with the Hubble Space Telescope, glows because of the ultraviolet radiation of the brightest of the four stars at its center.

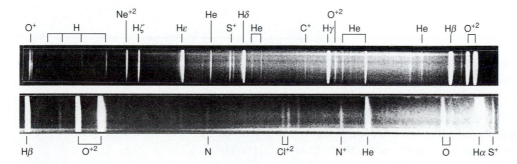

Figure 14.2
The Orion Nebula's spectrum, presented between 3,500 Å and 5,100 Å on the top and between 4,800 Å and 6,800 Å on the bottom, exhibits recombination lines produced by hydrogen, helium, and carbon, and collisionally produced lines of oxygen, nitrogen, neon, chlorine, and sulfur.

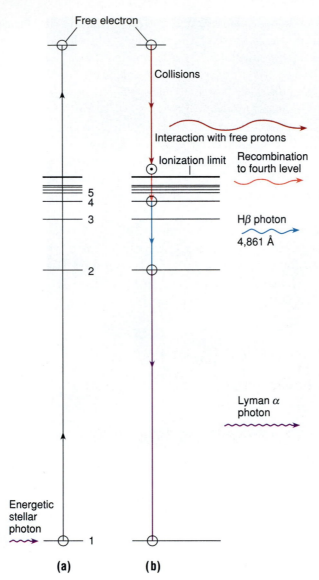

Free electron

Collisions

Interaction with free protons

Ionization limit

Recombination to fourth level

Hβ photon

4,861 Å

Lyman α photon

Energetic stellar photon

(a)   (b)

Figure 14.3
(a) A neutral hydrogen atom is ionized by an energetic ultraviolet stellar photon. The electron flies away at high velocity. **(b)** The electron first loses energy through collisions and close encounters with free protons in which it emits radio radiation (dark red photon). It then recombines (light red) with a different proton and lands on some energy level (here, the fourth). It can then jump to any lower level with the production of emission lines (here Hβ, denoted by blue, and Lyman α, in violet).

higher orbits from which they descend back to the ground state, producing powerful emission lines like those of doubly ionized oxygen at 5,007 and 4,959 Å.

From the strengths of the spectrum lines, we calculate that densities are commonly very low, only 100 to 1,000 atoms per cubic centimeter, better than the best vacuums produced on Earth. Yet the nebulae are still bright because of their great masses: the Orion Nebula has a mass several hundred times that of the Sun. Because of its low density, a diffuse nebula, like the solar corona, is not a blackbody. Kinetic temperatures are typically 10,000 K. The compositions are like those of the Sun and the stars: about 92% hydrogen (by number of atoms), 8% helium, and Sunlike proportions of other elements.

The diffuse nebulae are the remnants of O stars' births.

Though some of the galactic disk's blue O and B stars are found in open clusters, most of them populate looser groups called **OB associations.** The constellation Orion is actually one large OB association called Orion OB 1. Unlike the stars of an open cluster, an OB association's members are not gravitationally bound to one another. All the stars are seen to be moving away from a common center, presumably the place of their birth. All OB associations are therefore rapidly disintegrating. Because these massive stars do not get far from their birthplaces before they die and disappear, they cannot live very long, explaining their absence in so many open clusters (see Section 13.6.2). And because diffuse nebulae are associated with O and B stars, we conclude that the nebulae are the remnants of the stars' births.

The diffuse nebulae represent only the brightest portions of a more general interstellar gas in the Galaxy's disk. Light from distant stars must traverse interstellar space, the gas superimposing a variety of **interstellar absorption lines** (Figure 14.4a) that arise from atoms like calcium and sodium and from simple molecules (CH and CN). The ultraviolet (Figure

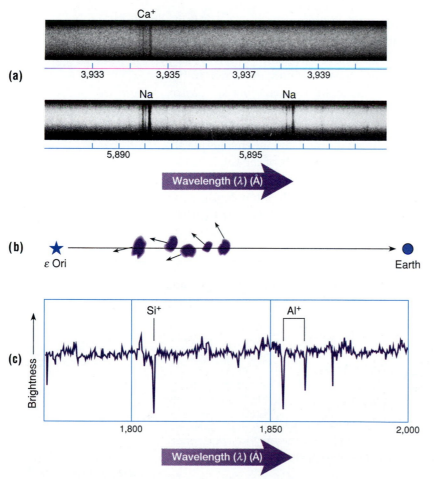

Figure 14.4

(a) Small sections of two very high dispersion (that is, greatly spread out) spectrograms of ϵ Orionis show multiple absorption lines of interstellar ionized calcium (Ca^+) (top) and neutral sodium (Na) (bottom). **(b)** We can see evidence for five separate clouds of interstellar matter along the line of sight, with radial velocities between $+3$ and $+27$ km/s. **(c)** A graphical spectrogram made by the *International Ultraviolet Explorer* satellite shows interstellar absorption lines of ionized silicon and aluminum.

14.4c) is especially rich in these interstellar absorptions, displaying lines from sulfur, carbon, and metals such as chromium, manganese, iron, copper, and zinc. The absorption lines shown in Figure 14.4a are split by the Doppler effect, demonstrating that the interstellar medium consists of discrete lumps of gas moving with velocities that differ by a few km/s (Figure 14.4b).

This interstellar gas is also revealed by a powerful emission line of neutral hydrogen with a wavelength of 21 cm. Protons and electrons have a property called *spin*, an atomic version of rotation. The magnetic fields associated with the orbit and spins cause the particles to interact. A proton and its attached electron may spin either in the same direction (parallel), or in opposite directions (antiparallel). As a result, the ground state of hydrogen actually consists of two very closely spaced orbits (Figure 14.5) that have slightly different energies, the parallel configuration the higher of the two. The electron can be shuffled from one of these orbits to the other by collisions with neighboring atoms. Once in the higher-energy orbit, it can also reverse direction by itself, dropping into the lower orbit with the release of a photon that contributes to a **21-cm line** that is found along any direction we look within the Galaxy's disk.

The combination of optical and radio observations reveals cool, discrete, *neutral hydrogen clouds* set within a warmer, lumpy, partially ionized *intercloud medium*, the ionization produced by the light of distant stars: these and other components of the interstellar medium are shown in Figure 14.6. Along any line of sight, there are about 10 clouds per kiloparsec. Typically a few pc across, the clouds have temperatures around 100 K and densities of a few tens to a few hundreds of atoms/cm^3. The denser ones are colder. Masses are several times that of the Sun, and taken together the neutral hydrogen clouds constitute roughly a quarter of the total mass of the interstellar medium. The warm intercloud gas has a temperature around 8,000 K but has a much lower density of only a few tenths of an atom/cm^3. It constitutes about half the mass of the interstellar medium. The clouds lie within about 100 to 150 pc of the galactic plane, but the warm gas spreads a few kpc into the halo.

The chemical compositions of these portions of the interstellar medium are odd. The abundances of lighter elements (such as oxygen and

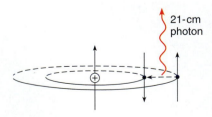

21-cm
photon

Figure 14.5

When hydrogen's electron is in the outer, higher energy suborbit of the ground state, it spins in the same direction as the proton. The electron can then change its spin direction and jump to the inner orbit by emitting a 21-cm photon. The difference in orbital radii is enormously exaggerated.

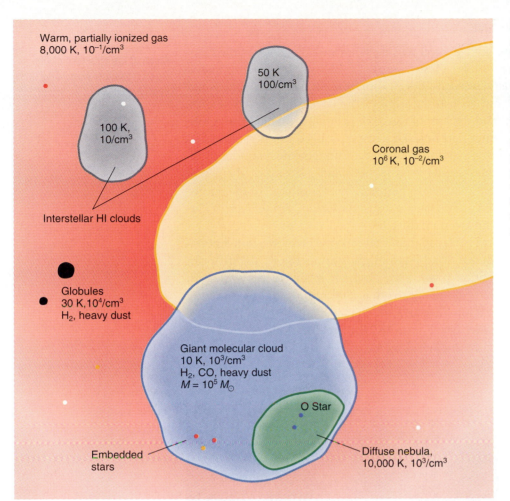

Warm, partially ionized gas
8,000 K, 10^{-1}/cm^3

50 K
100/cm^3

100 K,
10/cm^3

Coronal gas
10^6 K, 10^{-2}/cm^3

Interstellar HI clouds

Globules
30 K, 10^4/cm^3
H_2, heavy dust

Giant molecular cloud
10 K, 10^3/cm^3
H_2, CO, heavy dust
$M = 10^5 \, M_\odot$

O Star

Embedded
stars

Diffuse nebula,
10,000 K, 10^3/cm^3

Figure 14.6
A schematic view of the interstellar medium shows neutral hydrogen clouds (gray) enmeshed in a warm intercloud medium (red); a hot low-density coronal gas (yellow); small, dense, dusty globules (black); and giant molecular clouds (blue) filled with stars. Hot stars inside or at the cloud edges create diffuse nebulae (green).

nitrogen) relative to hydrogen are like those found in the Sun. However, the abundances of atoms like silicon and iron are below solar by a factor of 10 to 100. Something is removing these atoms from the gas and tying them into a less directly observable form. We will find them later.

A very different component of the interstellar medium is revealed by low-energy X rays and by ultraviolet observations of interstellar absorption lines from five-times ionized oxygen (O^{5+}), which requires a temperature of at least 300,000 K. The source, called the *coronal gas* because of its high temperature, has an extremely low density, near 10^{-3} atoms/cm^3. Such a high temperature cannot be produced by radiation from ordinary stars: it almost certainly represents the effects of blast waves from exploding high-mass stars. Over the aeons, the blast waves have interconnected, weaving a tapestry of tunnels within the gaseous medium. Although the coronal gas contains less than 5% of the interstellar medium's mass, it occupies perhaps half the interstellar volume.

14.1.2 Interstellar Dust

In any view of the Milky Way you will see what look like holes in the blanket of stars (Figure 14.7a). These gaps are caused by unilluminated gaseous clouds—*dark nebulae*—filled with **interstellar dust,** tiny solid grains that scatter and absorb the light of the background stars (Figure 14.7b). The smallest and densest of the dark nebulae are called **Bok globules** after the astronomer Bart Bok. They are typically a few parsecs across and have masses of 10 to 100 times that of the Sun. The most famous globules are the Horsehead Nebula (Figure 14.8) and a group of them that make the 5°-wide Coalsack, seen in Figure 3.6 just southeast of the Southern Cross. Much of the apparent structure of the Orion Nebula seen in Figure 14.1 is formed by intervening dust.

However, if a cloud is transparent and has one or more stars embedded within it (or if a dense cloud has a bright star next to it), the dust can scatter the stellar radiation toward the observer to produce a **reflection nebula** (see Figure 14.8), whose spectrum is a faithful reproduction of the embedded stars, absorption lines and all. The most famous reflection nebula enmeshes the Pleiades (see Figure 13.24b), whose hottest star, class B3 Merope, cannot ionize the surrounding gas.

Much of the structure of the Orion Nebula is formed by intervening dust.

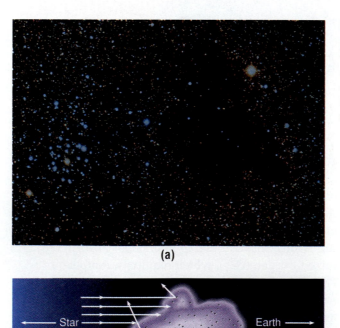

(a)

(b)

Figure 14.7
(a) Barnard 86, a Bok globule, is an optically opaque cloud of gas and dust. Note the open cluster to the left of it. **(b)** The diagram shows how it blocks the light of stars on the other side of it.

Figure 14.8
The Horsehead Nebula, a famous globule, protrudes into bright nebulosity that hangs below Zeta Orionis, the bright star at upper center and the left-hand star in Orion's belt. Up and to the left of the Horsehead is the bluish reflection nebula NGC 2023. Up and to the left of Zeta Orionis is NGC 2024, a diffuse nebula whose dust hides a huge number of new stars (see Figure 14.25).

The obscuring dust lies not just in the obvious globules but also in the cool clouds and the intercloud medium, and it is mixed everywhere with the gas. Its pervasive nature is revealed by a dark obscuring strip that runs down the spine of the Milky Way (Figure 14.9). Because the dust absorbs starlight, it is heated to temperatures of the order of 100 K and radiates at far infrared wavelengths. Gloriously recorded by the *Infrared Astronomical Satellite* (*IRAS*), the dust fills the plane of the Milky Way (Figure 14.10).

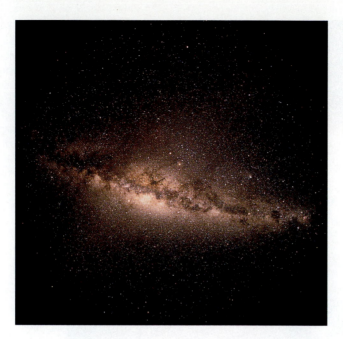

Figure 14.9
The brightest part of the Milky Way, the disk of our Galaxy, stretches from horizon to horizon, from the Northern Cross to the Southern Cross. Down the middle runs a dark strip, a band of obscuring dust that blocks the light of background stars.

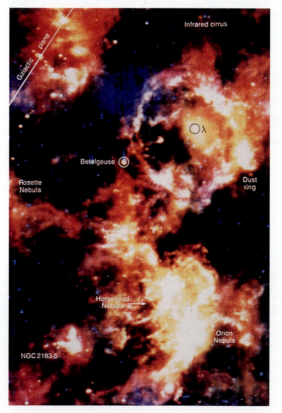

Figure 14.10
Orion takes on a different appearance when viewed by the *IRAS* satellite at far infrared wavelengths. All the stars but cool Betelgeuse have disappeared, replaced by masses of heated dust that fill the picture. The dust is concentrated toward the lower half of the constellation and is centered on the Orion Nebula. Toward the top is a huge bubble blown into the interstellar medium by the luminous O star λ Orionis. Thin filaments of dust are seen across the top of the picture.

Because the dust absorbs and scatters light, all stars beyond the local neighborhood appear to be fainter and consequently more distant than they actually are (an effect that must be accounted for when applying the method of spectroscopic distances). On the average, light is dimmed by about one visual magnitude per kiloparsec. Scattering and absorption are more efficient at shorter wavelengths than at longer ones, the degree of absorption very high at short wavelengths (making ultraviolet observations quite difficult) and going to zero at long wavelengths, making our view of the Galaxy clear if we observe in the long-wave infrared or radio spectral domains. The dust therefore reddens starlight much as the Earth's atmosphere reddens the Sun at sunset. From the way in which starlight is reddened and a theoretical knowledge of how light interacts with small particles, astronomers find that the dust grains have diameters from around a thousandth of a millimeter down to no more than a few Ångstroms, not much larger than complex molecules. Unlike familiar household dust, the grains would require a microscope to be seen.

The degree of interstellar absorption indicates that about 1% of the mass of the interstellar matter is dust. Only about one of the larger grains per cubic meter is needed to cause the observed interstellar absorption and reddening. The dust has an effect only because of the vast distances involved: the path from Earth to a reddened star 1 kpc away is over 10^{18} meters long.

Interstellar diamonds seem to be similar to those found in primitive meteorites.

A broad absorption feature in the infrared spectra of highly reddened stars reveals the presence of silicates, and the way in which the absorption changes in the ultraviolet reveals solid carbon grains made of graphite (Figure 14.11). Additional carbon is in the form of simple unstructured soot. Another absorption feature suggests that some 20% of interstellar carbon is in the form of tiny diamonds, created from carbon grains by compression from shock waves from exploding high-mass stars. These diamonds seem to be similar to those found in primitive meteorites (see Section 11.1), directly linking the stuff of interstellar space with the raw material of the Solar System. Other absorption features show that the grains are coated with ices.

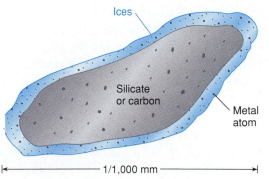

Ices

Silicate or carbon

Metal atom

|← 1/1,000 mm →|

Figure 14.11

An interstellar grain is expected to have an elongated silicate or carbon core. It is coated with a mantle of water, methane, ammonia, and other ices in which are embedded atoms of metals and other elements as well as simple and complex molecules.

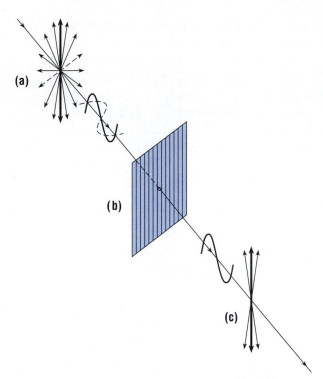

(a)

(b)

(c)

Figure 14.12
A beam of light proceeds from upper left to lower right. The arrows show the planes in which the light waves are oscillating. The heavy waves are up-and-down oscillations, the dashed one a side-to-side oscillation. **(a)** The light is randomly polarized (or unpolarized) and the waves oscillate in all directions. **(b)** The waves strike a polarizing filter that allows only a limited range of oscillation planes to go through. **(c)** The emerging light is now polarized.

In addition to dimming and reddening, the grains are also responsible for the *polarization* of starlight. Light (or any kind of radiation) is perfectly polarized when its waves oscillate only in a particular direction (Figure 14.12). Light can be polarized by passing it through a molecular filter that allows only a particular direction of oscillation to pass, and it is naturally polarized when it is reflected or scattered. The glare from a road surface is polarized light. Polarizing sunglasses block the polarized component of the light, allowing you to see better. We find that distant starlight is partially polarized, the grains selectively scattering a few percent more waves that oscillate in one direction than in others. For the light to be polarized, the grains must be both elongated and aligned with one another to produce a natural filter. The only explanation is that metal atoms within the grains must be affected by a weak magnetic field a hundred-thousandth the strength that of Earth's that runs through the Galaxy. The grains have apparently captured metal atoms from the interstellar gas, the process explaining the observed elemental depletions seen in interstellar clouds.

14.1.3 Interstellar Molecules

The radio spectrum of the interstellar medium is incredibly rich in molecular emission and absorption lines (Figure 14.13). As of 1996, over 100 different kinds of interstellar molecules and molecular ions were known (a sampling is given in Table 14.1), including such common compounds as

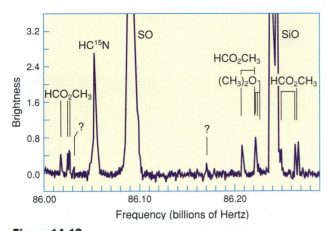

Figure 14.13
A short segment of the Orion Nebula's radio spectrum shows a remarkable number of interstellar molecules.

TABLE 14.1
A Selection of Interstellar Molecules

Two-Atom
CH (methylidine)
CN (cyanogen)
CO (carbon monoxide)
CS (carbon monosulfide)
H_2 (molecular hydrogen)
NO (nitric oxide)
OH (hydroxyl)
SO (sulfur monoxide)

Three-Atom
HCN (hydrogen cyanide)
HCO (formyl radical)
H_2O (water)
H_2S (hydrogen sulfide)
SO_2 (sulfur dioxide)

Four-Atom
HNCO (hydrocyanic acid)
H_2CO (formaldehyde)
HC_2H (acetylene)
NH_3 (ammonia)

Five-Atom
HCOOH (formic acid)
CH_4 (methane)
HC_3N (cyanoacetylene)

Six-Atom
$HCONH_2$ (formamide)
CH_3OH (methyl alcohol)
CH_3CN (methyl cyanide)
C_5H (pentynylidyne)

Seven-Atom
NH_2CH_3 (methylamine)
$HCOCH_3$ (acetaldehyde)
CH_2CHCN (vinyl cyanide)
CH_3C_2H (methylacetylene)

Eight-Atom
$HCOOCH_3$ (methyl formate)
CH_3COOH (acetic acid)

Nine-Atom
CH_3CH_2OH (ethyl alcohol)
$(CH_3)_2O$ (dimethyl ether)
CH_3CH_2CN (ethyl cyanide)
HC_7N (cyanotriacetylene)

Ten-Atom
NH_2CH_2COOH (glycine)

Eleven-Atom
HC_9N (cyanotetraacetylene)

Thirteen-Atom
$HC_{11}N$ (cyanopentaacetylene)

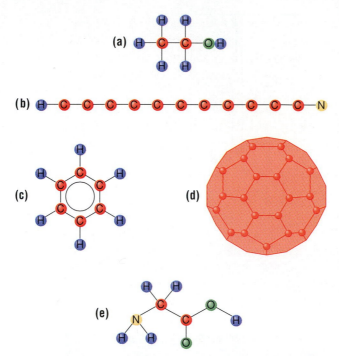

Figure 14.14
(a) Common ethanol, ethyl alcohol (CH_3CH_2OH), is abundant both on Earth and in interstellar space. A pair of carbon atoms form a bond. One carbon atom has three H atoms connected to it, the other two H atoms and an OH molecule. **(b)** A weird chain of carbon atoms makes the $HC_{11}N$ molecule. **(c)** A benzene ring can build with others to form complex polycyclic aromatic hydrocarbons that may be responsible for certain broad emission lines. **(d)** C_{60} is a soccer-ball-shaped molecule of 60 carbon atoms found in primitive meteorites. **(e)** Glycine is the first amino acid to be found in interstellar space.

molecular hydrogen (H_2), carbon monoxide (CO), water (H_2O), and ammonia (NH_3). Many are organic, based on carbon. Some, like CO, formaldehyde, acetylene, and both methyl (wood) and ethyl alcohol (Figure 14.14a) are familiar in everyday life. Others, like CH_9N and $HC_{11}N$ (Figure 14.14b), do not exist on Earth.

Interstellar chemistry can become remarkably complex. A pair of infrared emissions have been identified as polycyclic aromatic hydrocarbons (PAHs). These organic structures, built of benzene rings (Figure 14.14c), have strong odors, hence their name. They may be so large that they take on the form of tiny grains, blurring the distinction between the dust and the molecular gas. The carbonaceous chondrites (see Section 11.1) are effectively made of interstellar dust and contain C_{60} (Figure 14.14d). Most remarkable perhaps is the likely identification of interstellar glycine, NH_2CH_2COOH (Figure 14.14e), an amino acid. Amino acids, the foundations of proteins, are also found in primitive meteorites (see Section 11.1). This molecule (like the diamond dust described in Section 14.1.2) strengthens the intimate connection between the interstellar medium and the formation of our Solar System.

Molecules are easily broken apart. They require low temperatures and must be hidden from energetic starlight to survive, so they concentrate in dark, dusty clouds where temperatures may be only a few K. Nevertheless, we find that within a dark cloud almost all the hydrogen is tied up in hydrogen molecules, which constitute roughly a quarter of the interstellar mass. The abundances of other molecules are much lower and in line with the compositions of stars. From the line strengths, we find about one CO molecule for every 10,000 molecules of H_2. More complex molecules are considerably rarer: there is roughly one acetylene (HC_2H) molecule for every 10^{10} of H_2.

Extensive surveys of CO emission show that it—and the H_2 (which is difficult to observe)—are concentrated not just in the obvious Bok globules, but also in much more massive **giant molecular clouds** (GMCs) (Figure 14.15). The GMCs in turn are concentrated along the plane of the Galaxy, within the optical rifts of the Milky Way caused by the dark dust. Some 6,000 GMCs, with a huge range of radii and masses, are known. A typical GMC might be 100 pc across and contain over 200,000 solar masses. GMCs are the most massive structures in the Galaxy, and (because they contain nearly all the H_2) constitute a quarter or more of the interstellar medium's mass. The GMCs that pass close to the Earth are believed to disturb the Oort comet cloud (see Section 11.3.3) and send new comets into the inner Solar System. More important, they are the principal birthplaces of the stars, as we will see below.

Our picture of interstellar space (see Figure 14.6) is now fairly complete: we see cold giant molecular clouds, dense cold globules, and thinner cool neutral hydrogen clouds embedded in a warm gas tunneled through by hot coronal gas—all of it totaling around 10% of the stellar mass of the Galaxy.

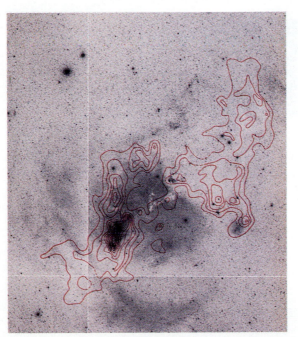

Figure 14.15

The contours show concentrations of carbon monoxide that outline a giant molecular cloud in Monoceros. They are superimposed on a photograph (presented as a negative) of the Milky Way. Diffuse nebulae are commonly seen at the edges of the clouds, demonstrating recent star birth from the cloud material. This nebula is probably on the forward edge of the cloud.

14.2 INFANT STARS

The interstellar medium, and the giant molecular clouds in particular, are the sites of star formation, where the cold raw material condenses to form the billions of stars that surround us. To build a theory of star formation, we first search for stars now in the process of creation, for **protostars** (from the Greek *proto*, "first").

The sky—and the HR diagram—are heavily populated with a great variety of **variable stars,** stars whose magnitudes change with time. These variations are caused not by eclipses within binary systems but by instabilities associated with single stars. Variations range from minutes and tenths of a magnitude for unstable white dwarfs to years and several magnitudes for the largest giants. We will encounter several different kinds of variables along the paths of stellar evolution. Associated with the dark, dusty clouds of the Milky Way are numerous F, G, K, and M variables called **T Tauri stars** after the first star found (single or double Roman letters are commonly used to indicate variability). On the HR diagram they lie up and to the right of the main sequence, among the subgiants. Typical of its class, T Tauri itself (Figure 14.16a) normally shines at apparent visual magnitude 11, but can unpredictably brighten to 10th or fade to 14th.

T Tauri stars gang together into associations that are not bound gravitationally and from which the stars are escaping, demonstrating that, like the O and B stars, they have a common and recent origin. Another indica-

The interstellar medium, and the giant molecular clouds in particular, are the sites of star formation.

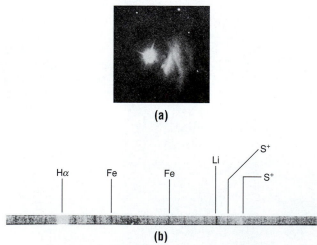

(a)

(b)

Figure 14.16

(a) T Tauri is surrounded by a small nebula associated with the star's birth. **(b)** The star's spectrum has emission lines of hydrogen and sulfur that reveal a surrounding low density gaseous cloud, and a strong absorption line of lithium that indicates extreme youth.

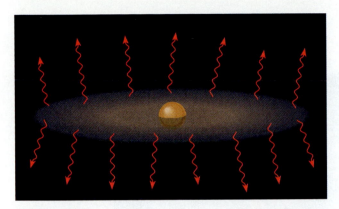

Figure 14.17
A T Tauri star is surrounded by a disk of dust heated by starlight that in turn radiates strongly in the infrared. If the Earth is not in the disk's plane, the star is easily seen.

tion of youth is found in their spectra (Figure 14.16b), which contain strong absorption lines of neutral lithium. This element is rare in the Sun and in most stars because it is easily destroyed by nuclear processes in hot, convecting, stellar envelopes. To have a high lithium abundance, a star must be young.

These variables have several other distinguishing characteristics. First, they are exceptionally bright in the infrared, which indicates surrounding clouds of heated dust, and their spectra display hydrogen and other emission lines that indicate accompanying gas. There is so *much* dust that if it were spherically distributed, it would hide the star. Since we can *see* T Tauri stars, the dust—and the gas—must be distributed in disks tilted to the line of sight, allowing us to see the majority of the stars (Figure 14.17).

Second, T Tauri stars are sometimes centered between two blobs of radiating gas called **Herbig-Haro objects** (Figure 14.18a). HH objects are produced when two jets of gas emerge in a **bipolar flow** from opposite sides of a T Tauri star in the direction perpendicular to the surrounding disk (Figure 14.18b). The bipolar flows, moving outward at speeds in excess of 100 km/s, shovel the local interstellar medium before them, lighting the compressed blobs with energy delivered from giant shock waves.

Third, T Tauri stars are also extraordinarily bright in the ultraviolet and X-ray spectral regions for their spectral classes. We can most easily explain this energetic radiation by matter spiraling toward the star from the surrounding dusty gaseous disk and violently crashing onto the stellar surface. The stars are apparently growing, accumulating mass from surrounding **accretion disks** (Figure 14.19). The rotation of the disk apparently produces a powerful magnetic field that directs some of the falling matter away from the star, ejecting it into the perpendicular bipolar flows

We have found real stellar infants that will develop into Sunlike dwarfs.

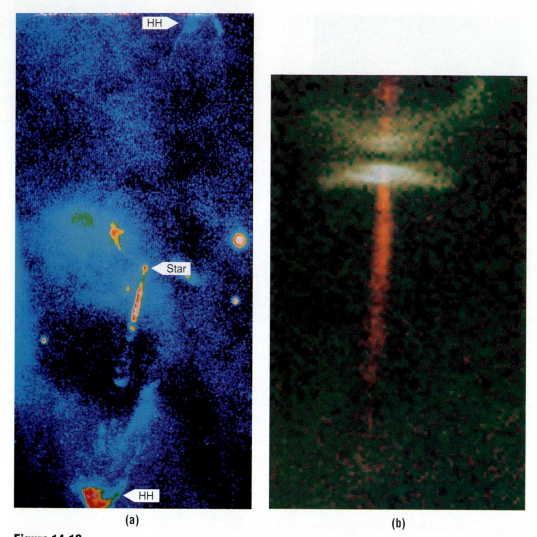

(a)   (b)

Figure 14.18

(a) HH 34, in the Orion Molecular cloud about 1° south of the Orion Nebula, consists of a set of two Herbig-Haro (HH) objects. They outline shock waves created by twin jets—bipolar flows—emanating from the star at center ramming into the surrounding interstellar medium. The upper jet is not visible because of foreground dust, but it is surely there. **(b)** The Hubble Space Telescope imaged HH 30's disk, from which emerges the characteristic jet. The edges of the disk are lit by reflection from the star buried inside.

that make the HH objects. All the evidence leads us to believe we have found real stellar infants in the process of growth, pre–main-sequence stars that will develop into Sunlike dwarfs.

We can enter the dark clouds with radio telescopes and see the disks and bipolar flows at even earlier stages of development. The interstellar ammonia molecule radiates its emission lines from particularly dense clouds of gas. Figure 14.20 shows the ammonia (yellow contours) outlin-

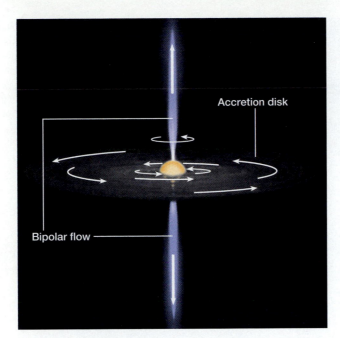

Figure 14.19
Mass falls from an accretion disk onto a T Tauri star that is still increasing its mass. A wind generated by the star escapes through the poles of the disk as a bipolar flow that will make Herbig-Haro objects beyond the confines of the drawing.

Figure 14.20
Yellow curves outlining ammonia radiation show that a thick disk surrounds an infrared star in the dark cloud Lynds 43. A bipolar carbon monoxide flow (blue) runs perpendicular to the disk. The dashed and solid blue lines show receding and approaching matter respectively.

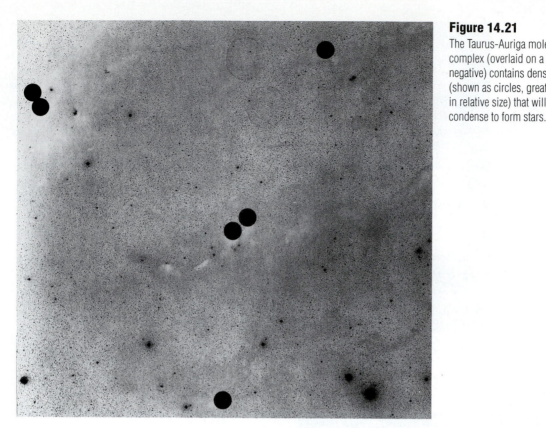

Figure 14.21
The Taurus-Auriga molecular cloud complex (overlaid on a photographic negative) contains dense cores (shown as circles, greatly exaggerated in relative size) that will ultimately condense to form stars.

ing a thick disk of matter. Perpendicular to the dense disk is a bipolar flow of carbon monoxide (blue contours), one jet tilted toward us, the other away. In both cases, the molecules trace the abundant molecular hydrogen.

We can trace the steps back even farther. Molecular clouds are highly fragmented. Figure 14.21 shows the locations of relatively compact knots or **dense cores** of matter typically a tenth of a parsec (20,000 AU) in radius revealed by ammonia radiation. They have densities of a few tens of thousands of H_2 molecules/cm^3, masses about that of the Sun, and are very cold, with temperatures of only about 10 K. Dense cores associate with T Tauri stars; they are the original blobs of matter out of which new stars will be born.

14.3 THE BIRTH OF SUNLIKE STARS

The observations allow us to construct a pathway that takes lower-mass (roughly solar-type) stars from conception onto the main sequence. Start with a rotating dense core of intensely cold molecular gas and dust. It may be one of many within a giant molecular cloud or by itself within a small

Bok globule. The core may have been created when the interstellar medium was compressed by a shock wave from an exploding star or by other means to be discussed later.

The core contracts under the force of its own gravity. The budding star's major problem is its angular momentum. As the core contracts, the conservation of angular momentum (see Section 5.2) makes the core spin ever faster, and if left alone it would ultimately tear itself apart. To form a star, the contracting core must rid itself of its angular momentum. The behavior of a cloud or core depends on the balance of forces within it. As gravity contracts the core, gas pressure pushes in the other direction. In addition, a few ions within the core (created by penetrating high-speed atomic particles called "cosmic rays" that we will examine in Chapter 16) grab the Galaxy's magnetic field (Figure 14.22a). As the core contracts, the field lines squeeze together and provide another outward pressure, slowing the contraction; the field lines also act like ropes tied to the Galaxy and gradually slow the rotation, the first critical act in removing angular momentum. If the core is dense enough, gravity ultimately wins the battle of forces. As the core shrinks, the magnetic field drags the ions to the outside, the inside becomes less charged, the magnetic field loosens its grip, and the core contracts ever faster.

If enough angular momentum has been removed by the magnetic field, a single star is now on its way to forming at the core's center. However, if the rotation speed is still too high, the contracting cloud may fragment into a *pair* of cores, allowing the formation of a binary. Angular mo-

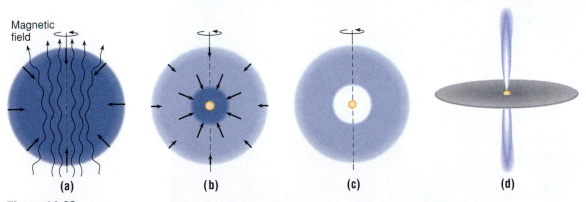

Figure 14.22
(a) A star first develops as a contracting dense core supported by gas and magnetic pressure, its rotation axis aligned along the field direction. The magnetic field also slows the rotation. **(b)** As neutral molecules and atoms fall inward, the magnetic field loosens its grip and the core contracts. **(c)** Infalling gas generates a shock wave that heats a protostar a few solar diameters across at the center, creating a clear zone within a surrounding dusty cocoon about the size of our Solar System. **(d)** Shrinkage makes the surrounding cloud spin faster and develop into a disk, the wind from the star creating a pair of opposing jets. (The figures are not to scale.) Fragmentation may also make a binary.

mentum then goes into orbital motion instead of spin. These cores may even split again, producing a quadruple or even a sextuple star.

Once enough angular momentum has been removed, the outer part of the core feeds matter into the denser center (Figure 14.22b). The rapidly collapsing cloud now produces a shock wave that creates heat and a genuine protostar with a radius a few times that of the Sun and a luminosity many times solar. The heat hollows the interior and creates a dusty shell comparable to the size of our planetary system (Figure 14.22c). The dust is heated to a few hundred Kelvins by the interior protostar and in turn radiates powerfully in the infrared (Figure 14.23). As a result of the conservation of angular momentum, the whole assembly of accreting matter flattens into a disk (Figure 14.22d) like the ones around the youthful objects

Figure 14.23
IRAS 16293-2422 (seen in the inset infrared image) is a cold (20-K) source 20 times more luminous than the Sun, a protostar buried in a dusty cocoon set within the Ophiuchus dark cloud. Note also in the inset the newly formed cluster.

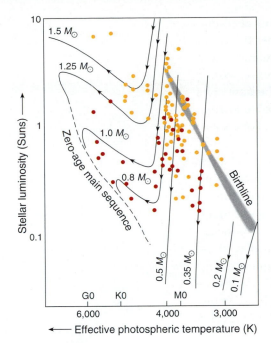

Figure 14.24

Protostars of different mass contract along evolutionary tracks (black lines) on the HR diagram, expressed here in terms of luminosity and effective temperature. They stabilize first at the birthline and then dim as T Tauri stars (yellow dots). As they shed their envelopes they turn into quieter T Tauri stars (red dots). When they are hot enough to fuse their internal hydrogen, they settle onto the zero-age main sequence (dashed line), their positions depending on mass.

shown in Figures 14.18 and 14.20. Bipolar flows and jets then create the HH objects.

The contracting protostar continues to accrete matter, and the temperature rises. After 10^5 to 10^6 years, the interior reaches a temperature of around a million K, and its small amount of natural deuterium (about one atom out of every 10^5 of ^1H) begins to fuse to helium. Convection continually sweeps fresh deuterium into the hot center, keeping the nuclear fire burning. This is the moment of truth. The star now begins to stabilize, to develop its T Tauri characteristics, and for the first time we can place it on the HR diagram at the *birthline* (Figure 14.24).

Theoretical calculations show how the luminosity of a protostar changes with effective (photospheric) temperature. We can plot graphs of these changes, as curves called **evolutionary tracks,** on the HR diagram (see Figure 14.24). From the birthline, the stars descend along nearly vertical tracks as they contract at nearly constant surface temperature. At some point, deuterium fusion dies down and the evolutionary tracks turn to the left as the protostars heat at their surfaces at roughly constant luminosity.

As the interiors heat by gravitational compression past the critical 10-million-Kelvin mark, true hydrogen fusion (starting with ^{1}H rather than deuterium) turns on, accretion and the T Tauri characteristics die away, contraction is halted, and the stars quickly enter the main sequence. With a full supply of fuel, the stellar clocks are set to "zero age," the new stars defining a **zero-age main sequence** (ZAMS). They are now set to burn at an allotted luminosity for a lifetime that both depend on mass. For a solar-type star the whole process has taken about 10 million years.

Protostars that are in thin parent globules or at the edges of their GMCs become visible to us in the early stages of their development and can be placed on the HR diagram along with the theoretical evolutionary tracks (see Figure 14.24). Others are in thicker dusty clouds and remain hidden from view. A look with an infrared camera into dusty regions associated with bright nebulae commonly reveals hundreds of new stars, many of which have probably already landed on the main sequence (Figure 14.25). In still other cases, intense radiation from nearby O stars evaporates the birthcloud to reveal dozens of new stars hatching like newborn chicks (Figure 14.26). Sometimes many stars form all at once in an open cluster; others may form with few neighbors and eventually orbit the Galaxy as single stars like our Sun or binaries like α Centauri, their places of origin long lost.

Figure 14.25
An infrared view into a diffuse nebula called NGC 2024 (seen in context in Figure 14.8) reveals dozens of newly formed stars hidden by the parent dust cloud.

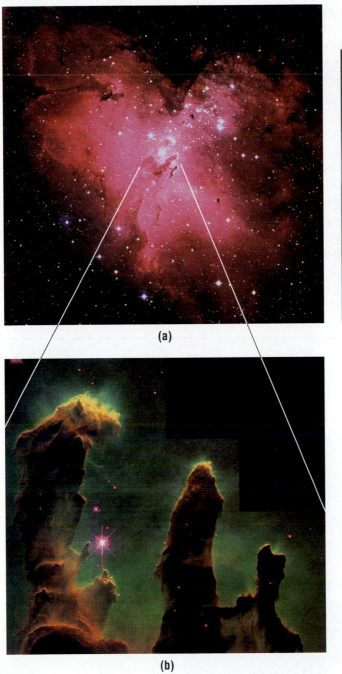

Figure 14.26
(a) The Eagle Nebula, M 16, contains three prominent shafts of dark dust. **(b)** The Hubble Space Telescope shows the tips of the shafts evaporating under the intense radiation from nearby hot stars. **(c)** The dissipation of the dust reveals stars that have been created within the shafts.

14.4 BIRTH UP AND DOWN THE MAIN SEQUENCE

T Tauri stars become main-sequence stars of classes roughly K through A with masses that range between about 0.75 and 3 $M_\odot$. The creation of higher-mass B stars is similar, involving higher-mass versions of the T Tauri variables. The formation of O stars is somewhat different, however. All stages proceed at a greatly increased rate. An O star is expected to begin hydrogen fusion even before it has stopped accreting matter. Its enormous luminosity hollows out a shell within the birth cloud, which the ultraviolet radiation lights up to produce a compact diffuse nebula. As the O star develops it sweeps away its surrounding cloud, and a fierce wind may compress the embracing interstellar medium into a bubble. Combined winds from OB associations may even be a significant factor in promoting star formation. The upper limit to stellar mass is really unknown. Massive stars are apparently so difficult for nature to make that the chance of any with masses greater than about 120 $M_\odot$ developing within a galaxy is effectively zero.

Stars of low mass, which are faint and hard to see, probably form in much the same way as did the Sun. The lower limit to the main sequence is imposed by interior temperature. Below 0.08 $M_\odot$, gravitational compression does not generate enough heat to raise the interior above the critical level of about 7 million K required for full hydrogen burning. Starlike

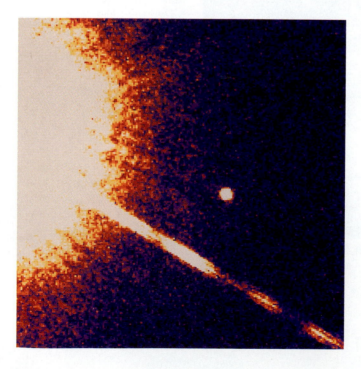

Figure 14.27

Though bright in the picture, Gliese 229 (at far left) is actually a dim red dwarf. Forty AU away from it (just right of center) is an odd faint body much cooler than an ordinary main-sequence star that appears to be the first known brown dwarf.

objects below this limit of 0.08 $M_\odot$ are called **brown dwarfs.** For a time, brown dwarfs should appear bright enough to see, especially in the infrared. Gravitational contraction heats them, producing radiation, and internal temperatures can exceed the million-degree limit required for the burning of their natural deuterium. Though astronomers have been searching for them for many years, and there are several good candidates, only one seems truly to measure up. Methane bands formed in the atmosphere of Gliese 229 B (Figure 14.27) yield a temperature of 1,100 K, which suggests a mass well below 0.08 $M_\odot$. Only a mass measurement made by observing the binary orbit will tell for sure. Even if the star's status is confirmed, however, it is clear that brown dwarfs are very rare. No one knows why.

14.5 PLANETS

Nature, however, makes bodies *much* less massive than stars, such as Jupiter and the other planets in our own Solar System. Jupiter is nowhere near massive enough to initiate thermonuclear fusion, even deuterium fusion, and it is not a brown dwarf. A brown dwarf is expected to be created whole as a result of contraction and accretion in a dense cloud; Jupiter and the other planets were accumulated from a vast number of primitive planetesimals and surrounding gas.

We can now finish the story begun in Chapter 11, where we saw that the planetesimals and planets are distributed *in a disk about the Sun.* The disk appears to be the remnant of the original accretion disk that must have formed around the protosun over 4.6 billion years ago. The dust in this disk accumulated to form larger pellets and finally planetesimals. More profoundly, the dust that created the bulk of the terrestrial planets and the cores of the Jovians came from interstellar space. *We are the products of other stars, our Earth a distillate not just of the solar nebula but of the dusty, long since dissipated interstellar cloud that gave us birth.*

If the Sun created planets from its spinning protostellar disk, other stars should have developed planetary systems too. Even Jupiter, Saturn, and Uranus have their own little sets of "planets," their satellites, further suggesting how common these systems might be. Can we find evidence for extrasolar planets? In 1984, *IRAS* satellite detected the infrared signatures of dusty disks surrounding Vega and Fomalhaut. Hubble Space Telescope images also reveal dusty gaseous structures around the young stars in the Orion Nebula (Figure 14.28). More telling is an edge-on disk 400 AU in radius that surrounds the main sequence A star β Pictoris (Figure 14.29). Though the disk is 10 times larger than the distance between the Sun and Pluto, remember that our Solar System extends vastly farther, into the Kuiper belt of comets that circulates in a thick disk about the Sun. We may

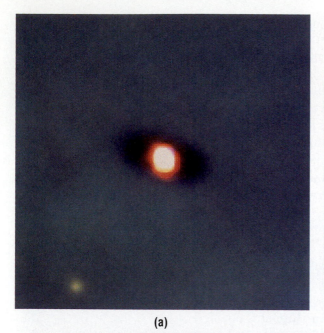

(a)

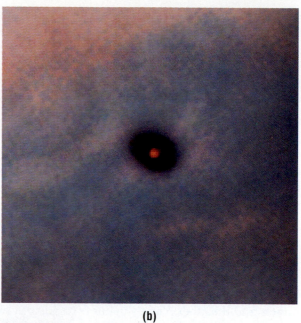

(b)

(c)

Figure 14.28

A deep, expanded look inside Figure 14.1 reveals several dusty disks seen in relief against the nebular background. The disks, several times the size of our planetary system, surround young stars similar in mass to the Sun and only a million or so years old. **(a, b)** The dark disks are more or less perpendicular to the line of sight, allowing us to see the stars at their centers. **(c)** An edge-on disk hides the central star. The bright spot at the center is starlight reflected from the dust at the disk's edge. Do the disks contain planets?

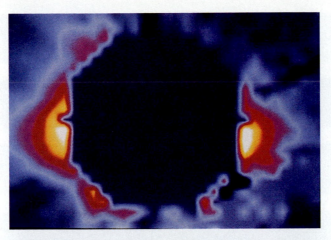

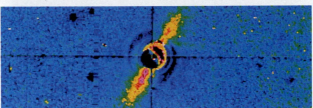

Figure 14.29
A disk of heated gas surrounds
β Pictoris (the star itself has been
eliminated from the picture) with
silicates shown in yellow, ices in red.
The extent of the disk (400 AU in
radius) is shown below.

be seeing dust and debris in β Pictoris's own Kuiper belt. Infrared observations show that the β Pictoris disk contains silicates and ices mixed with carbon, recalling the compositions of comets and asteroids in our own system. Moreover, the inner portion of the disk appears to be thinned out, suggesting the clearing of the dusty debris by the formation of planets.

We cannot yet actually see planets orbiting other stars. From α Centauri, Jupiter would shine at 22nd magnitude at most only 4 seconds of arc from a bright star whose glare would render it impossible to observe with our present technology. However, even though a planet may be invisible, it still has a gravitational tug on its parent star. Jupiter's mass is a thousandth the Sun's, effectively placing the center of mass of the Solar System 0.005 AU—the solar radius—from the solar center. The circumference of the solar orbit is therefore 0.03 AU ($2\pi \times 0.005$), about 4.7 million km, which the Sun traverses in Jupiter's sidereal period of 12 years (3.8×10^8 seconds). The Sun's orbital speed is thus 12 m/s. Seen from the outside, our Sun would have a variable radial velocity of up to 12 m/s, comparable to our current detection abilities.

Such radial velocity variations have clearly been seen for several stars, including 51 Pegasi, 47 Ursae Majoris, 70 Virginis, and ρ^1 Cancri, all nearby G dwarfs similar to the Sun. In effect, each star presents one-half of a spectroscopic binary (see Section 13.5.1), for which we observe only one

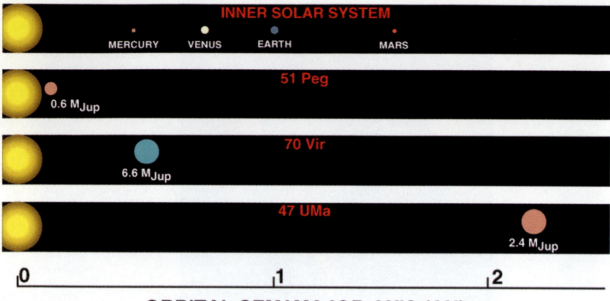

Figure 14.30
Three recently discovered planets and their minimum masses are compared with our own Solar System.

set of absorption lines periodically wobbling back and forth. However, we already know the mass of the parent star from the mass-luminosity relation. We can therefore determine a lower limit to the mass of the invisible companion (all we can do since we do not know the orbital tilt). Estimates of the most likely tilts powerfully suggest that the companions to these stars have masses a few times the mass of Jupiter (Figure 14.30). The surprise is their distribution: semimajor axes range from only 2.1 AU for 47 UMa down to a mere 0.05 AU for 51 Peg. How a Jupiter-like planet could form so close to its parent star is a mystery.

Our own planetary system seems to represent only one possibility out of many, and as we begin to explore other such systems we are surely in for surprises. Whatever we do find, it is becoming clear thet planets are a natural by-product of star formation. Direct sighting is only a matter of time.

> It is becoming clear that planets are a natural by-product of star formation.

14.6 LIFE

The ultimate question is not really about other planets but about what may be on them. We have few conjectures how life formed on Earth, but form it did, and quickly too. In Australia there are primitive fossils of organisms that were alive only a billion years after our planet was created. The Earth's waters, and much of its carbon, apparently arrived aboard crashing comets in the early days of the Solar System. Taking this view

BACKGROUND 14.1   Unidentified Flying Objects

The Search for Extraterrestrial Intelligence has the misfortune to be confused with the "sighting" of unidentified flying objects (UFOs). The two have nothing to do with each other. For centuries, people have been seeing things in the sky, lights and other objects, that they have not understood. Some then propose that because the objects are unexplained, they must be of extraterrestrial origin and must contain visitors from another planet.

UFOs fall into three major categories: natural phenomena that the viewer cannot explain, hoaxes, and delusions. The last of these cannot ever really be explained, as the evidence is not real. Hoaxes are remarkably easy to fabricate, and many have been perpetrated on an unsuspecting public. Photographs are simply not good evidence. All you need do is take an out-of-focus picture of a flying garbage can lid and you have a picture of a UFO you could probably get published in a supermarket tabloid. Figure 14.31a shows another example, an upside-down banana-split dish hung on a wire.

The natural phenomena taken for UFOs are far more interesting. Venus tops the list. As you drive along a highway the planet seems to follow you. Military pilots have tried to shoot it down, and a sheriff and his deputy in Ohio chased it into Pennsylvania, where it disappeared (most likely, the Sun came up). Stars have been mistaken for UFOs, as have been mirages (which are far more common than most people think), oddly shaped clouds, and lights seen through clouds (Figure 14.31b). Fireball meteors, with their long, persistent afterglows, have been mistaken for tubular spaceships.

To be sure, there *are* unexplained sightings. "Unexplained," however, does not mean "extraterrestrial." Those who champion an extraterrestrial origin of at least some unexplained UFOs have a burden of proof: they must show credible evidence. And to date, there is none. We have no reason to believe that UFOs fall into any category other than the three discussed above.

Real star travel is a massively difficult problem. The amount of energy required to send even a robot craft to the nearest star in a reasonable time, within a human lifetime, is staggering and far beyond our current capability. The propulsion problem may be solved sometime in the distant future. But if it is, there is no evidence that it will have anything to do with "unidentified flying objects."

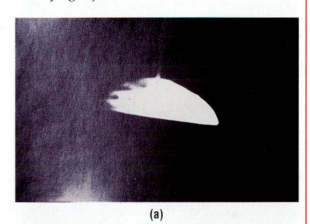

(a)

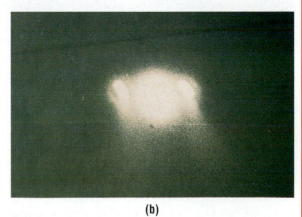

(b)

Figure 14.31
Photographic evidence of UFOs is not reliable. **(a)** An upside-down banana-split dish is hung on a wire. **(b)** The landing lights of an airplane are seen through fog.

one step farther, we might speculate that comets and meteorites, which contain complex molecules from interstellar space (including amino acids), also brought the building blocks out of which life eventually developed. If that was the case, late-arriving planetesimals must also have seeded the Solar System's other planets. However, only the Earth has had long-term abundant supplies of a near-universal solvent, liquid water, that might have allowed the seeds to grow into the more complex molecules that support life.

No matter what the speculation, however, the origin of life on our planet—and of ourselves—is shrouded in deep mystery. We have only one sample planet on which we *know* life formed, and that formation occurred in the dim past. If we cannot find further evidence in the Solar System for life and its origins, we must look elsewhere: to the stars. As planets may be a by-product of star formation, is life a by-product of planet formation? If it exists, can we detect it? We have not yet even been able to find extrasolar planets, so we are certainly not going to make observations of plants or insects. The search involves not just life, but *intelligent* life, something similar, however vaguely, to ourselves, something with which we might communicate even though the harboring planets are still invisible.

A means of estimating the possibilities of life elsewhere can be made with a version of the *Drake equation* (developed by radio astronomer Frank Drake). The probability (P) that a star harbors intelligent life is

$$P = f_p n_h f_l f_i f_t,$$

where f_p is the fraction of stars that have planets, n_h the number of planets such stars have in habitable zones, f_l the fraction of the planets that lie in a habitable zone that actually evolves life, f_i the fraction of these that evolves intelligent life, and f_t the fraction of planets on which life manages to endure (the ratio of the lifetime of a civilization to the age of the Galaxy).

We might expect that many if not most stars may have a planet in a habitable zone (f_p and n_h both close to 1). But the other factors are unknown. Even if these fractions are very small, however, the possibility of such life is increased by the vast number of stars in the Galaxy. Civilizations may teem around us, or the odds for the creation of life might be *so* low that we are alone.

Given the absence of data or even accurate speculation, all we—like scientists everywhere and in every discipline—can do is to go and look. This subject, the **Search for Extraterrestrial Intelligence** (SETI) has long been taken seriously, endorsed by the International Astronomical Union (the world organization of professional research astronomers), the U.S. National Academy of Sciences, and the National Commission on Space. Because of the immense distances between stars, SETI involves not direct personal contact but rather communication by radio. The task is to look into space to try to find some artificial signal, something periodic that is clearly not part of the cosmic radio-noise background.

The search is daunting, the reward measureless for both science and philosophy. Where do we start in the vastnesses of the sky and the radio spectrum? What stars do we look at, where do we tune our receivers? The first step was taken in 1960 by Frank Drake, whose Project Ozma at the National Radio Astronomy Observatory examined ϵ Eridani and τ Ceti. He chose radio frequencies near 21 cm on the assumption that if extraterrestrial engineers were broadcasting an intentional signal, they would place it near an obvious and recognizable spectral reference. He found nothing.

There have since been many, more sophisticated projects. One at Harvard called META (Mega-channel Extraterrestrial Assay) uses a 26-m radio telescope and a receiver that can scan 8.4 million radio channels at the same time. A second META observatory with two 30-m dishes operates in Argentina. The biggest effort was NASA's 10-year High Resolution Microwave Survey, which was to employ the 34-m dishes of the Deep Space Network used for tracking space probes, the giant Arecibo radio telescope, and other instruments. No longer supported by Congress, the project is now funded privately. Part of this program involves a survey of the entire visible sky, and another search concentrates on nearby solar-type stars. So far, however, despite the effort, we have found nothing.

The lack of data should not yet be discouraging. The Galaxy is enormous, and the signals, especially if they are passive (that is, not produced by a transmitter actively trying to make contact), may be weak. Is anyone else out there? We will never know if we do not look. Perhaps a century into the future we will have a better answer. So far, however, for all the grand speculation, the search for intelligence in the Universe is a science with but one datum: ourselves.

▶ KEY CONCEPTS

Accretion disk: A disk of gas revolving around a star from which matter flows onto the stellar surface.

Bipolar flow: Twin flows of gas emitted perpendicular to a rotating disk.

Bok globules: Small, opaque, relatively dense, dusty interstellar clouds, typically a few pc across and containing 10 to 100 solar masses.

Brown dwarfs: Bodies that contract directly from the interstellar medium with masses below the main-sequence cutoff of 0.08 $M_\odot$, too low to allow the proton-proton chain to operate.

Dense cores: Knots of matter in the interstellar medium, observable by ammonia radiation; the first observable step in star formation.

Diffuse nebulae: Bright clouds of interstellar gas ionized by stars at least as hot as type B1.

Evolutionary tracks: Graphs of luminosity against effective temperature (or M_V against spectral class) on the HR diagram.

Giant molecular clouds (GMCs): Massive dusty clouds (typically 200,000 $M_\odot$ and 100 pc across) made mostly of hydrogen molecules.

Herbig-Haro (HH) **objects:** Blobs of gas in the interstellar medium lit by shock waves from bipolar flows.

Interstellar absorption lines: Narrow absorption lines superimposed on the spectra of stars, produced by atoms and ions in the interstellar medium.

Interstellar dust: Small solid grains of silicates or carbon coated with ices and embedded with heavier atoms.

Interstellar medium: The lumpy mixture of gas and dust that pervades the plane of the Galaxy.

OB associations: Unbound groups of O and B stars found in the Galaxy's disk.

Protostars: Stars in the process of formation.

Reflection nebula: A bright cloud created by light scattered by dust from a star too cool to cause ionization (that is, cooler than type B1).

Search for Extraterrestrial Intelligence (SETI): The search for artificial radio signals that would indicate intelligence beyond Earth.

T Tauri stars: Active, young, variable pre–main-sequence stars.

21-cm line: A powerful radio line of neutral atomic hydrogen produced in the interstellar medium.

Variable stars: Stars whose magnitudes change with time.

Zero-age main sequence (ZAMS): The line on the HR diagram defined by new stars that are just beginning to fuse hydrogen into helium.

▶ **EXERCISES**

Comparisons

1. List the differences between diffuse and reflection nebulae.

2. How does the intercloud medium differ from the coronal gas?

3. Compare the galactic distributions of interstellar dust and interstellar carbon monoxide.

4. How do brown dwarfs differ from real stars and from planets?

Numerical Problems

5. What is the frequency of the 21-cm line?

6. What is the minimum frequency a photon can have and still ionize hydrogen from the ground state?

7. A protostar is 10 times more luminous than the Sun and has an effective temperature of 100 K. How big is its dust shell in AU?

8. A planet with 1 Jupiter-mass orbits 1 AU from a star with a mass of $0.5 \, M_\odot$. Assuming that the orbital plane lies in the line of sight, what is the star's maximum Doppler shift at 5,500 Å as a result of the planetary orbit?

9. What would be the approximate apparent magnitude of the Earth and the Earth's maximum angular separation from the Sun as seen from α Centauri?

Thought and Discussion

10. What processes produce the emission lines of diffuse nebulae?

11. What is described by the temperature of a diffuse nebula?

12. What observations provide evidence for the existence of a general gas in the interstellar medium?

13. What observations provide evidence for the existence of dust in the interstellar medium?

14. How is the 21-cm line created?

15. What is polarized light?

16. How do we know that dark regions in the Milky Way are really clumps of dust and not gaps in the stellar distribution?

17. We use carbon monoxide to trace the location of H_2 in the interstellar medium. Why would ethyl alcohol not make a very good tracer?

18. What is the effect of interstellar dust on the calculation of distances to stars?

19. What observations lead us to believe there are two different kinds of interstellar grains? What are they?

20. What evidence leads us to believe that there are magnetic fields in the Galaxy?

21. Why are interstellar molecules associated with dark clouds?

22. What are the various components of the interstellar medium? What fraction of the mass does each represent?

23. Why do we think that T Tauri stars are young?

24. What is the evidence for mass-loss in T Tauri stars and for disks and bipolar flows in these and other protostars?

25. How are Herbig-Haro objects produced?

26. Why does a dense core not collapse quickly?

27. How does deuterium fusion affect a protostar?

28. On the HR diagram, what is the "birthline"?

29. What happens inside a star that settles it onto the main sequence?

30. How might we locate planets in orbit around other stars?

31. Using the Drake equation, make your own estimate of the number of civilizations there might be in the Galaxy. Adopt values for the various factors and then multiply by the number of stars in the Galaxy. Explain your reasoning.

Research Problem

32. Brown dwarfs and extrasolar planets have proved elusive. From news notes in popular astronomy magazines, find instances in which discoveries have been publicized and later withdrawn. What led to the false discoveries?

Activities

33. On a dark moonless night in the summer, draw the Milky Way as you see it and outline the dark clouds.

34. Make a list in order of development of catalogued protostellar objects and their properties. Diagram the sequential development of the Sun up to the time it reaches the main sequence and develops planets.

Scientific Writing

35. Amateur astronomers are most interested in what they can see with their telescopes. Write a column for an amateur magazine that describes hidden wonders: interstellar gas, dust, and molecules that can only be seen by observations made in the infrared and radio.

36. Write an essay in which you solidly link the Sun's planetary system with the disks that are believed to exist about T Tauri stars. Summarize all the evidence from the Solar System that shows why we believe the planets formed in a disk, and summarize the evidence that T Tauri stars have disks.

THE LIFE AND DEATH OF STARS
The flow of stellar lives from the main sequence to death

15

We string the different kinds of stars together with theory, and now know that the Sun and stars will someday expire, their deaths providing sustenance for stellar birth.

15.1 THE MAIN SEQUENCE

The evolution of a star is ultimately the result of gravitational contraction as it tries to make itself as small as possible. The phases of a star's life are the result of pauses in the shrinkage or of the act of shrinking itself. A dense core contracts to a main-sequence star, which represents the first pause.

The Sun and other main-sequence stars are stable because the inward pull of gravity is balanced by the outward push of gas pressure. The solar luminosity depends on the internal temperature generated by gravitational compression and consequently depends on the solar mass. Stars of higher mass have higher internal temperatures and therefore greater luminosities; theory shows that they should brighten in accord with the general form of the observed mass-luminosity relation, ($L \propto M^{3.5}$). Thermonuclear fusion of hydrogen is not immediately responsible for the Sun's luminosity but for the 10-billion-year duration over which that luminosity can be sustained. Because main-sequence stars are made from the same material, they must also be supported by hydrogen fusion. The main sequence is thus a state of long-term hydrogen-burning stability.

Solar energy is produced by the proton-proton chain (see Figure 12.21). Energy can also be created by the **carbon cycle** (Figure 15.1), in which a ^{12}C nucleus is progressively struck by protons that turn it into various isotopes of nitrogen, carbon, and oxygen. At the end, four protons—hydrogen nuclei—are transformed into a ^{4}He nucleus and the ^{12}C is recovered. The carbon cycle is more temperature dependent than the p-p chain. At lower masses, among the K and M dwarfs, it is insignificant. Above about 2 $M_\odot$, however (near spectral class F0), the carbon cycle increasingly dominates.

The main-sequence lifetime of a star depends on the amount of fuel available divided by the rate at which it is consumed. The amount of fuel

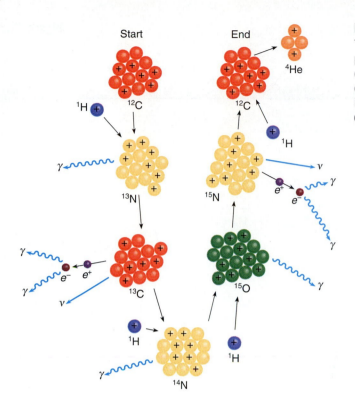

Start

End

^{1}H

^{12}C

^{13}N

γ

^{15}N

^{4}He

^{1}H

ν

γ

e^+

e^-

γ

γ

e^- e^+

γ

ν

^{13}C

^{15}O

γ

^{1}H

^{1}H

γ

^{14}N

Figure 15.1

The carbon cycle uses ^{12}C to fuse four protons into ^{4}He through the creation of isotopes of carbon, nitrogen, and oxygen. The original carbon nucleus is then recovered. Each reaction except the last produces gamma rays.

is proportional to the stellar mass (*M*) and the rate of nuclear burning is proportional to the luminosity (*L*). The lifetime (T) must therefore be proportional to *M/L*. But $L \propto M^{3.5}$, so

$$T \propto M/L = M/M^{3.5} = 1/M^{2.5}.$$

Higher-mass stars have dramatically lower lifetimes.

Higher-mass stars therefore have dramatically *lower* lifetimes. The Sun began with enough hydrogen fuel to live on the main sequence for 10 billion years. Vega, 2.5 times more massive than the Sun, will live only 1/10 as long, only a billion years. The M5 dwarf Proxima Centauri, with half a solar mass, will live 10 times as long as the Sun, 100 billion years. When we factor in the ratio of *burnable* mass (that located where the temperature is sufficiently high) to *total* mass and the structural changes that alter the size and mass of the nuclear-burning core with age, we see that a 120-$M_\odot$ O3 star will stay on the main sequence for a mere 3 million years. At the lower main-sequence mass limit of 0.08 $M_\odot$, stars will live for over a *trillion* years (Figure 15.2).

The differences among the HR diagrams of clusters (see Figures 13.24 and 13.26) are now more understandable. Stars of the entire main sequence may appear within a newly born cluster. The cluster's higher-mass stars evolve faster and disappear from the main sequence to become supergiants and giants. The older a cluster, the lower the main sequence's

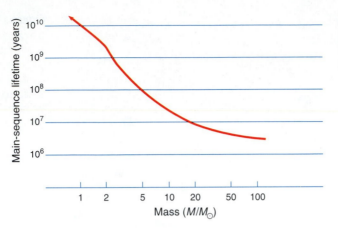

Figure 15.2
As the masses of stars increase, their main-sequence lifetimes rapidly decrease.

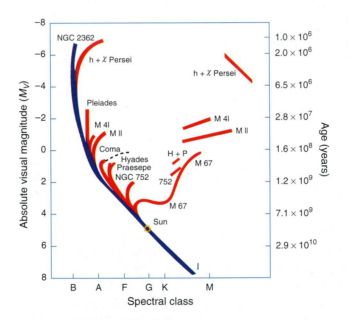

Figure 15.3
As clusters age, they lose their main sequences from the top down. The top of the main sequence (blue) for each cluster is indicated by where the giant branch (red) turns off. Ages that correspond to the tops of the main sequence are given at the right.

stopping point (Figure 15.3). From theory and the observed point where the main sequence ends and turns off to the giant or supergiant branch, we can derive the cluster's age.

The main sequence is a zone of stability because of its stars' abilities to adjust themselves internally as the percentage of hydrogen fuel declines. As fusion proceeds, four atoms of H are turned into one of He, and the number of atoms drops. The pressure of a gas is proportional to the number of atoms per unit volume, so the core compresses under the weight of the overlying layers. Temperature and density climb, and the rate at which the particles fuse offsets the diminishing fuel supply.

The compensation cannot be perfect, however, so main-sequence stars undergo slow evolutionary changes. At birth, when the Sun was on the zero-age main sequence, it was 30% fainter and nearly 8% smaller than it is now. Over the last 4.6 billion years it has "moved" on the HR diagram along an evolutionary track (that is, its graphical position in temperature and luminosity has changed), slowly becoming brighter, larger, and slightly hotter. By the time the internal fuel is gone, the Sun will be twice as bright and 50% larger. All main-sequence stars behave in somewhat the same fashion. As a result, the observed main sequence (see Figure 13.13) is a *band* with the zero-age main sequence (see Section 14.3) along its left-hand edge (Figure 15.4).

In 5 billion years we will receive as much energy from the Sun as Venus does now, and the Earth should be a greenhouse oven long before solar main-sequence life is finished. When life began on Earth some 3.5 billion years ago the amount of sunlight we received should have been

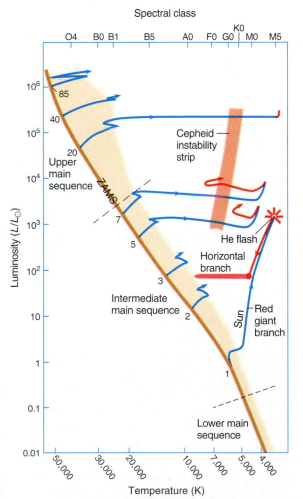

Figure 15.4

The main sequence is indicated in yellow and the evolutionary tracks for evolving stars in blue and red. Slight evolution on the main sequence makes it a band instead of a line. No lower-main-sequence star (0.08 $M_\odot$ to 0.8 $M_\odot$) has ever evolved away. From the intermediate main sequence (0.8 $M_\odot$ to 8 $M_\odot$), solar-type stars become cooler and brighter as they climb the red giant branch of their evolution with burned-out helium cores (blue). Solar-type stars terminate their ascents of the giant branch at the helium flash and then descend about halfway back down fusing helium (red), from which point low-metal stars spread out into a horizontal branch. From the upper main sequence, above 8 $M_\odot$, stars cool at roughly constant luminosity as they become supergiants. As luminous stars cross the instability strip they become Cepheid variables.

more like today's conditions halfway to Mars. Was the Sun more active in the past? Did a different terrestrial atmosphere provide more insulation? We do not know.

15.2 GIANT STARS

Globular clusters have the dimmest main-sequence upper cutoff (see Figure 13.26c) and therefore must be the Galaxy's oldest systems. Ages approach 15 billion years, a value that is identified with the age of the Galaxy itself. Such a main-sequence lifetime corresponds to a star of 0.8 solar mass. Stars below this limit, those on the **lower main sequence** (see Figure 15.4), have never had time to evolve away from it.

The stars of the **intermediate main sequence,** those between 0.8 $M_\odot$ and 8 $M_\odot$, produce the giants and eventually the white dwarfs. Intermediate-main-sequence evolution is exemplified by the Sun. In 5.4 billion years (10 billion years after the Sun's birth), all the hydrogen in the solar core will be converted to helium. Fusion will stop, removing a source of internal support, and the nearly pure helium core will finally be able to resume its contraction. Because of the huge gravitational energy released, the evolving core will heat, allowing hydrogen burning to expand into a shell around it (Figure 15.5a), for a time keeping the Sun very much alive.

For about a billion years following the exhaustion of the core's hydrogen, the solar luminosity will stay roughly constant. However, the new energy of the shrinking core will nearly double the size of the Sun's outer envelope. As a consequence, the solar surface will cool to spectral class K, near 4,500 K. The nuclear-reaction rates in the shell will become so intense that the Sun will brighten. As the core heats and compresses to ever-higher density, the Sun's outer parts will continue to balloon outward. For another billion-plus years the Sun will climb an evolutionary track called the **red giant branch** (see Figure 15.4), first becoming a subgiant and then a true red giant of class M with an effective temperature of only about 3,500 K.

The Sun will extend roughly to the orbit of Venus.

During this core-contraction phase, the Sun will become 2,000 times more luminous than it is today. The solar radius will increase to about 150 times its current value, and the Sun will extend roughly to the orbit of Venus, vaporizing little Mercury. Greater radius and luminosity will promote a wind from the star—a **stellar wind**—that blows a million times more strongly than the solar wind, and mass loss will become an important factor in evolution, the red-giant Sun losing as much as 20% of its initial matter back into space. At the same time, the solar core, now increased to half a solar mass, will shrink to only a few times the size of Earth. Its temperature will then be an astonishing 100 million K and its density will approach 1 million g/cm^3.

Under these conditions, the inert helium becomes a fuel for a new reaction and begins to fuse into carbon (Figure 15.6). Helium fusion is difficult. Two helium nuclei (α particles) collide and meld briefly to form ^8Be

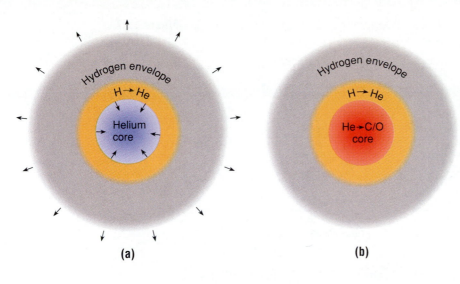

Figure 15.5
(a) As a star begins its evolution to gianthood, its quiet, shrinking helium core is surrounded by a shell of burning hydrogen encased in an expanding hydrogen envelope. **(b)** When the temperature is high enough, the helium core, surrounded by a shell of fusing hydrogen, begins to burn to carbon and oxygen. (The diagrams are not to scale.)

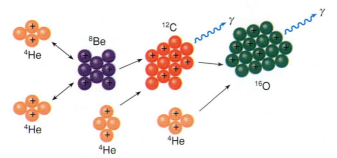

Figure 15.6
In the triple-alpha process, two ^{4}He nuclei—α particles—collide to produce ^{8}Be, which is struck by another α particle to produce ^{12}C before it falls apart. The carbon nucleus can subsequently capture another particle to make ^{16}O.

(beryllium), which falls apart in only 10^{-16} seconds. However, if a third helium nucleus can bang into the ^{8}Be just before it disintegrates, the reaction produces carbon (^{12}C) and a gamma ray. Following this **triple-alpha** (3-α) **process** an additional α particle collision can change some of the carbon to oxygen, with the creation of yet more energy.

In the Sun, the 3-α process will ignite violently in a *helium flash*, the huge new energy source terminating the ascent of the red giant branch (see Figure 15.4). The Sun will thereupon reverse direction on its evolutionary track and move about halfway back down the giant branch. Our star will now have reached another pause in its relentless attempt to contract, residing at this point on the HR diagram for about 10% of its hydrogen-burning lifetime, a K giant with a helium-burning core surrounded by a hydrogen-burning shell (Figure 15.5b). Stars more massive than the Sun evolve similarly, though above about 2 solar masses, the evolutionary tracks are somewhat different and helium ignition is quiet. Stars that exemplify the future of the Sun and other intermediate-main-sequence dwarfs can be identified by their orange color. Aldebaran, Arcturus, the brightest bowl stars in the Big and Little Dippers (α Ursae Majoris and β Ursae Minoris), and many others are helium-burning giants.

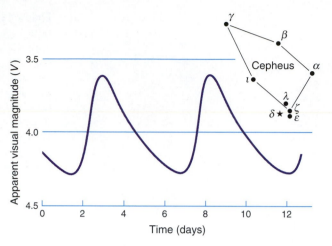

We can now explain many of the features of the observed HR diagram. High-mass stars are relatively rare. Most of those now evolving are in the neighborhood of 0.8 to somewhat over 1 solar mass. The combined effect of the evolution of these stars is the distinctive observed giant branch of Figure 13.13. The giants of a globular cluster are currently evolving from dwarfs of about 0.8 solar mass. As the stars fire their helium and come back down from the red giant tip, they spread out to the left into the horizontal branch (see Figure 13.26c) because of their low metal content (which makes their gases more transparent and increases their surface temperatures) and slight differences in mass caused by variations in mass-loss rates.

As the higher-mass stars, those above about 5 $M_\odot$, move across the HR diagram, they encounter a nearly vertical strip within classes F and G (see Figure 15.4) in which they become unstable and begin to pulsate (Figure 15.7), expanding and contracting at a regular pace. The pulsation causes them to vary in brightness. The first known example of such a star was δ Cephei, which varies between $V = 3.6$ and $V = 4.3$ over a 5.37-day period and gave its name to this class of **Cepheid variables**. (See Appendix 2 for a short list of naked-eye Cepheids.) The HR diagram's *Cepheid instability strip* crosses the globular cluster horizontal branch, where we find the lower-mass, metal-poor, short-period (about a day) *RR Lyrae stars* (see Figure 13.26c). As we will see in Chapter 17, Cepheids are crucial to the establishment of distances in the Universe.

15.3 THE SECOND ASCENT

Eventually, the giant's helium fuel is consumed and the nuclear fire once again goes out. The core loses its support and resumes its contraction. Internal compression generates more heat, making the star brighter and larger as it climbs upward on the HR diagram for the second time, now along the **asymptotic giant branch** (AGB) (Figure 15.8), an evolutionary

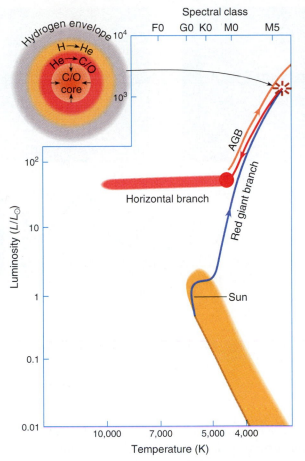

Figure 15.8

A solar-mass star ascends the asymptotic giant branch (AGB) (orange curve). It now has a dead carbon-oxygen core surrounded by a shell of helium that is fusing into carbon and oxygen, and another of hydrogen burning to helium (inset). Surrounding these is an immense hydrogen envelope that on this scale would fill roughly the average house. At the top of the AGB the star has lost almost all its envelope and has a mass of only about 0.6 $M_\odot$.

track that is roughly parallel to the red giant branch. At the center of an AGB star is a dead carbon and oxygen core that is surrounded by two shells (Figure 15.8 inset). The inner shell fuses helium to carbon and oxygen and the outer fuses hydrogen to helium, the two alternately switching on and off in an increasingly violent fashion.

As the star climbs the AGB its changing interior structure causes it to pulsate and vary in brightness (Figure 15.9). These stars are enormous compared to Cepheids, however, and they take much longer to vary. The first, found in the year 1574 in the constellation Cetus, so surprised its discoverer that he named it Mira (the "amazing one"). This star, also called *o* (Omicron) Ceti, normally reaches third magnitude at maximum, but can hit first. It then plunges to tenth, and for a good part of the year it is invisible to the naked eye, returning to third magnitude after 330 days. Some of these aptly named **long-period variables** take over 3 years to vary. (A short list of bright long-period-variables is given in Appendix 2.)

AGB stars become larger, brighter, and redder than they ever were on their first ascent of the HR diagram as ordinary red giants. The Sun will become 4,000 times brighter than it is now, and more-massive stars will

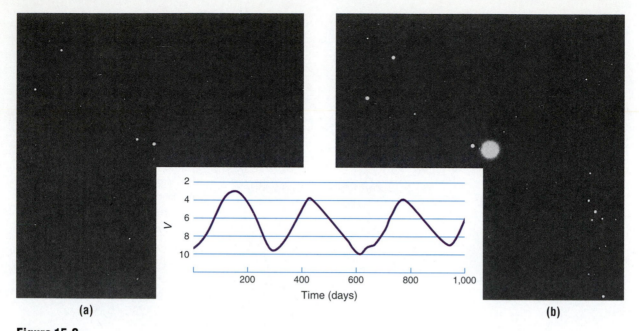

Figure 15.9
(a) At minimum, Mira is at 10th magnitude and can be seen only with a telescope. **(b)** At maximum, it can nearly dominate its constellation. Though periodic, Mira's light curve (inset) is not as regular as the light curves of Cepheids.

become even more luminous. Mira is now the size of the orbit of Mars and the Sun will come close to the current orbit of the Earth. Spectral classes can extend to M9 and temperatures can drop below 2,000 K.

During the AGB ascent, the intense heat causes a variety of nuclear re-actions that turn the star's interior into a chemical factory. The AGB star's envelope is in a state of deep convection, as are regions in and near the nuclear-burning shells. A few stars in the right mass range can become highly enriched in carbon made by the 3-α process and in a variety of other elements. As direct evidence of such nuclear processing, a few Mira variables exhibit spectral lines of the radioactive element technetium, whose half-life is so short that none is found naturally on Earth.

Mass loss rates on the AGB, pumped by Mira-type pulsations, can reach 10^{-5} $M_{\odot}$, 100 million times greater than the loss rate from the solar wind. By the end of this stage, the Sun will have lost almost half its mass back to space, and a 7-$M_{\odot}$ star will lose an astounding 80%. The decreas-ing solar mass will allow the Earth to spiral outward to just beyond the AGB-Sun's reach, probably saving us from complete fiery destruction. The outflowing winds of some of the AGB stars are enriched in carbon, he-lium, nitrogen, and elements produced by other nuclear reactions, the dying star now enriching the chemical content of the interstellar medium. In the cold of space above the stellar surface, some of the gas condenses

By the AGB stage, the Sun will have lost almost half its mass back to space.

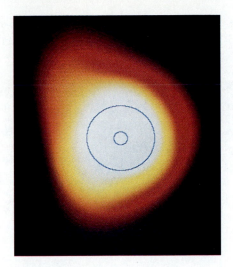

Figure 15.10
The shell of the carbon-rich long-period variable IRC+10 216 is imaged in radio radiation produced by carbon monoxide. The circles show the size of the invisible star and the inner edge of the dusty shell. The encircling cloud is loaded with organic molecules.

into dust that may become so thick that the star hides itself from view (Figure 15.10). The dust's composition will depend on that of the star: we will see silicate dust if the star is oxygen-rich (more oxygen than carbon), graphite dust if the star is carbon-rich. Once the dust grains are launched into interstellar space, they evolve, picking up ices and metal atoms from the interstellar gas. We now know the origin of the dark clouds of interstellar space: *luminous AGB stars provide all the dust needed to produce the globules and GMCs of the interstellar medium.*

15.4 PLANETARY NEBULAE AND WHITE DWARFS

The carbon-oxygen core of the future Sun, now over half a solar mass, will shrink to the size of Earth and the density will climb to around a million g/cm^3. In the world of the atom, we find that there is a minimum volume of space that can be occupied by identical particles moving at a specific velocity. When the density of a gas becomes so high that these minimum volume units are filled, the particles at a given velocity can be packed no closer, and the gas enters a state of **degeneracy.** We can always add particles to a degenerate gas, but only at ever-higher velocities. *Electron degeneracy* provides powerful pressure and support and keeps the star from contracting and heating any further. The result is a white dwarf, stabilized forever against contraction.

As the white dwarf forms inside the star, the last of the outer envelope is being cast off by the star's wind. The developing white dwarf core becomes progressively more exposed, the effective temperature climbs, and the star moves evenly to the left on the HR diagram (Figure 15.11), the specific evolutionary tracks depending on mass. When the effective tempera-

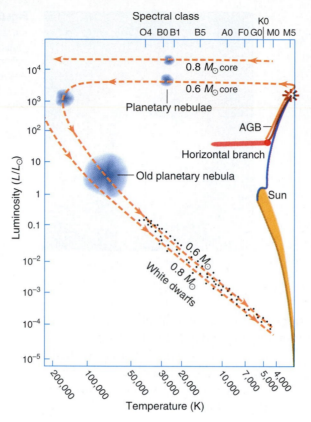

Figure 15.11
When mass loss has nearly exposed the core of the AGB star, the star tracks to the left on the HR diagram (dashed orange curve). When its effective temperature is high enough, it illuminates the surrounding gas to produce a planetary nebula. The core, a white dwarf in the act of revealing itself, eventually cools and dims. Higher-mass cores will be brighter when they leave the asymptotic giant branch and will become hotter, but when they cool as white dwarfs they are more gravitationally compressed and are smaller and dimmer.

ture hits about 25,000 K, the dying star begins to radiate enough ultraviolet radiation to ionize and light the surrounding matter that was blown away into space, producing a **planetary nebula** (Figure 15.12), which appears in the telescope as a gaseous ring or disk with a single blue star at its center. (The name refers only to the nebula's superficial visual resemblance to a planetary disk.)

At a maximum effective surface temperature that exceeds 100,000 K, the now-exposed stellar core finally begins to cool and dim. Some 50,000 years after it was created, the nebula, expanding at some 20 km/s, dissipates, leaving behind the white dwarf. As the white dwarfs created by aeons of stellar evolution cool and dim, they string out on the HR diagram in a long line far below the main sequence (see Figure 15.9). The coolest have effective temperatures of roughly 4,000 K. The time it takes to reach this point is comparable to the age of the Galaxy. Every white dwarf that has ever been created is still shining.

A large fraction of stars resides in binary systems. The more-massive main-sequence component of a binary star will be the first to evolve into a giant and then into a white dwarf. If the two are sufficiently close, tides

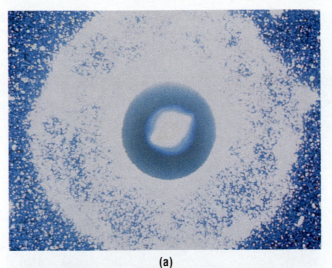

(a)

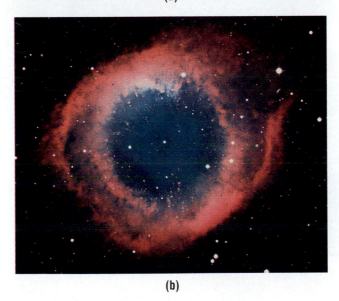

(b)

Figure 15.12

(a) A planetary nebula, like NGC 6826 shown here, is the inner compressed edge of a much larger low-density shell that resulted from eons of stellar mass loss. **(b)** The large, old, and spectacularly beautiful Helix Nebula, over 0.6 pc across, will soon be lost to interstellar space, leaving a white dwarf (centered on the nebula) behind.

raised by the white dwarf in its unevolved companion may cause mass to flow onto the white dwarf's dense surface. The fresh hydrogen, compressed and heated by crushing gravity, erupts in a thermonuclear explosion, hurling the newly deposited skin into space. From Earth we see a **nova** (from the Latin word for "new") blossom into the sky, the star brightening to an absolute magnitude of −8 to −10. Several novae are discovered each year. Every few decades one will erupt close to us and can reach first apparent magnitude or brighter (Figure 15.13a). After a few months the nova fades away, and a few years later, the blasted shell itself expands into view (Figure 15.13b).

No white dwarf heavier than 1.4 solar masses can exist.

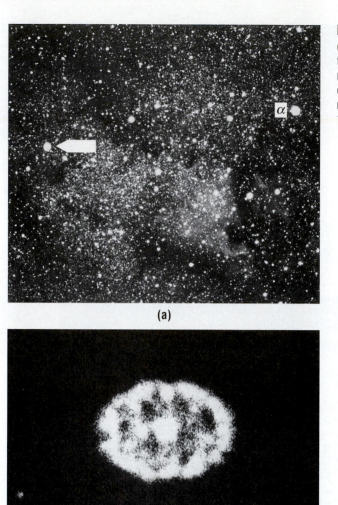

(a)

(b)

Figure 15.13
(a) Nova Cygni 1975 reached nearly first magnitude and for a brief time rivaled Deneb (α Cygni). **(b)** A cloud of debris is seen expanding around a nova that exploded in Hercules in 1934 (Nova Herculis 1934).

The higher the mass of the original star, the greater the mass of the remnant. Most white dwarfs are by-products of near-solar-mass stars and have masses around $0.6\ M_\odot$. Because the number of stars declines upward along the main sequence, more-massive white dwarfs are rarer. Only one (Sirius B) is known that approaches even $1\ M_\odot$. As the mass of a white dwarf increases, so does its interior compression. As a result, the degenerate electrons are forced to move faster. At some point their velocities approach that of light. The theory of relativity shows that at 1.4 solar masses, degenerate electrons can no longer provide support and the star must collapse under the force of gravity. This **Chandrasekhar limit** (named after Subramanyan Chandrasekhar, its discoverer) is 1.4 solar masses. No heavier white dwarf can exist, as confirmed by observation.

15.5 SUPERGIANTS AND THE ROAD TO DISASTER

In spite of their great masses and luminosities, the stars of the **upper main sequence,** those with masses greater than 8 $M_{\odot}$, at first behave similarly to those of the intermediate main sequence. The 20-$M_{\odot}$ star in Figure 15.4 begins life as a brilliant 07 dwarf, one that may light a diffuse nebula. By the time its core is depleted of hydrogen and its main-sequence lifetime ends, it has cooled enough so that any remaining birth cloud becomes a reflection nebula. As its helium core contracts, the star tracks to the right at nearly constant luminosity, becoming first a hot blue supergiant then a cool red helium-burning supergiant of class M. By that time, it has expanded to a radius of some 5 AU, about 1,000 times that of the Sun, comparable with the orbit of Jupiter. Examples of such red supergiant stars are Betelgeuse in Orion and Antares in Scorpius.

The full range of upper main sequence evolution is shown in Figure 15.14. Stellar winds are of paramount importance, mass loss rates reaching 10^{-5} $M_{\odot}$/year. As a result, O stars begin to whittle themselves away. As the more luminous stars evolve into blue supergiants, they can become surrounded by thick nebulae (Figure 15.15a) and can powerfully affect

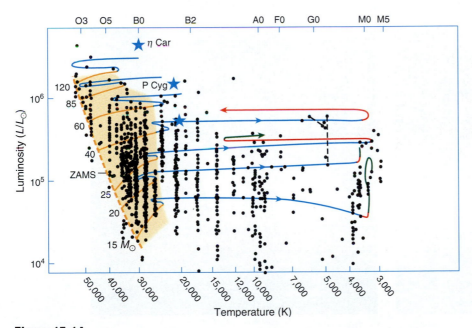

Figure 15.14
Evolutionary tracks show high-mass stars evolving within the main sequence and then away from it, becoming first blue, then red, supergiants. The various phases of evolution are: hydrogen burning (yellow), helium core contraction (blue), helium burning (red), and carbon burning (green). Observed stars are represented by black dots. In the upper range of masses, powerful winds prevent the evolution to the red supergiant stage, stars such as Eta Carinae and P Cygni burying themselves in their own ejecta.

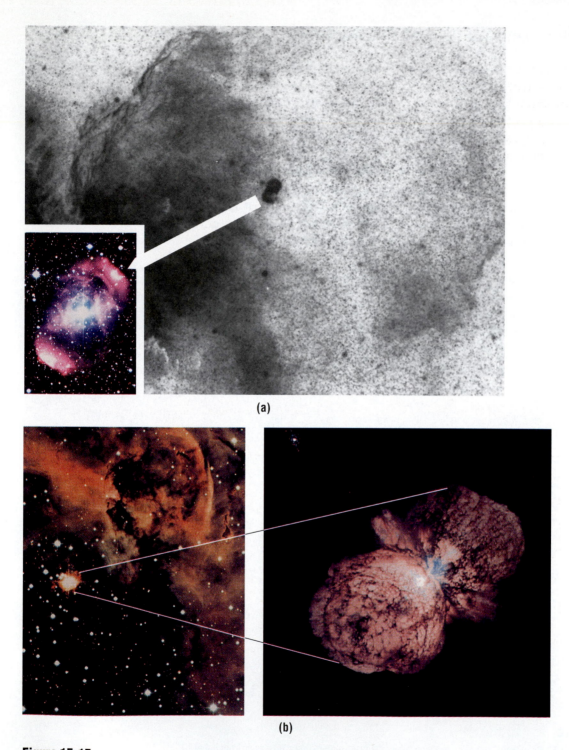

Figure 15.15

(a) The evolving supergiant HD 148937 in Ara is surrounded by its own ejecta, NGC 6164-5 (inset); it is inside a wind-blown bubble that in turn is in an immense volume of ionized gas, a huge diffuse nebula. **(b)** Massive Eta Carinae, seen within the stars and nebulae of the southern Milky Way (left), is enmeshed in gas and dust flowing outward in two opposing streams (right).

their surroundings by blowing and lighting bubbles in the interstellar medium. Above 40 $M_\odot$, evolution is so severely affected by increasing mass-loss rates that the stars cannot make it all the way across the HR diagram and remain blue supergiants. Here we find some of the oddest characters in the Galaxy. About the time of the American Civil War, Eta Carinae (Figure 15.15b) was the second brightest star in the sky. It then underwent a long episode of powerful winds, mass-loss rates reaching 0.1 $M_\odot$ per year, and blanketed itself completely in obscuring dust and dropping below naked-eye vision. We think there is a 100-solar-mass star at its heart.

Intermediate-main-sequence evolution ends with the creation of a degenerate carbon/oxygen (C/O) core. Upper-main-sequence stars, however, have cores larger than the Chandrasekhar limit and *cannot become white dwarfs*. Once the helium runs out and a C/O core is developed, it contracts and heats (Figure 15.16a), allowing nuclear burning to continue

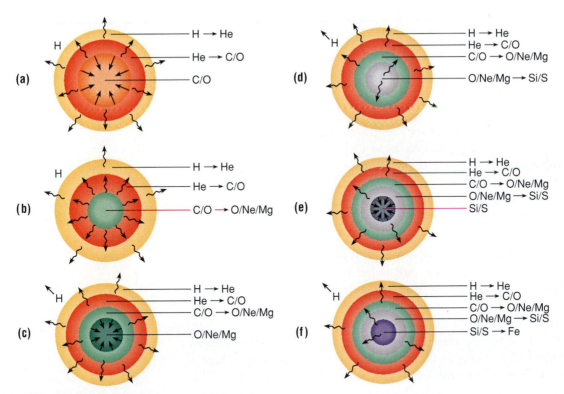

Figure 15.16
Stages of nuclear burning in supergiant cores are **(a)** a contracting C/O core; **(b)** C-burning to oxygen, neon, and magnesium; **(c)** a contracting O/Ne/Mg core; **(d)** burning of O, Ne, and Mg to a mixture of silicon and sulfur; **(e)** a contracting Si/S core; **(f)** Si-burning to iron. These zones are surrounded by shells of previous burning stages and by vast hydrogen envelopes. The diagrams are not to scale; the contraction stages are successively smaller.

to more advanced stages. At a temperature near 1 billion K, contraction is halted as carbon begins to fuse to oxygen, neon, and magnesium (Figure 15.16b), the core still surrounded by helium- and hydrogen-burning shells. When carbon is exhausted at the center, the nuclear fire shuts down and the O/Ne/Mg core contracts under the inexorable force of gravity (Figure 15.16c). At 2 billion K, the oxygen, neon, and magnesium fuse to a mixture of silicon and sulfur (Figure 15.16d), the star beginning to look like a nuclear onion. When the Si/S core starts to contract (Figure 15.16e), and the temperature approaches 3 billion K, the silicon starts reacting with itself to form iron (Figure 15.16f).

Each stage of nuclear burning releases less energy, and as a result, each supports the star for a shorter period of time. An O7 star of 20 solar masses takes 8 million years to use up its initial hydrogen. Helium burning then takes about 10% of the main-sequence lifetime, or only about a million years. Once the carbon starts to fuse, it is gone in about 10% of the He-burning lifetime, or a mere 100,000 years. What happens next is astonishing. Over a solar mass of oxygen burns in only 20 years, and silicon-burning is completed in a week. Clearly, some monumental event is about to take place.

15.6 SUPERNOVAE

In 1572, a "new star," studied by the great Tycho Brahe, blazed forth in the Milky Way in Cassiopeia. It reached a spectacular apparent magnitude of −4 and rivaled Venus. Even a hundred days after the event, it was as bright as Vega, and it did not fade from sight until 1574. Only 30 years later, in 1604, another spectacular event graced Ophiuchus. Studied by Kepler, it outshone Jupiter and reached apparent magnitude −2.5, finally disappearing from view after 14 months.

Supernovae are among the most violent events in the Universe.

A similar event took place in the Andromeda Galaxy (see Figure 13.23) in 1885. In 1924, Edwin Hubble calculated the galaxy's distance (now known to be 750,000 pc), and we learned that the so-called "new star" had reached an amazing absolute magnitude of −17, over 500 times brighter than an ordinary nova and only 10 or so times fainter than the whole galaxy itself! These **supernovae** (Figure 5.17) were clearly among the most violent events in the Universe. From their brilliance and duration, there is no doubt that Tycho's and Kepler's stars were supernovae. Supernovae are rare. Over the course of this millennium, only five have been seen (in the years 1006, 1054, 1181, 1572, and 1604) and none in our Galaxy since the invention of the telescope.

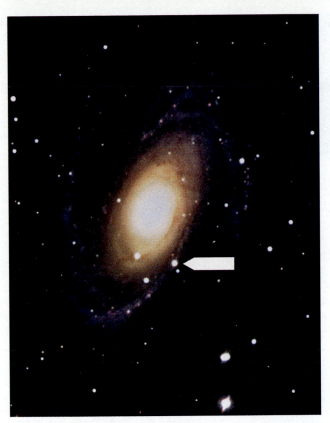

Figure 15.17
A supernova gleamed within the galaxy M 81 in Ursa Major in March 1993. From Earth it appeared at 10th magnitude, but from the distance of the galaxy (1.4 million pc) and the magnitude equation we find that it reached an absolute visual magnitude of −16.

15.6.1 Core Collapse

The iron core of an evolving supergiant is the end of the peaceful evolutionary line: no energy can be gained by iron's fusion. To fuse iron into heavier nuclei *requires* energy. Once silicon burning ceases, the iron core contracts and heats, but the core can no longer support itself by further burning. The result is a catastrophic collapse at speeds up to a quarter that of light (Figure 15.18a). In *less than a tenth of a second* the radius drops from 1,000 km to less than 50 km, and within a few seconds more down to a mere 10–20 km, releasing a spectacular amount of energy, 99% of it in the form of neutrinos. At the moment of collapse, the star generates a power output (the rate of energy released per second) *comparable to that of all the stars in the observed Universe combined.* The temperature now approaches 200 billion K. The density is over 10^{12} g/cm^3, that of the nucleus itself, a million times greater than that in a white dwarf. Protons and electrons merge into neutrons, the reactions producing the outbound neutrinos. Like the electrons in a white dwarf, the neutrons become degenerate, their outward pressure finally halting the collapse.

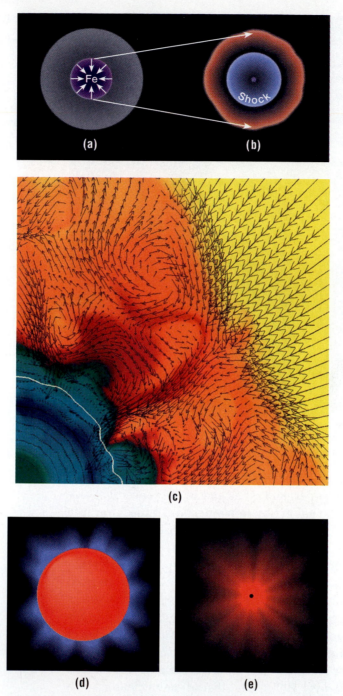

Figure 15.18
The stages of supernova start with
(a) the collapse of the iron core and
the generation of **(b)** a shock wave
(light circle) trapped within it. **(c)** In
the center, a computer simulation of a
supernova's core only 120 km across
a few hundredths of a second after the
collapse shows matter falling in from
the envelope (yellow), the shock wave
where the matter hits atoms broken
into protons and electrons (orange),
and neutrons (green and blue). The
matter inside the neutron sphere is
being pushed on by a wall of neutrinos (wavy white line), which will
shortly help release the shock and
blast the envelope outward. **(d)** After
the neutrinos (blue) have begun the
explosion process inside, they exit the
star, whose surface (red) has not yet
been affected. **(e)** The shock wave follows, finally tearing the star apart.

During the collapse, part of the imploding core rebounds and produces a shock wave that attempts to rip through the core's outer layers but
is not strong enough to do so (Figure 15.18b). The density of the core is so
great, however, that the gas is actually opaque to the newly-forming neutrinos, particles that are ordinarily so non-interactive that they can fly

right through the Sun without stopping (see Section 12.4). Acting together in the supergiant's internal cauldron, the neutrinos give the shock a mighty shove that releases it, and it roars outward to tear the star apart (Figure 15.18c). As the density drops, the neutrinos, having done their job, fly free (Figure 15.18d), almost immediately escaping the still-calm surface of the dying star. A few hours later, the shock wave makes it to the stellar surface and a supernova erupts into the sky (Figure 15.18e), its brilliance —typical absolute magnitudes are –18—announcing that another massive star has died. The supernova might take a full year to drop back to its original supergiant absolute magnitude, from which, over a period of many more years, it will fade to near or total invisibility.

Supernova 1987a (Figure 15.19), which exploded from a 20-solar-mass B3 Ia supergiant in a small companion Galaxy to ours called the Large

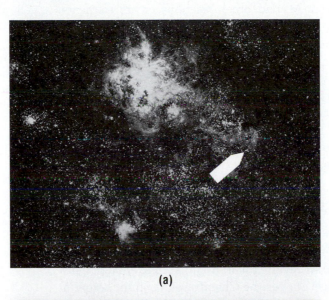

(a)

(b)

Figure 15.19
In **(a)** "before" and **(b)** "after" pictures, we see Supernova 1987A erupting near the Tarantula Nebula from a B3 Ia supergiant in the Large Magellanic Cloud.

Magellanic Cloud, provided stunning proof of the theory of these **core-collapse supernovae.** Three hours before it was seen visually, 11 neutrinos coming from the direction of the Large Magellanic Cloud hit the Japanese neutrino detector (see Section 12.4). At the same time, eight more hit another neutrino detector in the United States. The number caught indicated that neutrinos passed through the Earth at a rate of 10 billion/cm²/s, approximately that expected from a collapsing core at the distance of the Large Magellanic cloud, 50,000 pc away. The neutrinos directly revealed the core collapse before the shock wave had reached the stellar surface and anyone saw the star brighten. The delay is about that expected from the size of a B supergiant.

Within the shock wave of a developing supernova, nuclear reactions go berserk. About half a solar mass of the expanding debris is converted into radioactive nickel, which decays into iron: such a decay was actually observed in Supernova 1987a. Myriad other reactions create elements heavier than lead all the way to plutonium and perhaps even beyond. The by-products of this nuclear burning are then blasted into space, further enriching the interstellar gas.

As descendants of massive stars, core-collapse supernovae must be found where we find O stars and their supergiants, in the spiral arms of galaxies. Sometime in the next million or so years someone may see Antares, Betelgeuse, or the best candidate of all, Eta Carinae, explode. Since we have no way of knowing their current internal constructions, we might even see one go off tonight.

15.6.2 Binary Systems

We also see supernovae erupt in the *halos* of other galaxies, where there are no massive stars. Like novae, supernovae may be caused by interactions within binary systems. A massive white dwarf might accrete enough matter from a companion to push it over the Chandrasekhar limit before a nova can fling the accumulated mass into space. The high temperature and density produced by the collapse would cause runaway nuclear reactions, the star going off in a massive thermonuclear explosion, the C/O core fusing to nickel and then back to iron, resulting in another load of enriched debris delivered into the interstellar medium. Supernovae of this type are even brighter than their core-collapse counterparts, typically reaching $M_V = -19$, 4 billion times brighter than the Sun. From their light curves (reconstructed from old observations), Tycho's and Kepler's stars are thought to have been of this kind.

15.6.3 Significance

Supernovae have been observed in our Galaxy about once every 200 years. However, many more must have been hidden from view behind thick clouds of obscuring interstellar dust. The best estimate of the actual rate is

Supernovae can easily have created all the iron in the Universe.

roughly one every 25 years. Over the age of the Galaxy (around 15 billion years), perhaps a quarter billion or more supernovae have occurred. If we average all the types, each blast kicks about 10 $M_\odot$ back into space for a total yield of over a billion $M_\odot$, about 1% of the stellar mass in the Galaxy. More important, this material is highly enriched in the by-products of fusion reactions. Supernovae can easily have created all the iron in the Galaxy, including that in the Sun and the Earth, and almost all the other heavy elements as well. The silicon in rocks, the metal in your chair, and the gold in your ring came from supernovae.

15.6.4 Supernova Remnants

Supernova explosions ought to be marked by clouds of expanding debris. Just off the eastern horn of Taurus, near ζ Tauri, we find a fuzzy filamentary gas cloud called the Crab Nebula, or M 1 (Figure 15.20a). It lies at the location of the "Guest Star" observed by the Chinese in 1054. Rich in heavy elements, this **supernova remnant** is almost certainly the result of a core-collapse supernova. Doppler shifts in its spectrum lines show that it is expanding at 1,500 km/s, or nearly 50 billion km/yr. Measurement of the yearly *angular* rate of expansion, the angle subtended by 50 billion km

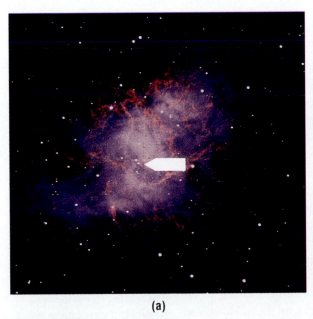

(a)

(b)

Figure 15.20
(a) The Crab Nebula, M 1, is the remains of the Chinese "Guest Star" of 1054. The arrow points to the collapsed core of the supernova, now a pulsar, or rotating neutron star. **(b)** The Vela supernova remnant, 4° across (it would almost fill the bowl of the Big Dipper) is a blast wave sweeping up and heating interstellar matter.

(see MathHelp 2.1), gives a distance of about 2,000 pc. At its peak the exploding star must have had an absolute visual magnitude of about −16.

A number of supernova remnants decorate the visual sky. In the southern hemisphere constellation Vela lies a vast network of filaments that mark the death of a star about 100,000 years ago (Figure 15.20b). Unlike the young Crab, which is filled with gas, the Vela Remnant (typical of older remnants) is an empty shell. We are looking not at exploded debris but at the powerful expanding shock wave of the supernova sweeping up interstellar matter. X rays from older supernova remnants reveal matter within the hollow shells that reaches 10^7 K and is the origin of the hot coronal gas that threads throughout interstellar space (see Section 14.1.1 and Figure 14.6).

Radio observations offer a better view, allowing us to see about 150 galactic supernova remnants. The radio spectrum reveals synchrotron radiation (see Section 9.2.1 and Figure 9.2) created by fast electrons spiraling in magnetic fields at speeds near that of light. The Crab's field is related to the parent star, and its synchrotron radiation is visible through the optical spectrum and even into the X-ray spectral region.

The compressing blast waves of supernovae are an important factor in inducing star formation, stellar death begetting stellar birth. Meteorites show isotope ratios—including the daughter products of plutonium—expected in supernovae, demonstrating that our existence may have been linked to a supernova that began the condensation of the Sun's birth cloud.

15.7 NEUTRON STARS AND PULSARS

A core-collapse supernova forms a dense ball of self-supporting degenerate neutrons with roughly a Chandrasekhar mass and a radius of about 10 km—a **neutron star.** Though predicted in the 1930s, the existence of neutron stars was not confirmed until 1967, when graduate student Jocelyn Bell, working with a new radio telescope, recorded a series of pulses of radio radiation spaced 1.337011 . . . seconds apart (Figure 15.21a) coming from the constellation Vulpecula. Radio signals in space are refracted and dispersed by the electrical properties of the interstellar medium, longer waves traveling more slowly. The farther the pulsar and the denser the interstellar medium, the greater the difference between pulse arrival times at any two frequencies. By measuring the difference, and by assuming a density for interstellar gas, the distance to this **pulsar** (for *puls*ating *r*adio source) was estimated at about 500 pc. The signals came too fast to be Cepheid-like pulsations. The only real possibility was rotation. However, for a body to rotate that quickly it would have to be small, no more than a few tens of km across. It had to be a spinning neutron star.

The argument was clinched with the discovery of a pulsar in the Crab Nebula with a remarkable period of only 0.0316 . . . seconds (Figure 15.21b). It pulses not only in the radio spectrum, but in the optical and X-

The neutron star in the Crab Nebula is a star of more than a solar mass 20 km across spinning over 30 times a second.

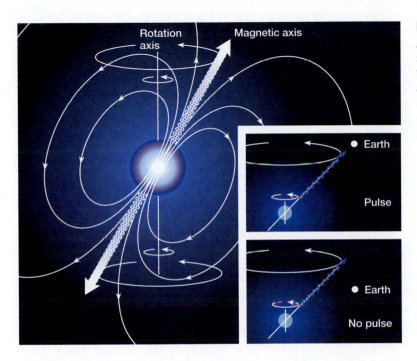

Figure 15.21
(a) Pulses at two frequencies for the first known pulsar are shown from the discovery paper. Those at 80.5 MHz arrive 0.2 second after those at 81.5 MHz, the delay allowing a measure of distance. **(b)** To the eye, the light from the Crab pulsar appears continuous. It actually flashes 30 times per second, appearing to turn "on" (left) then "off" (right) a 60th of a second later.

Figure 15.22
A pulsar's magnetic field spins in space around its rotation axis. Radiation is beamed along the magnetic axis, and if the Earth is in the way (upper inset), our telescopes will pick up pulses; otherwise they will not (lower inset).

ray spectral regions as well. We at last could see the stellar end product of a core-collapse supernova, a star of more than a solar mass, some 20 km across, spinning over 30 times a second! Over 400 of these bizarre objects are now known. The pulses are complex and pulse activity may cease for long intervals only to resume right on the beat. Long-term observations show that pulse periods lengthen as rotation slows, the Crab's declining by 0.0001% per day.

The radiation mechanism most likely originates in a magnetic field that has been concentrated in harmony with the stellar density to a strength about 10^{12} times that of the Earth's field (Figure 15.22). Like most astronomical magnetic fields, stellar or planetary, a pulsar's field axis is tilted relative to its rotation axis. The magnetic field creates a powerful electric field that accelerates electrons out along the magnetic poles. The result is a pair of oppositely directed beams of tightly focused radiation. The pulsar acts like the beacon of a lighthouse. If the Earth is in the right position, a beam will sweep by us and we will see a burst of radiation (Figure 15.22 upper inset). If not, we see nothing (lower inset). The beams can somehow be suppressed temporarily, but since they are controlled by rotation, they are exactly on the beat when they come back.

Pulsars spin fast because conservation of angular momentum speeds up the supernova core as it collapses. Since the pulses are driven by rotation, radiation must suck energy from the star, which therefore slows with time. The Crab pulsar, only 940 years old, is still highly energetic and is rotating quickly, so it even produces X-ray pulses. As a pulsar slows, it becomes incapable of producing high-energy radiation, and by the time the period is measured in seconds, the little star is bright only in the radio spectrum. After millions of years, when the period lengthens to 10 seconds or so, the star fades from sight.

Supernovae produced by white dwarfs in binaries may annihilate themselves, so no pulsar is produced. In most core-collapse supernovae, the pulsar is likely to be aligned in the wrong direction, and therefore we cannot see it. On the other hand, the pulsars we do see long outlive their expanding supernova remnants, so most of them are bare. The 18 we see inside supernova remnants (which include both the Crab and the Vela Remnant) are about what we expect to find.

Just as there is a maximum mass that can be supported by degenerate electrons (1.4 $M_\odot$), so is there a maximum mass that can be held up by degenerate neutrons. Though a good theory does not yet exist, the limit is estimated to be about 3 $M_\odot$. Masses of pulsars derived from binary companions confirm the neutron-star limit: none has been found greater than 3 $M_\odot$. For reasons no one understands, most pulsars with estimated masses actually hover near the 1.4 $M_\odot$, the Chandrasekhar limit.

As a result of binary interactions, some pulsars have been forced to spin at rates of *hundreds* of times per second. One of these exhibits variations in pulse times produced by orbiting bodies —*planets*—of only a few Earth masses that are pulling the pulsar back and forth along the line of sight. The planets were apparently created from debris of the companion star after it was destroyed by the pulsar's intense radiation. The importance of the discovery is the demonstration that, given a source of matter for their assembly, orbiting planets will apparently form almost anywhere.

15.8 BLACK HOLES

If the mass of a collapsing supernova core were beyond the neutron degeneracy limit of 3 $M_\odot$, it would no longer have any known means of support. We now examine the possibility—indeed, the probability—that stars can go into total collapse and disappear, pulling the fabric of the Universe over their heads.

15.8.1 Theory

Throw a ball up; gravity brings it down. But if you could throw it fast enough, at the Earth's escape velocity of 11.2 km/s, the ball would not return. Now do an experiment in your mind. Build a machine that shrinks the Earth and therefore increases the acceleration of gravity and escape velocity (v_{esc}) at its surface. From Section 5.2, v_{esc} is proportional to $\sqrt{M/R}$. When the Earth is a quarter its present diameter, v_{esc} doubles to 22.4 km/s, and when the Earth is as dense as a neutron star, with a radius of only 50 meters, v_{esc} is 3,997 km/s. At a radius of 9 mm, $v_{esc} = 299{,}793$ km/s, the speed of light. Light can no longer escape; to an outside observer, the Earth *disappears from view*. It has become a **black hole**.

When you threw the ball upward, it lost energy of motion as gravity slowed it down. Shine a flashlight toward the zenith. The light too must work against gravity, but it must always move at c. Since $E_{photon} = h\nu$ (see Section 6.2), light loses energy by becoming redder. Gravitational reddening of light is observed in the laboratory and in the spectra of white dwarfs and the Sun. Assume a friend is orbiting the Earth and observing your flashlight. As the Earth shrinks, the beam reddens, shifting to the infrared and then to the radio. When $v_{esc} = c$, the light red-shifts to infinity and disappears from view.

For ease in understanding, these explanations invoke a Newtonian view of gravity. The German physicist Karl Schwarzschild introduced the modern view in 1916 with the first exact solution of Einstein's equations involving gravity. Einstein's theory considers gravity as a curvature of spacetime (see Section 5.7). If you shine your flashlight horizontally, the beam will bend to Earth (Figure 15.23a), as starlight does when it passes near the Sun. Earth's gravity is so weak that such bending is undetectable. But if the Earth were shrunk sufficiently, the bending would be so severe that the light could go into a spiral orbit before escaping (Figure 15.23b); if gravity were further increased, the light would fall back to the surface (Figure 15.23c).

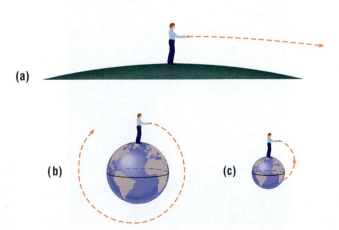

(a)

(b)

(c)

Figure 15.23

This figure and the next two make a continuous sequence that shows the development of a black hole. **(a)** A horizontally directed light ray will bend imperceptibly to Earth because of the curvature of spacetime. **(b)** The Earth has been shrunk, and the curvature is great enough to put the light nearly into orbit. **(c)** The curvature is so large that the light falls back to the surface and cannot escape. These diagrams are schematic and not drawn to any scale.

As the Earth is made smaller yet, only light within an *exit cone* around the zenith would be able to leave (Figure 15.24a); the rest of it curves back. As the planet shrinks, the cone becomes smaller (Figures 15.24b and c). When the radius is 9 mm, only a vertical ray can escape (Figure 15.24d). If

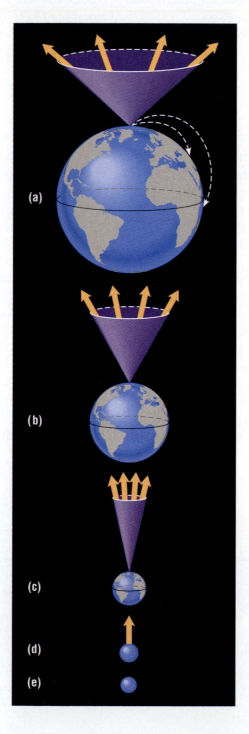

Figure 15.24
The sequence of Figure 15.23 continues. **(a)** Only light within the exit cone can escape. Outside the exit cone, all light falls back to the surface. **(b, c)** The Earth becomes smaller, and the cone shrinks. **(d)** The Earth is just over 9 mm in radius, and only light directed vertically escapes. **(e)** The radius is 9 mm, spacetime at the surface is infinitely curved, nothing escapes, and the Earth is a black hole. The diagrams are not to scale.

the body shrinks even more, spacetime folds back on itself, *all* light is returned, and the beam and the Earth disappear from view (Figure 15.24e). The light still travels at c, but it will not emerge from an Earth of radius less than 9 mm because of the infinite distortion of spacetime.

Every physical body—not just the Earth—has a radius, R_{bh}, at which it will disappear. Since $v_{esc} \propto$ (is proportional to) $\sqrt{M/R}$, $v_{esc}^2 \propto M/R$, and $R \propto M/v_{esc}^2$. For a black hole, we just substitute the speed of light, c, for v_{esc}. Since c is a constant, $R_{bh} \propto M$. The Sun's mass is 333,000 times that of the Earth, so the Sun's black hole radius is 333,000 times 9 mm, or 3.0 km. R_{bh} at the neutron star limit of 3 $M_\odot$ is 9 km.

Under these conditions, a body should contract to a point, a **singularity,** leaving a sphere of radius R_{bh} around it called the **event horizon** (Figure 15.25). Nothing—neither light, mass, nor information—can return from within it. We can know nothing of actual conditions within the event horizon, and the laws of physics as we know them have no meaning. In

Neither light, mass, nor information, can return from within a black hole.

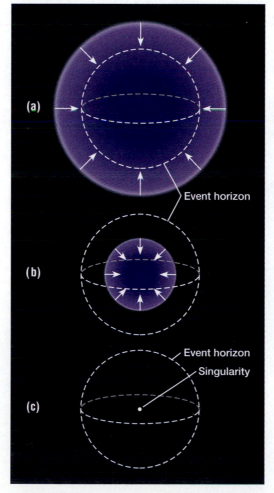

Event horizon

Event horizon
Singularity

Figure 15.25
(a) In an expanded view of Figure 15.24d, the dashed curves represent the event horizon. The body is not yet a black hole, as part of its contracting mass is outside its event horizon. **(b)** The body has contracted within the event horizon and is no longer visible. **(c)** The body has contracted to a point, or singularity.

spite of its strangeness, however, a black hole is a simple object. The only property required of Schwarzschild's black hole is mass, and the theory can be expanded to include rotation. Any other property the object had before it became a black hole is crushed out of existence.

Black holes are popularly misconceived. Poorly written stories and films give the idea that they are inherently dangerous, that they will suck in matter—including you—if you venture too close. The black hole is no different gravitationally from any star. If the Sun were to become a black hole, it would disappear from sight, but the curvature of spacetime and the gravitational acceleration at the distance of the Earth would remain precisely the same. An orbiting body can fall in only if some outside force dissipates its orbital energy. Orbiting a black hole is no more dangerous than orbiting the Sun.

15.8.2 Observation

We have yet actually to detect a black hole, but their existence would explain a variety of bizarre phenomena. Though black holes radiate no energy, we should be able to find them by their effects on their surroundings. Cygnus X-1 (Figure 15.26) is a powerful source of X rays. It has been identified with a B0 supergiant—one-half of a spectroscopic binary (see Section 14.5)—whose spectrum lines shift back and forth in wavelength showing that another, but invisible, body is orbiting it. The mass of the B0 star is found by comparing it with theoretical evolutionary tracks. From the orbital velocity of the B star, we find the companion's mass to be

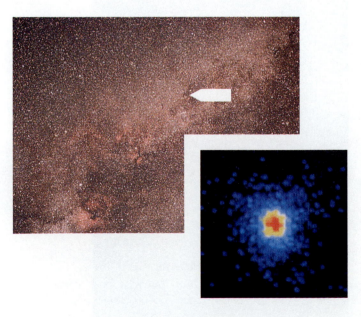

Figure 15.26
Cygnus X-1 is identified with HDE 226868, a B0 supergiant located here within its constellation (note Deneb at lower left). The star uncharacteristically radiates powerful X rays (inset).

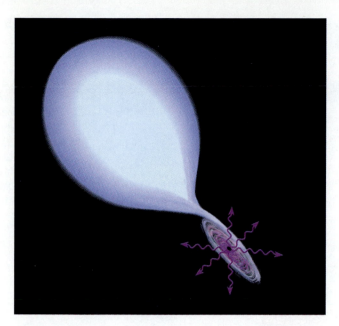

Figure 15.27
Cygnus X-1 appears to be a binary system. The black hole (black dot) raises tidal distortions in the blue supergiant, causing matter to flow from the supergiant into an accretion disk. As the matter spirals from the accretion disk into the black hole (violet) it heats and radiates X rays.

greater than 10 $M_\odot$ and possibly as large as 15 $M_\odot$. Any normal star with that high a mass would be visible. The companion is too massive to be a neutron star, so we believe it to be a black hole. Its gravity raises tides so severe that mass is transferred from the supergiant into an accretion disk around the black hole (Figure 15.27). As the matter loses angular momentum by friction and spirals inward, it heats and radiates X rays. Nearly a dozen other candidates may harbor black holes between 3 $M_\odot$ and 15 $M_\odot$. More evidence for the existence of black holes will be presented in Chapter 17.

Astronomers are generally convinced that black holes exist. If they do, then we have finally found the ultimate in contraction: some stars, as they are born out of the mists of interstellar space, are destined to shrink all the way down to points.

15.9 THE CYCLE OF EVOLUTION

We can now fit intermediate- and high-mass stars into the full cycle of stellar evolution (Figure 15.28). Giants and supergiants manufacture helium, carbon, and nitrogen in their interiors by the cycles of nuclear-energy generation, make heavier elements by the other nuclear processes, and transfer them to the surface by convection. Mass loss through winds (producing planetary nebulae, shells around evolved O stars, and the like) strip the hydrogen envelopes from the stars and spread the chemical byproducts of evolution into interstellar space. Supernovae create and eject

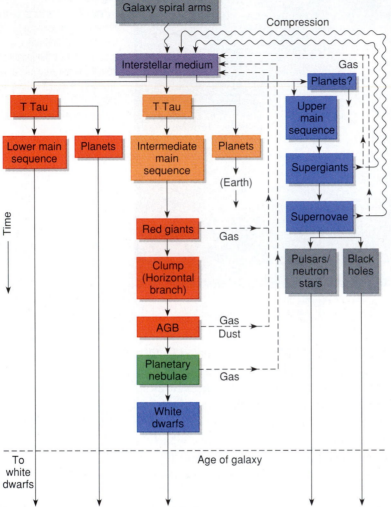

Figure 15.28

In the cycle of stellar evolution, each generation of stars depends on those that came before for its heavy elements and for its formation. Time runs downward, the flow of stellar evolution indicated by black arrows. (The relative speeds of evolution are not to scale.) Dashed lines show the return of matter to the interstellar medium and wavy lines show the origin of compressive forces that can initiate star formation.

yet more heavy elements, including all the galaxy's silicon, iron, and heavy atomic nuclei, as stellar evolution continually enriches the interstellar medium.

AGB stars make dust and spew it into the interstellar medium, creating the seeds out of which interstellar grains and molecules—and eventually stars and planets—will grow. Supernovae (as well as the Galaxy's spiral arms, to be discussed in the next chapter) act as triggers that compress the interstellar medium and initiate star formation. The characteristics of stars and planets therefore depend on earlier stellar generations in a tightly interlocked cycle. We now begin to suspect why the stars of the disk are richer in metals than are those of the halo. The globular clusters are the oldest stars known, and since they are in the halo, the halo must be old as well. The open clusters of the galactic disk are not as old as the globular clusters, and therefore the disk appears to be younger than the halo. The processes of stellar evolution continually enrich the interstellar medium with heavy elements, so stars born later are—at least in some general sense—richer in metals than stars born earlier. Earth is not isolated from the stars and the Galaxy, but is a part of them. We are a distillate not just of the solar nebula or the interstellar medium but of all of stellar evolution.

▶ KEY CONCEPTS

Asymptotic giant branch (AGB): The stage of stellar evolution in which intermediate-mass stars grow to become giants for the second time; these stars have contracting carbon-oxygen cores.

Black hole: A mass with a radius so small that the force of its gravity keeps light from escaping, so the body cannot be seen.

Carbon cycle: The fusion of four atoms of hydrogen into one of helium with carbon as a nuclear catalyst.

Cepheid variables: F or G giant or supergiant stars that vary in brightness with regular periods.

Chandrasekhar limit: A limit of 1.4 $M_\odot$, above which a star cannot be supported by degenerate electrons; thus no white dwarf can have a mass greater than 1.4 $M_\odot$.

Degeneracy: A state in which the particles (protons, neutrons, or electrons) of a gas at a given velocity cannot get any closer; in a star, degeneracy provides an outward stabilizing pressure.

Event horizon: The surface surrounding a black hole where the escape velocity equals the velocity of light.

Intermediate main sequence: The main sequence between 0.8 $M_\odot$ and 8 $M_\odot$; these stars evolve into white dwarfs.

Long-period variables: Giant pulsators with periods over about 100 days and large visual magnitude ranges.

Lower main sequence: The main sequence between 0.08 $M_\odot$ and 0.8 $M_\odot$; no lower-main-sequence star has ever had time to evolve away from the lower main sequence.

Neutron star: A collapsed star smaller than about 3 $M_\odot$, made of degenerate neutrons.

Nova: A thermonuclear explosion on the surface of a white dwarf of matter flowing from a main-sequence star.

Planetary nebula: A shell of illuminated gas around a hot star; the last part of the ejected wind of an AGB star lit by the old nuclear-burning core.

Pulsar: A rotating neutron star that beams energy along a tilted magnetic field; it is visible only if a beam hits Earth.

Red giant branch: The stage of stellar evolution in which intermediate-mass stars grow into giants for the first time; these stars have contracting helium cores.

Singularity: A mass that has been shrunk to a point.

Stellar wind: A flow of mass from the surface of a star.

Supernovae: Outbursts that involve the destruction or near-destruction of a star; **core-collapse supernovae** are produced by supergiants; other supernovae are produced by white dwarfs that exceed the Chandrasekhar limit.

Supernova remnant: The expanding cloud of debris caused by the explosion of a supernova.

Triple-alpha (3-α) process: The fusion of three atoms of helium into one of carbon.

Upper main sequence: The main sequence above 8 $M_\odot$; its stars are too massive to evolve to white dwarfs.

▶ **KEY RELATIONSHIP**

Radius of a black hole's event horizon:

$$R_{bh} \propto M_\odot.$$

▶ **EXERCISES**

Comparisons

1. Compare the three domains of the main sequence. What are they like now? What will happen to the stars within them?

2. What is the essential difference between a giant and a supergiant?

3. Compare the interiors of first- and second-ascent giant stars.

4. What distinguishes Cepheids from long-period variables both observationally and theoretically?

5. What are the similarities and differences between **(a)** planetary and diffuse nebulae; **(b)** diffuse nebulae and supernova remnants?

6. What is the difference between a nova and a supernova?

7. Compare the two ways in which we think supernovae are created.

8. Compare the limits to electron and neutron degeneracy.

9. Compare the dimensions of white dwarfs, neutron stars, and black holes.

10. How does a black hole's singularity differ from its event horizon?

Numerical Problems

11. Using the average mass-luminosity relation, what are the expected lifetimes of stars of 10 $M_\odot$ and 0.1 $M_\odot$? Why would the actual lifetimes differ from your answer?

12. About how many times farther away can you see a typical supernova than an ordinary nova?

13. What is the radius of the event horizon for a black hole with as much mass as the most massive star in the Galaxy?

Thought and Discussion

14. What is meant by a main-sequence star?

15. Why do high-mass stars die sooner than low-mass stars?

16. If it is the evolutionary fate of a star to shrink under the force of gravity, why do stars become giants? What part of the star is shrinking?

17. What proofs do we have of the change of one element into another inside stars?

18. What is the evidence for stellar mass loss?

19. What is a degenerate gas and why is it important?

20. How are planetary nebulae made?

21. What is meant by the Chandrasekhar limit?

22. How do we know that supergiants are related to high-mass main-sequence stars?

23. What effects do stellar winds have on high-mass stars and their surrounding environments?

24. What evidence supports the core-collapse hypothesis of supernova explosions?

25. Why do pulsars and supernova remnants not correlate very well?

26. What effects do supernovae have on the interstellar medium?

27. Why do young pulsars spin so fast and why do they slow with time?

28. What effect does the discovery of planets orbiting a pulsar have on our estimates of life in space?

29. How does gravity work to produce a black hole?

30. Summarize the evidence for the existence of black holes.

Research Problems

31. From old astronomy textbooks in your library, describe some early views of stellar evolution.

32. There are many kinds of variable stars other than those described in this chapter. What are their natures? Where do they fall on the HR diagram?

33. Using photocopies of illustrations in magazines and books, assemble an atlas of supernovae that have taken place in other galaxies. On the basis of their locations within the galaxies, suggest what types they might be.

Activities

34. Construct your own light curve of δ Cephei by watching it with the naked eye or binoculars over several light cycles. Use ζ ($V =$ 3.35), ε ($V = 4.19$), and λ ($V = 5.04$) Cephei as comparisons (see Figure 15.7).

35. Use a telescope to observe the effects of stellar evolution. Find a giant star. Then find an AGB star (use Appendix Table A2.5 or the predictions of the maxima of long-period variables in *Sky & Telescope* magazine). Finally, locate at least one of the three planetary nebulae found in the Messier list in Appendix Table A2.1. Describe what you see and reflect on its evolutionary significance.

36. List the known pause and contraction phases found in stellar evolution and the reasons for the known pauses in contraction.

Scientific Writing

37. Some science fiction magazines have articles about real science. Write such an article that explains how past and future life on Earth may be affected by the evolution of the Sun.

38. Write a script for a lecture to a high school chemistry class that explains the origins of the chemical elements.

THE GALAXY

The local assembly of stars and the structure of the Milky Way

16

Stars, nebulae, molecular clouds, and collapsed objects interact both gravitationally and by cyclic evolution. All these objects are interlocked in a grand picture, not only of stellar evolution, but of the evolution of the Galaxy as a whole.

16.1 THE GRAND DESIGN

None of the great variety of astronomical bodies—from planets to pulsars—exists in isolation. To understand their natures we must see them in the unifying context of the Galaxy. Some of the Galaxy's character has been revealed through the discussions of stars in Sections 13.6, 14.1, and 15.9. In the simplest view, our Galaxy (Figure 16.1) consists of a disk made

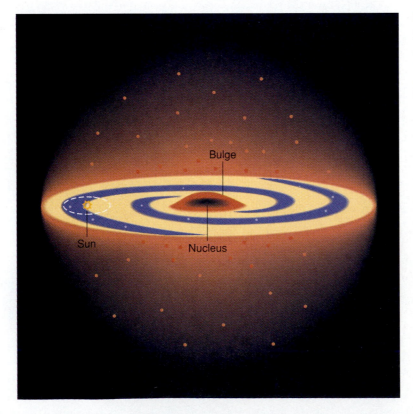

Figure 16.1

A schematic model of the main portion of our Galaxy (which contains 90% of the stars) shows a disk (in yellow) containing spiral arms studded with blue O and B stars. Surrounding the disk is a sparsely populated halo (in red) with globular clusters that merges with the disk in a central bulge. The Sun is located 8 kpc from the center, roughly two-thirds of the way out. In reality, the Galaxy does not have a sharply defined edge but fades out to a much larger extent than shown here.

of billions of stars enclosed in a sparsely populated spherical halo that merges with the disk into the central galactic bulge (Table 16.1). The disk includes the Sun and is easily visible to us at night as an encircling band of light, the Milky Way (see Section 13.6.1, Figure 3.9, and Figure 13.22).

Disk stars orbit the galactic center on roughly circular paths; halo stars move on elliptical orbits that can be highly inclined to the plane of the disk (see Figure 13.21). The galactic disk contains stars of all spectral classes, but from a distance appears bluish because its light is dominated by the massive and brilliant young blue O and B stars (Figure 16.2; see also Figures 13.23 and 15.17). The halo contains only lower mass older stars and their evolved red giants. Open clusters and associations belong to the galactic disk, whereas the densely packed globular clusters relate to the halo. The halo is generally deficient in metals compared with the disk, though in the bulge the metal content can exceed solar. The interstellar medium pervades the Milky Way and the galactic disk, and therefore the disk is the birthplace of new stars. The halo has little gas and dust and cre-

TABLE 16.1
Characteristics of the Galaxy's Disk and Halo

Galactic Disk	*Galactic Halo*
Blue O and B stars, interstellar medium, supergiants, and all other classes	Red dwarfs and giants
Open clusters	Globular clusters
Metal rich	Metal deficient
Star formation and young stars	Old stars only
Both kinds of supernovae	White dwarf supernovae

Figure 16.2
The disks of galaxies like ours are bluish as a result of their blue O and B stars (see also Figures 13.23 and 15.17). The red regions in the disk are diffuse nebulae lit by the O stars. The central bulges of such galaxies are redder than the disks, as they contain fewer O stars and greater proportions of evolved red giants.

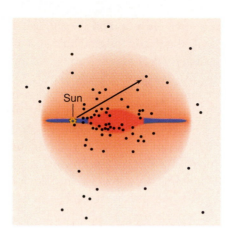

Figure 16.3
If we look at an angle relative to the galactic plane (red arrow), not much dust intervenes and we can easily see most of the Galaxy's globular clusters. We can measure their distances, and since the center of the whole collection coincides with the galactic center, we can find the distance from it to the Sun. (Adapted from "Globular Clusters," by Ivan R. King from *Scientific American*, June 1985, page 80. Copyright © 1985 by Scientific American, Inc. All rights reserved.)

ates no new stars. The halo, which contains the ancient globular clusters, is the oldest part of the Galaxy.

To understand the Galaxy, we need to know its size. William Herschel made the first attempt by counting the number of stars seen in different directions. By assuming that all stars have the same luminosity, he estimated their relative distances, and thereby showed their disklike distribution. However, Herschel did not know about absorption of starlight by interstellar dust, which severely limited his view in the disk and caused him to place the Sun near the center of its own local neighborhood (see Figure 16.1).

The breakthrough in measuring the Galaxy's size was made at Mount Wilson around 1918 by Harlow Shapley. He looked away from the dusty galactic plane at the halo's globular clusters (Figure 16.3) and made the logical assumption that the center of the whole system of globulars should coincide with the center of the Galaxy. From modern HR diagrams of globular clusters, spectroscopic distances of the halo's individual RR Lyrae stars (see Section 15.2)—whose system must also coincide with the Galaxy's center—and other methods, we adopt a distance of the Sun to the galactic center (R_0) of 8 kpc. The Sun lies only about 15 pc above the central plane of the galactic disk.

There is no clearly defined edge to the Galaxy, only a gradual diminution in the number of stars and in the density of the interstellar medium. About 90% of the light of the galactic disk falls within 12.5 kpc of the center, a distance that is taken as the traditional radius and which places the Sun about two-thirds of the way outward from the center (as in Figure 16.1). We will see, however, that the Galaxy really extends much farther.

16.2 THE GALACTIC DISK

Galaxies like ours are dominated by disks that contain the vast majority of the visible mass. The disk is the location of most of the action and where we first focus attention.

16.2.1 Structure

A closer look at the galactic disk shows that it is complex and separated into layers (Figure 16.4). The thinnest portion, about 200 pc thick, contains the Galaxy's gas and dust. As a result, it also contains the youngest stars, including the blue O and B stars. This young disk is readily apparent in other galaxies (especially if they are seen edge-on) as a dark band of dust, and in our own Galaxy as the central rift of the Milky Way (Figure 16.5). A thicker disk roughly 700 pc thick encompasses older stars, those born over the past several billion years.

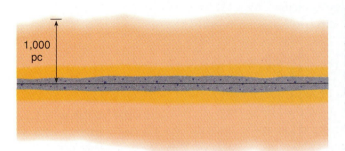

Figure 16.4

The disk of the Galaxy is made of layers that gradually thin out in the perpendicular direction. The thinnest and densest layer (gray) is made of dust and gas and is studded with young blue O and B stars. A thicker layer (yellow) embraces most of the Galaxy's stars, and a more sparsely populated thick layer (orange) has a somewhat lower metal content.

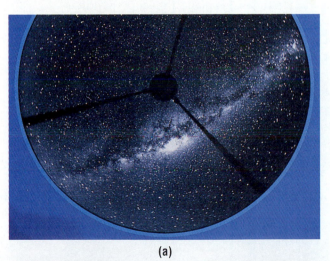

(a)

(b)

Figure 16.5

(a) A wide-angle view of the Milky Way centered on the galactic nucleus bears a remarkable resemblance to **(b)** NGC 891, a disk galaxy seen edge-on. Both structures exhibit a central bulge and the dark dusty band of the thinnest part of the disk.

Spreading outward is a much thicker disk roughly 2,000 pc thick that is something of a bridge to the halo, since its metal content is about half that of the Sun and the young disk at the center. Together these components contain 90% of the Galaxy's stars, the number dropping quickly in the direction perpendicular to the disk's plane.

16.2.2 Rotation

Further exploration of the disk's structure, and ultimately of the mass of the Galaxy, requires a knowledge of the disk's rotation. We first need to know the velocity of the Sun in its orbit about the galactic center. We encountered a similar problem in determining the direction of solar motion through the surrounding stars (see Section 13.1). This time we are observing objects outside the galactic disk that do not participate in galactic rotation, such as distant globular clusters and RR Lyrae stars (Figure 16.6). Although they all have their own orbital motions around the galactic center, those toward which the Sun is moving directly will *on the average* appear to be approaching, and those in the opposite direction, receding. The average observed speed of approach or recession will equal the solar orbital speed.

We find that the Sun is moving in the disk's plane almost perpendicular to the line to the galactic center (in the direction of Cygnus) at about 240 km/s (see Figure 16.6). The solar orbit, however, is somewhat elliptical. What we need to know is the velocity of a body in a *circular* orbit at the distance of the Sun from the galactic center, which is that of the average orbital velocities of the local disk stars. Since we know both the solar orbital velocity and the velocity of the Sun relative to the local stellar swarm (see Section 13.1), we find the circular velocity to be 220 km/s.

The circular orbital velocity at the Sun's distance from the galactic center is about 220 km/s.

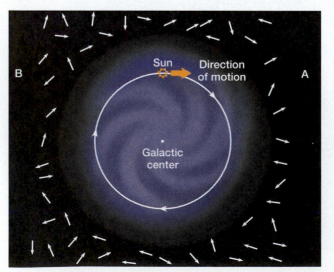

Figure 16.6
The Sun is moving in the direction of the yellow arrow. Objects not participating in galactic rotation are moving along randomly oriented orbits (white arrows). Those in the A direction will on the average appear to be approaching the Sun at the solar velocity; those in the B direction will on the average appear to be receding.

The next task is to measure the circular orbital velocities at different distances from the galactic center, which we do with radio observations that can penetrate the obscuring interstellar dust. At 21 cm, we can see neutral interstellar atomic hydrogen all the way across the disk. For simplicity, assume the Sun is on a circular orbit. In Figure 16.7a, a radio telescope is pointed into the disk at an angle of 35° from the galactic center (indicated by the red dashed line). It receives 21-cm radiation from neutral hydrogen clouds lying all along the line of sight. Each cloud is at a different distance from the galactic center; each has a different radial velocity and a different Doppler shift. The result is the complex combined spectrum line in Figure 16.7b.

Project the circular orbital velocities of the Sun and the clouds onto the line of sight in Figure 16.7a (as indicted by the arrowheads on the

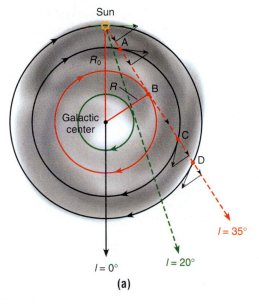

(a)

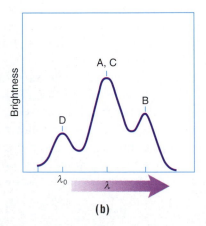

(b)

Figure 16.7

(a) A radio telescope looks into the galactic disk at an angle (l) of 35° from the galactic center (red arrow). Clouds A, B, C, and D and the Sun are assumed to have circular orbits. The clouds' radial velocities depend on the projections of the orbital velocities (dark arrowheads) along the line of sight. **(b)** The telescope records a complex 21-cm line radiated by clouds moving at different radial velocities: cloud D radiates at the rest wavelength, λ_0; cloud B, at the tangent point to the line of sight, has the greatest radial velocity, from which we can find the orbital speed at distance R from the galactic center; clouds A and C have equal intermediate velocities. Observation at an angle of $l = 20$° (dashed green arrow) gives the rotation velocity at a different distance from the center.

dashed red line). The radial velocities of the clouds are the *differences* in the clouds' projected orbital velocities and the projected orbital velocity of the Sun. Cloud D, on the solar orbit, moves along the line of sight with the same speed as the Sun. It therefore appears to be at rest, radiating the 21-cm line at the rest wavelength of λ_0. However, greater proportions of the orbital velocities of clouds A and C will be projected along the line of sight, and their spectra will be Doppler-shifted toward longer wavelengths.

Cloud B lies at the tangent point to its orbit, and *all* its orbital velocity is along the line of sight. Radiation from B therefore has the greatest Doppler shift and the longest wavelength. Because we know how the solar velocity is directed along the line of sight, we can find the actual orbital velocity of cloud B. The Sun, the galactic center, and cloud B define a right triangle. Since we know the distance to the galactic center, R_0, we can compute the distance of cloud B from the center and the radius of its orbit, R. Observations along different directions (such as the dashed green line in Figure 16.7a) allow us to determine speeds at different distances from the galactic center.

This method does not work for the outer part of the Galaxy. But since there is less dust in that direction, we can use radial velocities of diffuse nebulae and molecular clouds coupled with spectroscopic distances of their embedded O and B stars. This technique allows us to probe the outer tenuous disk more than 20 kpc from the center, far beyond the traditional radius of 12.5 kpc. The final result of these studies is the **galactic rotation curve** (Figure 16.8), which gives circular velocity graphed against distance from the center.

16.2.3 Spiral Arms

Embedded within the disks of other galaxies are sets of graceful, winding **spiral arms** (Figure 16.9) made of OB associations, diffuse nebulae, and associated giant molecular clouds. We can apply the method of spectroscopic distances to our Galaxy's O and B stars and plot their positions to establish local structure and the locations of our own Galaxy's arms outside the solar orbit, where the dust is thinner.

To find the arms beyond our local neighborhood, but inside the solar orbit, we use the galactic rotation curve. As we saw in Figure 16.7, the radial velocities of the neutral hydrogen and giant molecular clouds, which mark the arms, depend on their distances along a line of sight. Once the rotation curve is known, we can calculate the clouds' distances from their radial velocities. When we chart the clouds' positions, the spiral arms pop into view (Figure 16.10). From the available data—including the known

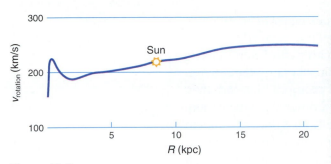

Figure 16.8
The rotation speed of the Galaxy quickly increases from the galactic center, then levels off with a slight, but steady, rise.

Figure 16.9
M 51 in Canes Venatici displays a beautiful set of spiral arms in its disk, outlined with blue O and B stars and red diffuse nebulae. A small companion galaxy below M 51 raises a tide in its larger neighbor, distorting one of the spiral arms.

Figure 16.10
(a) The first detailed map of the distribution of interstellar neutral hydrogen showed the existence of galactic spiral arms. (Velocities could not be defined in the blacked-out wedge.) **(b)** Giant molecular clouds also outline spiral structure.

(a)

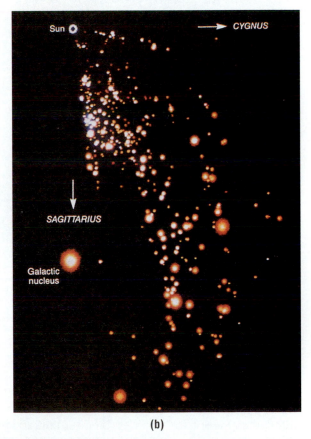

(b)

distances to individual objects—we can construct maps of the Galaxy (Figure 16.11).

The spiral arms (which constitute only a small part of the Galaxy's total mass) cannot be permanent collections of stars, because the rotation of the Galaxy would cause them to wind up and disappear. Instead, they appear to be waves of density. The waves rotate around the Galaxy more slowly than do the stars, so that the stars continually pass through them in the direction of rotation, slowing down temporarily in response to the increased gravitational field. The effect is not unlike the behavior of cars on a highway slowing for an accident. The vehicles keep moving, but a wave in which they are close together propagates toward the rear of the line.

Portrait of the Milky Way

Figure 16.11

In the inset at lower left, we see a simplified map of the Galaxy's spiral arms. The large *Portrait of the Milky Way*, by Jon Lomberg, shows the Galaxy as viewed over the top of the Perseus arm and the Sun, which is near the next spiral arm inward from the Perseus arm (identified in the inset). The *Portrait* is as realistic a view of the Galaxy as possible, placing objects of known distance in proper context.

The cars move through the wave, then speed up. The Galaxy's density waves are a product of the interstellar medium. Some disturbance causes a clumping of interstellar gas. The increased gravity produces further clumping away from the disturbance, which causes even more clumping, and the whole pattern moves outward through the Galaxy in response to the rotation.

The density waves may cause the initial compression of interstellar material that leads to star formation. Stars develop within the arms, which are subsequently lit by the brilliant, ephemeral, blue O and B dwarfs and blue supergiants that give the arms their characteristic color. These massive stars evolve and explode, leading to further compression and more star formation. As time proceeds, the lower-mass stars leave the arms and populate the general disk.

Density waves in the Galaxy's disk may cause the initial compression of interstellar matter that leads to star formation.

16.2.4 The Magnetic Field and Cosmic Rays

Polarization of starlight caused by aligned interstellar grains (see Section 14.1.2) indicates a magnetic field, a millionth or so the strength of the Earth's field, permeating the disk. It is undoubtedly related to the Galaxy's rotation and the movement of ionized matter. The field controls **cosmic rays**, high-speed atomic nuclei that can weakly ionize dense cores (see Section 14.3). They are observed directly with high-flying balloons and spacecraft, and indirectly when they smash into atoms in the Earth's atmosphere, producing atomic debris that rains to the ground. Cosmic rays consist mostly of protons and helium nuclei, but nuclei as heavy as bismuth have been seen. With velocities near that of light, they can carry astonishing amounts of energy. At maximum, a cosmic ray particle—a mere atomic nucleus only 10^{-13} cm across—can carry the energy of a professionally pitched baseball!

Cosmic rays are almost certainly blasted out of supernovae, the atomic nuclei accelerated to high velocities by riding supernova shock waves. Once these charged particles have been injected into interstellar space, the galactic magnetic field bends their paths around the Galaxy, a tiny fraction crashing into Earth.

16.3 THE HALO AND BULGE

Like the spiral arms, the halo and bulge of our Galaxy are best appreciated by looking at those of other galaxies (Figure 16.12). Except for its globular clusters, the galactic halo is hard to see, sparsely populated as it is with only about 2% the number of stars in the disk. There is little interstellar gas and dust. The halo is old and metal-deficient: the amount of iron relative to hydrogen is only about 1% that found in the Sun and in the galactic disk. Though the halo's stars and globular clusters orbit the Galaxy's center, there is little or no rotation of the system as a whole.

Figure 16.12
M 104, the Sombrero Galaxy, seen nearly edge-on, displays a prominent, thin, dusty disk. The galaxy's faintly visible halo is filled with globular clusters seen as fuzzy, almost starlike, images. The huge bulge is illuminated by bright red giants.

Like the disk, the halo is structured. There are two sets of globular clusters. One occupies the full spherical halo to distances far beyond the traditional galactic radius; the other lies within a much flatter volume with a thickness of only about 4 kpc. The inner globulars have a higher metal content, a third that of the Sun. We now see even more of the intriguing trend discussed in Section 15.9: as we approach the disk from the halo, the metal contents of the stars tend to increase; this pattern is a key to understanding something of the evolution of the Galaxy as a whole.

Our own bulge (see Figures 16.1 and 16.11), roughly a kiloparsec in radius, is visible through only a few "windows" in the thick intervening interstellar dust. It consists primarily of old stars, its characteristic color provided by red giants. The region where the halo and disk merge, it has some of the characteristics of both as well as unique characteristics of its own. Containing about 10% of the Galaxy's stars, the bulge has a huge range of metal abundances. Like the extended halo, it has metal-deficient stars. As the inner extension of the disk, however, it also contains stars with remarkably high metal contents, up to ten times solar. Such metal enhancement appears to be the result of rapid star formation and evolution that have caused great amounts of heavy elements to be deposited in interstellar space.

16.4   THE MYSTERIOUS NUCLEUS

Buried at the center of the bulge, 6° west of γ Sagittarii (Figure 16.13) in the thickest part of the Milky Way's central dust lane, is Sagittarius A, one of the strongest radio sources in the sky. Because of its compactness, it was immediately identified with the true center of the Galaxy, the **galactic nucleus**. At a distance of 8 kpc, the nucleus and its environs are hidden in the visual spectrum by 30 magnitudes of dust absorption (only one visual photon out of a trillion gets through the haze). In the infrared and radio spectral regions, however, where the view clears, astronomers find extra-

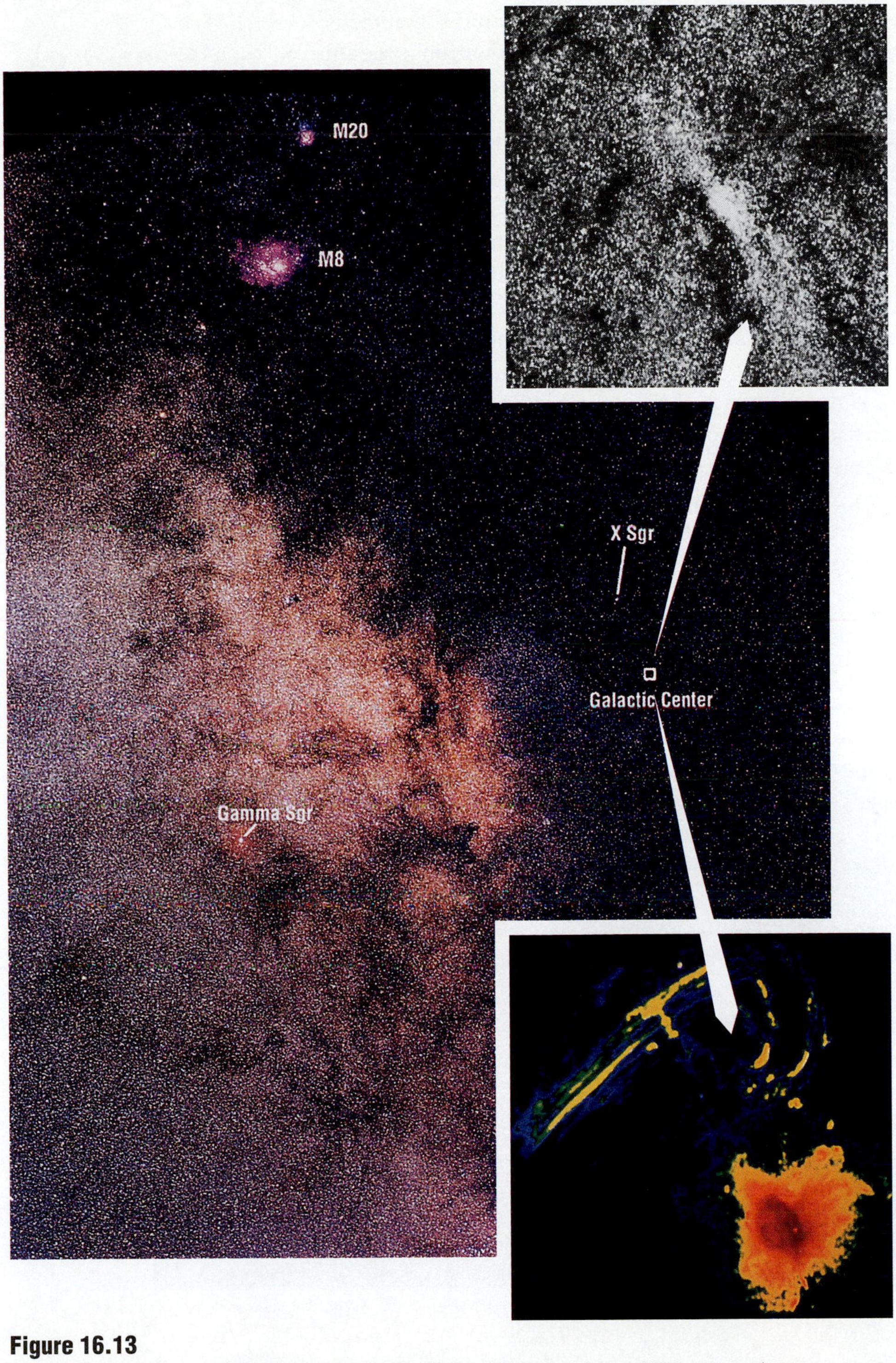

Figure 16.13

The small box to the west of the dense star clouds of Sagittarius, 25 minutes of arc and 60 pc across, shows the location of the galactic center and Sagittarius A. (upper inset) An infrared view within the box reveals masses of stars hidden by interstellar dust. (lower inset) A radio view centered a bit to the northeast shows streamers of gas set within a relatively strong magnetic field. Ionized gas (red) hides the actual nucleus.

ordinary complexity. Infrared observations of the box around the nucleus in Figure 16.13 (upper inset) show dense masses of stars within a region only 60 pc across. A radio view (lower inset) shows streamers of gas that emit synchrotron radiation, revealing a magnetic field a thousand times the strength of that in the outer Galaxy.

Within the inner 2 pc of the Galaxy (Figure 16.14a), we see a tilted spiral whose matter may either be falling into the center or shooting from it. Above the spiral's center is a pointlike radio source, Sagittarius A* (Figure 16.14b). Very Long Baseline Interferometry shows Sagittarius A* to be no more than 4 ten-thousandths of a second of arc, or 3 AU, across—about the size of the orbit of Mars. Sagittarius A* is apparently the actual nucleus of the Galaxy.

Infrared spectra of stars and gas in the Galaxy's central regions show that orbital speeds increase steadily toward the center. Limits on distance from the center and Kepler's generalized third law suggest that the motions are controlled by a mass of perhaps 1 or 2 million $M_\odot$. Infrared observations, however, do not show a sufficient number of stars to account for this mass. If it is concentrated inside Sagittarius A*, we have little recourse but to believe that the galactic nucleus contains a massive black hole and that Sagittarius A* is its bright, radiating accretion disk. The disk is created and sustained by stars that are pertubed in their orbits by their

We have little recourse but to believe that the galactic nucleus contains a massive black hole.

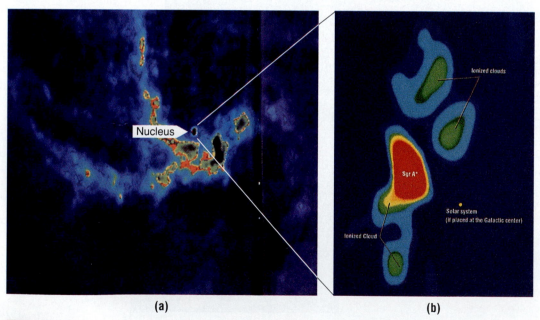

(a)   **(b)**

Figure 16.14

(a) The center of the Galaxy, Sagittarius A, consists in part of a gaseous spiral 2.5 pc across. Sagittarius A*, a pointlike source buried within Sagittarius A, appears to be the actual nucleus. **(b)** The view is only 0.05 pc across. Sagittarius A*, within the elongated red area, is seen surrounded by clouds of ionized gas.

neighbors, wander too close to the black hole, and are tidally disrupted, their torn shards entering the madly circulating mass that surrounds the dark central body. The existence of such a "monster in the middle" is uncertain, and we may yet find that a dense cluster of O and B stars is responsible. But we have opened a fascinating possibility, evidence for which will grow in later chapters.

16.5 THE MASS OF THE GALAXY AND DARK MATTER

Any body in the Galaxy orbits the galactic center in response to the total mass inside that orbit (including the mass in the halo). The Sun and the inner Galaxy—that portion inside the solar orbit—can therefore be considered a two-body system, and we can apply Kepler's generalized third law to measure the inner Galaxy's mass. From the circular velocity at the Sun's distance from the center, we find an orbital period of 225 million years. From the simple formula $M_{\text{inner Galaxy}} + M_{\text{Sun}} = a^3/P^2$, where M, a, and P are respectively in solar masses, AU, and years (see Section 13.5.1), the mass of the inner Galaxy is $9 \times 10^{10}\ M_{\odot}$. Because the Galaxy's disk outside the solar orbit contains only about 20% of the galactic light, and because the halo has a low stellar population, the mass of visible stars and interstellar gas and dust in the whole Galaxy is around $10^{11}\ M_{\odot}$. From the average stellar mass of about 0.5 $M_{\odot}$, the Galaxy contains around 200 billion stars.

However, the Galaxy's rotation curve clearly indicates a severe problem. If all the mass of the Galaxy were effectively inside the solar orbit, the orbital velocities of stars and neutral hydrogen clouds farther from the center would be lower than ours: Mars, for example, moves more slowly around the Sun than does the Earth. Since 90% of the light of the Galaxy is inside a radius of 12.5 kpc, we expect some kind of drop in rotation speed at large distances from the galactic center in accord with Kepler's laws of motion (red curve in Figure 16.15). However, the speeds do not drop. The rotation curve remains flat or even rises a bit to at least a radius of 20 kpc.

To keep speeds high, each orbit outside the Sun's must contain increasingly more matter. There is $2\frac{1}{2}$ times as much mass inside 20 kpc as

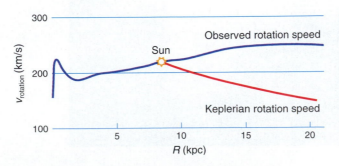

Figure 16.15
The observed rotation curve of Figure 16.8 is shown by the blue line. If nearly all the mass of the Galaxy were inside the solar orbit, stars outside the orbit would, according to Kepler's laws, move more slowly (red curve).

there is inside the solar orbit, yet only 20% of the galactic light resides outside the solar orbit. We cannot see the matter that causes the high velocities at large galactic radii, and it is therefore called **dark matter**. Do not confuse dark matter with dark nebulae or globules, whose compositions we know. Dark matter can be "seen" only by its gravitational effects.

From investigations of the velocities of stars in the direction perpendicular to the disk, which are affected by the disk's gravitational attraction, we know that only a small fraction of the dark matter can reside there. The majority of the dark matter, or even all of it, seems to be in a greatly extended halo called the **dark matter halo** (Figure 16.16). Analyses of the velocities of other galaxies close to ours, those under the influence of our Galaxy's gravity, indicate that the dark matter halo may have a radius of 100 kpc or more and that it and the extended disk may contain 10^{12} solar masses, 10 times the galactic mass interior to the solar orbit: dark matter may constitute 90% of the Galaxy's mass! The traditional Galaxy described at the beginning of this chapter is therefore seen to be only a small part of the overall structure.

We do not know the nature of dark matter, though several possibilities have been proposed. The Galaxy may be filled with dim, low-mass main-sequence stars, brown dwarfs (see Section 14.4), or white dwarfs. If they are responsible for the dark matter, there must be extraordinary numbers of them. Yet to date only one brown dwarf has been identified (and even that may be something else); and Hubble Space Telescope studies show that, if anything, the halo has *fewer* dim stars than expected near the end of

Dark matter may constitute 90% of the Galaxy's mass.

Figure 16.16
The traditional disk-halo structure of the Galaxy seems to be surrounded by a huge halo filled with mysterious dark matter.

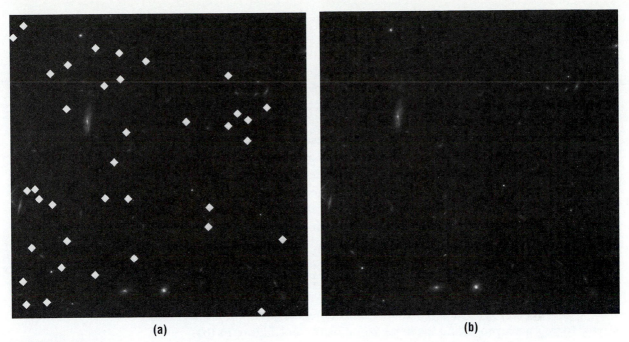

(a)   (b)

Figure 16.17
The Hubble Space Telescope took an extraordinarily deep look into the galactic halo. In the left-hand version of the image **(a)**, the diamonds simulate the number of faint red stars that would be seen if they were the explanation of dark matter. **(b)** The unadulterated view shows *fewer* stars than expected rather than more: about all we see are fuzzy images of distant galaxies far outside the confines of our own system.

the main sequence rather than more (Figure 16.17). Other researchers have proposed massive interstellar "grains" the size of bowling balls, myriad black holes, and even vast numbers of exotic subatomic particles, these to be examined in Chapter 19.

16.6 THE ORIGIN AND EVOLUTION OF THE GALAXY

The most important parameter in the study of the development of the Galaxy is obviously its age and the ages of its components. The main-sequence cutoffs of the oldest open clusters indicate that they are roughly 7 to 10 billion years old. Since open clusters belong to the galactic disk, they give us its age. However, the looseness of open clusters makes them prone to destruction. Less-massive stars can gravitationally encounter more-massive ones and gain enough energy to be hurled out of the cluster. The effect is enhanced by tidal forces induced by the Galaxy as a whole, which can stretch a cluster and cause some of the less-massive stars to flee. An open cluster then slowly evaporates. It is therefore quite possible that none has lived for the full lifetime of the disk. An age derived from such clusters must therefore be a lower limit, although one probably not far from the truth.

We identify the ages of globular clusters with that of the Galaxy itself and adopt an average of 15 billion years.

The globular clusters of the halo, however, have measured ages between about 13 and 17 billion years, showing that the halo is older than the disk. Since they are not so subject to disruption, we identify their ages with that of the Galaxy itself. There is a real difference among globular cluster ages (showing that they were made over a period of time), but the age analysis also contains uncertainties, so we adopt an average of 15 billion years.

Metal content correlates crudely with age (a concept first developed in Section 15.9), showing the Galaxy to be a changing, evolving structure. The ancient halo and its globular clusters are low in metals, the younger disk relatively high. The Galaxy appears to have contracted from a huge, somewhat spherical structure into the disk at the same time that stellar evolution pumped heavy elements into the star-forming interstellar medium.

We believe that somewhere around 15 billion years ago the gaseous mass that composed the Universe fragmented into billions of clouds that contracted under the force of their own gravity (Figure 16.18a). One of them was to become our Galaxy. Within a short time, the gas of the cloud started to fragment, producing the first generation of stars (Figure 16.18b). Study of the origin of the Universe (see Chapter 19) strongly indicates that the initial contracting cloud was made only of hydrogen, helium, and a tiny amount of lithium; so, therefore, was the first stellar generation. Historically, the stars of the galactic disk are called Population I, those of the halo Population II, so the stars of this earliest generation are called **Population III**. As Population III's more massive stars evolved into supergiants and giants, they created the first heavy atoms, which were spread through the interstellar medium by stellar winds and explosions and incorporated into the next stellar generation.

This second generation had a low metal content, and we identify it with the most metal-poor stars of the galactic halo. We thus encounter the weakest link in the chain of reasoning that leads to the present. Where are the remaining low-mass stars of Population III, those with no metals? Even the ancient globular clusters have metal contents about 1% that of the Sun. Among halo noncluster stars we find several with metal abundances all the way down to about 10^{-4} solar, but we never find zero.

Such stars may simply be very rare and hard to locate. It would not have taken many Population III stars to seed the Galaxy with the metals we see incorporated into the halo stars. The higher-mass component of Population III would long ago have burned away, leaving only a few dim dwarfs still on the main sequence. Alternatively, early conditions may have allowed the formation of only massive stars that died long ago. In either case, the compositions of halo stars are consistent with the synthesis of atoms in an earlier generation, so we are fairly confident that some kind of Population III once existed.

As the Galaxy continued to contract and to make stars, it also divided into large lumps of material that fragmented into yet more stars to make

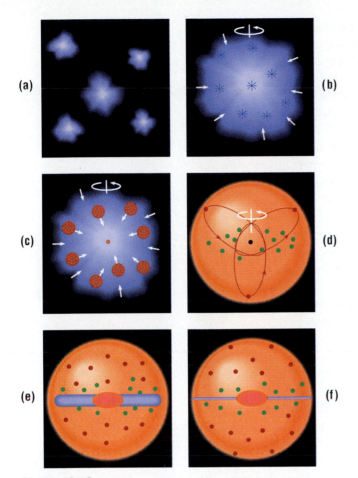

Figure 16.18

(a) The Galaxy was likely created from a blob of hydrogen and helium fragmenting out of the early intergalactic medium. **(b)** The initial generation of stars, Population III (blue), seeded the gas with the first heavy atoms. **(c)** The contracting cloud then produced the ancient halo component of stars and globular clusters (red), which were given elliptical orbits toward the galactic center. Conservation of angular momentum caused the interstellar medium to begin to flatten. **(d)** The next generation of globulars (green) was then distributed in a flattened halo and had higher metal abundances, which in turn contracted further to **(e)** the thick portion of the stellar disk (light blue). **(f)** Five billion years after initial formation, the interstellar medium had collapsed to the thin disk with its O and B stars (blue), leaving the globulars in elliptical orbits.

the globular clusters we see today (Figure 16.18c). The first generation of globulars had few metals (those produced by Population III stars). Since the Galaxy was rapidly contracting, these assemblies had strong components of velocity toward the galactic center. The result is that the halo globulars today have highly elliptical orbits. As they evolved, they partially disrupted and populated the halo with at least some of the noncluster stars we now see.

The Galaxy continued to contract, and conservation of angular momentum began to flatten the dusty interstellar gases, the part that was making stars (Figure 16.18d). The next generation of globular clusters thus had a higher metal content, was distributed in a more flattened fashion, and had less-elliptical orbits. The metal content of the Galaxy continued to increase as contraction continued. A few billion years after formation, the interstellar medium had spun itself into an even flatter disk (Figure 16.18e), making the thick portion of the galactic disk shown in Figure 16.4, and finally into the current thin disk (Figure 16.18f) with its young O and B stars, leaving behind the globular clusters and the single stars of the galactic halo. The Sun, which came along 10 billion years after the formation of the Galaxy and maybe 5 billion years after the development of the disk, is a relative newcomer.

This theoretical picture, pleasing as it is, presents some severe problems. The relations among stellar orbits, ages, and metal abundances are complex. Galactic evolution was apparently highly chaotic. Though there does seem to be an age spread among globular clusters, the inner globulars—those that are more flatly distributed—have about the same measured ages as those of the outer halo. We do not know how Population III stars could have formed without interstellar dust generated by stars, why the halo took so long to collapse into the disk, or why globular clusters are no longer being made. It also seems more and more likely that some of our Galaxy's stars and globular clusters have come from other galaxies in different states of evolution that have collided and merged with ours. Most important, we do not know the nature of dark matter, or how it and the dark matter halo have influenced star formation and galactic evolution.

Think, however, not of our ignorance, but of what we have learned. As we stand under the stars, admiring the Milky Way, we can now ponder its meaning and its powerful relation to the Earth and to ourselves. The Galaxy is a vast recycling engine, its magnetic field and density waves instrumental in the formations of stars. Stellar death promotes stellar birth by providing both compression and ionizing cosmic rays. The heavy atoms out of which we are made came from aeons of stellar evolution and mass loss. We now know we are not only citizens of our town or even of the world, but are truly residents of the Galaxy. We are no more separate from it than we are from our own planet.

▶ **KEY CONCEPTS**

Cosmic rays: High-energy atomic nuclei accelerated by the Galaxy's magnetic field; possibly produced in supernovae or by their collapsed remnants.

Dark matter: Unilluminated mass, detectable only through its gravitational effect; its content is unknown.

Dark matter halo: The volume of space filled with dark matter, encompassing the traditional disk and halo of the Galaxy.

Galactic nucleus: The energetic center of the Galaxy; it is quite likely a black hole.

Galactic rotation curve: The rotation velocity of the Galaxy graphed against distance from the galactic center.

Population III: The original population of stars, made only of hydrogen and helium, that seeded the Galaxy with the first heavy elements; no representatives are known.

Spiral arms: Arms in the galactic disk that wind outward from near the galactic center, contain O and B stars and clouds of interstellar matter, are sites of star formation, and are produced by density waves.

▶ **EXERCISES**

Comparisons

1. Compare the metal abundances and ages of the stars of the galactic disk, the galactic halo, and the galactic bulge.

2. Why are spiral arms bluish? Why is the bulge reddish?

3. Compare different-wavelength views of the galactic nucleus.

4. What is the difference between Sagittarius A and Sagittarius A*?

Numerical Problems

5. Show that the rotation period of the Sun around the galactic center is about 225 million years.

6. How many times has the Sun orbited the Galaxy since the Sun's birth?

7. What is the mass of the Galaxy within 12.5 kpc of the center assuming a flat rotation curve?

8. About how many stars are in the galactic bulge?

Thought and Discussion

9. Why can we *not* use the Galaxy's O stars to find the distance to the galactic center?

10. Why are the old stars of the galactic halo found within the young galactic disk?

11. How can observations of the 21-cm radio line be used to find the rotation curve of the Galaxy? Show why the tangent method will not work outside the solar orbit.

12. How would you use O and B stars to define the Galaxy's local spiral arms?

13. Describe three processes that can compress the interstellar medium to aid in the creation of new stars.

14. What are the effects of the Galaxy's magnetic field?

15. What are cosmic rays and how are they detected?

16. Why do we think there is a black hole in Sagittarius A*?

17. Summarize the evidence for dark matter in the Galaxy.

18. What are some possible candidates for dark matter?

Research Problem

19. When did astronomers become aware of the galactic nucleus and its bizarre properties? Find properties of the galactic center not described in this book.

Activities

20. Use a pair of binoculars to locate the galactic center. Sketch the stars you see in the region and draw in the nucleus.

21. Make a three-dimensional scale model of the Galaxy that includes the halo and disk, globular clusters, and the galactic nucleus.

Scientific Writing

22. A philanthropist is willing to support galactic research, a subject in which you are an expert. Write a proposal to obtain the funds, explaining the unknown aspects of the Galaxy and the research that must yet be done.

23. You wish to communicate your enthusiasm about the Milky Way to as large a group of people as possible. Write a nontechnical article for a newspaper's Sunday supplement in which you describe the beauty and nature of the Milky Way and its significance to us. Include two illustrations with captions.

PART

IV

GALAXIES AND
THE UNIVERSE

GALAXIES

Other galaxies and their relation to our own

17

Stars are organized into large, separated systems called **galaxies.** We have examined our own Galaxy in detail; we now look at the vast number of others and assemble the units that make the Universe.

17.1 DISCOVERY

With the development of the telescope, astronomers began to find vast numbers of small fuzzy patches of light. Some were readily resolved into individual stars and thus revealed as nearby clusters, but others defied resolution. In the nineteenth century, a few, the diffuse and planetary nebulae, were found to have emission-line spectra, and so had to be gaseous. A fraction, however, were neither resolvable *nor* gaseous.

As early as 1750, the English astronomer Thomas Wright suggested that our Galaxy is not limitless but has boundaries, and that the unresolved nebulae are comparable but distant collections of stars. Visual observations (Figure 17.1) and then photography disclosed that many had spiral structure and spectroscopy revealed absorption lines, demonstrating that the "spiral nebulae" are indeed made of stars. Yet the significance of the spiral nebulae remained controversial, the puzzle they presented deepening when in 1917 V. M. Slipher of Lowell Observatory found from spectral shifts that most of them are moving away from us at high speed.

The controversy peaked with the *Great Debate* that took place in 1920 between Heber D. Curtis of Lick Observatory and Harvard's Harlow Shapley. Curtis took the "Many Galaxies" view. The apparent brightnesses of novae in Andromeda's M 31, one of the mysterious nebulae (see Figure 13.23), indicated a then-amazing distance of 150 kpc. The structure, and presumably the other spiral nebulae, must therefore be huge. Curtis also noted that the nebulae are moving away from us at speeds that clearly release them from our Galaxy's gravitational grasp, so they could not belong to us. Shapley, however, championed the concept of one "Big Galaxy," believing the spiral nebulae to be nearby, within our own system.

Three years later Edwin Hubble (Figure 17.2) settled the issue. With the 100-inch at Mount Wilson, he resolved M 31 and its nearby neighbor

Figure 17.1
Spiral structure in what would later be recognized as spiral galaxies was first seen by Lord Rosse in Ireland with his new 72-inch reflector around 1845. Compare his drawing of M 51 with the modern photograph in Figure 16.9.

Figure 17.2
Edwin Hubble's research on galaxies changed our view of the Universe.

M 33 in Triangulum into stars (Figure 17.3). If they were anything like the stars of the solar neighborhood, M 31 and M 33 had to be far away and comparable in nature to our Galaxy. Recognition of familiar kinds of stars whose absolute visual magnitudes were known from galactic studies allowed Hubble to estimate M 31's distance at 300 kpc. The debate was over: the mysterious "nebulae" were true galaxies at great distances. Our Galaxy is not unique, but belongs to a family of billions of others that extends as far as we can see.

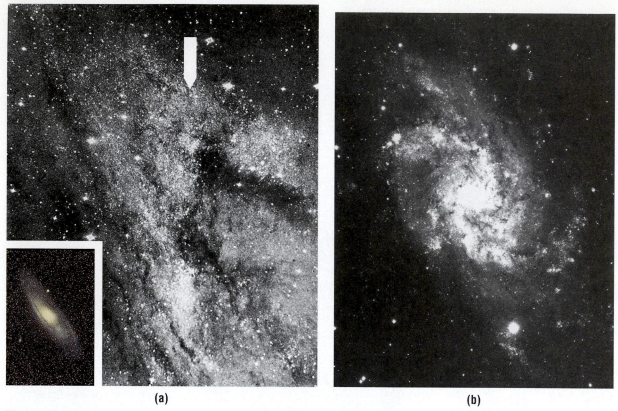

(a)   **(b)**

Figure 17.3
(a) The outer disk of M 31 swarms with millions of stars and a handful of Cepheid variables (one indicated by the arrow), from which its distance can be found. The entire galaxy is shown in the inset.
(b) M 33, about the same distance of M 31, is also easily resolved into stars.

17.2 KINDS OF GALAXIES

To begin to understand galaxies, we must classify their great variety of forms. The original **Hubble classification** has three major groups and several subgroups. The simplest major class is composed of **elliptical galaxies** (E). They have smooth outlines that appear as ellipses (Figure 17.4) and—unlike our Galaxy—have no disks. Hubble subclassified elliptical galaxies according to their apparent shapes by following E with a number. A round system (Figure 17.4a) is E0. NGC (which stands for *New General Catalogue*) 3377 in Figure 17.4b is E6. Any galaxy more elongated than E7 (for which the long axis is 3.3 times the length of the short axis) has lost its elliptical shape and shows evidence of a disk.

An elliptical galaxy's apparent shape depends on its three-dimensional structure and its orientation. A spherical galaxy must appear as E0. However, an elongated system, like that in Figure 17.4c, can take a variety of apparent forms. If we view it end-on, it appears as E0. However, if we look at it sideways, we see it as E6. If we assume that elliptical galaxies are

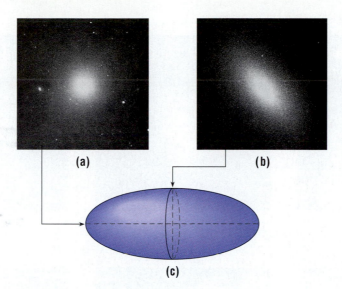

Figure 17.4
The elliptical galaxies **(a)** NGC 4636 in Coma Berenices and **(b)** NGC 3377 in Leo have different apparent shapes and are respectively classified as E0 and E6. **(c)** Both could in fact have the shape of this drawing, but we may be viewing them from different angles, as indicated by the arrows; NGC 4636 may, however, be spherical.

oriented randomly in space, we find that the most common have actual shapes about midway between these two.

Elliptical galaxies generally contain no O or B stars, active star formation, or cool, star-forming gas and dust. What gas they have appears to be largely a hot, chemically enriched by-product of winds from evolved stars and supernova explosions. In that sense they are similar to the halo of our own Galaxy except that the star density (the number of stars per cubic parsec) is *much* greater. The metal abundance may—especially in large ellipticals—also be much higher, making them more like our Galaxy's bulge. Nearby elliptical systems can be resolved into a grainy structure, into millions of bright red giants that give the ellipticals a characteristic reddish glow. Stars like the Sun and dimmer dwarfs farther down the main sequence are present, but they are too faint to be seen individually.

Ellipticals are not nearly so captivating as **spiral galaxies**, those with thin bluish disks and complex spiral arms comparable to those in our own Galaxy. There are two kinds. **Normal spirals** (S), like our Galaxy, have disks and arms that emerge from a central bulge (Figure 17.5a). Sa systems have nearly circular arms that wind outward slowly, the arms of Sc galaxies open out quickly, and those of Sb are in between. The more open the arms, the smaller the bulges, and the more we see the arms break into knots and clumps made of O and B associations and complexes of diffuse nebulae. The more open spirals have more gas and dust, and consequently star formation proceeds at a greater pace. The distribution in types is continuous, and intermediate classes like Sab are common. M 31 and our own Galaxy are Sb. About 20% of spirals are **barred spirals** (SB) (Figure 17.5b), in which the arms branch almost perpendicularly from a straight bar punched through the bulge like a knitting needle through a ball of yarn. They divide into SBa, SBb, and SBc, in parallel with the classification of normal spirals. Although difficult to detect, all the spirals are expected to

Elliptical galaxies generally contain no O or B stars, active star formation, or cool, star-forming gas and dust.

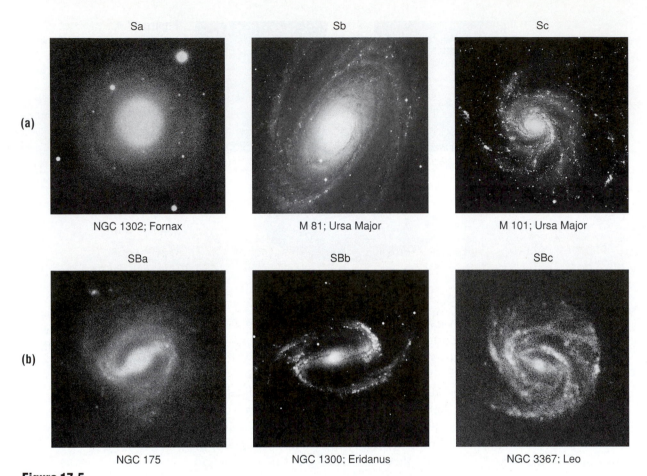

(a)   Sa   Sb   Sc

NGC 1302; Fornax   M 81; Ursa Major   M 101; Ursa Major

(b)   SBa   SBb   SBc

NGC 175   NGC 1300; Eridanus   NGC 3367; Leo

Figure 17.5

(a) Normal spiral galaxies (S) show disks and spiral arms. **(b)** The arms of barred spirals (SB) emerge from a bar through the nucleus. A spiral's subclass (Sa, Sb, Sc, or SBa, SBb, SBc) depends on the openness of its arms.

have surrounding halos that merge with the disks to create the central bulges. Between the ellipticals and the spirals is a set of "SO" galaxies that display obvious disks but minimal dust lanes and no spiral arms. (The SB0 barred version has a bar but no projecting arms.)

The different major types of galaxies are summarized by Hubble's **tuning fork diagram** (Figure 17.6). The diagram does *not* represent an evolutionary sequence, only a classification scheme. The tines of the tuning fork close again at the far right as they merge into various kinds of chaotic systems of stars and interstellar matter called **irregular galaxies** (Irr) that do not fit into the classes described above. The Large and Small Magellanic Clouds (Figure 17.7) make good examples. These are small gravitational companions to our own Galaxy, are visible to the naked eye from the southern hemisphere, and were first reported to Europe by the remnants of Ferdinand Magellan's crew. The Large Magellanic Cloud (LMC) was the site of Supernova 1987a (see Section 15.6.1).

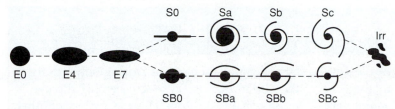

Figure 17.6
Hubble's classification sequence has the shape of a closed tuning fork. Ellipticals merge smoothly into either spirals or barred spirals that become more open and then blend into chaotic irregulars. There is no regular evolution between kinds.

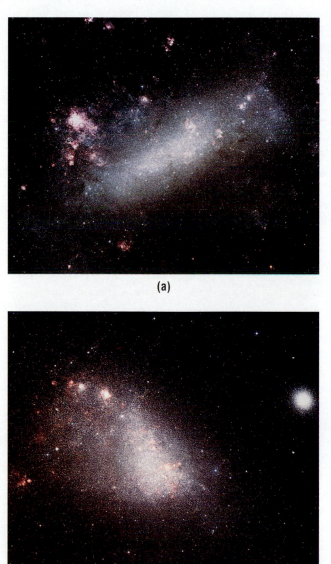

(a)

(b)

Figure 17.7
Our Galaxy has a pair of close companions, both irregular galaxies, **(a)** the Large Magellanic Cloud (LMC) and **(b)** the Small Magellanic Cloud (SMC). The great globular cluster 47 Tucanae appears just to the right of the SMC.

BACKGROUND 17.1   Observe a Galaxy

Galaxies extend to the visible limit of the Universe, and you might be tempted to think of them as exotic things to which you cannot very well relate. The truth is quite the opposite. Galaxies may be distant, but they are luminous and can easily be seen. Start with the naked eye and on a clear moonless night look for M 31 in Andromeda (see Figures 1.12 and 13.23). (Its position is marked on the star maps in Appendix 1.) A view with binoculars in a dark sky discloses an amazing sight, M 31's disk stretching over a degree. The Large and Small Magellanic Clouds are not visible from the continental United States, but they can be seen with the naked eye from the tropics, and the LMC is visible even from Hawaii.

Although the Triangulum spiral M 33 (see Figure 17.3b) is not ordinarily classed as a naked-eye object, you can in fact see it if conditions are right (see the star maps). You need an absolutely clear, dust-free sky with no Moon and no artificial lighting. Locate it first with binocu-

lars, then pull them away. Do not look directly at the object, but use *averted vision*, as the eye is more sensitive just off center. A variety of other galaxies are also visible with binoculars, including M 81 (see Figure 17.5a, center). With even a small telescope you can roam through the Virgo cloud at random and find a variety of smudges that belong to the cluster.

It is often disappointing to people that telescopic images do not look like the photographs that grace so many books. A telescopic view of M 31 still shows you only the bulge and the inner disk. However, with an instrument of 12-inch aperture, you can view the spiral arms of the Whirlpool Nebula, M 51 (see Figure 17.1).

The excitement of observing galaxies lies not so much in viewing detail but in realizing that you can encompass the light of billions of stars in one sight and that you are looking back into time at light that left the systems before human beings ever walked the Earth.

17.3 DISTANCES

To learn the physical natures of galaxies, to measure radii and masses, we need to know their distances. To find distance, we try to resolve the Galaxy into stars or other recognizable objects whose luminosities (that is, whose absolute magnitudes) we have found from cluster main-sequence fitting (see Section 13.6.2) or from other means. We then measure the apparent magnitudes of these **distance indicators,** apply a correction for absorption of light by dust, and determine distance from the magnitude equation (see Section 13.2). Good, and commonly used, indicators are O and B stars, RR Lyrae stars (see Section 15.2), novae, and the brightest planetary nebulae (see Section 15.4), whose maximum absolute magnitudes are about the same from one galaxy to the next.

The most important distance indicators, however, are Cepheid variable stars. In 1912, Henrietta Leavitt of Harvard examined the Cepheids in the Small Magellanic Cloud (SMC). Because all the stars within each of the clouds have about the same distance, the clouds make superb natural laboratories that allow astronomers to compare relative stellar properties. Leavitt found that the longer-period Cepheids are also the brighter. When she graphed the average apparent magnitudes of Cepheids against the logarithms of their periods (Figure 17.8a), she saw the resulting relation to be nearly a straight line. Since all the Cepheids in the SMC have the same distance, their average *absolute* magnitudes must depend on period in the same way. There are a few Cepheids in open clusters in our own Galaxy. From main-sequence fitting, we can find the clusters' distances and the Cepheids' absolute magnitudes. Because we know the periods of the Galaxy's Cepheids, we can place them on Figure 17.8a and find the absolute magnitudes that correspond to the observed apparent magnitudes. Adding the Cepheids of the LMC, we finally establish the **period-luminosity relation** in terms of absolute visual magnitude (Figure 17.8b).

The period-luminosity relation follows from the simple fact that the more luminous Cepheids are larger, so it takes them longer to pulsate. The relation allows us to find the distance of any Cepheid. We measure its period and average apparent magnitude, read the absolute magnitude from the period-luminosity relation, and apply the magnitude equation. The difference between the absolute and apparent magnitudes of the Cepheids

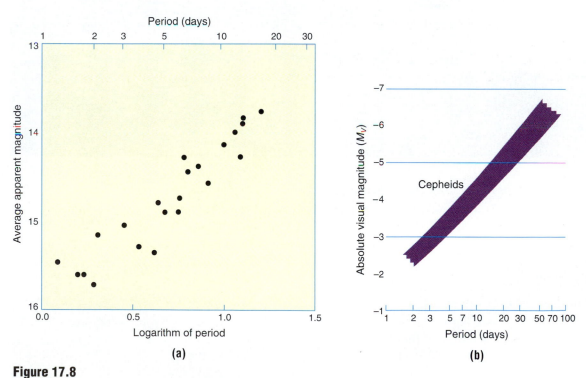

(a)   (b)

Figure 17.8

(a) Henrietta Leavitt showed that the average apparent magnitudes of the Cepheids in the Small Magellanic Cloud correlate with the logarithms of their periods. The actual periods are shown across the top.
(b) Measurement of the distance of the Large Magellanic Cloud allows us to find how *absolute* visual magnitude is related to period.

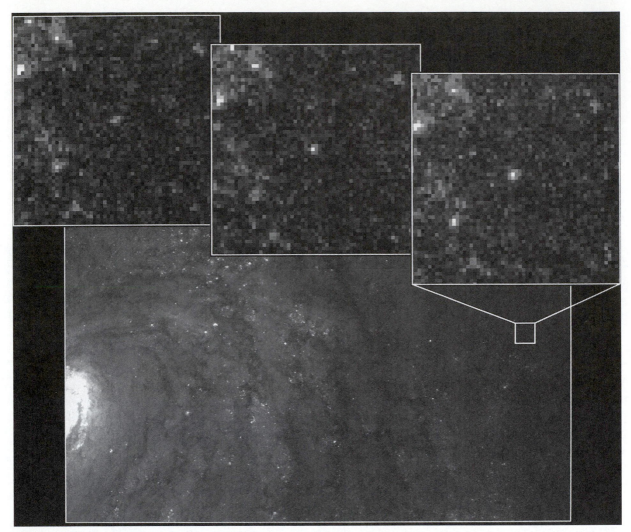

Figure 17.9
Hubble Space Telescope observations of Cepheids (like the one seen varying in the insets) showed the galaxy M 100 (a prominent spiral in the Virgo Cluster) to be 15.6 Mpc away.

in the <u>Magellanic Clouds</u> also gives us the clouds' distances. From this method and from spectroscopic distances, we find the LMC to be 52 kpc away, the SMC 58 kpc. Hubble used Cepheids in his first determination of the distance of M 31 (see Figure 17.3a), now measured at 690 kpc. The supergiant Cepheids are so brilliant they can be seen over 15 million parsecs (15 *megaparsecs*) away (Figure 17.9), allowing us to begin to step across the vast reaches of the Universe.

Beyond about 20 Mpc, we can no longer see a galaxy's individual stars. However, we can still find its globular clusters (Figure 17.10a),

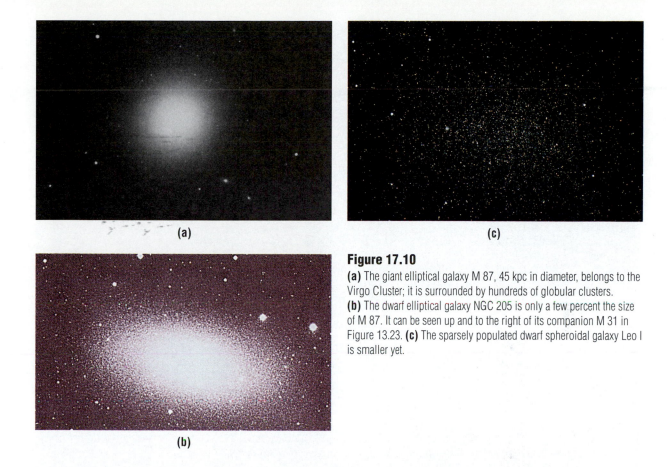

(a)

(c)

(b)

Figure 17.10
(a) The giant elliptical galaxy M 87, 45 kpc in diameter, belongs to the Virgo Cluster; it is surrounded by hundreds of globular clusters. **(b)** The dwarf elliptical galaxy NGC 205 is only a few percent the size of M 87. It can be seen up and to the right of its companion M 31 in Figure 13.23. **(c)** The sparsely populated dwarf spheroidal galaxy Leo I is smaller yet.

whose absolute magnitudes are inferred from galactic studies. Globulars are so bright they can be observed to 100 Mpc. Moreover, they are also seen in elliptical galaxies, which, since they lack star-forming disks, do not contain youthful luminous distance indicators such as O and B stars and supergiant Cepheids.

As distances increase, even the globulars become difficult, if not impossible, to see. White dwarf supernovae (see Section 15.6.2) are potentially among the most important distance indicators, since they are thought to reach a common absolute visual magnitude and can be seen over enormous distances. They are relatively rare, however, and can be used only for the few galaxies in which they appear. A more general method is to determine typical absolute magnitudes of different classes of galaxies. (We can calculate absolute magnitudes of large objects like galaxies by assuming that all their light comes from a starlike point.) For example, M_V for the Andromeda Galaxy is -21.1. If we find a faraway Sb spiral similar to M 31, its distance can be estimated from its apparent magnitude and the magnitude equation.

TABLE 17.1
A Collection of Galaxies

The Local Group[a]

Name	α[b] (h)	(m)	δ[b] (°)	(')	Class	Distance (kpc)	Diameter[c] (kpc)	Mass[c] ($M_\odot$)	Absolute Visual Magnitude (M_V)	v_r (km/s)
Milky Way	—		—		Sb	—	24	2×10^{11}	−21.1	—
LMC	5	24	−69	46	Irr	50	6.5	1×10^{10}	−17.7	−270
SMC	0	52	−72	53	Irr	55	2.9	2×10^{9}	−17.7	168
Draco	17	20	+57	54	E3/d[d]	70	0.3	1×10^{5}	−8.5	—
Ursa Minor	15	14	+67	07	E5/d	70	1	1×10^{5}	−9	—
Sextans	10	14	−01	37	d	80	—	—	−10	—
Carina	06	42	−50	57	d	85	—	—	−9	—
Sculptor	01	00	−33	41	E3/d	90	1	3×10^{6}	−12	62
Fornax	2	40	−34	34	E3/d	170	2	2×10^{7}	−13	40
Leo I	10	08	+12	21	E3/d	230	1	4×10^{6}	−11	—
Leo II	11	13	+22	10	E0/d	230	1	1×10^{6}	−9.5	—
Phoenix	01	51	−44	28	Irr	390	—	—	−10	—
NGC 6822	19	44	−14	50	Irr	500	2	3×10^{8}	−15.6	−40
IC 1613	01	04	+02	04	Irr	660	2	3×10^{8}	−14.8	−240
M 31	00	42	+41	13	Sb	690	20	3×10^{11}	−21.1	−275
NGC 147	00	33	+48	27	E5	690	1	1×10^{9}	−14.8	−250
NGC 185	00	39	+48	17	E3	690	1	1×10^{9}	−15.2	−300
NGC 205	00	40	+41	38	E5	690	2	8×10^{9}	−16.3	−240
M 32	00	42	+40	49	E2	690	1	3×10^{9}	−16.3	−210
M 33	01	33	+30	36	Sc	720	10	1×10^{10}	−17.8	−190
IC 10	00	20	+58	45	Irr	1,260	2	1×10^{10}	−15.3	−343

Prominent Galaxies Not in the Local Group

Name	Constellation	α[b] (h)	(m)	δ[b] (°)	(')	Class	Apparent Visual Magnitude (V)	Distance (Mpc)	Diameter[c,e] (kpc)	Mass[c] ($M_\odot$)	v_r (km/s)
M 81	Ursa Major	09	55	+69	04	Sb	8.3	1.4	8.3		−43
M 83	Hydra	13	37	−29	52	Sc	8	4.7	16	1.2×10^{11}	+520
NGC 5128	Centaurus	13	25	−42	59	E0	8	5	20	2×10^{11}	+526
M 101	Ursa Major	14	03	+54	20	Sc	8	5.4	35	1.6×10^{11}	+231
M 82	Ursa Major	09	55	+69	41	Irr	9.0	5.2	16	—	+210
M 51	Canes Venatici	13	30	+47	15	Sc	7.4	7.7	27	4×10^{10}	+470
M 87	Virgo	12	31	+12	23	E0	10	16	45	4×10^{12}	+1,260
M 104	Virgo	12	40	−11	38	Sa	8	20	49		+1,130

[a]Nine additional dwarf spheroidals and irregulars are excluded from the list.

[b]See the introduction to the star maps in Appendix 1.

[c]Diameters and masses relate to the optically bright parts of the galaxies and do not encompass much dark matter; given only for comparison with our own Galaxy.

[d]"d" denotes dwarf spheroidal.

[e]Long axis.

This method is limited because any class of galaxy has a great range in size and absolute magnitude, the spirals, for example, ranging from the magnificent examples seen in Figure 17.5 to vast sprawling massive systems so large and dim they are barely detectable. The ellipticals are the most extreme. At one end we find brilliant, massive **giant ellipticals** that considerably outshine our own (see Figure 17.10a). From these we descend in luminosity through smaller **dwarf ellipticals** like M 31's companions M 32 and NGC 205 (Figure 17. 10b), to the dim **dwarf spheroidals** (Figure 17. 10c) that are tens of thousands of times fainter than our Galaxy and not much more luminous than big globular clusters. Moreover, at great distances it becomes difficult even to classify galaxies. As a result, estimates of large distances by this simple method are subject to considerable error.

Once the distances to galaxies are known, we can easily derive their diameters in kiloparsecs from their angular sizes. Distances and diameters for a variety of galaxies out to about 50 Mpc are given in Table 17.1.

17.4 CLUSTERS OF GALAXIES

Galaxies have a powerful tendency to gather together. The irregular Magellanic Clouds (see Figure 17.7) orbit our Galaxy, and M 31 is similarly accompanied by M 32 and NGC 205. Multiples like Stephan's Quintet (Figure 17.11) are common. The real organization of the Universe, how-

Galaxies have a powerful tendency to gather together.

Figure 17.11
Four galaxies of Stephan's Quintet epitomize small groups of galaxies and their interactions.

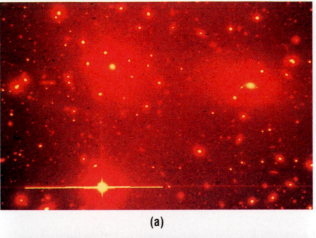

(a)

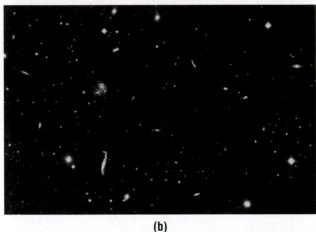

(b)

Figure 17.12
(a) The rich Coma Berenices Cluster is highly concentrated at the center and dominated by a pair of elliptical galaxies. **(b)** The Hercules Cluster is a spread-out system with many spirals.

ever, involves gravitationally bound **clusters of galaxies** (Figure 17.12) that range from poor groups of a dozen or so members to rich assemblies like the great Virgo Cluster (named after its constellation of residence), which has more than 1,000 galaxies and is dominated by the giant ellipticals M 87, M 84, and M 86. With a distance of about 16 Mpc (determined from Cepheids like the one in Figure 17.9) and an angular extent of some 12°, the Virgo Cluster is roughly 4 Mpc across. Especially prominent clusters are listed in Table 17.2.

Galaxy clusters exhibit considerable variety. The rich Coma Berenices Cluster (Figure 17.12a) contains mostly elliptical systems: highly concentrated toward the center, it is also dominated at its center by giant elliptical galaxies. Others, like the Hercules Cluster (Figure 17.12b), contain mostly spirals and are not centrally condensed. We belong to a sparsely populated cluster, the **Local Group** (Figure 17.13 and Table 17.1). It is dominated by two great spirals, our Galaxy and M 31, each of which has numerous small companions. M 33 (see Figure 17.3b) lies near M 31, and a few small assemblies hover well outside the boundaries of Figure 17.13.

TABLE 17.2
Some Important Clusters of Galaxies

Name	Distance	Characteristics
	(Mpc)	
Local Group	...	Small cluster; bright spirals
Virgo	16	Rich, loose cluster; ellipticals
Coma Berenices	100	Rich cluster; ellipticals
Hercules	145	Loose cluster; many spirals
Corona Borealis	300	Rich cluster; ellipticals
Hydra	800	Rich cluster, ellipticals

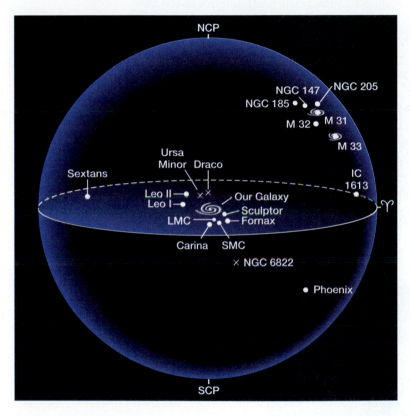

Figure 17.13
The main portion of the Local Group of galaxies is placed within the celestial sphere, with our Galaxy at center. Galaxies noted by **X** are in the foreground; the others are in the background.

Clusters allow us to extend the distance scale. Since we know the distance of the Virgo Cluster, we know the absolute magnitudes of its galaxies. If we find another cluster with a similar structure, we assume that its brightest galaxies have the same absolute magnitudes as Virgo's. We can then find the cluster's distance from its galaxies' apparent magnitudes and the magnitude equation. The method allows us to estimate the distances of clusters that are more than a billion parsecs away.

With the distances known, we can assess the numbers of the different kinds of galaxies per unit volume and find that 60% of galaxies are ellipticals and 20% are spirals. Of the ellipticals, the vast majority are dwarf ellipticals or dwarf spheroidals. Clusters of galaxies, and the Universe at large, appear to be dominated by impressive spirals and giant ellipticals, but most of the systems are the faint dwarfs and irregulars that we cannot even see once the distances are sufficiently large.

17.5 DYNAMICS AND MASSES

A galaxy's mass is a powerful factor in controlling its development. To find the mass of a spiral galaxy is surprisingly easy. Because the galaxy in Figure 17.14a is rotating, one side is coming at the Earth relative to the average speed and the other side is moving away. Spirals are filled with diffuse nebulae that radiate emission lines from which we can find velocities from the Doppler effect. We simply measure the velocities outward along the disk relative to the velocity of the central bulge and correct them for the galaxy's tilt to the line of sight. The result, presented as a graph, is the galaxy's rotation curve (Figure 17.14b). The rotation velocity at any distance from the center lets us calculate the mass interior to that radius, just as it does for our own Galaxy (see Section 16.5).

Elliptical galaxies are more difficult to evaluate because they exhibit little rotation as a unit. However, the stars of any galaxy orbit the center at different velocities that result in different Doppler shifts. If we examine the spectrum of a galaxy's combined light, we find that the absorption

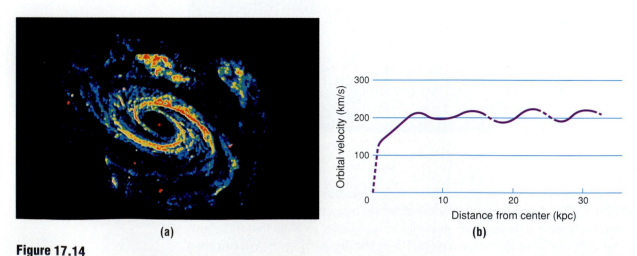

(a) (b)

Figure 17.14
(a) M 81 is a rotating spiral galaxy tilted to the line of sight. The parts colored blue are approaching relative to the central bulge, the parts colored red receding. **(b)** The rotation curve of NGC 2998 is, except for some wiggles, relatively constant outward from the center, implying considerable dark matter.

lines are smeared out or broadened. From the degree of broadening we can infer the spread of stellar speeds, from which we can (with the aid of theory) calculate the strength of the gravitational field and the mass.

Table 17.1 presents masses appropriate to the galaxies' visually bright interior portions, those brightly filled with stars, the *luminous masses*. They are given only for comparison with the stellar mass of our Galaxy. Our Galaxy's mass is exceeded by that of M 31; M 87, the giant elliptical in the Virgo Cluster, has over 10 times the luminous mass of our system. At the other extreme, the irregulars and dwarf spheroidals have luminous masses only a hundred-thousandth of ours. Although there are more small galaxies than large ones, their combined mass is not too significant. The vast majority of the stellar mass in the Universe belongs to the big spirals and ellipticals.

However, the amount of luminous mass—illuminated matter—is only a part of the story. Rotation curves derived both from visual spectra and from the 21-cm line of neutral hydrogen usually do not fall with distance from the center as expected from Kepler's laws, but instead remain constant even as the luminosities of the galaxies, and the masses of stars and interstellar matter, drop nearly to zero (see Figure 17.14b). As in our own Galaxy (see Section 16.5), there is a great deal of dark matter that cannot be seen or accounted for. Observations of the tidal effects that spirals have on their small satellite galaxies suggest dark matter halos that extend over 100 kpc outward and contain 10 times as much dark matter as illuminated.

We find comparable amounts of dark matter in elliptical galaxies. Some ellipticals (Figure 17.15) are enmeshed in large X-ray–emitting halos of hot gas at temperatures of over 10^6 K. The gas itself does not have much mass, but its very existence means that it must be trapped in the grip of a powerful gravitational field, from which the mass of the system can be cal-

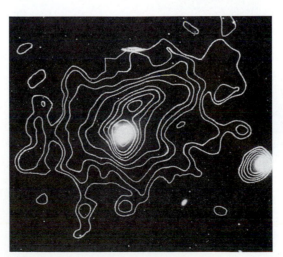

Figure 17.15

This view of the central region of the Virgo Cluster is only 50 minutes of arc across, yet it contains two massive ellipticals (M 84 at right, M 86 near the center) and half a dozen bright spirals. M 87 lies off the picture to the lower left. The cluster has a linear extent 30 times larger than shown here. The contour lines indicate the strength of an X-ray-emitting halo around M 86 from which the mass of the galaxy can be found.

culated. M 87 (see Figure 17.10a) may have a total mass of 3×10^{13} solar masses, over 10 times the amount found in its stars.

Clusters of galaxies tell the same story. In clusters, individual members orbit a common center of mass. The larger the cluster's total mass, the faster the galaxies must move in orbit and the greater the spread of observed radial velocities, from which the cluster's total mass can be found. Once again, the clusters typically contain 10 times the combined luminous masses of the individual galaxies. We cannot see the dark matter, and we do not know what composes it. Confoundingly, a few galaxies seem to have hardly any at all.

17.6 GALACTIC NUCLEI

The nucleus of our own Galaxy may contain a massive black hole (see Section 16.4); we might be more convinced if other galaxies exhibited evidence for them too. M 31 is so close that an inner region only 2 pc across can be resolved. The velocities around this core imply a mass 10^7 times that of the Sun, strongly suggestive of a black hole.

Several similar systems have been observed. Velocity measurements made a mere 20 pc from the center of M 87 with the Hubble Space Telescope reveal a rotation speed of 550 km/s (Figure 17.16). Within that tiny volume lie between 2 and 3 billion solar masses. We know of nothing that could cause that kind of compaction other than a central black hole. Indeed, it is beginning to appear likely that most—or even all—galaxies above a certain mass have them. Most of these central black holes sit there quietly. Some, however, are responsible for the spectacular phenomena to be described in Section 17.8.

Within 20 pc of the center of M 87 lie between 2 and 3 billion solar masses.

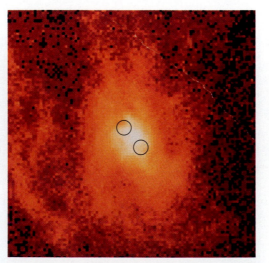

Figure 17.16
This view of the inner 1% of M 87 (see Figure 17.10a) is only 300 pc across. The circles indicate where the galaxy was observed with the Hubble Space Telescope to have a rotation speed of 550 km/s. Such a speed only 20 pc from the center suggests that the nucleus is a black hole of 2 to 3 billion solar masses.

17.7 THE ORIGINS OF GALACTIC FORMS

Why are there so many different galactic forms? The answer—or answers—is far from clear. One possibility is that the initial clouds from which the galaxies developed may have had different values of angular momentum. The clouds that had slow rotations or low angular momentum fell in on themselves to produce ellipticals and irregulars and never had enough spin to make disks and spiral arms. As a result, star formation proceeded quickly. These galaxies therefore exhausted much of their interstellar material in their youths, and winds and explosions resulting from stellar evolution blew the rest of it away. The clouds that formed the spirals, on the other hand, were spinning more quickly and followed the sequence that led to the formation of our Galaxy (see Figure 16.18). It is not at all clear, however, why some galaxies have bars and others do not. Computer simulations show that bars seem to develop naturally as a result of gravity and the conservation of angular momentum. The real question may be: Why do many galaxies *not* have bars? Or do they? Perhaps when they are looked at more closely most galaxies will be found to have bars. Another controlling factor is mass. Smaller systems take on the forms of dwarf ellipticals and irregulars. Moreover, too much dark matter apparently suppresses spiral arms.

Collisions and mergers appear to play a powerful role. Sprinkled across the sky are hundreds of *peculiar galaxies,* ones that do not fit standard categories. In most cases, they are pairs of **interacting galaxies** that affect each other tidally or even undergo collisions (Figure 17.17). Stars, however, do not collide. Compared to the distances between them, they are essentially points that rarely come close to one another. But the separation between our Galaxy and M 31 is only 30 times the galaxies' diameters (depending on where you establish the vaguely defined "edges"). In rich clusters, galaxies are even closer. The odds of a galactic collision are therefore high. In such an event, the stars pass by one another with no direct hits. In spirals, however, the interstellar clouds are large enough to bump, sweeping gas from the colliding systems as they pass through one another.

Figure 17.17a (as well as Figure 17.11) shows that the interaction can raise huge tides in the systems and throw long streamers of matter into intergalactic space. Galaxy interactions may be a reason for gas-poor spirals within galaxy clusters and may be instrumental in making the giant ellipticals that lie at the hearts of many rich clusters. Much of the gas removed falls to the center of the cluster, where it can be consumed by one or more of the massive central galaxies, which then grow at the expense of the others. The disturbances produced in such encounters can trigger intense bursts of star formation (Figure 17.17b). The evolution of our own Galaxy may have been profoundly influenced by such collisions and mergers (see Section 16.6).

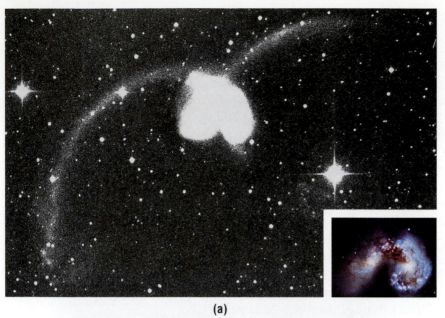

(a)

Figure 17.17
(a) NGC 4038 and 4039 in Corvus are caught in the act of collision, the tidal forces spreading gas and stars into great long streamers. **(b)** The interaction in the center of the system, viewed here through the Hubble Space Telescope, promotes rapid star formation within the tidally distorted spiral arms.

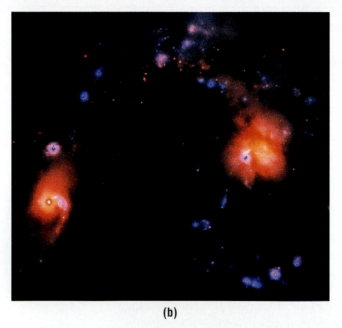

(b)

If the circumstances of the encounter are right, the galaxies can merge. NGC 6240 (Figure 17.18) is probably the combination of two spirals. Much of the matter may also fall to the center to feed a black hole that may have been at the nucleus of one or both of the galaxies. NGC 6240 might harbor a black hole with a mass of 100 billion solar masses! After the merger, a pair of spirals may fall in a heap (Figure 17.19), resulting in an elliptical galaxy.

Figure 17.18
NGC 6240 is the result of the merger of a pair of spirals. Its center may harbor a black hole of 100 billion solar masses.

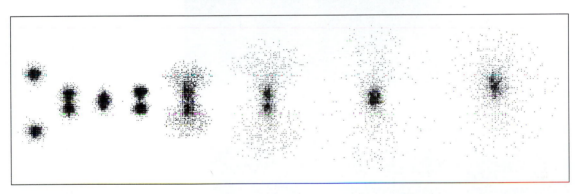

Figure 17.19
Computer-simulated spiral galaxies collide head-on. As a result of gravitational forces, the two merge into a large elliptical.

17.8 ACTIVE GALAXIES

Sprinkled in among the countless galaxies is a variety of **active galaxies**, those that have bright nuclei associated with great energy production or energetic outflows of matter. *Seyfert galaxies* (Figure 17.20a) constitute about 1% of the spirals and are generally confined to those with more tightly wound arms. Their central regions have bright starlike nuclei and display strong emission lines. A large fraction has broad, spread-out hydrogen lines that indicate high gas velocities, up to 10,000 km/s. Powerful continuous spectra underlying the emissions extend through the ultraviolet into the X ray. The continuous spectra have no absorption lines, so are not produced by stars (which *do* have absorption lines).

The brightest Seyfert galaxies radiate up to a trillion solar luminosities from a region a few thousand AU wide.

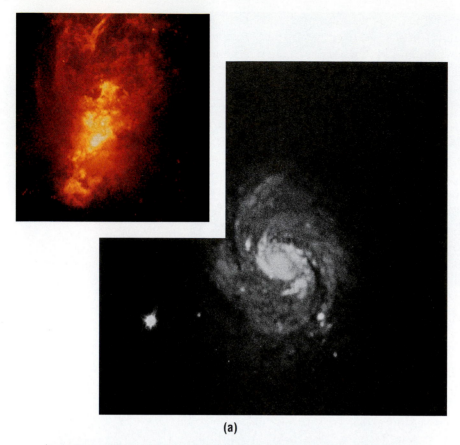

(a)

Figure 17.20
(a) Within the central bulge of the Seyfert galaxy NGC 1068 is a brilliant nucleus. The inset, an image made with the Hubble Space Telescope, shows the nucleus surrounded by tubular clouds of ionized gas.
(b) A Seyfert galaxy is probably powered by a mass-accreting black hole at its center that is successively surrounded by a rapidly spinning disk containing high-velocity gas that produces broad hydrogen emission lines, a ring of dust, and a more slowly rotating (low-velocity) disk that radiates narrow emission lines. Radiation can escape perpendicular to the disk to illuminate interstellar gas, as seen in the inset to (a).

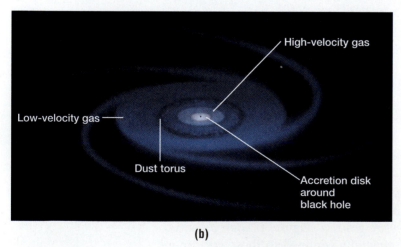

(b)

The broad lines vary in brightness with periods on the order of weeks. For a body to vary in a coherent fashion, its parts must communicate—exchange information—with one another. The light-travel time across the emitting region must therefore be less than the time it takes for the body to change its brightness: if the body varies with a period of a day, then it can be no larger than one light-day (the distance light travels in a day) across.

The broad emission lines come from regions only a few light-weeks—a few thousand astronomical units—wide. The continuous spectra can vary on time scales of minutes! The brightest systems radiate up a trillion solar luminosities, all from a region that may be at most a few thousand AU wide.

Seyfert nuclei are too small and bright to be energized by stars and thermonuclear fusion. However, we have already established the likelihood of massive black holes at the centers of galaxies. A good explanation for Seyfert activity suggests that large amounts of matter—disrupted stars and interstellar gas—fall erratically into an energetic accretion disk around a black hole of 10^7 or so solar masses (Figure 17.20b). The radiation from the accretion disk produces the bright continuum and ionizes the gas clouds around it. The broad hydrogen lines are produced by gas rapidly orbiting within a few parsecs of the black hole. Surrounding this region is a doughnut, or torus, of dust that radiates in the infrared. Radiation from the accretion disk escapes through the poles of the disk, ionizing and illuminating interstellar gas in the perpendicular direction (Figure 17.20a inset). We see a continuum of this kind of behavior from the bright Seyferts down to galaxies like our own.

Radio galaxies are commonly found among the ellipticals. Most galaxies, ours included, emit much more optical radiation than radio. The opposite is true for radio galaxies. M 87 (see Figure 17.10a) is a superb example. A straight jet of matter that appears to be about 2.5 kpc long, roughly 10% of the galaxy's radius, streams from its nucleus (Figure 17.21). The jet radiates powerfully at radio wavelengths and is seen even in the X-ray part of the spectrum. Radio spectra show that the jet is producing synchrotron radiation (see Section 9.2.1), so it must be related to magnetic fields and energetic electrons moving near the speed of light. The jet is not smooth but broken into knots. Proper-motion studies with radio telescopes suggest that the knots are shock waves moving outward through an ionized gas at nearly half the speed of light, implying enormous generating energy. There is also a faint counterjet pointing in the opposite direction.

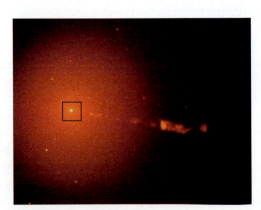

Figure 17.21

M 87's bright jet pours from the brilliant galactic nucleus in this near-infrared image taken with the Hubble Space Telescope. Knots are regularly spaced from the center outward. The black square shows the size of the inner region illustrated in Figure 17.16.

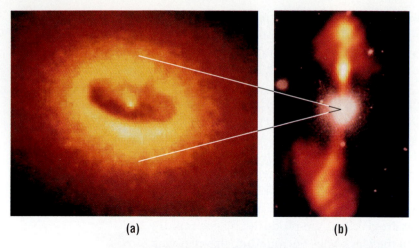

Figure 17.22
(a) The Hubble Space Telescope reveals a dusty accretion disk only 100 pc across within the elliptical galaxy NGC 4261. **(b)** An expanded view shows an optical image of NGC 4261 at the center combined with radio images that disclose 12-kpc-long jets flowing away from the galaxy perpendicular to the disk.

(a)   (b)

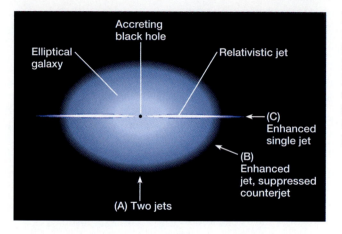

Figure 17.23
If an active elliptical galaxy is viewed from the side (A), both jets will be visible. If viewed from an angle (B), the jet coming toward the observer will be enhanced, the one going away suppressed. If viewed head-on (C), the observer sees the jet as a brilliant point set against the galaxy.

The phenomenon is even better revealed by the elliptical galaxy NGC 4261 (Figure 17.22a), in which we see a dusty disk surrounding a bright spot at the center. Mass apparently spirals from the outer disk into a compressed hot accretion disk immediately surrounding the supposed black hole. It is thought that magnetic fields twisted along the rotation axis cause mass to be expelled perpendicular to the disk in a tight bipolar flow (Figure 17.22b).

The imbalance in the brightnesses of M 87's jets is explained by orientation. Relativity theory shows that radiation from a beam of matter moving toward the observer at nearly the speed of light will be greatly amplified, whereas that from matter rapidly moving away will be suppressed. Therefore, M 87's central black hole is so oriented that the jets lie more or less along the line of sight (Figure 17.23, view B) and are consequently much longer than they appear in Figure 17.21. NGC 4261's jets, however, are more nearly perpendicular to the line of sight (Figure 17.23, view A), consistent with the orientation of the dust ring in Figure 17.22. This *relativistic beaming* reaches an extreme in the *BL Lacertae objects* (Figure 17.24),

Figure 17.24
BL Lacertae (once thought to be a variable star) consists of a brilliant variable stellar nucleus surrounded by the faint stellar component of a giant elliptical galaxy.

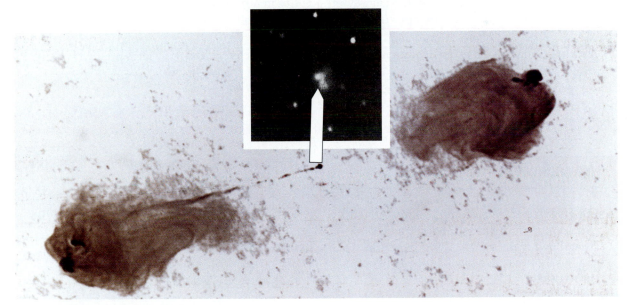

Figure 17.25
A radio image of Cygnus A shows a jet emerging from the left side of the bright spot at the center and blossoming into a huge radio lobe. The lobe on the other side is fed by an invisible jet. (inset) An optical closeup of the center shows a peculiar distant elliptical galaxy.

in which brilliant starlike nuclei lie at the centers of faint giant elliptical galaxies, an amplified relativistic jet coming straight at the observer (Figure 17.23, view C).

In some radio galaxies, the jets extend for vast distances, each ending in huge lobes illuminated in the radio spectrum by the synchrotron mechanism. Cygnus A (Figure 17.25) is a brilliant celestial radio source identi-

Jets from active galaxies can reach distances of megaparsecs from the galactic nuclei.

fied with a modestly faint, distant galaxy. Radio observations show a prominent, relativistically beamed jet pointing straight from the galaxy into one of the lobes. There must be one coming from the other side as well, forming a galactic bipolar flow. The jets shovel intergalactic material in front of them, finally braking to a halt within chaotic lobes, much in the way that Herbig-Haro objects (see Section 14.2) are formed by bipolar flows from new stars. These structures are among the largest in the Universe, some reaching distances of *megaparsecs* from the galactic nuclei. The jets are the straightest structures known and can extend on a perfect line from the lobes right down into the galactic cores.

If these theories are correct, an active galaxy requires a significant input of raw material to feed the black hole. In giant ellipticals that reside in clusters, like M 87, the nourishment for the black hole may be mass stripped from other galaxies in collisions, the same mass that made the galaxy and the black hole so big to begin with (see Section 17.7). In other systems, the collision itself might send fresh matter (and, as in NGC 4261, dust) into the center. A large amount of mass that begins to fall into the central black hole might suddenly trigger Seyfert activity in a previously quiet galaxy. Several Seyferts are apparently the products of collisions as well.

Normal galaxies have an insufficient mass flow to be active, thus starving "the monster in the middle." Yet some galactic nuclei can be so active as to be visible across much of the Universe, a subject into which we now expand our sight.

▶ KEY CONCEPTS

Active galaxies: Galaxies with bright, variable nuclei sometimes associated with bipolar jets; they are probably caused by matter falling into massive black holes from accretion disks.

Clusters of galaxies: Gravitationally bound groups of galaxies.

Distance indicators: Objects with known absolute magnitudes that can be used to find galaxy distances.

Elliptical galaxies (E): Galaxies without disks and spiral arms, ranging in mass and luminosity from **giant ellipticals** through **dwarf ellipticals** to **dwarf spheroidals.**

Galaxies: Self-contained collections of mass that include stars, interstellar matter, and dark matter.

Hubble classification: A system of classifying galaxies that branches from ellipticals through the two kinds of spirals to irregulars; the scheme is depicted by the **tuning fork diagram.**

Interacting galaxies: Galaxies that interact by tides or collision.

Irregular galaxies (Irr): Galaxies with little structure.

Local group: The sparse cluster that contains the Galaxy.

Period-luminosity relation: The correlation between the periods of Cepheids and their average absolute magnitudes, which allows derivation of distances.

Spiral galaxies: Galaxies displaying disks, spiral arms, and a large youthful component; divided into **barred spirals** (SB) (those with central bars) and **normal spirals** (S) (no bars).

EXERCISES

Comparisons

1. What are the similarities and differences among Sa, Sb, and Sc spirals? Between normal and barred spirals?

2. Compare the kinds of stars found in spiral and elliptical galaxies.

3. How do irregular galaxies differ from S0 galaxies?

4. Compare the extreme forms of elliptical galaxies.

5. Compare different kinds of clusters of galaxies.

6. List the similarities and differences between **(a)** active and nonactive galaxies; **(b)** radio galaxies and Seyfert galaxies.

Numerical Problems

7. What is the diameter of a galaxy 5 minutes of arc across and 10 Mpc away?

8. What do you expect the apparent magnitudes of the brightest stars in M 32 to be? State your assumptions.

9. If the Sun were in M 31, what would its apparent magnitude be?

10. What fraction of the mass of the Local Group is in **(a)** M 31; **(b)** the combined spirals?

11. Select a Cepheid variable from Appendix Table A2.5 and estimate its distance from the period-luminosity relation.

12. The magnitude of the nucleus of a distant galaxy varies with a period of five days. What is the maximum size of the emitting region in AU?

Thought and Discussion

13. Why would an E0 galaxy not necessarily be spherical?

14. Describe the steps that allow astronomers to measure the distance to the Virgo Cluster of galaxies.

15. How can astronomers derive the masses of spiral and elliptical galaxies?

16. List the evidence for dark matter in galaxies.

17. What probably produces the giant elliptical galaxies that we find at the centers of many galaxy clusters?

18. How do we know that galaxies collide? What are the major results of galaxy collisions?

19. In what ways might elliptical galaxies form?

20. How does orientation affect the jets from active galaxies? In what orientation will an observer see both jets equally well?

21. What do we think is the driving force behind active galaxies?

Research Problem

22. Examine astronomers' concepts of the distance of M 31 over the twentieth century. Find these distances and plot them against the year. Explain why measurements of the distance kept increasing up to about the middle of the century.

Activity

23. If you have access to a telescope and a good amateur star atlas in your library, spend an evening trying to find as many galaxies as you can. Sketch each and use your library to find photographs for comparison.

Scientific Writing

24. Write a two-page pamphlet in which you show how the distance scale of the Universe is established, beginning with the nearest stars. Indicate possible sources of error.

25. Write a one-page description for an astronomy course in which you showcase galaxies to interest potential students. Use two photocopied illustrations.

THE EXPANSION AND STRUCTURE OF THE UNIVERSE

Large-scale motions and the distribution of matter in the Universe

18

We now examine the motions of galaxies and their large-scale organization in three-dimensional space as we begin to explore **cosmology,** the study of the origin, construction, and evolution of the Universe.

18.1 THE VELOCITY-DISTANCE RELATION

In 1912, Vesto M. Slipher of the Lowell Observatory measured the first Doppler shift and radial velocity for a spiral nebula. M 31, the Andromeda Galaxy, was seen to be approaching at an unprecedented speed of 300 km/s. By 1925 he had observed the spectra of 41 nebulae (by then known to be galaxies). All but three exhibited **redshifts** (z) in their spectra (Figure 18.1), the spectral lines shifted toward the red, toward longer wavelengths. The redshift is specified by

$$z = (\lambda_{observed} - \lambda_{emitted}) / \lambda_{emitted,}$$

where $\lambda_{observed}$ is the observed wavelength of a spectrum line and $\lambda_{emitted}$ is the wavelength as emitted by the body, or the true or *rest* wavelength, the wavelength that would be observed if the body were not moving. Interpreting the redshifts as Doppler shifts, for which

$$v_r = c(\lambda_{obs} - \lambda_{rest}) / \lambda_{rest} = cz$$

(see Section 13.1), Slipher's results implied that the galaxies were moving away at speeds approaching 1,800 km/s, far greater than any of our Galaxy's stars. Only M 31, its companion M 32, and M 33 were seen to be approaching.

By the end of the decade, Edwin Hubble had distance estimates for two dozen galaxies. When Slipher's velocities were graphed against them

The more distant the galaxy, the faster it is receding.

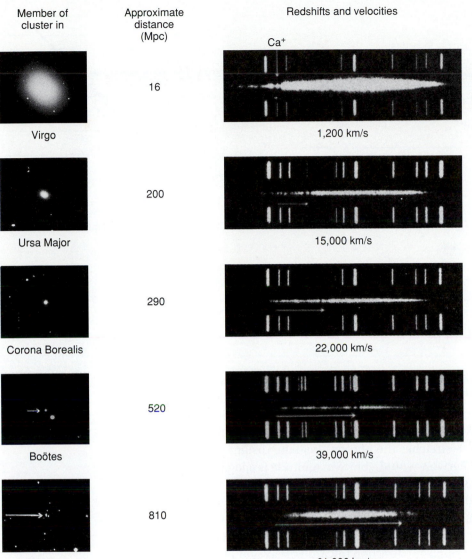

Member of cluster in	Approximate distance (Mpc)	Redshifts and velocities
Virgo	16	Ca⁺
		1,200 km/s
Ursa Major	200	15,000 km/s
Corona Borealis	290	22,000 km/s
Boötes	520	39,000 km/s
Hydra	810	61,000 km/s

Figure 18.1

The brightest elliptical galaxies of five clusters exhibit redshifts in their spectra as seen by the positions of the absorption lines of ionized calcium. The diminishing angular sizes of the galaxies show that distance increases with increasing redshift and velocity.

(Figure 18.2a), they were seen to increase together along a straight line: the more distant the galaxy, the faster it was receding (see Figure 18.1). Only within the Local Group do orbital motions make a few galaxies come toward us. By 1936, Hubble's colleague, Milton Humason, had extended this **Hubble relation** to a redshift of $z = 0.13$, which corresponds to 40,000 km/s and a distance of roughly half a billion parsecs.

Modern astronomers have extended the Hubble relation vastly farther. To avoid problems of interpretation, Figure 18.2b is based almost purely on observational data: the redshifts (z) of 82 clusters are plotted

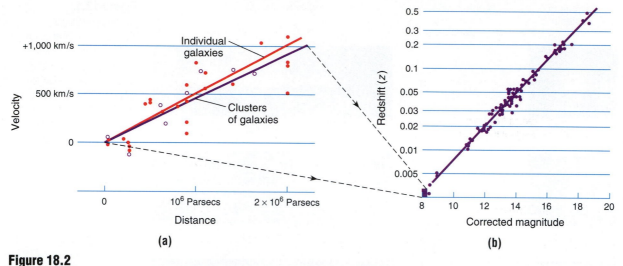

Figure 18.2

(a) When Edwin Hubble plotted the velocities of galaxies (red symbols) and their clusters (violet symbols) against distance, he discovered the expansion of the Universe. **(b)** The original diagram is a tiny portion of this more modern plot in which redshift climbs with increasing magnitude, and therefore with increasing distance, to a redshift of at least $z = 0.5$.

against the apparent magnitudes of their most luminous galaxies. These galaxies are assumed to have similar absolute visual magnitudes (see Section 17.4), so apparent magnitude becomes a measure of relative distance. The only corrections involve those for absorption by dust and for the effect of the redshift, which stretches the photons, making distant galaxies look fainter than they otherwise would. (Since magnitude is a logarithmic scale, z must be plotted similarly.)

Up to a redshift of 0.5 the observational data still fall along a straight line. Equally important, the observed clusters are found in all directions from the Earth. It does not matter if we look outward toward Virgo or Coma Berenices or in the opposite direction toward Fornax and Eridanus. We now have sufficient evidence to assume with some safety that the Hubble relation is a property of the Universe at large and is *isotropic*, or independent of direction. Everywhere we look, galaxies are flying away from us according to the same law.

Figure 18.2 shows that velocity increases in direct proportion to distance; a galaxy twice as far away recedes at twice the speed. The system of galaxies therefore appears to be *expanding*. Our Galaxy seems to be at the center, but that is an illusion. All clusters of galaxies are getting farther away from *one another*. In Figure 18.3, galaxy B is twice as far from our Galaxy as galaxy A, and galaxy C is four times as far. A, B, and C are respectively moving away from our Galaxy at speeds of 1,000 km/s, 2,000 km/s, and 4,000 km/s (black arrows). But *our* Galaxy appears to be moving away from the people in galaxy B at 2,000 km/s in the opposite di-

Figure 18.3

Our Galaxy appears to be at the center of an expanding system of galaxies (black arrows) that are shown to be moving away from us at specific speeds in km/s (black numbers). An observer in galaxy B also sees the galaxies receding (red arrows), and would also seem to be at the center of the expansion, from which the galaxies recede at the speeds indicated by red numbers.

rection. To find the speeds of A and C relative to B, subtract B's arrow from the others. You find that A moves away from B at 1,000 km/s, and C—which is twice as far away—departs at double that value, 2,000 km/s (red arrows). The speeds seen from galaxy B show the same effect as seen from our Galaxy. You would see exactly the same Hubble relation from any of the galaxies in Figure 18.3 as you see at home; in fact, you would see the same expansion *from any point in the Universe*. There is no favored position. Humanity has now received the ultimate displacement from the center of attention.

18.2 THE MEANING OF THE REDSHIFT

Even before Hubble disclosed the natures of galaxies, mathematical physicists were developing theories of the structure of the Universe using the equations of Einstein's new general theory of relativity (see Section 5.7). There are three simple possibilities: the Universe can expand, contract, or do neither. Einstein solved his equations in 1917 to obtain a universe that does neither, that is, a *static universe*. To prevent contraction and to counter the attractive force of gravity provided by the mass within it, he postulated a **cosmological force** (for which there is no laboratory evidence) whose strength *increases* with distance. However, at the same time the Dutch mathematician Willem de Sitter found a solution for an **expanding universe.** Hubble's discovery of the expansion of the system of galaxies then fit into a theoretical framework already in place.

The solutions of general relativity show that the clusters of galaxies are not moving away *into* space but *with* space, in a smooth motion called the **Hubble flow.** Clusters travel with the flow rather like chips of wood in a stream. The distinction is profound and has vast implications. Space itself is expanding, carrying the clusters along with it. However, neither the galaxies nor their clusters are getting larger, because they are held together by gravity.

Look at a two-dimensional analogy. Several people hold a sheet of rubber containing a grid of lines and colored pins representing galaxy

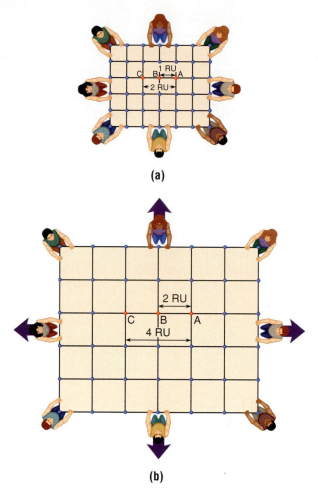

(a)

(b)

Figure 18.4
(a) A rubber sheet with a grid is stuck with colored pins. Pins A, B, and C are each one "rubber unit" (RU) apart. As the sheet is stretched, the pins ride along with it and are separated.
(b) The sheet has doubled in size, B is 2 RU from A, and C is 4 RU from A. An observer on A (and on every other pin) would see an expanding "rubber universe."

clusters (Figure 18.4a). Pins A, B, and C are each one "rubber unit" (RU) apart. The people step back and the sheet expands (Figure 18.4b), carrying the clusters with it. A and B are pulled to 2 RU apart, but B and C are *also* now 2 RU apart, so C is 4 RU from A. To move to those separations in the same amount of time, the velocity of C relative to A has to be twice that of B. An observer on A, or on *any other pin*, sees an expanding two-dimensional rubber universe.

What, however, is expanding? The grid lines in Figure 18.4 represent coordinates in the Universe that move apart with the pins. By specifying which lines the pins are on, we specify location. Mathematically, the pins are not moving because they stay at the same coordinates. It is the coordinate framework, or *space*, that is expanding, carrying the pins with it. During the expansion, pin B emits a photon in the direction of pin A. While the photon travels, "rubberspace" expands, stretching the wave. When the wave arrives at pin A, the photon's wavelength is longer than it was and the light is redshifted. The amount of redshift (z) depends on how long the photon has been in flight, or on the distance between pins. The observer on A (and again at any other point) sees increases in recession velocities in proportion to distance, or a rubber-Hubble relation.

This redshift is therefore *not* a Doppler shift, which occurs when objects change their coordinates relative to each other with time. In the expanding rubber universe, the redshift is caused by the expansion of the coordinate framework itself. The wavelength of the photon observed at pin A depends on the change in the "size" of the Universe, more properly the **scale of the Universe** (the relative distance between the grid lines) during the time the photon was in flight. The redshift therefore gives the fractional increase in the size (scale) of the Universe between now and when the light was emitted. If $z = 0$, the photon was emitted close by and the Universe could not significantly have changed its scale during its flight. However, if we observe a distant galaxy with $z = 0.5$, the grid-line distance has increased by 50%, so the Universe is 1.5 times larger now than it was when the photon was emitted.

Because the redshift is *not* a Doppler shift, we cannot properly use the Doppler formula to relate redshift to velocity. Instead, we must use an equation derived from the theory that provides our description of the Universe, the general theory of relativity. However, there is a snag: relativity shows us that spacetime can be curved, and that gravity is the result of spacetime curvature caused by the presence of mass (see Section 5.4). Any formula relating redshift and velocity *depends on the exact nature of the overall spacetime curvature of the Universe*, a subject to be dealt with in Chapter 19. An example, from a model of the Universe devised by Einstein and de Sitter in 1932, is shown by the red curve in Figure 18.5. Fortunately, any of these equations become nearly identical to the Doppler formula (the blue line in Figure 18.5) for low z (below about 0.3 or so), allowing us to make simple calculations for nearby galaxies. Note from the relativistic curve

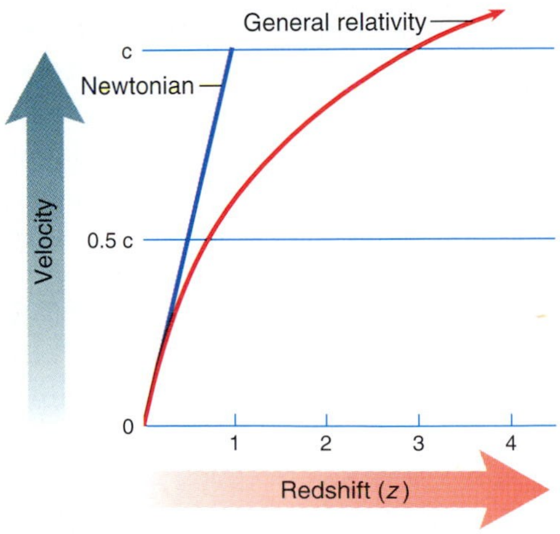

The redshifts of distant galaxies are not Doppler shifts.

Figure 18.5
According to Newtonian theory and the ordinary Doppler formula (blue line), the relation between velocity and redshift is a straight line, and a body with $z = 1$ recedes at the speed of light. The relation between z and velocity is different in an expanding universe, in which we must adhere to the principles of general relativity. The red curve shows the relation derived from a model of the Universe developed by Albert Einstein and Willem de Sitter. As z increases, the expanding Universe can exceed the speed of light. The Doppler formula can still be applied at low z, however.

(the one derived from relativity theory) that the expansion of the Universe *can exceed the speed of light*. Such an expansion violates no law, because it involves the expansion of space itself and not the objects within it.

18.3 THE HUBBLE CONSTANT

The most important quantity describing the expansion of the Universe is its rate, the **Hubble constant** (H_0), found by dividing the recession velocity of a cluster of galaxies v by its distance D, or

$$H_0 = v/D.$$

Because the relation between expansion velocity and distance is a straight line, all clusters give the same value. And since the redshift-velocity relation within the expanding Universe is close to that for the Doppler effect for low z (well under 1), astronomers can apply the Doppler formula to the redshifts of relatively nearby galaxies to find the velocities.

In spite of the apparent simplicity of the procedure, the actual value of H_0 has been elusive. Hubble's first diagram (see Figure 18.2a) gave an expansion rate of 500 km/s/Mpc: that is, for every increase in distance of 1 Mpc, the speeds of galaxies were seen to go up by 500 km/s. His distances, however, were wrong. Modern measurements, though much better, are still plagued by uncertainty. In addition, *real* Doppler shifts contaminate the smooth Hubble flow: galaxies within a cluster move about as a result of gravitational orbits, and there are significant gravitational effects that cause whole clusters to move relative to one another.

These gravitational motions have proven quite complex. Galaxies in the direction of the great Virgo Cluster have slightly lower redshifts than they do in the other direction, indicating that our Galaxy and the Local Group are "falling" toward Virgo. The Local Group is apparently part of a cluster of clusters, a **supercluster,** centered on the massive Virgo system. The falling speed is hard to measure; estimates range from 150 to 300 km/s. However, since the Hubble flow is moving us away from the Virgo Cluster at 1,200 km/s, "falling" simply means that we are not moving away quite as quickly as we would if there were no gravitational attraction: we will never arrive. Redshifts of galaxies examined over larger volumes of space show a variety of other streaming motions not related to the Hubble flow that are very confusing and that we are only beginning to sort out. We and the Virgo cloud are "falling" relative to the Hubble flow at some 600 km/s toward a massive supercluster in the direction of Hydra and Centaurus, while on a yet larger scale we seem to be moving toward Lepus at yet another 600 km/s.

Different distance methods and velocity corrections give measurements for the Hubble constant that range from 30 km/s/Mpc to as high as 100 km/s/Mpc. A thoughtful average of all determinations, including

Hubble Space Telescope measures of the distances to M 100 in Virgo (see Section 17.3 and Figure 17.9) and M 96 in Leo, gives 75 km/s/Mpc. That value will be adopted here, but keep in mind that it is still uncertain.

18.4 THE BIG BANG AND THE AGE OF THE UNIVERSE

Los Angeles and San Francisco are 600 km apart. You plan to drive between them at a steady 50 km/h. You calculate the time (t) the trip will take by dividing distance (D) by your speed (v),

$$t = D/v,$$

so it will take 600 km/(50 km/h) = 12 hours. If you now want to drive at the same speed from Los Angeles to Klamath Falls, Oregon, a distance of 1,200 km, the trip will take 24 hours. But if you double your speed to 100 km/h, you can get to Klamath Falls in the same time of 12 hours.

Similarly, you can figure how much time has elapsed since galaxy A in Figure 18.3 and our Galaxy were together (their mass in contact even if the galaxies had not yet formed) by dividing galaxy A's distance from us by its recession velocity. If you calculate the elapsed time since galaxy B and ours were together, you get the same answer: galaxy B has twice the distance and twice the velocity. Since velocity increases in direct proportion to distance, it does not matter which galaxy you use: all give the same answer, suggesting that at one time all the observed matter in the Universe—from which the observed galaxies were made—was concentrated into a small, hot, dense volume and dispersed in a sudden event called the **Big Bang**. The elapsed time since the Big Bang can then be defined as the **age of the Universe** (t_0), more properly called the **Hubble time**. Note, however, that this event was *not* an explosion in which matter, and subsequently galaxies, began to fly through space. Instead, space itself suddenly expanded, carrying the decompressing matter along with it.

The Hubble constant is defined as velocity divided by distance, and age is defined as distance divided by velocity; one is the reciprocal of the other, or

$$t_0 = 1/H_0.$$

The distances of galaxies are measured in megaparsecs, and velocities in kilometers per second. To find t_0 we must use consistent units. Since there are 3.09×10^{19} kilometers in a megaparsec, the Hubble constant can be written as $H_0/3.09 \times 10^{19}$ (km/s)/km = 3.24×10^{-20} H_0 (km/s)/km. The Hubble time, t_0, is thus $t_0 = 3.09 \times 10^{19}/H_0$ seconds = $978 \times 10^9/H_0$ years. If H_0 is 50 km/s/Mpc, $t_0 = 18.6$ billion years, and if $H_0 = 100$ km/s/Mpc, $t_0 = 9.8$ billion years. Roughly speaking, the Hubble time falls somewhere between 10 and 20 billion years, with a most likely value of 13 billion for $H_0 = 75$ km/s/Mpc.

The observed matter in the Universe was concentrated into a small, hot, dense volume and dispersed in the Big Bang.

This discussion ignores the powerful drag of gravity, which causes the expansion to slow over time and has the effect of shortening the age of the Universe for given values of H_0 and t_0. It also leaves out other age determinations. The measured ages of the oldest globular clusters are greater than some of the estimated ages of the Universe or even of t_0. The Universe, however, cannot be younger than its oldest stars. These critical matters will be discussed in Chapter 19.

18.5 THE DISTRIBUTION OF GALAXIES

Beyond a few tens of megaparsecs, direct distance measurement becomes very difficult because the galaxies cannot be resolved into recognizable objects. If we look much farther, we cannot even classify the galaxies and must rely on the brightest members of large clusters for distance estimates. That procedure, however, does not serve to locate the myriad faint galaxies and smaller clusters. But once the Hubble relation has been established, we can turn the problem around and use the Hubble constant to *find* distance.

With the spectrograph, we measure a galaxy's redshift. If the redshift is relatively small, so that there is little or no difference between the redshift-velocity curves as shown in Figure 18.5, we can convert the redshift to velocity with the Doppler formula. Because $H_0 = v/D$, a **redshift distance** is defined as $D = v/H_0$, and since, from the Doppler formula, $v = cz$,

$$D = cz/H_0.$$

As an example, if we observe a galaxy moving away with a redshift, z, of 0.2, and if $H_0 = 75$ km/s/Mpc, the distance is 800 Mpc. As long as a spectrum can be obtained, we can locate galaxies in three-dimensional space. We can avoid the uncertainty in H_0 and in the curvature of spacetime by expressing distances simply in terms of z. The major difficulty involves the inevitable deviation of the motions of individual galaxies from the smooth Hubble flow. Gravitational influences will make galaxies go either faster or slower than expected, leading to errors in distance.

The *local supercluster* can be examined in three dimensions better than any other such assembly because we can easily see the faint galaxies within it. It is spread all over the sky, allowing us to probe its three-dimensional structure with considerable accuracy. Its effect is found in the remarkable concentration of galaxies in the northern celestial hemisphere centered on the constellation Virgo; a good amateur star map shows them all over that part of the sky. A three-dimensional map made both from directly determined and redshift distances reveals that the Virgo Cluster is indeed a center of attraction (Figure 18.6). Our own Local Group is but a spur in a much larger collection of galaxies related to the massive system in the center. However, though the local supercluster is a region over

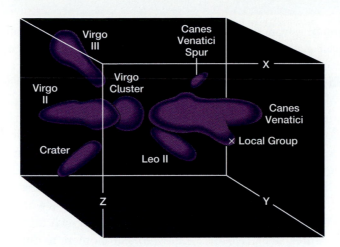

Figure 18.6
This three-dimensional representation shows the local supercluster centered on the Virgo Cluster. The Local Group is at its edge.

which gravity has a strong influence, enough to produce significant variations in the Hubble flow, *it is still expanding*, the individual clusters and spurs getting farther apart.

An early belief about the Universe was that its galaxies are uniformly (or homogeneously) distributed, and that if we were to sample different locations we would find the same average number of galaxies per unit volume. The only deviations from this uniformity were believed to be local, caused by clustering or some level of superclustering. The observations show otherwise. The first hints of the real distribution of galaxies go back to the middle 1970s, when astronomers began finding *voids*, huge regions 100 Mpc or more across containing barely any galaxies. A number of groups of astronomers then mounted massive programs to map the Universe on a grand scale. The goal is not only to find how galaxies are grouped and distributed, but also to see if we can actually find a scale of size over which the Universe really *does* become homogeneous, a scale of distance over which the hierarchical groupings (successive groupings within groupings) stop and over which the number of galaxies per unit volume stays the same.

So far, we have redshift distances of over 50,000 galaxies, and the goal is many times higher. Figure 18.7 shows the locations of galaxies with depth—outward from the Earth—within arcs parallel to the celestial equator in the mid-northern and mid-southern celestial hemispheres, each 135° long and 5° wide. The irregular distribution is obvious. We see great voids more than 50 Mpc across outlined by filaments of galaxies up to 200 Mpc long. The clusters are only the most intense concentrations within the filaments. The most prominent northern filament is the Great Wall, a long string of galaxies that extends from one side of the diagram to the other at a distance of between 50 and 150 Mpc and that contains the massive Coma Berenices Cluster. The southern hemisphere shows a similar structure called the Southern Wall.

The distribution of galaxies has a spongelike texture.

The technological revolution in astronomical instrumentation over the past two decades has had its greatest impact on cosmology. The pioneers of the subject labored heroically to obtain the data they needed: it took Slipher 13 years to assemble radial velocities for a mere 41 bright (by today's standards) galaxies. The only available detectors were photographic plates. Even today, photography is a slow way of collecting photons; the plates in use in the 1920s were a hundred times slower.

With extraordinary patience and dedication, astronomers ran some exposures over two or more nights. The camera shutter would be closed at the end of the first night and the plateholder sealed against stray daylight, to be opened on the same object the next night. In attempting a redshift record in 1928, Milton Humason, Hubble's talented observing colleague, exposed the spectrum of NGC 7619 for 45 hours over five consecutive nights, during which time he had to keep guiding the telescope on the object. Flexure in the telescope and changes wrought by temperature fluctuations resulted in a poor spectrum, but one in which the absorption lines of ionized calcium were identifiable, allowing Humason to measure a redshift of 3,780 km/s, close to the modern value.

The same spectrum can be obtained with much better quality with a CCD in less than 10 minutes. Provided the objects were above the horizon, all of Slipher's work could now be done in a single night! Such technological ability has allowed astronomers to determine the redshifts of tens of thousands of galaxies and to map the Universe. Even so, the science of observational cosmology is still young. For all our telescopic and detector capability, we have still not been able to nail down the exact value of the expansion rate of the Universe, the Hubble constant—testimony to the extreme remoteness of the objects and the difficulty of the subject.

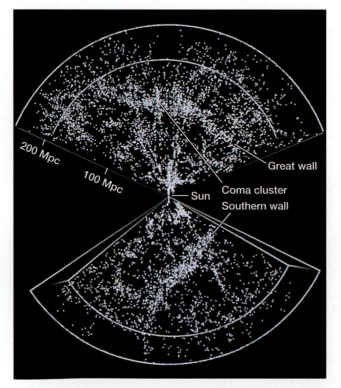

Figure 18.7
The distances of galaxies are plotted in 5°-wide strips each 135° long in the mid-northern (top) and mid-southern (bottom) celestial hemispheres. The northern hemisphere's Great Wall, which runs from side to side, is centered on the huge Coma Cluster. A similar Southern Wall is seen in the southern hemisphere. Another survey extending twice as far shows similar structure.

The concept of clustering and superclustering does not adequately describe the way galaxies are distributed. What we actually find is a texture more like that of a sponge, the voids surrounded by connected sheets that contain the gravitationally bound clusters of galaxies. The Hubble flow then expands the sizes of both sheets and voids, stretching them, and leaving the clusters together as knots in the connective tissue. We would like to extend such maps as far as we can see and are now building telescopes and spectrographs dedicated to the task. In the meantime, deep probes within narrow areas of the sky reveal similar structures to distances of 1.3 billion parsecs (over 4 billion light-years) and clumpings again nearly 200 Mpc across. The Doppler motions of the visible galaxies described in Section 18.3 (movements within the Hubble flow) suggest structures at least half again as large.

We do not know where the hierarchy stops. However, surveys of radio emission from distant galaxies *do* demonstrate a remarkably smooth distribution, as do CCD images of faint blue galaxies (Figure 18.8). Spectrograms of the brightest (and presumably nearer) ones reveal redshifts up to 0.8, showing that over distances measured in billions of parsecs the galaxy distribution may indeed be uniform.

Distances to these faint blue galaxies are great enough that, because of the time it takes light to travel, we look considerably back into the past. The faintest of these galaxies is billions of light-years away, so we see the systems as they *were* billions of years ago; we see them when they were *young*. Their notable blue colors are different from the colors of the nearby galaxies, revealing the effects of galaxy evolution. The redshifts are so high that ultraviolet light is shifted into the blue and visual parts of the spectrum, and the high level of UV radiation attests to rapid rates of star formation.

The visible Universe could contain upward of a trillion galaxies.

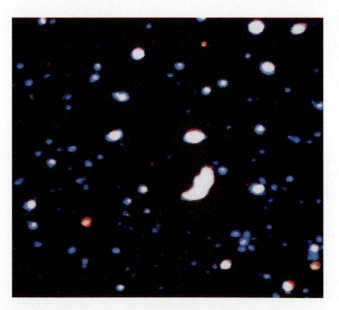

Figure 18.8
Hundreds of faint blue galaxies crowd a square only 2.2 minutes of arc—1/14 the angular diameter of the full Moon—on a side.

Figure 18.9

The extraordinary *Hubble Deep Field* image displays roughly 1,500 galaxies, the vast majority never before seen, within an area of the sky only 2.5 minutes of arc across. Almost every image in the picture is a galaxy. Although some faint galaxies may be relatively nearby, many are exceedingly far away. With some imagination you can make the picture come alive in three dimensions. At such great distances we view the Universe at a younger time, allowing us to see galaxy evolution at work.

The number of galaxies in the visible Universe and a deeper glimpse of their evolutionary histories are both dramatically revealed by the *Hubble Deep Field* image taken in 1995 (Figure 18.9). The *Hubble* astronomers pointed the *Space Telescope* toward a tiny region of the sky near the Big Dipper that, from the ground, appeared nearly blank, and stared at it for 10 days in four wavelength bands. The resulting combined image, only 2.5 minutes of arc across (1/10 the angular diameter of the full Moon), reached nearly to 30th magnitude and shows some 1,500 galaxies. Multiplying this view over the whole sky yields a galaxy count of more than *10 billion*. The galaxies we see in the Hubble deep field are strung out in three dimensions, from their apparent magnitudes ranging from about 1 to 3 billion parsecs away. If we could examine the Universe twice as far out, we could conceivably find 10 times as many, and these are only the brightest. The visible Universe could contain upward of a trillion (10^{12}) galaxies, and this number does not include the huge numbers of large faint systems and dim dwarfs. If their average population is 10 billion (10^{10}) stars, the Universe accessible to us could contain $10^{12} \times 10^{10} = 10^{22}$ stars!

The *Deep Field* is a time machine, distant galaxies (on the average, the fainter ones) younger than the brighter, closer ones, and all much younger than the galaxies that surround us. We may be looking back as far as 10 billion light-years. While other ground-based surveys have probed the sky to galaxies almost as faint and distant, none can approach *Hubble's* incredible resolving power. In the great distance we see galaxy forms that differ from the familiar normal spirals and ellipticals and that may be predecessors to the more familiar local systems. This and other observations also suggest that spirals were more common in the distant past, consistent with their mergers at a later time into ellipticals. However, only redshift data—and a great deal of research—will sort out the clues to the flow of evolution buried within this great mass of data. Such evolution of galaxies with time clearly complicates our ability to derive distances from apparent magnitudes, as the galaxies we see had different—and unknown—absolute magnitudes in the distant past.

18.6 QUASARS

Redshift distances of extreme forms of active galaxies allow us to probe even farther into the Universe. In the 1950s and 1960s, great effort was spent cataloguing new radio sources and attempting to identify them with optical counterparts. One, 3C 48 in Triangulum (Figure 18.10a), was iden-

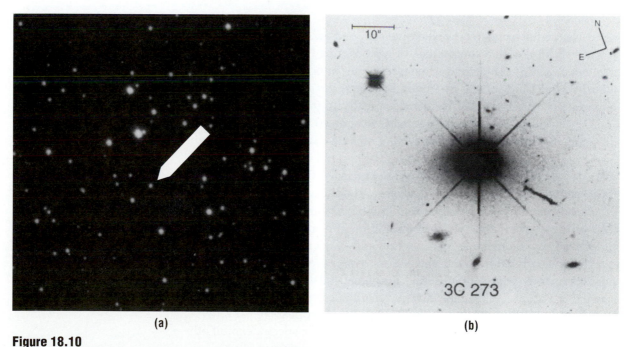

(a)   (b)

Figure 18.10
(a) An image of the quasar 3C 48 shows that it is pointlike. **(b)** 3C 273, the brightest and closest quasar (seen here in a negative image), is the source of a short jet. The faint halo is probably the quasar's dim host galaxy, from which the black hole at the center draws mass.

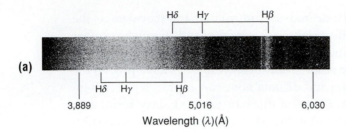

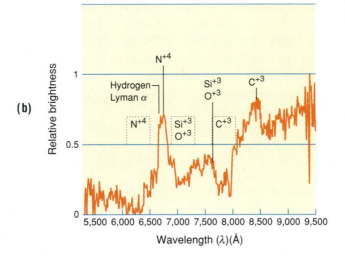

(a)

(b)

Figure 18.11

(a) The spectrum of 3C 273 (top) shows three hydrogen Balmer lines, Hβ through Hδ, redshifted by 16%, $z = 0.16$. The positions the lines would have if the body were at rest are shown below. **(b)** Quasar Q0051-279, $z = 4.43$, recedes so fast that the hydrogen Lyman α line at 1,216 Å is shifted into the red part of the spectrum.

tified with what appeared to be a blue star, and another, 3C 273 in Virgo (Figure 18.10b), with a short streak of light coming from yet another blue star. Spectra of 3C 48 displayed emission lines at odd and unidentifiable wavelengths. Lines of the hydrogen Balmer series redshifted by a then-remarkable 16% ($z = 0.16$) were subsequently found in the spectrum of 3C 273 (Figure 18.11a). A look back at 3C 48 showed that its lines were hydrogen as well, but with $z = 0.37$, a value so high that no one had thought of the redshift as an explanation. These objects, and others quickly found, were definitely *not* stars and were dubbed **quasars,** short for "quasi-stellar radio sources." Soon after the discovery of quasars, optical surveys in which astronomers examined particularly blue "stars" revealed that over 99% of the quasars emit no radio radiation. Nevertheless, *quasar* is still used as a general term that includes the radio-quiet objects.

All quasar redshifts are large. Values over 1 are common, and many quasars have z greater than 2, the ultraviolet redshifted into the visual spectrum (Figure 18.11b). The current record is $z = 4.897$. For this quasar the Lyman α line of hydrogen, normally at 1,216 Å, appears red at 7,149 Å. In addition to the strong hydrogen emission lines, the spectra exhibit others from elements like carbon (see Figure 18.11b) superimposed on a continuous spectrum. These emissions must be produced by low-density gas that presumably surrounds a central source. The continuum can be observed into the ultraviolet and the X-ray parts of the spectrum as well as into the infrared. In a few cases, the continuum even reaches into the radio, where it produces the classic quasars.

Some quasars, like 3C 273 (see Figure 18.10b), are surrounded by faint material and/or have streaks of light—jets—coming from them, but the bright central cores *all* appear stellar. Very long baseline interferometry shows the cores to be unresolved down to 0.0001 second of arc (roughly the angular size of a pinhead in Los Angeles as observed from New York). They are erratically variable over periods of years (Figure 18.12a) and over intervals as short as 10 days (Figure 18.12b), so they must therefore be smaller than 10 light-days, or about 2,000 AU.

There are opposing explanations for these objects. They may be taking part in the Hubble flow and are therefore at **cosmological distances,** distances that span a large portion of the Universe and that relate to cosmology. If the redshifts do not exceed a few tenths, we can readily determine redshift distances. The nearby quasar 3C 273 ($z = 0.16$) recedes at

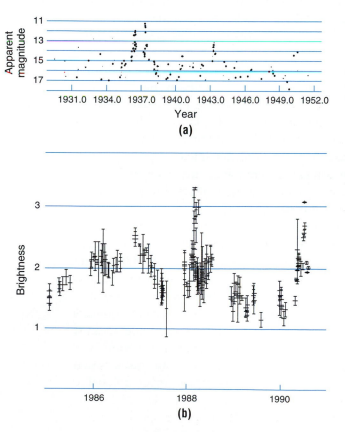

Figure 18.12
(a) The light curve of the quasar 3C 279, reconstructed from old photographic plates, shows considerable variation as well as powerful outbursts over a 20-year interval. **(b)** A short-term examination of 3C 273 reveals variations over a mere 10 days. (Adapted from "The Quasar 3C 273," by Thierry J.-L. Courvoisier and E. Ian Robson from *Scientific American*, June 1991, page 54. Copyright © 1991 by Scientific American, Inc. All rights reserved.)

40,000 km/s and (for $H_0 = 75$ km/s/Mpc) is 530 Mpc away; 3C 48 recedes at about 91,000 km and is roughly 1.2 billion parsecs (nearly 4 billion light-years) away. At large z we need to know the degree of the Universe's spacetime curvature (a subject for Chapter 19) to derive distances from z. However, the velocity of Q0051-279 (see Figure 18.11b) must be very high, approaching that of light.

From its redshift distance and apparent visual magnitude of 12.9, the absolute magnitude of 3C 273 is -26.1, making it a phenomenal 2×10^{12} times more luminous than the Sun and 100 times more luminous than our whole Galaxy. Other quasars are similar, and if we take nonvisual radiation into account they are more luminous yet. One with $z = 4.7$ is estimated to be more than 10,000 times brighter than our Galaxy, all this energy pouring from a volume not all that much larger than our Solar System.

Such huge calculated luminosities have suggested to some astronomers that quasars simply cannot be at cosmological distances, that instead they must be nearby, and their luminosities therefore much smaller. Perhaps they are "local," small, high-speed "bullets" ejected from the core of our Galaxy in some major calamity. For a number of reasons, however, nearly all astronomers believe that quasars are indeed at cosmological distances. If they were local, we would expect some to be associated with other galaxies, and we would then also expect to see blueshifts from those ejected *toward* us. But none is found. The cosmological hypothesis is supported by some quasars that are allied with other galaxies or clusters of galaxies with the same redshift. In addition, the spectra of quasars commonly exhibit absorption lines that are also redshifted, but with z always less than or equal to the redshifts displayed by the emission lines. These absorptions are interpreted as originating in clouds of low-density gas—even galaxies—lying along the line of sight to the distant but brilliant sources of continuous emission. Any cloud that lies in front of a quasar must be closer to us and receding more slowly as a result of the Hubble flow, and its redshift will be less.

Furthermore, the faint clouds of fuzz that surround many quasars have the same redshifts as the quasars themselves. The quasars therefore seem to be the brilliant nuclei of galaxies, the unresolved fuzz of the galactic disks or halos. Quasars without the accompanying fuzz are thus also likely to be galactic nuclei whose surrounding galaxies are either too faint to be seen or have not yet developed. The similarity between most quasars and Seyfert galaxies (see Section 17.8) is striking. If we could make a Seyfert's nucleus a little brighter and the disk a little fainter, it would appear as a quasar. Comparable evidence is provided by some quasars that are associated with radio lobes, showing there is another subset related to

One known quasar is estimated to be over 10,000 times brighter than our Galaxy.

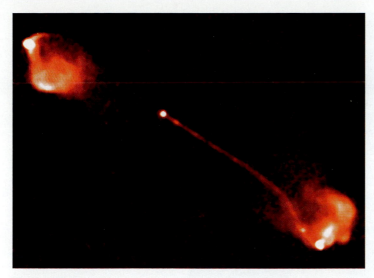

Figure 18.13
The quasar 3C 175 ($z = 0.77$) appears as a classic double radio galaxy with twin lobes and a relativistically beamed jet pointing into one of them.

elliptical radio galaxies (Figure 18.13). Moreover, the jet of 3C 273 (see Figure 18.10b) is reminiscent of the one emanating from M 87.

Quasars therefore are not different kinds of objects but active galaxies with hyperactive nuclei. Seyferts and radio galaxies are essentially quasars in which the surrounding galaxy can easily be seen. The power source of the quasar is interpreted as a massive black hole accreting matter from its surroundings, which may be the galaxy that contains it or mass being stripped from some kind of companion. The emission lines come from ionized clouds of matter orbiting outside the black hole's accretion disk. The models for quasars are still far from perfect, but at least we now have a rational working explanation for them.

There are huge numbers of quasars. Brighter than 23rd magnitude, we find about 100 per square degree, implying a total of around 4 million (Figure 18.14). At 28th magnitude, near the limit of observation, there should be 30 times as many. Still, this number is only a fraction the count of normal galaxies. We actually see significant numbers of quasars only when we look back in time to when the Universe was young. However, if we look back far enough, beyond about $z = 2$, the count begins to drop off. Quasars are rare above $z = 3$ and almost unknown above $z = 4$. More-distant quasars, those with the higher redshifts, tend strongly to be the more luminous.

Quasars are not different kinds of objects but active galaxies with hyperactive nuclei.

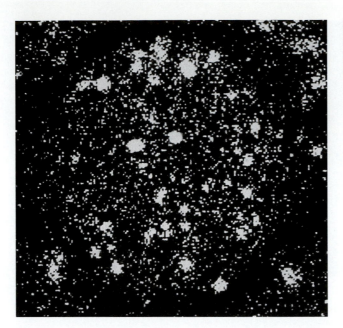

Figure 18.14
More than 40 quasars are seen in an X-ray view of the sky less than a degree across.

We may be witnessing an evolutionary sequence that is telling us how at least some galaxies were created. At the highest redshifts, the farthest back in time we can look, we see the first, most brilliant, quasars being born. The act of formation may have sent vast amounts of matter inward to feed growing black holes. As we proceed to lower redshifts and distances, we look less and less far back into time, the number of quasars (after first increasing) drops off, and the objects we see are progressively older. As the quasars aged, the amount of infalling mass dropped and the activity dimmed. At the same time, star formation began to create visible galactic halos and disks. Some quasars faded into Seyfert and radio galaxies. Others perhaps turned into more ordinary systems, those like our own, with relatively inactive nuclei. Behind the vast dust clouds of the Milky Way, there may be a nearly dead quasar at the center of our Galaxy.

18.7 GRAVITATIONAL LENSES

In 1979, astronomers at the University of Arizona discovered two quasars in Ursa Major only 6 seconds of arc apart (Figure 18.15a). More surprisingly, the quasars' spectra are identical, as are their redshifts of 1.4136. A true binary can quickly be ruled out: the components would be somewhat different, and we would expect a velocity difference because of orbital movement.

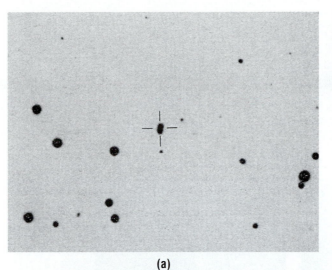

Figure 18.15
(a) A photograph taken with a wide-angle telescopic camera shows the first-known double quasar in Ursa Major. **(b)** A magnified view with a larger telescope reveals an intervening giant elliptical galaxy (arrow), which has gravitationally focused and split the light of a single distant quasar.

(a)

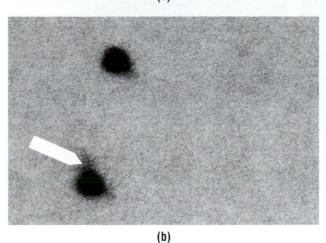

(b)

The double quasar is produced by an intervening **gravitational lens** that gravitationally deflects the light and focuses it, causing a distant *single* quasar to *appear* double. Deep imaging of the double quasar revealed the lensing object (Figure 18.15b), a giant elliptical galaxy situated between the two quasar images. With $z = 0.36$, it is considerably closer than the quasar.

Figure 18.16a shows how such a gravitational lens works. A quasar, a point source of radiation, is positioned behind an extended galaxy. The total mass that affects a light ray increases as we proceed from the galaxy's center to its edge. As a result, the rays that pass farther from the center are increasingly deflected, and the light is focused. An observer on Earth

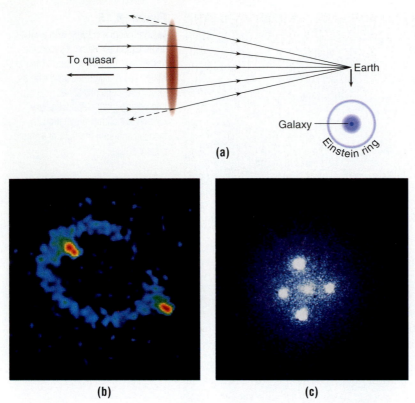

(a)

(b)   **(c)**

Figure 18.16
(a) A distant single quasar lies directly behind an intervening mass. The farther away from the center the light hits, the more mass it "sees" and the more it is deflected. The observer then views the quasar as a lensed Einstein ring around the galaxy. **(b)** This radio image of a gravitational lens is a fair approximation to the ideal Einstein ring. **(c)** The Hubble Space Telescope views an Einstein cross, a configuration in which an intervening galaxy at the center of the image splits the light of a distant quasar into several components.

would see light coming from all around the galaxy in the form of an *Einstein ring* (Figure 18.16b). If the alignment is shifted slightly, the ring becomes a pair of asymmetrically placed arcs, one considerably brighter than the other. If the alignment is offset a little more, the arcs become more like actual point images of the source, and the placement becomes more asymmetrical (see Figure 18.15). Because the intervening galaxy is extended, we might also see multiple images like the *Einstein cross* (Figure 18.16c). Depending on alignments, the brightness of the quasar can also be amplified. Such gravitational lenses would not be possible if the lensed quasars were not at great, and probably cosmological, distances.

Gravitational lenses are of considerable importance. In clusters of galaxies, intervening dark matter lenses the light of distant quasars and

Figure 18.17
Dark matter in a galaxy cluster lenses more-distant galaxies into arcs.

galaxies into numerous arcs (Figure 18.17), allowing the amount to be assessed and the distribution to be mapped. Dark matter up to 50 times as abundant as luminous matter has been found in this way. On a much smaller scale, astronomers have examined huge numbers of stars in the Magellanic Clouds for lensing effects that might be produced by intervening dark matter—or planetary-sized bodies—in our own Galactic halo. And though a few lensing events were seen, there were not enough to make up the amount of dark matter expected, in line with the similar conclusion drawn from *Hubble* observations about the lack of red dwarfs in the halo (see Figure 16.17). Dark matter remains deeply mysterious.

The quasars provide a bridge from the galaxies to the distant reaches of the Universe, which we now enter, and perhaps even begin to understand.

▶ KEY CONCEPTS

Big Bang: A sudden expansion of space 10 to 20 or so billion years ago that is believed to have produced the present Universe.

Cosmological distance: A large distance that relates to the expanding Universe and thus to cosmology.

Cosmological force: A force suggested by Einstein that increases with distance.

Cosmology: The study of the structure, origin, and evolution of the Universe.

Expanding universe: The steady growth of space that generates the **Hubble flow,** which increases distances between clusters of galaxies.

Gravitational lens: A body whose gravity causes relativistic distortion or amplification of the light from a more distant body.

Hubble constant (H_0)**:** The rate at which the Universe expands.

Hubble relation: The linear correlation between redshifts and distances of galaxies.

Quasars: "Quasi-stellar radio sources," bright, starlike objects with high redshifts; some are strong radio emitters; they are probably distant, young versions of active galaxies.

Redshift (z): The general spectral shift to longer wavelengths (redward) observed from receding galaxies.

Scale of the Universe: The concept that represents relative distances in the Universe or the distances between coordinate grid lines; the scale of the Universe increases with time as the Universe expands and clusters of galaxies get farther apart.

Supercluster: A cluster of clusters of galaxies.

KEY RELATIONSHIPS

Hubble time (age of the Universe):

$t_0 = 1/H_0$ if H_0 is in consistent units
$= 978 \times 10^9/H_0$ years if H_0 is in km/s/Mpc.

Redshift:

$$z = (\lambda_{observed} - \lambda_{emitted})/ \lambda_{emitted}.$$

Redshift distance:

$$D = v/H_0 \text{ (for small } z\text{).}$$

EXERCISES

Comparisons

1. Compare the natures of, and relations among, clusters, superclusters, and walls of galaxies.

2. Compare quasars with other kinds of active galaxies.

Numerical Problems

3. What is the approximate redshift (z) of the Hydra Cluster, shown in Figure 18.1? What are the observed wavelengths of the galaxy's ionized lines, which are normally at 3,934 Å and 3,968 Å?

4. We measure $z = 2.4$ for a distant receding source. What is the recession velocity from **(a)** the Doppler formula; **(b)** the general relativity relation of Einstein and de Sitter?

5. A distant galaxy has $z = 0.75$. How many times bigger is the Universe than it was at the time the light left the galaxy?

6. What is the Hubble time if $H_0 = 60$ km/s/Mpc?

7. What is the distance to a galaxy receding at 15,000 km/s for $H_0 = 50, 75,$ and 100 km/s/Mpc?

8. What is the angular diameter of a quasar that has a physical diameter of 1,000 AU and is a billion parsecs away?

9. What would you expect the redshift of the quasar in the previous question to be? Explain your assumptions.

Thought and Discussion

10. What is the evidence that the Universe is expanding?

11. Why are we not actually at the center of an expanding Universe?

12. What is meant by the Hubble constant (H_0)?

13. Why do some galaxies exhibit spectral blueshifts? Why are there deviations from the smooth Hubble flow?

14. Why is our Galaxy not expanding with the Universe at large?

15. How does the redshift of a distant galaxy differ from a Doppler shift?

16. Relate the Local Group to the Virgo Cluster and to the Local Supercluster.

17. What is the Great Wall? Why is it important to our understanding of the Universe?

18. List the defining characteristics of quasars.

19. What is the evidence that quasars are at large distances?

20. What is the most likely origin of the absorption lines seen in the spectra of quasars?

21. What do gravitational lenses demonstrate about quasars?

22. What evidence suggests that quasars evolve into normal galaxies?

Research Problems

23. Use your library to find a textbook on astronomy written before 1925. Comment on the views of the Universe presented there and contrast them with those held today. What observations might make the present textbook appear out of date in another 70 years?

24. Examine science and astronomy magazines from about 1963 on and summarize the changes that have taken place in astronomers' views of quasars as new data were acquired.

Activities

25. Construct your own Hubble diagram from Table 17.1, using different colors for Local Group and non–Local Group galaxies.

Show how the two divisions of the table give different results, and explain why data for the Local Group confuses the results for non-Local Group Galaxies.

26. Create your own "gravitational lens" using optics in place of gravity. Hold the bowl of a wine glass toward your face and point the stem at a small, bright light. The wine glass is the galaxy and the light the quasar. You will see a ring of light around the stem similar to the Einstein ring. Then displace the center of the glass slightly from the light and you will see a pair of arcs that become progressively more pointlike. Make notes describing what you see.

Scientific Writing

27. A politician is publicly critical of astronomers, writing that with all the instruments we have and the research money that has been spent we should at least have some firm answers about the nature of the expanding Universe. Answer the criticism, describing how much we have learned in a short time and explaining why the problems are difficult.

28. There still remains debate about the interpretation of quasars as cosmological objects. Write an essay supporting the cosmological hypothesis in which you include one or two captioned diagrams.

THE UNIVERSE

How the Universe is constructed, how it began, and what its fate may be

Gathering all our knowledge, we attempt to create theories about the Universe, the grand concept that incorporates everything, to understand its structure, origin, and evolution.

19.1 INFINITY?

In spite of the seeming complexity of cosmology, you can make a simple cosmological observation from your back yard. The nighttime sky is dark. A search for the reason directs us toward the comprehension of the Universe.

We adopt some assumptions and see where they lead us. Assume the Universe to be **infinite** in size. Infinity means never-ending: no matter how large the number, you can always add to it. Assume also that the Universe is populated uniformly with stars and galaxies and that it is static in time and space, so that nothing moves or evolves. Though stars may seem like points from Earth, they are not. Every star in the Universe has an angular diameter that, no matter how small, is greater than zero. Given our assumptions, any line of sight must therefore encounter the surface of some star (Figure 19.1). Even if our line of sight slips between the relatively nearby stars (those represented in Figure 19.1), it will finally land on some other, more distant star. The sky should therefore be covered with the overlapping apparent disks of stars and should be as bright as the surface of the Sun. It is not. Discussion of this apparent contradiction can be traced to Edmund Halley and maybe even Kepler. Among the many who examined it was the nineteenth century's H. Wilhelm Olbers, and it has been known as *Olbers' paradox* ever since.

At least one of the initial assumptions must be wrong. The expanding Universe plays a small role: the Hubble expansion reddens radiation from distant stars and reduces the energies of their photons. The principal reason for nighttime blackness, however, is that the Universe is not static in time, that stars do not live forever. To see a sky filled with stars we would

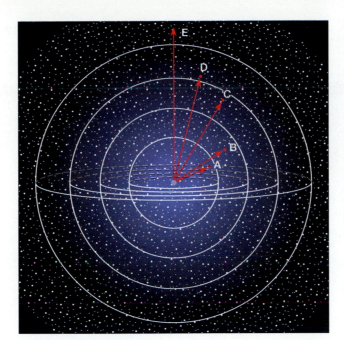

Figure 19.1
From the Earth (the blue dot at center), we look through ever-increasing distances into an infinite Universe uniformly filled with galaxies, hence stars. Lines of sight A, B, C, and D end at the surfaces of stars. Line of sight E gets through the part of the Universe shown here, but since the Universe is infinite, even E must eventually hit a star. Since all lines of sight end in stars, the sky must be about as bright as the surface of the Sun. Olbers' paradox is that the sky is not bright, but dark.

have to look much farther in light-years than the ages stars can have. Stars therefore do not fill the sky and the night is dark. This resolution of Olbers' paradox allows us to eliminate one combination of possibilities. The dark night sky, however, does not tell us if the Universe is infinite in space or time. So we dig deeper.

19.2 COSMOLOGICAL PRINCIPLES AND THE COSMIC BACKGROUND RADIATION

To attempt to understand the Universe more fully, we postulate three principles, the first leading to the other two in turn. We live near one edge of a small cluster of galaxies on the edge of a much larger supercluster. The location of the Earth is not special, a concept called the **Copernican Principle** after the man who first displaced humanity from the center of attention (Figure 19.2). We also live in an expanding Universe filled with galaxies as far as we can see. We would see the same kind of expansion from any location, and over distance scales of a few hundred megaparsecs the lumpy distribution of galaxies smooths out. These observations lead to a broader concept, the **Cosmological Principle,** which states that the Universe must look essentially the same from *any* vantage point, that there are *no* special places. The extension of our observations through the Cosmological Principle means that the Universe can have neither edge, boundary, nor center.

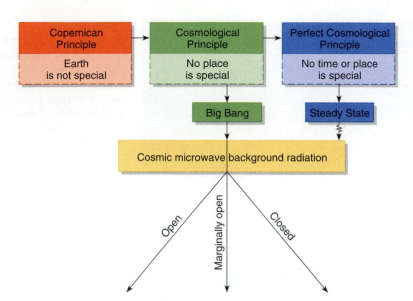

Figure 19.2
The chart summarizes properties of three fundamental principles. Cosmic background radiation disallows the Perfect Cosmological Principle but allows the Cosmological Principle and the Big Bang. Three possible structures for the Universe—open, closed, or marginally open—branch off from the basic concept.

The Cosmological Principle addresses space. What about time? A universe with neither beginning nor end is defined by the **Perfect Cosmological Principle,** which states that the Universe is the same from all vantage points in space *and* time. According to the Perfect Cosmological Principle, the Universe will look the same no matter from *where* you look and no matter *when*, whether infinitely long ago or in the future. We know the Copernican Principle to be correct. What about the others?

Big Bang theory, developed by George Gamow and his colleagues in the late 1940s, incorporates a beginning to the Universe as well as its evolution, maintains the Cosmological Principle, but contradicts the Perfect Cosmological Principle. An alternative, which embraces the Perfect Cosmological Principle, is the *Steady State* theory, proposed in 1948 by Hermann Bondi, Thomas Gold, and Fred Hoyle. The Steady State theory also incorporates the expanding Universe. Clusters of galaxies move farther apart over time, and the spaces between them enlarge. To keep the Universe looking the same, new matter, which condenses into new galaxies, must continually be created within the spaces. We cannot rule out such creation with laboratory tests. To fill the space left by the expansion of the Universe requires that three or four new atoms appear per year per cubic kilometer. A cubic kilometer of air at sea level contains 10^{37} atoms, so the few new ones would be completely undetectable.

How can we discriminate between the theories, between the two Principles? Observing at a wavelength of 10 cm, Arno Penzias and Robert Wilson of Bell Laboratories discovered in 1965 that no matter where they pointed their radio telescope, there was always a minimum, uniform level

The cosmic microwave background fills all space with blackbody radiation near a chilling 3 K.

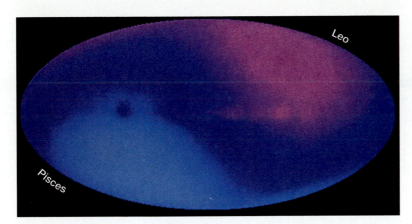

Figure 19.3
The average spectrum of the cosmic background radiation is a blackbody at 2.735 K. The colors of this all-sky map made by *COBE* show a slight change in temperature caused by the Doppler motions of the Sun, Galaxy, and Local Group, red warmer and approaching, blue cooler and receding.

of radiation unassociated with any of the usual galactic or extragalactic radio sources. They found this **cosmic microwave background radiation** everywhere, filling all space. Additional measurements showed that this radiation has the characteristic spectrum of a blackbody near a chilling 3 K. The *Cosmic Background Explorer* satellite (*COBE*) later refined the average value to 2.735 K.

The cosmic background radiation is remarkably isotropic (Figure 19.3), the temperature a mere 0.003 K higher in the direction of Leo and 0.003 K cooler oppositely. This variation is interpreted as a Doppler shift (which shifts the blackbody curve and gives the illusion of temperature change) caused by the combined motions of the Sun, Galaxy, Local Group, and local supercluster relative to the large-scale homogeneous structure of the Universe, all of which make the Sun seem to move toward Leo at about 600 km/s. When these motions are accounted for, the background radiation evens out.

Cosmic background radiation had been predicted in 1948 by Gamow's collaborators as the cooled remnant of the hot fireball created in the Big Bang. The Universe therefore appears to be changing with time. This conclusion is strongly supported by the observed evolution of galaxies and quasars as we look to distant and therefore younger reaches of the Universe (see Sections 18.5 and 18.6). We therefore exclude the Perfect Cosmological Principle and the Steady State theory, and accept the Cosmological Principle and the Big Bang as basic premises.

19.3 THE STRUCTURE OF THE UNIVERSE

There are several possible structures for the Universe that incorporate the Big Bang. These relate to the question of whether the Universe will ever come to an end.

19.3.1 Models

Throw a ball upward. If its velocity is less than the escape velocity, it will return, and its path is *closed*. If the velocity is greater than the escape velocity, however, the ball will never return, and the path is *open*. If the two velocities are equal, the ball will coast to a stop after an infinite time when it has reached an infinite distance from Earth, and the path is *marginally open*.

The expansion of space must similarly be affected by gravity, which acts to slow the expansion over time. Imagine a huge shell surrounding the Earth that contains distant galaxies (Figure 19.4a). The Earth is merely a reference: from the Cosmological Principle, there is no actual center, and any point will do. As space expands, the surface of the shell has an escape velocity that depends on the mass of matter enclosed within its spherical volume and on its size, or on the average density of its contained matter. Figure 19.4b shows three possibilities. If the shell's outbound velocity is

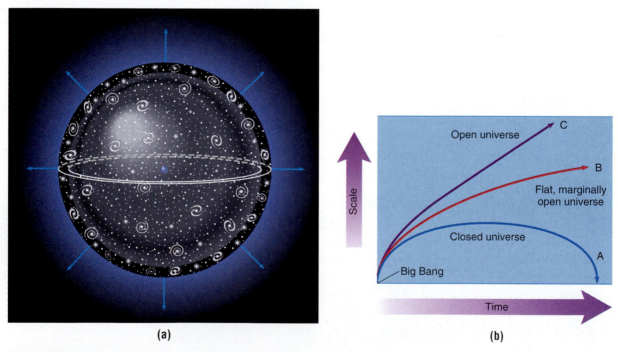

(a)   (b)

Figure 19.4

(a) A distant shell containing galaxies surrounds the Earth (blue dot) and represents the expanding Universe. The expansion speed is slowed by the gravity of all the mass interior to the shell. **(b)** Curves A and C represent the way in which space expands for the two Friedmann-Lemaître universes (where the expansion is represented by the scale of the Universe, which is proportional to the distances between clusters of galaxies). In A, gravity will stop the expansion and eventually bring the closed universe back to a collapse. In C, the open universe expands too fast to be brought back and will always expand. The Einstein-de Sitter universe (B) is barely open, bringing the rate of expansion to zero only after an infinite amount of time.

less than the escape velocity, it will slow to a halt in a *finite* (less than infinite) time, and we live in a **closed universe.** The Universe will thus someday begin to shrink, its *scale* (see Section 18.2) decreasing rather than increasing. It is fated eventually to crunch back to the dense structure from which it came. How long that will take depends on the density of matter. If, however, the outbound velocity is *greater* than the escape velocity, the shell and the Universe will expand forever and we live in an **open universe.** Finally, there is the dividing case of a **marginally open universe,** which will coast to a stop, but only after an infinite amount of time has elapsed; this universe will never contract.

These arguments are based on a Newtonian view of gravity. Similar concepts are derived from general relativity. The characteristics of closed and open expanding **Friedmann-Lemaître universes** were developed by Alexander Friedmann and the Abbé Georges Lemaître in the 1920s. The relativistic solution for the **Einstein-de Sitter universe** (see Section 18.2), the dividing case that comes to a stop only in infinite time, was found in 1932. Gravity, which controls the rate of expansion and the degree of closure, is the result of curved spacetime. Each of the three possible universes described is therefore associated with a unique kind of spatial curvature that controls the expansion and determines whether the Universe is open, closed, or marginally open.

Think of collapsing your three-dimensional world into a two-dimensional one. Instead of living in a universe in which you can go forward and back, sideways, and up and down, pretend you are living on a surface with no up or down. One of the basic postulates (unprovable assumptions) of geometry is that "only one straight line may be drawn through a point parallel to another line." Thus, parallel lines stay the same distance apart even if drawn to infinity. This statement is true only if your two-dimensional space is *flat* (Figure 19.5a), or *Euclidean*, after the ancient Greek inventor of plane geometry. In such a world, the angles of a triangle add to 180°.

This surface, however, need not be flat, but might be bent into other shapes through a third dimension of which you are unaware. Your two-dimensional world might be curved *negatively* or *hyperbolically* like a saddle (Figure 19.5b) that curves outward to an infinite distance. You could sense your world's curvature since you could then draw *more* than one line through a point parallel to a given line and the angles of a triangle would sum to *less* than 180°. Your two-dimensional surface might also be curved *positively*, closing back on itself in the equivalent of a sphere (Figure 19.5c). If so, then *no* line could be drawn through a point parallel to a given line, as the lines must inevitably converge, and a triangle's angles would add to *more* than 180°.

Now scale these pictures back to your real three-dimensional world. Your space could be Euclidean with unique parallel lines, or three-dimensional space could be bent either positively or negatively. Though the human mind cannot actually visualize such curvature, you could still

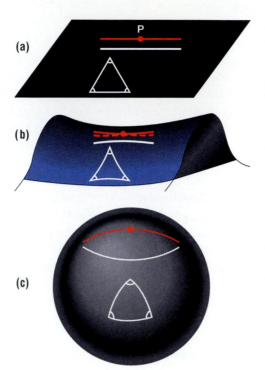

Figure 19.5

Our world of three dimensions is represented here by a variety of two-dimensional surfaces. **(a)** The surface is flat and Euclidean, and the Universe is marginally open. One line (red) can be drawn through a point parallel to another line, and the angles of the flat triangle sum to 180°. **(b)** The two-dimensional space is curved negatively, like a saddle. The Universe is open, more than one line can be drawn parallel to a given line, and the inwardly bowed triangle's angles add to less than 180°. **(c)** Space is positively curved and folds back on itself. There are no parallel lines, and the angles of the outwardly bowed triangle add to greater than 180°.

discover your world's shape by noting that *more* than one line could be drawn parallel to another, demonstrating hyperbolic curvature. Or you might find that *no* parallel lines exist, demonstrating that your universe is positively curved or spherical. You could also tell the shape of your universe by summing the angles of a triangle. In practice, however, the degree of curvature is so large that differences from true flatness—if they exist—are not locally detectable over the scale of the Earth, or perhaps over even a billion or more parsecs. A variety of tests *is* available, however, as we will see.

The three kinds of curvature represented in Figure 19.5 make the terms *open* and *closed* more meaningful. If the Universe is destined to expand forever and is open, it is also hyperbolic, infinite, and geometrically open. If it is closed, it is spherical and finite. If the Universe is borderline between the two cases and marginally open, space is flat and Euclidean, but still infinite.

If the Universe is open or flat (and infinite), it clearly has no center. Such a conclusion about a closed universe is not so obvious. Figure 19.6 represents a spherical universe in three dimensions, where our three-dimensional space is again represented by the sphere's two-dimensional surface. The surface of a sphere is closed and finite: there is only a certain amount of space, the sphere's surface area. Yet this world is unbounded and has no edge. A person could travel forever among the galaxies without encountering an impassable barrier, and an astronaut traveling in a

The Universe is not expanding into anything; it is everything and is merely getting larger.

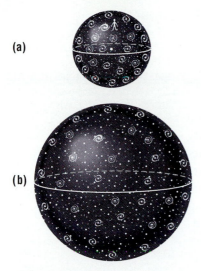

(a)

(b)

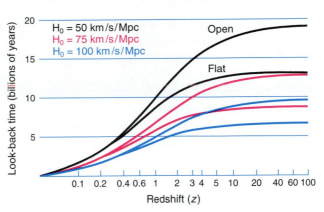

Figure 19.7

Look-back times (used to represent distances) in billions of years for three values of H_0 are indicated by different colors. The upper and lower curves for each refer respectively to open and to flat (marginally open) universes. We cannot accurately give cosmological distances unless we know both the Hubble constant and the curvature of the Universe.

Figure 19.6

(a) The Universe (assumed to be closed) is represented by the surface of a three-dimensional sphere; its people are two-dimensional and unaware of a third dimension. Their universe is the sphere's surface, which is closed and finite but is unbounded and has no center. **(b)** As the sphere expands, its clusters of galaxies become farther apart.

straight line would return to the starting point. The surface of a sphere has no center, a term that has meaning only if three dimensions are considered. In our real world, a closed universe is finite and has only a certain volume, but is unbounded and has no center. Moreover, a traveler moving in a straight line will return to the point of origin. As the sphere in Figure 19.6 expands and its size (or scale) increases, its galaxies get farther apart, but the two-dimensional world does not exist outside the sphere. Likewise, the real closed Universe is not expanding *into* anything; it *is* everything and is merely getting larger, carrying the galaxies along with it and making their clusters recede from one another. The same is true for open or flat universes. Space does not exist outside the Universe.

19.3.2 Distances and Ages

The curvature of space presents a problem in expressing a cosmological distance. A common measure of such a distance is the time required for an object's light to get to us, which is also the time to which we look in the past to see it. Such a distance can therefore be called the **look-back time.** The difficulty with this straightforward concept is that the relation between the observational quantity—the redshift—and the look-back time depends on the model assumed for the Universe. It is shown in Figure 19.7

for fully open and flat universes and for three values of H_0. As an example, for a redshift of $z = 4.9$ (the current quasar record), the look-back time for a flat universe is 93% of the way back to the Big Bang, at which time the Universe was less than a billion years old. However, any such distance given in billions of parsecs or light-years can only be approximate, since we do not yet know the degree of curvature.

Galaxies recede faster with increasing distance, in accordance with the Hubble law. At some distance L (the *Hubble length*), the recession velocity equals that of light, or $L = c/H_0 = ct_0$ (where t_0 is the Hubble time, as described in Section 18.4). Beyond the Hubble length, galaxies and quasars are now receding faster than light. Such speeds are possible because the distant objects are not moving *through* space but are being carried along *with* space (see Section 18.2). In a flat Einstein-de Sitter universe, the redshift of a galaxy now a Hubble length away is $z = 3$. As we move outward, z continues to increase. At some point z becomes infinite, we can see no farther, and we have reached the **horizon of the Universe** (Figure 19.8). In the Einstein-de Sitter version, it lies at a distance of twice the Hubble length. We have no knowledge of the Universe beyond the horizon, since no light has yet had time to reach us. As time proceeds, the Hubble constant drops under the backward pull of gravity, and the horizon expands into the Universe, allowing us to see a greater volume of space. At the highest observed values of z we look back in time to see youthful quasars,

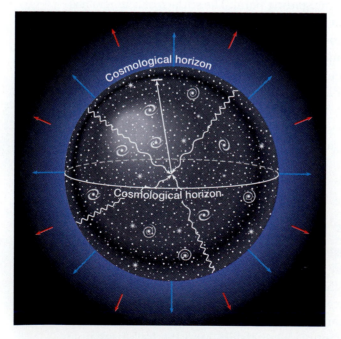

Figure 19.8
Light coming from the horizon of the Universe is redshifted to infinity. As the Universe expands (red arrows), the horizon expands even faster (blue arrows), bringing new sights into view.

their number rapidly thinning out. If we could watch for billions of years we would see those quasars age to galaxies, their places continuously taken by additional youthful quasars as the horizon expands outward.

The age of the Universe as a whole also depends on the assumed model. Assuming there is no expansive cosmological force, the Hubble time, $t_0 = 1/H_0$ (see Section 19.3), is the age of the Universe only if there is no gravity and consequently no mass. In a real universe that has mass, gravity slows the expansion, and the Universe was once expanding faster than it is today. The time since the *actual* birth of the Universe, the real age t, must therefore be smaller than t_0.

The ratio t/t_0 depends on the degree of gravitational drag. For an empty universe, $t/t_0 = 1$ (line A in Figure 19.9). Even for an open, negatively curved Friedmann-Lemaître universe (curve B), which does not have enough mass to cause the Universe to begin to contract, t/t_0 is reasonably close to 1, and the age of the Universe is approximated by $1/H_0$. For the marginally open, flat Einstein-de Sitter universe (curve C), t is two-thirds of t_0, and for a closed, spherical, positively curved Friedmann-Lemaître universe (curve D), which has sufficient mass to cause a collapse, t/t_0 can be considerably less.

In an open universe with little matter, if $H_0 = 50$ km/s/Mpc, t could be close to $1/H_0$, or 19.6 billion years. In the flat case, if $H_0 = 50$ km/s/Mpc, the true age is 2/3 of 19.6 billion years or 13.1 billion. If the Universe is flat and H_0 is 100 km/s/Mpc, t_0 and t respectively equal 9.8 and 6.5 billion years. For our adopted value of $H_0 = 75$ km/s/Mpc, the two numbers are 13.0 and 8.7 billion years. Assuming the Universe to be either open or flat, we have restricted the true age to between about 6.5 and 20 billion years, to some astronomers an embarrassingly large range.

We have restricted the age of the Universe to between about 6.5 and 20 billion years.

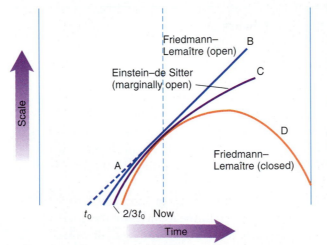

Figure 19.9
The curves show how the expansions of different models of the Universe are affected by gravity. An empty universe, or one with very little mass (A), has an age of $t_0 = 1/H_0$. The open Friedmann-Lemaître universe (B) is younger than t_0, the flat, marginally open Einstein-de Sitter universe (C) has an age 2/3 of t_0, and a fully closed universe (D) is younger yet.

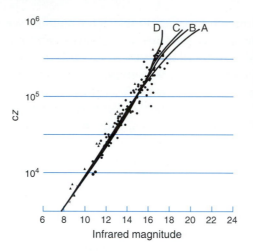

Figure 19.10
Curves A, B, and C respectively show theoretical Hubble relations for open, flat, and closed universes, and curve D shows a relation for a flat universe that includes the evolution of galaxies. The observational points cannot discriminate curvature.

19.3.3 Observational Tests

There are several ways to tell which model might be correct. We see galaxies where they used to be: the farther away they are, the less time gravity has had to slow the expansion. The degree of drag depends on how much mass is pulling on them, and consequently on the curvature, or model, of the Universe. Figure 19.10 compares different predictions with an observational Hubble diagram expressed as a redshift-magnitude relation. Curves A, B, and C respectively describe open, flat, and closed universes that assume no evolutionary effects (that is, distant galaxies have the same absolute magnitudes as those nearby), and curve D is a flat universe that takes into account the evolution of galaxies, that they had different absolute magnitudes in the distant past (see Section 18.5). The data lean toward evolution, but are not good enough to discriminate curvature.

A second method relies on local measurements. The degree of gravitational drag and the curvature of the Universe depend on the amount of mass enclosed within a sphere (see Section 9.3.1 and Figure 19.4a), and therefore upon the average density of matter (ρ). If ρ is low, the Universe is open and hyperbolic; if ρ is sufficiently high, the Universe is closed and spherical. For any value of H_0 there will be a **critical density** (ρ_c) that gives a marginally open, flat, Einstein-de Sitter universe. For H_0 equal to 50, 75, and 100 km/s/Mpc, ρ_c is respectively 5×10^{-30} g/cm^3, 10^{-29} g/cm^3, and 2×10^{-29} g/cm^3 (ρ_c is proportional to H_0^2). The curvature of the Universe can thus be measured by **Omega** (Ω), the ratio of the true density to the critical density: $\Omega = \rho/\rho_c$. If Omega is less than 1, the Universe is open; if greater, it is closed; if exactly 1, it is flat. Omega is probably the most sought-after number in modern astronomy.

To evaluate Ω, sum the masses of the constituents of a large and representative region of space and divide by the volume to measure the average density. From the luminous masses of galaxies, we find a small value for Ω, only about 0.01 (though we are still adding to the census, still discovering faint dim galaxies that will certainly alter this value). However, when we include gravitationally observed dark matter (see Sections 16.5 and 17.5), whose mass exceeds that of luminous matter by a factor of at least 10, Ω climbs to 0.3 or so, and if the most favorable numbers are chosen (a dangerous game), Ω might even approach 1.

A third and crucial test compares the age of the Universe with the ages of the oldest stars, those in the Galaxy's oldest globular clusters. For any given value of H_0 there is a range in ages that depends on the Universe's curvature (see Figure 19.9). If H_0 is known, the ages of the globulars (ignoring the time it took to make stars after the Big Bang) will give the true age of the Universe and consequently t/t_0 and the degree of curvature. The ages of the oldest globulars (there is a range, and some may well be younger) appear to lie between about 13 and 17 billion years (see Section 16.6). The lower limit excludes any value of H_0 over about 75 km/s/Mpc (Figure 19.11). That is, for our chosen value of $H_0 = 75$ km/s/Mpc, there is overlap between the age of the Galaxy and the age of the Universe only if the Universe is both fully open and the oldest globulars are 13 billion years old or younger. To obtain a flat, marginally open universe, we must accept a Hubble constant as low as 50 km/s/Mpc. If H_0 is eventually confirmed to be in the higher range and if we determine on other grounds that the Universe is flat or close to it, we will have to revise stellar evolution theory (an ongoing process) or revive Einstein's old idea of the cosmological force (see Section 19.2). An expansive force pushing on spacetime would make the Hubble constant higher than expected for a given age; then the Universe could be old enough to fit the globulars and H_0 could *still* be high.

As yet, none of the tests provide a firm answer. Density measurements suggest that the Universe is either open or flat and marginally open, the latter only if we adopt an extreme value for the average density. Instead of

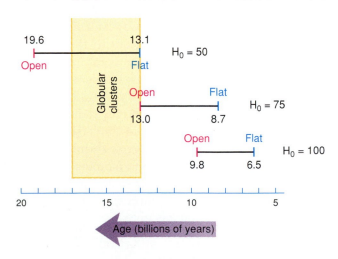

Figure 19.11

Three horizontal lines illustrate the allowable age ranges for universes with different Hubble constants from fully open (red) to marginally closed and flat (blue). The yellow box represents the currently estimated range of age of the oldest globular clusters, which allows Hubble constants only from about 75 km/s/Mpc on down.

being disappointed, however, we might be elated. We can determine an age for the Universe in two completely independent ways, by measurement of the expansion rate of the Universe and through the principles of stellar evolution. They give the same answer within a factor of 2, something we would not expect were we not on the right track.

19.4 THE ORIGIN AND EVOLUTION OF THE UNIVERSE

We now have some idea of the structure of the Universe, even if we do not know its fate. How did it come to be? The theories take us back to the beginning of time.

19.4.1 Back to the Beginning

Cosmology is still an enormously uncertain science, and many of our ideas are likely to be found incorrect. The current set of observations, however, lead to a theory that we hope will at least point us in the right direction.

The theory states that the photons of the cosmic background radiation were created in the Big Bang. As the Universe aged and expanded, they stretched along with space, their energies steadily decreasing. As the Universe enlarged its scale, it therefore also proportionately cooled. Doubling the scale (analogous to doubling the distance between the coordinate grid lines in the "rubber universe" in Figure 18.4) cuts the temperature in half.

Look first at the simplest possible case, that of a smoothly developing Universe in which temperature and density change in a constant, regular manner. Figure 19.12 shows the temperature of the Universe plotted against time in seconds. As we proceed backward in time and the scale of the Universe shrinks, the Universe becomes hotter and denser (density depending on the inverse cube of the scale). For most of the way back to the beginning, temperature changes are slow. Even looking out to the most distant quasar, with z nearly 5, the cosmic background temperature is still only a little over 10 K. But as we regress to earlier times, temperature and density mount. When the Universe was only a million years old, T was roughly 1,000 K. There could then have been no galaxies, only radiation and matter. At 200 seconds, conditions within the whole Universe were like those inside a star. We can in principle watch the Universe shrink backward to a *singularity* (see Section 15.8.1), when density and temperature both become infinite. What we see today is the result of steady expansion and cooling. Observations of carbon absorption lines superimposed on quasar spectra by distant cold intergalactic gas give a cosmic background temperature of 8° at $z = 1.8$ (a look-back time of around 12 billion years), just what we would expect.

At an age of 200 seconds, conditions within the whole Universe were like those inside a star.

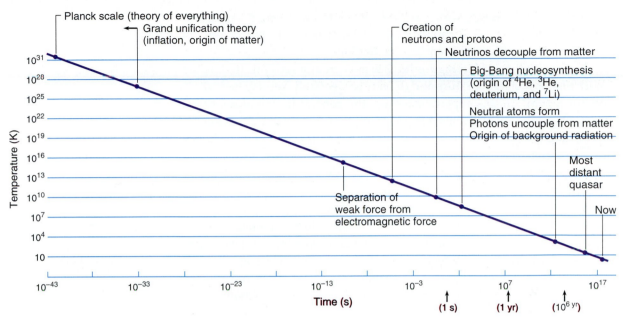

Figure 19.12
The history of the Universe as it is understood today is shown by plotting temperature against time in seconds with various milestones indicated by dots.

Such extreme conditions may have had the effect of once unifying the four forces of nature: gravity, the electromagnetic and nuclear forces (see Section 6.1), and the weak force, the one involved with radioactivity (see Section 7.1). The forces are intimately related to—in a sense carried by—different kinds of particles. The electromagnetic force, for example, is related to the photon. Changes in the electromagnetic force within a body (or between two bodies) can therefore generate photons, and the body is then said to radiate. The photon has no mass, so it can fly at the speed of light to infinity. The electromagnetic force therefore behaves according to an inverse square law. The strong force is theoretically related to massless particles called *gluons*. But because gluons interact with each other, the force acts over only ultrashort ranges, comparable to the size of the atomic nucleus. The weak force is related to exotic particles called W and Z. These are extremely heavy (in an atomic sense), and their range is smaller yet. (Though gravity has so far defied this concept of carrier particles, it may be related to *gravitons*.)

If temperature and particle energies are sufficiently high, W and Z lose their mass and behave as photons. Electromagnetism and the weak force then become one, the *electroweak force*, reducing the number of forces to three. The Universe had the required temperature of 10^{16} K in the early moments of the Big Bang when it was a mere 10^{-12} seconds old (see Figure 19.12). At a temperature of 10^{27} K, that which prevailed within 10^{-33}

seconds of the Big Bang, we expect the strong force to join. Such **Grand Unified Theories** (GUTS) cannot now be experimentally tested, and in fact we may never have the required technology.

The end of the road in Figure 19.12 is not actually zero but 10^{-43} s, the *Planck time,* when temperature and density respectively approached 10^{33} K and 10^{92} g/cm^3. In this superdense state, the general theory of relativity, our standard theory of gravity, no longer applied. Near that time, we speculate that gravity may have been united with the other forces in a **Theory of Everything.** The four forces would then have been in perfect *symmetry*.

The interactions of gravity in a Theory of Everything are related to a concept called *supersymmetry*. In a theory of supersymmetry, each kind of atomic particle has a counterpart that has different, opposing properties. For example, the fast-moving massless photon, which interacts strongly with matter, therefore is expected to have a slow-moving massive counterpart, called the *photino,* which does not. The photino is but one of a collection of *weakly interacting massive particles* (*WIMPS*). They have never been detected, but represent one possible component of dark matter, providing perhaps enough mass to bring Ω to 1.

19.4.2 The Early Universe

A steadily decelerating expansion after the Big Bang, as implied by Figure 19.12, runs into four main problems that must be solved. First is the *horizon problem*. To establish the constancy of the cosmic background radiation in all directions requires that all parts of the Universe must have been in some kind of contact with one another. Yet the radiation we observe from one direction is beyond the horizon of the radiation coming from the other. As we progress backward in time and the size of the Universe shrinks, the horizon shrinks even faster, and the problem gets worse. How did communication take place? Second is the *flatness problem*. The value of Ω is now not that far from 1. If in the early days of the Universe Ω was even slightly greater than 1, the Universe would have collapsed almost instantly; if slightly below, it would have expanded so quickly that no galaxies could have formed. To be even close to 1 now, it needed to be within 10^{-60} of 1 within the first moments, a not very likely balancing act. The *matter-antimatter problem* is next. Why is the Universe made almost entirely of only normal matter? Why not equal amounts of matter and antimatter, normal matter with charges and other properties reversed (see Section 12.4). And last is the *lumpiness problem*. The expanding matter of the Big Bang must have had variations in density that produced galaxies. Can we find them buried within the seeming smoothness of the cosmic background radiation? How and when did they develop?

In the first 10^{-35} seconds, the Universe could have expanded in scale by a factor of $10^{1,000,000,000,000}$.

These problems cannot be solved by a simple Big Bang that evolved smoothly and simply from the beginning. They can, however, be addressed by a new theory that has dramatically modified the very early moments of the Big Bang, the theory of **inflation.** It is important to remember here that these theories are not mere speculation, but have solid mathematical foundations; their purpose is to explain the actions of nature. Return now to the beginning. According to inflationary theory, most of the initial energy of the Universe was not concentrated in any of the four basic force-fields, but in another field that took on values from place to place but had no direction associated with it. Though this field held enormous energy, it contained almost no place-to-place variations. The energy field would therefore have been unnoticeable and would have appeared as a kind of unstable vacuum. (Think of being on a high plateau; you are unaware of altitude unless the plateau also contains mountains or you come to a steep slope down to the valley.) As the Universe began its expansion, and after gravity broke from the other forces, the energy of the field began to decline and was pumped into that of the expansion. The density of a universe filled with normal particles would have dropped very quickly. However, during the initial expansion, the field's energy decayed exceedingly slowly, and as a result could keep working on the expansion. The Universe therefore expanded much more quickly than expected under the original Big Bang theory. Inflation may have begun as early as 10^{-43} seconds after the creation of the Universe and may have lasted only 10^{-35} seconds. However, during this short stage we find that the Universe could have expanded in scale by as much as a factor of $10^{1,000,000,000,000}$ (1 followed by a quadrillion zeros)! The energy of this special field also provided the intense initial heat that we see today as the background radiation. After the energy of the field had declined as much as possible, the Universe then coasted outward according to the standard conception of the Big Bang, that seen in Figure 19.12.

Inflation solves the horizon and flatness problems. Before the inflationary period, the Universe was small enough to allow all its parts to communicate, to have contact with one another. Rapid inflation separated the parts, but they had already established the smoothness we now see. The inflation was so great that we see only a tiny portion of the Universe, and local space was flattened to the point that Ω effectively equals 1. (If you live on a large curved surface and cannot see very far, your world will look flat.) If this inflationary picture is correct, then luminous matter must account for only 1% of the total matter in the Universe: 99% is some kind of dark matter. However, even gravitational measures of dark matter can account for only 10% to 30% of the mass required to flatten the Universe. Another large factor is thus yet even to be found.

Energy and mass can be converted back and forth into each other ($E = mc^2$). As the temperature dropped at the end of the GUTS era, the energy decayed into mass, and spanning the time of the separation of the weak force from the electromagnetic force (10^{-11} seconds after the Big Bang), into electrons and **baryons,** a collective name for protons, antiprotons, and neutrons. Nature, however, is not perfectly mirror-symmetric. The laws that describe the Universe are slightly different when nature is reflected in a mirror. Matter and antimatter are therefore slightly different in ways other than reversed properties. As a result, the new Universe produced slightly more protons than antiprotons. The protons and antiprotons collided and annihilated each other, creating huge numbers of gamma rays. This reaction can go both ways. Two energetic gamma-ray photons can collide to make a pair of protons and antiprotons. But as the temperature dropped, gamma-ray collisions were no longer sufficiently energetic to make the particles. Matter and antimatter then continued to annihilate each other until only the excess protons were left. The theory accounts for the lack of antimatter and the observed ratio of protons to photons.

Collisions between the gamma rays could still produce electrons and positrons, but eventually, by about one second after the beginning, the lowered temperature caused this reaction to stop as well, leaving a slight excess of normal negative electrons. At about this time, the density of the Universe had dropped low enough to make it transparent to neutrinos (produced in abundance by reactions involving the weak force) that then were free to fly through space unimpeded. Solar experiments indicate that neutrinos may have a tiny amount of mass (see Section 12.4); so many were created in the Big Bang that they might provide some of the dark matter needed to flatten the Universe.

At an age of three minutes and a temperature of 10^9 K, the Universe began to take on more of its present character, and our theories appear to be on firmer ground. Deuterium was created by the collisions of protons and neutrons; as the temperature dropped, deuterium could no longer be broken apart by collisions and could survive. Deuterium atoms reacted with one another and with protons and neutrons to make helium, both ^{3}He and ^{4}He. The chain built to lithium (^{7}Li), but by then the density and temperature had dropped too low to allow anything else, leaving only the basic building blocks of Population III stars (see Section 16.6). Theory predicts an initial He/H ratio of about 0.08 (by number of atoms), just what we see in the Galaxy's ancient halo population.

The abundances of deuterium, ^{3}He, and ^{7}Li are very sensitive to the density of baryons. From Big Bang theory we can calculate these abundances in terms of what the baryonic density of the Universe ought to be *now* (Figure 19.13). From the observed abundances we find a maximum current mean density of about 3×10^{-31} g/cm^3. This value is several times larger than that found from luminous matter alone, confirming the exis-

The nuclear reactions in the early Universe built to lithium (^{7}Li), leaving only the basic building blocks of Population III stars.

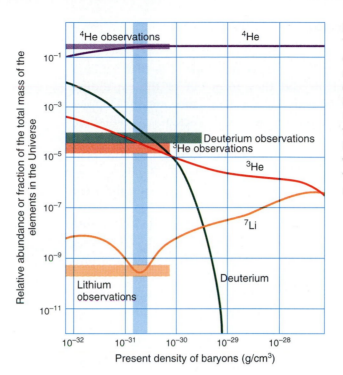

Figure 19.13
The colored lines show the abundances of ^2H, ^3He, ^4He, and ^7Li (by mass) predicted to have been formed during the initial phases of the Big Bang plotted against the density the Universe should now have (calculated for $H_0 = 50$ km/s/Mpc). The thick horizontal bars are the observations, which cross the predictions within the vertical bar, which indicates the current density.

tence of dark baryonic matter, presumably the dark unknown stuff that flattens galaxy rotation curves. However, it still yields Ω somewhat under 0.1. If Ω is actually 1, as suggested by the inflationary picture, baryons—what we think of as normal matter—constitute at best only a small percentage of the dark matter. The additional mass must therefore be *non-baryonic*, in the form of other kinds of particles—perhaps WIMPS or neutrinos.

At this time the density of the Universe was still so high that photons interacted with protons and electrons to keep the matter ionized. About 100,000 years after helium was formed, when the temperature was a few thousand K, the density dropped to the point where electrons could combine with protons and other nuclei to form permanent neutral atoms. Photons no longer were strongly absorbed by matter to keep it ionized, and the Universe became transparent to light. This was the critical moment of the creation of the cosmic background radiation. Since that time, the scale of the Universe has increased by over 1,000 and the background temperature (which is inversely proportional to the scale) has dropped by a like factor to 2.7 K. When we look into space with our radio telescopes, we can see at most only to the moment of the onset of transparency, to a redshift of about 1,000. As the horizon of the Universe expands, could we live long enough, we would see new quasars and then galaxies emerge from the murk of this opaque background.

19.4.3 The Formation of Galaxies

One of the challenges to theory—the lumpiness problem—remains to be addressed. The first galaxies appear to have developed less than a billion years after the Big Bang. How did they form out of the expanding medium? After removal of confusing sources of radiation and the Doppler effect from the *COBE* observations of the cosmic background radiation, the map of Figure 19.3 breaks into a pattern of fluctuations that change the average temperature by no more than 10^{-5} K (Figure 19.14). They appear to be density ripples that developed immediately after the Big Bang, variations in mass and gravity that ultimately produced galaxies and their clusters and superclusters (Figure 19.15).

However, the amount of baryonic matter in the Universe, including *both* the luminous and dark matter discussed in Sections 16.5, 17.5, 18.7 (at least the amount inferred from most observations), cannot by itself produce sufficient gravitational attraction to allow the ripples to grow. The ripples may instead be markers for much larger fluctuations in the distribution of gravitational *seeds* that can cause the baryonic matter to accumulate (Figure 19.16). These seeds may be the same things required to bring Ω to 1, if indeed Ω *is* actually 1.

There are two broad candidates for such seeds. First, they could be produced during inflation. The vacuum is not really empty. Energy is constantly appearing and disappearing in the form of quickly vanishing particles, a concept confirmed by laboratory experiment. The energy fluctuations in the vacuum could have been expanded to very large proportions by inflation, producing density, and therefore gravity, fluctuations. Second, the breaking of GUTS symmetry into the strong and electroweak

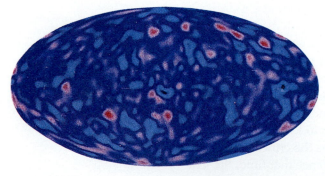

Figure 19.14

This all-sky map of background radiation from the *COBE* satellite, with Doppler effects removed, shows temperature fluctuations of a mere 10^{-5} K that could have given rise to the lumpy structure of the Universe.

forces, which has been likened to the freezing of water, introduces *defects* into spacetime that are conceptually akin to the cracks in an ice cube. The defects, regions where GUTS is still in effect, have mass and gravity. They have never been observed, but may make the necessary seeds that quickly attracted baryonic matter.

The fluctuations in the *COBE* background may be observations of the first actions of these seeds. Whatever the seeds' origins, the fluctuations

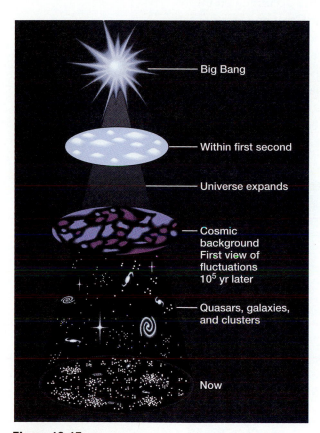

Figure 19.15
The fluctuations in the cosmic background radiation had their start in the first second following the Big Bang. They grew into the structures we see today.

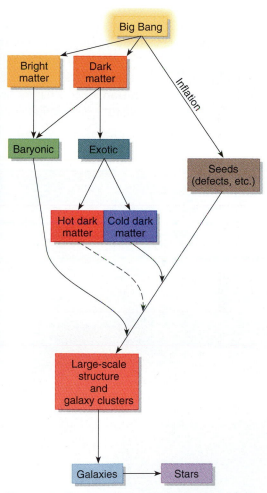

Figure 19.16
Galaxies may have been created from the Big Bang through seeds in the form of defects and vacuum fluctuations that accumulate dark and baryonic matter into large-scale structures (like the Great and Southern walls), galaxies, and stars.

grow as a result of the attraction of baryons. Their growth is especially fast if they can attract not only baryons but non-baryonic dark matter, which may be of two different types. Neutrinos, which were made in great abundance in the Big Bang and move at or near the speed of light, are **hot dark matter.** An alternative is slower **cold dark matter,** which consists of exotic particles, perhaps the WIMPS. We have no idea which candidate or combination of candidates may actually be present, or for that matter, whether we are on the right track at all!

Given that matter *can* accumulate from the Big Bang, there are two more possibilities. Either galaxies formed first and then accumulated into clusters, superclusters, and larger structures (like the Great and Southern walls seen in Figure 18.7), or the larger structures formed first and then collapsed into galaxies. Computer simulations of the latter scenario show structures that are remarkably similar to those observed (Figure 19.17).

Sections 18.5 and 18.6 outline at least some of the pathways of continuing development and evolution, once galaxies are formed. Some of the new galaxies are born as brilliant quasars that subsequently fade into ac-

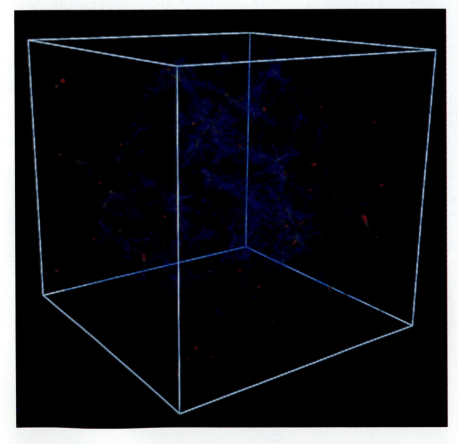

Figure 19.17

The box shows a local volume of space 150 Mpc on a side produced by a computer simulation in which baryonic matter (protons and neutrons) was accumulated under the action of a combination of hot and cold dark matter. Red represents high-density regions, blue low-density regions. This distribution is reminiscent of the real one seen in Figure 18.7.

tive galaxies and then perhaps into systems like our own. Others are born as the faint blue star-forming galaxies seen in Figure 18.8 that may eventually fade to become like some of the large dim systems now seen nearby. Imaging with the Hubble Space Telescope back to times when the Universe was only 15% of its present age reveal ragged-looking disrupted systems that gradually develop into the older spirals and ellipticals we see nearby at low redshift (Figure 19.18). Atop these scenes lie continuing collisions and mergers that produce ellipticals from spirals, giant ellipticals at the centers of massive clusters, and our own complex Galaxy.

Figure 19.18
The Hubble Space Telescope looks backward in time—to earlier days—from left to right. From right to left we therefore see the possible development of galaxies with time, ellipticals across the top and spirals in the middle and bottom. The ellipticals seem to have developed their forms early. The spirals seem to have taken longer. As time proceeds, the spirals develop ragged star-forming arms that grew into the elegant arms of today.

19.5 WE AND THE COSMOS

In spite of the apparent successes of current Big Bang theory, it is important to realize that it is still a *partial* theory. Parts of it are successful and other parts are not. However, a decade or two ago, it was even less successful. The Big Bang provides a theoretical framework that seems to work, one on which astronomers and cosmologists continue to build.

Yet we must bear in mind that every generation has thought it understood the basic aspects of the Universe, and every generation has been wrong in some crucial regard. Even if we are on the right track, there is a profusion of unknowns. We do not have agreement on the value of the Hubble constant. Various measurements and estimates of Omega run from well under 0.1 to 1. We do not know whether the Universe is infinite or finite, open or closed, or even the nature of the majority of its matter: if Ω is truly 1, then some 99% of the Universe's mass is shrouded in some form of mystery. Rapid developments over the past few years, particularly of inflationary cosmology, also show us that even the basic notions of Big Bang theory are in a continuous state of change. The inflationary universe (that brought about by inflation), for example, may be self-reproducing, some cosmologists envisaging an eternally growing tree of universes consisting of branches that produce yet more branches, each out of sight of the other. The Universe might contain an infinite number of big bangs, the chain going on forever.

Our ignorance is hardly confined to cosmology. We have yet to establish a full inventory of the Solar System and do not even know how our own Sun affects the Earth and its climate. We do not understand the deficit of solar neutrinos, nor even know the mass of a neutrino. With satellites we observe sudden and powerful gamma-ray bursts blossom all over the sky. No one knows what they are. Theories have ranged from events in the Oort Comet cloud through neutron star activity in an extended galactic halo to collisions between neutron stars and/or black holes at the fringes of the observable Universe. Yet none of these problems need be discouraging. Science, after all, is about wonder, about delight in mystery, about exploration of the unknown. Were there no unknown, there would be no science. Look first to what we *have* learned; then look to the excitement of learning yet more.

In the end, as remarkable as this grand cosmic structure is, we should perhaps remember that all of it can exist—can be comprehended—within our own minds, and all of it was necessary to create us. Galaxies developed only because of the way in which the Big Bang proceeded. We are here because of the way in which the stars evolved, building heavy elements out of light and throwing them to the stellar winds out of which the Sun and Earth and humanity were ultimately created. As we stand outside among the sparkling stars, we realize that all the Universe is our birthplace and our rightful domain.

The Big Bang provides a theoretical framework on which astronomers and cosmologists continue to build.

Figure 19.19
We journey home from distant reaches of the Universe.

KEY CONCEPTS

Baryons: Collectively, protons, antiprotons, and neutrons.

Closed, open, and **marginally open universes:** Respectively, a positively curved, spherical universe that folds back on itself; a negatively curved, hyperbolic universe that does not close back (both of these models are **Friedmann-Lemaître universes**); and a flat, Euclidean **Einstein-de Sitter universe.**

Copernican Principle: The Earth is not special; extended to the **Cosmological Principle** (no place in the Universe is special); extended further to the **Perfect Cosmological Principle** (there are neither special places nor times in the Universe).

Cosmic microwave background radiation: All-pervading blackbody radiation near 2.7 K, cooled from the Big Bang fireball.

Grand Unified Theories (GUTS): Theories combining the electromagnetic, weak, and strong forces.

Horizon of the Universe: The maximum distance we can look, which is determined by the age and structure of the Universe.

Hot and **cold dark matter:** Respectively, fast-moving dark matter (for example, neutrinos), and slow-moving dark matter (possibly WIMPS) that can cause the formation of galaxies and may help bring Ω to 1.

Infinite: Never-ending.

Inflation: A super-rapid expansion of the Universe, that may have occurred just after the Big Bang.

Look-back time: The time it has taken for light to reach us from a distant source.

Omega (Ω): The ratio of the true average density of the Universe to the **critical density,** the density needed to give a marginally closed universe.

Theory of Everything: A theory that combines all the forces of nature.

KEY RELATIONSHIPS

Age of an empty universe:
$$t = t_0 = 1/H_0.$$

Age of a flat universe:
$$t = 2/3\ t_0 = (2/3)/H_0.$$

EXERCISES

Comparisons

1. Compare the Copernican Principle, the Cosmological Principle, and the Perfect Cosmological Principle.

2. Describe the differences in fates of open and closed universes.

3. Compare the forces in Grand Unified Theories and in a Theory of Everything.

4. Distinguish between hot and cold dark matter.

Numerical Problems

5. What is the critical density if $H_0 = 60$ km/s/Mpc?

6. What should be the actual age of a marginally open universe with a Hubble constant of 35 km/s/Mpc?

7. When the Universe doubles in size 13 billion years from now, what will be the temperature of the cosmic microwave background?

Thought and Discussion

8. What does Olbers' paradox tell us about the Universe?

9. What evidence suggests that the cosmological force might exist?

10. Which theory of the Universe derives from the Perfect Cosmological Principle? What observations negate it? What does this theory require?

11. What is the origin of the cosmic microwave background radiation? Why does its temperature drop as the Universe ages?

12. What deviations from perfect uniformity are seen in the cosmic background radiation? What causes them?

13. Which property of quasars negates the Steady State theory?

14. How do parallel lines behave in **(a)** Euclidean space; **(b)** in spaces that are not Euclidean?

15. How does the curvature of the Universe relate to whether it is open or closed?

16. Why can the Universe have no center?

17. Explain why the actual age of the Universe is related to the curvature of space.

18. Summarize the evidence for the existence of the Big Bang.

19. What is Omega? How does it relate to the structure of the Universe? How can we find its value?

20. Define the horizon of the Universe.

21. Which problems presented by the Big Bang are solved by inflation?

22. What is meant by broken symmetry with regard to the forces of nature?

23. What possible roles do weakly interacting massive particles play in the Universe?

24. Name the isotopes made in the Big Bang. Why were heavier isotopes not made? What do they tell us about the early conditions in the Universe?

25. Identify the candidates for the seeds that can accumulate mass for galaxy formation.

Research Problem

26. Using your library, examine and list opinions that have appeared in the press on the reality of the Big Bang.

Activities

27. From the figures in the text, construct a detailed timetable of the sequence of events in the theory of the Big Bang that includes the idea of inflation. Include the approximate temperature at each point. Comment on the importance of each event along the way.

28. List the problems and uncertainties of modern cosmology; then list some observations that might resolve them.

Scientific Writing

29. The theory of the Big Bang is sometimes ridiculed in the press. Write a rebuttal letter to the editor of a newspaper demonstrating that the Big Bang is a valid theory that predicts and describes many of the properties of the observed Universe.

30. Write a five-page history of our view of the Universe; discuss how it has changed since Aristotle's time, and speculate on how it might change in the future.

STAR MAPS

The following six maps locate the constellations and brighter stars. The first shows the north polar region, the next four are seasonal equatorial maps, and the last shows the south polar region. Stars and other objects are located on each map according to their *right ascensions* and *declinations*. Declination is the sky's version of Earth's latitude, right ascension of Earth's longitude. To find a star's right ascension and declination, draw a great circle (called an *hour circle*) from one celestial pole to the other through the star. Declination is the arc measured along the hour circle north or south of the celestial equator to the star. Right ascension is the arc between the vernal equinox and the point of intersection of the hour circle and the celestial equator, measured to the east. Right ascension is measured not in degrees but in time units, in hours, minutes, and seconds of time: 1 hour = 15° and 1° = 4 minutes. (See Figure A1.1.)

Declinations are indicated on the maps along a central hour circle. Right ascensions are noted around the peripheries of the polar maps and

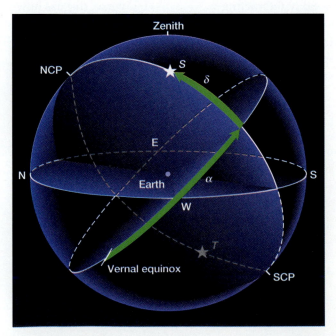

Figure A1.1

Right ascension (α) is measured counterclockwise from the vernal equinox to the intersection of the hour circle and the celestial equator. Declination (δ) is measured along the hour circle north and south of the celestial equator. The right ascension of star *S* is about 6 hours (or 90°); its declination is about 35°N.

along the celestial equator for the equatorial maps. The polar maps show the sky from the poles to 50° north or south declination; the equatorial maps show it between 60°N and 60°S declination.

Stars are generally selected to indicate constellation positions and outlines. They are generally shown through fourth magnitude, although the census here is not complete. A few fainter stars are included to indicate the locations of obscure constellations. All 88 constellations are represented, although several are indicated by only their brightest stars. A small number of non-stellar objects—clusters, nebulae, and galaxies—are also included.

Colors are indicated for the 48 brightest stars, those through magnitude 2.0, down to the brightness of Polaris. The colors are greatly exaggerated and are assigned according to spectral class as follows:

Blue—Class O through B5
Light blue—B6 through B9
White—Class A (all stars fainter than magnitude 2.0 are also assigned white)
Pale yellow—Class F
Yellow—Class G
Orange—Class K
Red—Class M

The maps show the broad outline of the Milky Way, though much of its intricate detail is omitted. The galactic equator, the mid-line of the Galaxy, is marked with *galactic longitude* (similar to right ascension but measured along the galactic rather than along the celestial equator) starting at the galactic center in Sagittarius.

Complete paths of precession of the North and South Celestial Poles are shown on Maps 1 and 6. Partial paths are shown for the NCP on Maps 2 and 5 and for the SCP on Maps 3 and 4; these paths do not appear as circles because of the distortions inherent in mapping a sphere onto a plane.

The times of year given around the edges of the polar maps and along the tops and bottoms of the equatorial maps indicate the appearance of the sky at approximately 8:30 P.M. (20^h30^m) local time. To use the north polar map, face north and rotate the map so that the current month appears at the top (in the southern hemisphere use the south polar map similarly). The celestial pole will have an elevation in degrees equal to your latitude. To use the equatorial maps in the northern hemisphere face south and line up the current month with the celestial meridian (to use them in the southern hemisphere, face north and turn them upside down). The point of intersection between the celestial equator and the meridian will have an elevation in degrees equal to 90° minus the latitude.

For each hour past 20^h30^m, shift or rotate the map one hour of right ascension (15°) to the west, that is, align an hour circle that is one additional hour to the east. For every 2 hours past 20^h30^m add one month to your current month. For example, if it is March 15 at 20^h30^m, you would align "March" on Map 4 with the celestial meridian. If it is 2^h30^m, set "June" (Map 5) on the meridian.

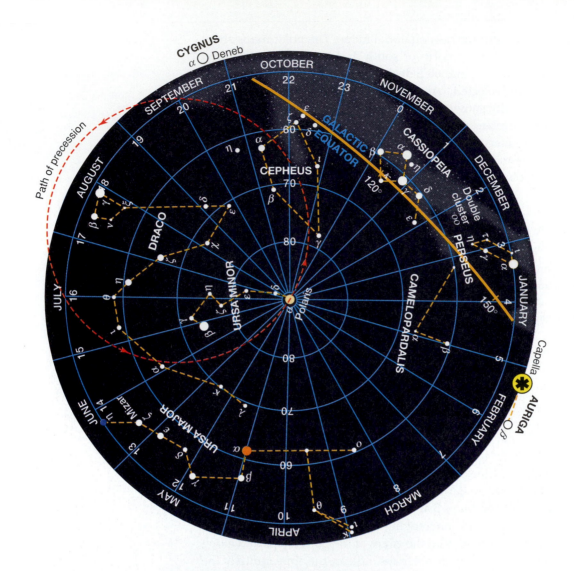

SCALE OF MAGNITUDES

Map 1 The North Polar Constellations

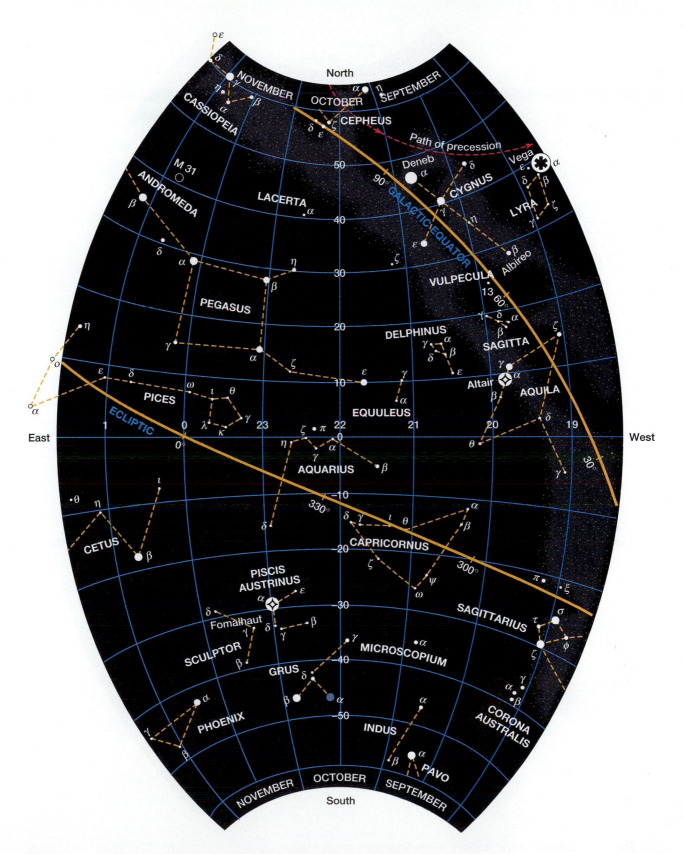

North

SEPTEMBER

OCTOBER

NOVEMBER

CEPHEUS

CASSIOPEIA

ε
δ
η γ
α β

α η

δ ε ζ

Path of precession

50

Deneb
α

90°

GALACTIC EQUATOR

Vega
α

CYGNUS

LYRA

δ β
γ ζ

M 31

ANDROMEDA
β

40

α

ε

ζ

VULPECULA

Albireo
β

13 60°

LACERTA

δ
α

30

β

η

PEGASUS

γ

DELPHINUS

γ δ
β

SAGITTA
γ δ
β

SAGITTA

ζ

α

γ
δ β

γ

ζ

η

o o

ε δ

ω ι θ

PICES

α

ζ

10

ε

γ

EQUULEUS

α

Altair
α

AQUILA

δ

ECLIPTIC

0

1

λ κ

γ

23

π

22

ζ

0

21

β

20

δ

19

East

η ζ

γ

AQUARIUS

β

θ

γ

West

30°

ι

θ η

β

δ

−10

330°

γ

ι θ

α

β

CAPRICORNUS

CETUS

β

ζ

−20

ψ

300°

PISCIS
AUSTRINUS

α ε

ω

SAGITTARIUS

π
ξ

δ

γ δ β

Fomalhaut
β

−30

σ
τ

SCULPTOR

β

MICROSCOPIUM
α

ζ φ

GRUS
δ

−40

γ

α

γ
α β

β

α

INDUS

CORONA
AUSTRALIS

PHOENIX
α

−50

β

α

PAVO

γ

β

NOVEMBER   OCTOBER   SEPTEMBER

South

Map 2 The constellations of Northern Autumn, Southern Spring

Map 3 The Constellations of Northern Winter, Southern Summer

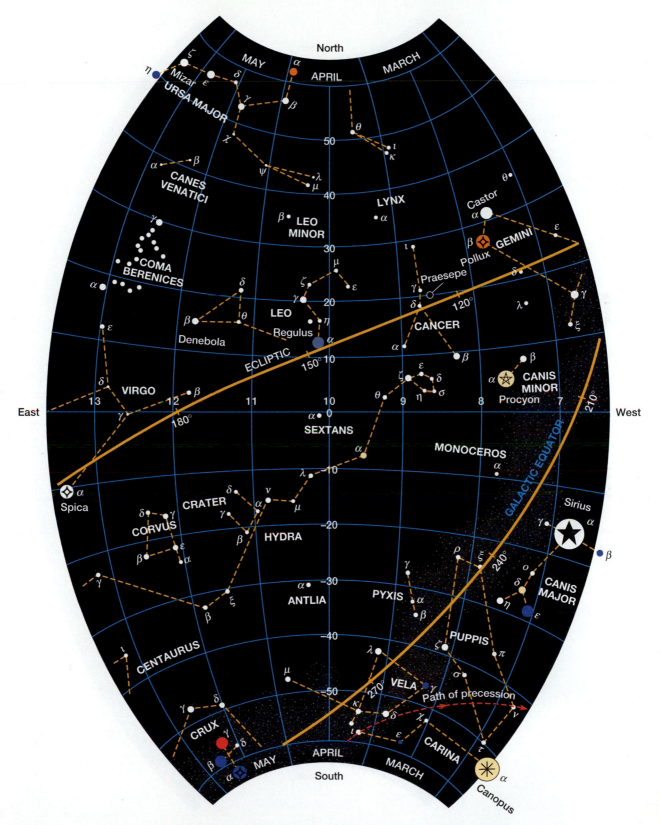

Map 4 The Constellations of Northern Spring, Southern Autumn

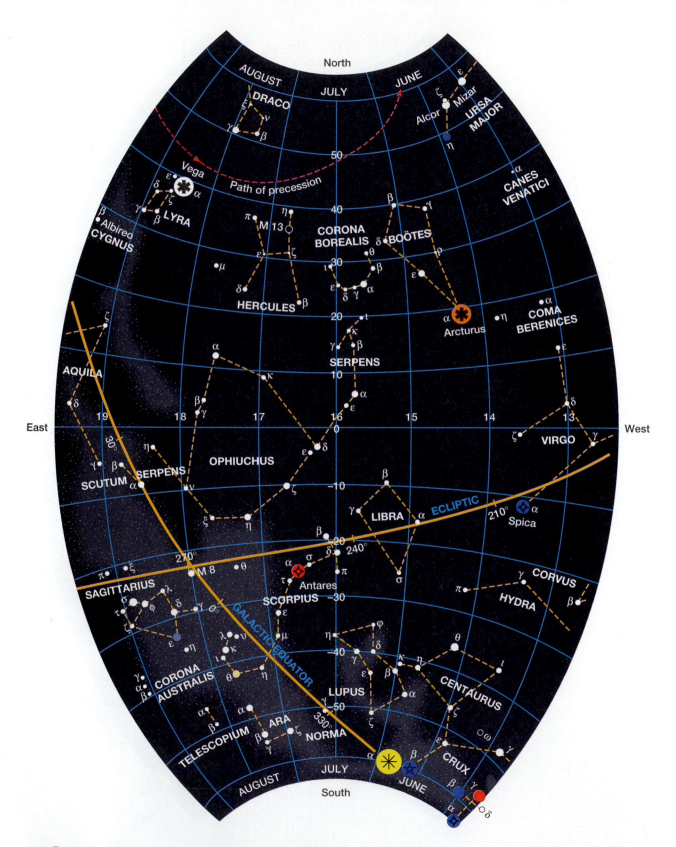

Map 5 The Constellations of Northern Summer, Southern Winter

Map 6 The South Polar Constellations

ASTRONOMICAL OBJECTS

The Messier Catalogue is a list of 103 clusters, galaxies, and nebulae compiled by Charles Messier between 1781 and 1784 and extended to 109 objects in 1786 by Pierre Méchain. The Catalogue is the fundamental list of such objects for beginning observers and contains some of the most famous and lovely of celestial sights.

TABLE A2.1
The Messier Catalogue

Messier Number	NGC	Constellation	α^a	δ^a	Angular Size	Description
1	1952	Taurus	05 35	+22 01	5′	*Crab Nebula*, remnant of supernova of 1054
2	7089	Aquarius	21 33	−00 50	12′	Globular cluster
3	5272	Canes Venatici	13 42	+28 25	19′	Bright globular cluster; binocular object
4	6121	Scorpius	16 24	−26 31	23′	Globular cluster; binocular object
5	5904	Serpens	15 19	+02 05	20′	Globular cluster
6	6405	Scorpius	17 40	−32 12	26′	Open cluster; easy binocular object
7	6475	Scorpius	17 54	−34 49	50′	Magnificent open cluster; naked-eye object
8	6523	Sagittarius	18 04	−23 23	1°	*Lagoon Nebula;* bright diffuse nebula; naked-eye object
9	6333	Ophiuchus	17 19	−18 31	6′	Globular cluster
10	6254	Ophiuchus	16 57	−04 06	12′	Globular cluster; binocular object
11	6705	Scutum	18 51	−06 18	12′	Open cluster; striking in telescope
12	6218	Ophiuchus	16 47	−01 57	12′	Globular cluster; binocular object
13	6205	Hercules	16 42	+36 30	23′	*Great Cluster in Hercules;* magnificent globular cluster; barely naked-eye; easy in binoculars
14	6402	Ophiuchus	17 38	−03 15	7′	Globular cluster
15	7078	Pegasus	21 30	+12 09	12′	Globular cluster
16	6611	Serpens	18 19	−13 46	8′	Open cluster
17	6618	Sagittarius	18 21	−16 12	40′	*Omega Nebula; Horseshoe Nebula;* diffuse nebula
18	6613	Sagittarius	18 21	−17 08	7′	Open cluster
19	6273	Ophiuchus	17 03	−26 15	5′	Globular cluster
20	6514	Sagittarius	18 03	−23 02	30′	*Trifid Nebula;* diffuse nebula
21	6531	Sagittarius	18 05	−22 30	12′	Open cluster
22	6656	Sagittarius	18 36	−23 55	17′	Bright globular cluster; binocular object
23	6494	Sagittarius	17 57	−19 01	27′	Open cluster
24	——	Sagittarius	18 18	−18 29	1.5°	Star cloud in Milky Way; naked-eye object

TABLE A2.1
The Messier Catalogue (continued)

Messier Number	NGC	Constellation	α^a	δ^a	Angular Size	Description
25	IC4725	Sagittarius	18 32	−19 15	35′	Open cluster; binocular object
26	6694	Scutum	18 45	−09 25	9′	Open cluster
27	6853	Vulpecula	20 00	+22 43	6′	*Dumbbell Nebula;* planetary nebula
28	6626	Sagittarius	18 25	−24 52	15′	Globular cluster
29	6913	Cygnus	20 24	+38 32	7′	Open cluster
30	7099	Capricornus	21 40	−23 11	9′	Globular cluster
31	224	Andromeda	00 43	+41 15	1° × 2°	*Great Nebula in Andromeda; Andromeda Galaxy;* spiral galaxy; naked-eye object
32	221	Andromeda	00 43	+40 51	3′	Elliptical galaxy; companion to M 31
33	598	Triangulum	01 34	+30 38	1°	*Triangulum Spiral;* spiral galaxy; binocular object, just visible to naked eye
34	1039	Perseus	02 42	+42 47	30′	Open cluster
35	2168	Gemini	06 09	+24 21	30′	Open cluster; easy binocular object
36	1960	Auriga	05 36	+34 08	16′	Open cluster
37	2099	Auriga	05 52	+32 33	24′	Open cluster; binocular object
38	1912	Auriga	05 29	+35 51	18′	Open cluster
39	7092	Cygnus	21 32	+48 26	32′	Open cluster
40	——					Does not exist
41	2287	Canis Major	06 47	−20 44	32′	Open cluster; binocular object
42	1976	Orion	05 35	−05 25	1°	*Orion Nebula;* bright diffuse nebula; easy binocular object
43	1982	Orion	05 35	−05 16	10′	Diffuse nebula at northern edge of Orion Nebula
44	2632	Cancer	08 40	+20 00	1.5°	*Beehive* or *Praesepe Cluster;* open cluster; naked-eye object
45	——	Taurus	03 48	+24 06	2°	*Pleiades; Seven Sisters;* open cluster; obvious naked-eye object
46	2437	Puppis	07 42	−14 49	27′	Open cluster
47	2422	Puppis	07 37	−14 29	25′	Open cluster; naked-eye object
48	2548	Hydra	08 14	−05 47	30′	Open cluster; binocular object
49	4472	Virgo	12 30	+07 59	4′	Elliptical galaxy
50	2323	Monoceros	07 03	−08 20	16′	Open cluster
51	5194	Canes Venatici	13 30	+47 11	10′	*Whirlpool Nebula;* spiral galaxy
52	7654	Cassiopeia	23 26	+61 35	13′	Open cluster
53	5024	Coma Berenices	13 13	+18 10	14′	Globular cluster
54	6715	Sagittarius	18 55	−30 28	6′	Globular cluster
55	6809	Sagittarius	19 40	−30 56	15′	Globular cluster
56	6779	Lyra	19 17	+30 02	5′	Globular cluster
57	6720	Lyra	18 54	+33 01	1.2′	*Ring Nebula in Lyra;* planetary nebula
58	4579	Virgo	12 38	+11 48	4′	Spiral galaxy
59	4621	Virgo	12 42	+11 39	3′	Elliptical galaxy
60	4649	Virgo	12 44	+11 33	4′	Elliptical galaxy
61	4303	Virgo	12 22	+04 28	6′	Spiral galaxy
62	6266	Ophiuchus	17 01	−30 07	6′	Globular cluster
63	5055	Canes Venatici	13 16	+42 01	6′	Spiral galaxy
64	4826	Coma Berenices	12 58	+21 41	6′	Spiral galaxy
65	3623	Leo	11 19	+13 07	6′	Spiral galaxy
66	3527	Leo	11 20	+13 01	6′	Spiral galaxy
67	2682	Cancer	08 51	+11 51	18′	Open cluster; one of oldest known

TABLE A2.1
The Messier Catalogue (continued)

Messier Number	NGC	Constellation	α^a	δ^a	Angular Size	Description
68	4590	Hydra	12 39	−26 45	9′	Globular cluster
69	6637	Sagittarius	18 31	−32 21	4′	Globular cluster
70	6681	Sagittarius	18 43	−32 18	4′	Globular cluster
71	6838	Sagitta	19 54	+18 47	6′	Globular cluster
72	6981	Aquarius	20 53	−12 33	5′	Globular cluster
73	6994	Aquarius	20 59	−12 38	——	Four stars
74	628	Pisces	01 37	+15 47	8′	Spiral galaxy
75	6864	Sagittarius	20 06	−21 56	5′	Globular cluster
76	650-1	Perseus	01 42	+51 34	1′	Planetary nebula
77	1068	Cetus	05 47	−00 02	2′	Spiral galaxy
78	2068	Orion	05 47	+00 03	7′	Diffuse nebula
79	1904	Lepus	05 24	−24 31	8′	Globular cluster
80	6093	Scorpius	16 17	−22 59	5′	Globular cluster
81	3031	Ursa Major	09 56	+69 04	13′	*Great Spiral in Ursa Major;* spiral galaxy; binocular object
82	3034	Ursa Major	09 56	+69 42	7′ × 2′	Irregular galaxy
83	5236	Hydra	13 37	−29 52	9′	Spiral galaxy
84	4374	Virgo	12 25	+12 53	3′	Elliptical galaxy
85	4382	Coma Berenices	12 25	+18 11	3′	Elliptical galaxy
86	4406	Virgo	12 26	+12 56	4′	Elliptical galaxy
87	4486	Virgo	12 31	+12 23	3′	*Virgo A;* active elliptical galaxy
88	4501	Coma Berenices	12 32	+14 25	6′ × 3′	Spiral galaxy
89	4552	Virgo	12 36	+12 33	2′	Elliptical galaxy
90	4569	Virgo	12 37	+13 09	6′ × 3′	Spiral galaxy
91	——	——	——	——	——	Erroneous discovery; unexplained
92	6341	Hercules	17 17	+43 08	12′	Globular cluster
93	2447	Puppis	07 45	−23 52	18′	Open cluster
94	4736	Canes Venatici	12 51	+41 07	5′	Spiral galaxy
95	3351	Leo	10 44	+11 42	3′	Spiral galaxy
96	3368	Leo	10 47	+11 49	7′ × 4′	Spiral galaxy
97	3587	Ursa Major	11 15	+55 02	3′	*Owl Nebula;* planetary nebula
98	4192	Coma Berenices	12 14	+14 54	8′ × 2′	Spiral galaxy
99	4254	Coma Berenices	12 19	+14 25	4′	Spiral galaxy
100	4321	Coma Berenices	12 23	+15 49	5′	Spiral galaxy
101	5457	Ursa Major	14 03	+54 21	22′	*Pinwheel Galaxy;* spiral galaxy
102	——	——	——	——	——	Same as M 101
103	581	Cassiopeia	01 33	+60 41	6′	Open cluster
104	4594	Virgo	12 40	−11 37	7′ × 2′	*Sombrero Galaxy;* spiral galaxy
105	3379	Leo	10 48	+12 35	2′	Elliptical galaxy
106	4258	Canes Venatici	12 19	+47 18	20′ × 6′	Spiral galaxy
107	6171	Ophiuchus	16 32	−13 03	8′	Globular cluster
108	3556	Ursa Major	11 12	+55 41	8′ × 2′	Spiral galaxy
109	3992	Ursa Major	11 58	+53 22	7′	Spiral galaxy

$^a\alpha$ (right ascension) and δ (declination) are explained in the introduction to the star maps in Appendix 1.

TABLE A2.2
Some Visual Binaries

Name	α^a h m	δ^a ° ′	Magnitudes (V)	Spectral Classes	Separation (″)	Remarks
γ And	02 04	+42 20	2.26–4.84	K3 II–B8 V+A0 V	10	Gold, blue
γ Ari	01 54	+19 18	4.75–4.83	A1p–B9 V	7.8	Both white
ϵ Boo	14 45	+27 05	2.70–5.12	K0 II–A2 V	2.9	Orange, white
ζ Cnc	08 12	+17 39	5.44–6.20	F9 V–G9 V	0.9	1968, $P = 60$ yr
ζ Cnc A–C[b]			6.01	G5 V	5.8	Triple star
α Cen	14 40	−60 50	−0.01–1.33	G2 V–K1 V	8.7	1946, $P = 81$ yr
α Cru	12 27	−63 06	1.58–2.09	B0 IV–B1 V	4.4	A is spectroscopic binary
β Cyg	19 31	+27 58	3.08–5.61	K3 II + B0 V–B8 V	34	Gold, blue
ν Dra	17 32	+55 11	4.87–4.88	A4 Vm–A6 V	62	Both white
α Gem	07 35	+31 54	1.58–1.59	A2 Vm–A1 V	1.8	Both spectroscopic binaries
α Gem A–C[b]			9.5		73	Spectroscopic binary
γ Leo	10 20	+19 50	2.61–3.80	K1 III–G7 III	4.4	
α Lib	14 50	−16 01	2.75–5.80	A3 IV–F4 V	231	Naked eye
ϵ Lyr	18 44	+39 39			208	Naked eye
ϵ^1 Lyr	18 44	+39 40	5.06–6.02	A4 V–F1 V	2.8	
ϵ^2 Lyr	18 44	+39 37	5.37–5.71	A8 V–F0 V	2.2	
σ Ori	05 39	−02 36	4.1–5.1	O9.5 V–B0.5 V	0.2	Multiple
σ Ori A–C[b]			8.79	A2 V	11	
σ Ori A–D[b]			6.62	B2 V	13	
σ Ori A–E[b]					42	
θ^1 Ori	05 35	−05 23	6.73–7.96	O7–B0 V	8	The Trapezium; many more components make a small cluster
θ^1 Ori A–C[b]			5.13	O6	12	
θ^1 Ori A–D[b]			6.70	B0.5 V	21	
ζ UMa[c]	13 24	+54 55	2.27–3.95	A1 V–A1 Vm	14	Both spectroscopic binaries
γ Vir	12 42	−01 27	3.65–3.68	F0 V–F0 V	4.7	

[a] α (right ascension) and δ (declination) are explained in the introduction to the star maps in Appendix 1.

[b] Multiple star. Data are given for the second named component (C, D, or E) in relation to the first (A).

[c] Has the same proper motion as Alcor; quintuple star.

TABLE A2.3
Some Bright Clusters

Naked-Eye Open Clusters

Name	Catalogued As	Constellation	α^a h m	δ^a ° ′	Diameter (min′)	Distance (pc)	Magnitude V^b
h Per[c]	NGC 869	Perseus	02 19	+57 07	29	2,200	7
χ Per[c]	NGC 884	Perseus	02 22	+57 05	29	2,200	7
Pleiades	M 45	Taurus	03 47	+24 06	110	138	3
Hyades	——	Taurus	04 26	+15 51	330	45	4
——	M 35	Gemini	06 08	+24 21	28	850	8
Praesepe	M 44	Cancer	08 40	+20 01	95	160	6
Coma Berenices		Coma Berenices	12 25	+26 10	120	90	5
Jewel Box	κ Crucis	Crux	12 53	−60 18	10	1,500	7
——	M 6	Scorpius	17 40	−32 12	14	490	7
——	M 7	Scorpius	17 54	−34 38	80	240	7

Bright Globular Clusters

Name	NGC	Constellation	α h m	δ ° ′	V^d	Distance (pc)	Remarks
47 Tuc	104	Tucana	00 24	−72 08	4.0	4,300	Naked eye
ω Cen	5139	Centaurus	13 26	−47 36	3.7	6,080	Naked eye; magnificent
M 3	5272	Canes Venatici	13 42	+28 26	6.4	10,000	
M 13	6205	Hercules	16 42	+36 28	5.9	7,400	Barely naked eye
M 22	6656	Sagittarius	18 36	−23 55	5.1	5,200	Contains planetary nebula
M 15	7078	Pegasus	21 30	+12 08	6.4	11,300	Contains planetary nebula

[a] α (right ascension) and (declination) are explained in the introduction to the star maps in Appendix 1.
[b] Of brightest star.
[c] Together, the Double Cluster in Perseus.
[d] Collective magnitude of all stars.

TABLE A2.4
Prominent Diffuse Nebulae

Popular Name	Constellation	Messier Number	NGC	α^a h m	δ^a ° ′	Distance (pc)	Associated with
California	Perseus	——	IC499	04 00	+36 00	400	ζ Per
Orion	Orion	42	1976	05 35	−05 25	420	θ¹ Ori
Rosette	Monoceros	——	2224	06 37	+05 00		Open cluster NGC 2237
Carina	Carina		3324 +3372	10 43	−59 00		η Carinae, HD 93129A
Trifid	Sagittarius	20	6514	18 03	−23 02	1,000	
Lagoon	Sagittarius	8	6523	18 04	−24 23	1,100	Open cluster NGC 6530
Eagle	Serpens	16	6611	18 18	−13 45	2,900	Open cluster of same name
Omega	Sagittarius	17	6618	18 21	−16 13	900	Open cluster IC 4707
North America	Cygnus	——	7000	20 45	+30 42	——	——

[a] a (right ascension) and (declination) are explained in the introduction to the star maps in Appendix 1.

TABLE A2.5
Prominent Variable Stars

Cepheids

Star	α^a h m	δ^a ° ′	Period (days)	Magnitude (V)	Spectral Class
T Vulpeculae	20 51	+28 13	4.44	5.44−6.0	F5 Ib−G0 Ib
FF Aquilae	18 58	+17 21	4.47	5.8−6.2	F5 Ia−F8 Ia
δ Cephei	22 29	+58 23	5.37	3.6−4.3	F5 Ib−G2 Ib
Y Sagitarii	18 21	−18 52	5.40	5.40−6.1	F6 I−G5 I
X Sagitarii	17 47	−29 50	7.01	4.24−4.8	F5 II−G9 II
η Aquilae	19 52	+00 59	7.18	3.7−4.5	F6 Ib−G2 Ib
W Sagitarii	18 04	−29 35	7.59	4.30−5.0	F2 II−G6 II
S Sagittae	19 55	+16 37	8.38	5.5−6.2	F6 Ib−G5 Ib
β Doradus	05 34	−62 30	9.84	3.8−4.7	F6 Ia−G2 Iab
ζ Geminorum	07 04	+20 35	10.15	3.68−4.16	F7 Ib−G3 Ib

Long-Period Variables

Name	α^a h m	δ^a ° ′	Period (days)	Magnitude (V)	Spectral Class[b]
Mira (o Cet)	02 19	−03 01	332	2.0−10.0	M5 IIIe−M9 IIIe
R Horologii	02 54	−49 55	404	4.7−14.3	M7 IIIe
R Leporis	04 59	−14 49	432	5.5−11.7	C6 IIe
U Orionis	05 55	+20 11	372	5.4−12.6	M6.5 IIIe
R Carinae	09 32	−62 45	309	3.9−10.5	M4 IIIe−M8 IIIe
R Leonis	09 47	+11 28	312	5.2−10.5	M8 IIIe
S Carinae	10 09	−61 31	149	5.4−9.9	K5 IIIe−M6 IIIe
R Hydrae	13 29	−23 15	390	4.5−9.5	M7 IIIe
χ Cygni	19 50	+32 54	407	3.3−14.2	S6 IIIe−S10 IIIe
R Aquarii	23 43	−15 20	389	5.8−12.4	M5 IIIe−M8.5 IIIe

[a]a (right ascension) and (declination) are explained in the introduction to the star maps in Appendix 1.

[b]Spectral classes S and C have about the same temperatures as class M but are respectively moderately and heavily enriched in carbon.

TABLE 2.6
Famous Twentieth-Century Novae

Nova	Variable-Star Name	V_{max}	$M_{V(max)}{}^a$
Persei 1901	GK Per	+0.2	−8.5
Aquilae 1918	V 603 Aql	−1.1	−9.2
Pictoris 1925	RR Pic	+1.2	−7.4
Herculis 1934	DQ Her	+1.4	−5.6
Puppis 1942	CP Pup	+0.4	−9.1
Cygni 1975	V 1500 Cyg	+1.8	−9.5

[a]Corrected for dimming by interstellar dust (see Chapter 14).

GLOSSARY

A stars: White stars with strong hydrogen lines and temperatures of 7,400 to 9,900 K.

AAAOs: Asteroids in the Amor, Apollo, or Aten families, all of which come close to the Earth's orbit.

Aberration of starlight: The shift in the apparent position of a star caused by the motion of the Earth relative to the velocity of light.

Absolute magnitudes (*M*): Apparent magnitudes that celestial bodies would have at a distance of 10 pc.

Absolute zero: The lowest possible temperature, achieved when all possible heat has been removed from a body.

Absorption lines: Dark spectrum lines, or gaps, indicating lack of radiation at specific wavelengths in a continuous electromagnetic spectrum.

Acceleration (*A*): A change in velocity, that is, a change in speed or direction.

Acceleration of gravity (*g*): The acceleration of a falling body in a gravitational field.

Accretion disk: A disk of gas revolving around a star or other body from which matter flows onto that body.

Achondrites: Chondrite-like meteorites that have no chondrules.

Active galaxies: Galaxies with bright, variable nuclei sometimes associated with bipolar jets, probably caused by matter falling into massive black holes from accretion disks.

Active Sun: The collective phenomena of the solar magnetic cycle.

Aitken basin: The deepest and one of the largest basins on the Moon; on the backside, near the Moon's south pole, and not visible from Earth.

Alpha particles: Helium nuclei.

Altitude (*h*): The angular elevation of a body above the horizon.

Amino acids: The chemical building blocks of proteins and of life, found in meteorites.

Ancient constellations: The 48 constellations handed down by the ancient Greeks.

Angle of incidence: The angle between an approaching light ray and the perpendicular to the surface it will strike.

Angle of reflection: The angle between a light ray and the perpendicular to the surface from which it will be reflected.

Angle of refraction: The angle between a refracted light ray and the perpendicular to the surface from which it has been refracted.

Angular diameter: The angle formed by lines projecting from the viewing point to opposite sides of a body.

Angular momentum: For an orbiting body, mass times velocity times orbital radius.

Annular eclipse: An eclipse of the Sun in which the Moon is too far from the Earth to cover the Sun completely, leaving a visible ring of sunlight.

Antarctic circle: The parallel of latitude, 66.5°S, below which it is possible to have a midnight Sun between September 23 and March 21.

Ante meridiem (A.M.): Before noon.

Antenna: That part of a radio receiver electrically excited by radiation.

Antimatter: Particles with reversed charges.

Aphelion: The point of greatest distance between a planet and the Sun.

Aphrodite Terra: The great equatorial rise on Venus.

Apogee: The point in the lunar orbit farthest from Earth.

Apollo program: The spaceflight program that landed human beings on the Moon.

Apparent magnitudes (*m*): Magnitudes of objects as measured by their appearance in the sky; apparent visual magnitudes are apparent magnitudes measured by eye or a yellow-green detector.

Arctic circle: The parallel of latitude, 66.5°N, above which it is possible to have a midnight Sun between March 21 and September 23.

Asterisms: Small named portions of constellations, or stellar groupings that extend over constellation boundaries.

Asteroid belt: A zone of debris that lies mostly between the orbits of Mars and Jupiter.

Asteroids: Small bodies that orbit the Sun mostly between Mars and Jupiter; concentrated in a main belt between 2 and 3 AU from the Sun; different kinds relate to meteorite classes.

Astronomical horizon: The great circle defined by the intersection between the celestial sphere and a plane at the observer's feet perpendicular to the line to the zenith.

Astronomical unit (AU): The semimajor axis of the Earth's orbit around the Sun and the average distance between the Earth and the Sun.

Asymptotic giant branch (AGB): The stage of stellar evolution in which intermediate-mass stars grow to become giants for the second time; these stars have contracting carbon/oxygen cores.

Atmosphere: The gases that surround a planet or satellite.

Atom: The basic unit of the chemical elements.

Atomic number: The number of protons in an atomic nucleus; the atomic number defines the chemical element.

Atomic weight: The number of protons plus neutrons in an atomic nucleus.

Aurora: Light in the upper atmosphere caused by the interaction between the terrestrial and solar magnetic fields and the solar wind; aurora borealis is seen in the north, aurora australis in the south.

Autumn: The period in the northern hemisphere between September 23 and December 22, when the Sun is moving southward in the southern celestial hemisphere (the period between these dates is spring in the southern hemisphere).

Autumnal equinox: The point on the celestial sphere where the ecliptic crosses the celestial equator as the Sun moves south; the crossing takes place about September 23.

Averted vision: A visual observing technique in which the observer does not look directly at an object but uses the more light-sensitive periphery of the eye.

Axis: a line about which a body rotates or moves.

B stars: Blue-white stars with strong hydrogen lines and neutral helium lines, and temperatures of 9,900 to 28,000 K.

Balmer series: A series of hydrogen absorption or emission lines that have the second energy level as their common bottom level.

Barred spirals: Spiral galaxies in which the arms come from the ends of a bar through the galaxies' centers.

Baryons: Collectively, protons, antiprotons, and neutrons.

Basalts: Fine-grained, metal-bearing silicates.

Basin: A depression in a planetary crust.

Bayer Greek letters: The Greek letters assigned to stars, usually in order of brightness within a constellation.

Belts: In Jupiter's atmosphere, strips of dark, lower-altitude, higher-temperature clouds.

Big Bang: A sudden expansion of space 10 to 20 or so billion years ago that is believed to have produced the present Universe.

Binary stars: Double stars with gravitationally bound components.

Bipolar flow: Twin flows of gas emitted perpendicular to a rotating disk.

Birthline: The line on the HR diagram, above and to the right of the main sequence, where protostars first become visible.

BL Lacertae objects: Bright nuclei of active elliptical galaxies from which a jet is coming directly at the observer.

Black hole: A mass with a radius so small that the force of its gravity keeps light from escaping, so the body cannot be seen.

Blackbody: A body that absorbs all the radiation that falls upon it and radiates an energy spectrum that depends only on temperature.

Blackbody (Planck) curves: Graphs of the brightnesses of blackbodies plotted against wavelength.

Bok globules: Small, opaque, relatively dense, dusty interstellar clouds, typically a few parsecs across, containing 10 to 100 solar masses.

Brown dwarfs: Bodies that contract directly from the interstellar medium, but with masses below the main sequence cutoff of 0.08 M., too low to allow the proton-proton chain to operate.

Bulge: The central, thick region of a spiral galaxy where the disk and the halo come together.

Caldera: A depression caused by the collapse of a portion of a volcano into an underground chamber emptied of magma.

Callisto: The outermost Galilean satellite of Jupiter.

Caloris: The major basin on Mercury.

Camera mirror: The mirror in a spectrograph that focuses dispersed light onto a detector.

Canals: Illusory linear features on Mars caused by the eye's linking of real features.

Carbon cycle: The fusion of four atoms of hydrogen into one of helium using, with carbon as a nuclear catalyst.

Carbonaceous chondrites: Primitive chondrite meteorites with a high carbon content.

Cassegrain focus: A configuration of a primary mirror and a curved secondary mirror that extends the focal length and sends the light back through a hole in the primary of a reflecting telescope.

Cassini division: A resonance gap that divides Saturn's A and B rings.

Celestial: Pertaining to the sky.

Celestial equator: The great circle on the celestial sphere above the Earth's equator, equidistant from the celestial poles.

Celestial meridian: The great circle through the celestial poles and the zenith.

Celestial poles: The points of rotation of the celestial sphere that lie above the north and south poles of the Earth.

Celestial sphere: The apparent sphere of the sky.

Celsius (centigrade) temperature scale (°C): A scale in which water freezes at 0°C and boils at 100°C.

Center of mass: The point at the mutual focus of two orbiting bodies.

Centers of activity: Magnetic regions of the solar surface that contain sunspots and associated phenomena.

Cepheid instability strip: A nearly vertical strip in the HR diagram among class F and G giants and supergiants in which Cepheid variable stars are found.

Cepheid variables: F and G giant and supergiant variables with regular periods.

Ceres: The largest and first-discovered asteroid.

Chandrasekhar limit: A limit of 1.4 $M_\odot$, above which a star cannot be supported by degenerate electrons, the maximum mass of a white dwarf.

Charge-coupled device (CCD): An electronic imaging device.

Charon: Pluto's satellite.

Chondrites: Stony meteorites with small round inclusions known as chondrules.

Chondrules: Small round inclusions found in stony meteorites.

Chromosphere: The thin reddish emission-line layer of the Sun that lies between the photosphere and the corona.

Chryse Planitia: The "Plains of Gold", a plain in the Martian northern hemisphere.

Circle: A conic section derived by slicing a cone with a plane parallel to the cone's base; an ellipse of zero eccentricity in which both foci are at the center.

Circumpolar stars: Stars that do not set on their daily paths.

Closed universe: A four-dimensional, positively curved, spherical universe that folds back on itself; parallel lines meet, and the angles of a triangle sum to more than 180°.

Clusters: Gravitationally bound groups of stars.

Clusters of galaxies: Gravitationally bound groups of galaxies.

Coalsack: A famous Bok globule located near the Southern Cross (Crux).

Cold dark matter: Slow-moving dark matter that can cause the formation of galaxies; possibly WIMPS.

Collimating mirror: The mirror in a spectrograph that makes converging light rays from the telescope parallel to one another.

Coma: The bright cloud of ions and dust that surrounds a comet's nucleus.

Coma cluster: A rich cluster of galaxies in Coma Berenices, about 100 Mpc away.

Comet: A fragile interplanetary body with a nucleus of dirty ice that is heated by the Sun to produce a surrounding coma and tails of gas and dust.

Comparative planetology: The science of comparing planets and their varied conditions, using one to learn about another.

Comparison spectrum: A spectrum with lines of known wavelength produced in a spectrograph to calibrate the wavelengths of spectrogram of a celestial object.

Conic sections: The curves (circle, ellipse, parabola, and hyperbola) defined by the intersections of a cone and a plane.

Conjunction: The position in which a planet (as viewed from Earth) is in the same direction as the Sun (or in which two celestial bodies are aligned with each other).

Conservation of angular momentum: The concept that in a closed system total angular momentum is always constant.

Constellations: Named patterns of naked-eye stars; 88 are now officially recognized and used.

Continental drift: The motion of continents across the mantle of the Earth.

Continents: Raised portions of the Earth's crust made out of light rock.

Continuous radiation: Radiation that has a spectrum with no breaks, gaps, or sudden changes in brightness.

Convection: The up-and-down circulation of a heated fluid.

Copernican Principle: The Earth is not special.

Copernican system: A system of circular, heliocentric, planetary orbits.

Copernicus: A recent large crater on the Moon.

Core: The metallic (nickel/iron) core of a terrestrial planet, the ice and rock core of a Jovian planet, or the inner portion of a star within which nuclear reactions are produced.

Core-collapse supernovae: Supernovae produced by the collapse of iron cores that have developed in high-mass stars.

Corona (solar): The hot (2-million-K) outer envelope enclosing the Sun; seen during a total solar eclipse.

Coronae (Venus): Large, volcanically uplifted areas with fractured centers, surrounded by concentric fractures.

Coronal gas: A very hot gas in the interstellar medium created and heated by supernova blasts.

Coronal holes: Gaps in the solar corona where there is little hot gas and from which the solar wind blows most strongly.

Coronal mass ejection: A bubble of gas released from the solar corona by magnetic activity; the process is responsible for the aurorae.

Cosmic: Pertaining to the cosmos, to the sky, or to the Universe.

Cosmic Background Explorer: An Earth-orbiting satellite designed to observe the cosmic microwave background radiation.

Cosmic microwave background radiation: All-pervading blackbody radiation near 2.7 K, cooled from the Big Bang fireball.

Cosmic rays: High-energy atomic nuclei accelerated by the Galaxy's magnetic field; probably produced in supernovae.

Cosmological distance: A large distance that relates to the expanding Universe and thus to cosmology.

Cosmological force: A force, suggested by Einstein, that increases with distance.

Cosmological Principle: There is no special place in the Universe.

Cosmology: The study of the structure, origin, and evolution of the Universe.

Cosmos: The sky or Universe.

Crab Nebula: The remnant of the supernova of 1054 in Taurus.

Craters: Pits caused by impacts (or, in the case of volcanic craters, by explosions).

Crescent moon: A lunar phase in which we see less than half the illuminated side of the Moon; seen when the Moon makes less than a right angle with the Sun.

Critical density: The density of matter needed to close the Universe.

Crust: The top, thin, light layer of a terrestrial planet or satellite.

Cygnus X-1: The best-known black hole candi

Daily paths: The apparent paths taken by celestial bodies as the Earth rotates.

Dark matter: Unilluminated mass, detectable only through its gravitational effect; its content is unknown.

Dark matter halo: The volume of space filled with dark matter, encompassing the traditional disk and halo of the Galaxy.

Decameter bursts: Bursts of radio radiation from Jupiter, caused by the action of a ring of electrical current that connects the planet and Io.

Defects: Places in the Universe where the symmetry of forces did not break.

Degeneracy: A state in which the particles (protons, neutrons, or electrons) of a gas at a given velocity cannot get any closer; in a star, degeneracy provides an outward stabilizing pressure.

Dense cores: Knots of matter in the interstellar medium, observable by ammonia radiation; the first observable step in star formation.

Density: Mass per unit volume, measured in kg/m^3, or g/cm^3.

Density waves: Gravitational accumulations of matter that spread into a spiral pattern as a result of orbital revolution.

Deuterium: Hydrogen with a neutron in its nucleus; ^2H.

"Diamond ring": The visual effect at the moment of the first appearance of the Sun after a total solar eclipse when the corona is still visible.

Differential rotation: Rotation in which different parts of a body rotate with different speeds, and therefore with different periods.

Differentiation: The separation of a planet's interior into layers of different composition.

Diffraction: A phenomenon that spreads light outward in all directions around barriers.

Diffraction disk: The central image produced by diffraction.

Diffraction grating: A plate with narrowly spaced lines that disperses light to produce spectra.

Diffuse nebulae: Bright clouds of interstellar gas ionized by stars at least as hot as type B1.

Dipole: A field with two poles.

Dispersion: The spreading of radiation into a spectrum by refraction or diffraction.

Distance indicators: Objects with known absolute magnitudes used to find distances; the term is most commonly applied to objects in other galaxies.

Doppler effect: The observed shift in wavelength or frequency caused by relative radial motion.

Drake equation: An equation consisting of several factors which, when multiplied, give the probability of intelligent life in the Universe.

Dry ice: Frozen carbon dioxide.

Dust: Small, solid grains with a variety of chemical compositions.

Dust tail: The diffuse fan-shaped tail of a comet caused by sunlight scattered from dust released by the nucleus.

Dwarf elliptical galaxies: Small elliptical galaxies.

Dwarf spheroidal galaxies: The smallest kind of elliptical galaxies.

Dwarfs: Stars of the main sequence on the HR diagram.

E0 galaxy: An elliptical galaxy with a circular outline.

E7 galaxy: The most elongated form of elliptical galaxy.

Earth: Our world, the third planet from the sun.

Earthquake waves: Compressional (P) and transverse (S) waves sent through the Earth by the shocks of earthquakes.

Earthquakes: Vibrations caused by the slippage of the Earth's crust along faults.

Eccentricity: The degree of flattening of an ellipse (the center-to-focus distance divided by the semimajor axis), ranging from 0 for a circle up to (but not including) 1.

Eclipse of the Moon: The passage of all or part of the Moon through the Earth's shadow.

Eclipse of the Sun: The passage of the Moon across the Sun (that is, of the lunar shadow across the Earth).

Eclipsing binaries: Binary stars whose components eclipse each other.

Ecliptic: The apparent path of the Sun through the stars.

Effective temperature: The temperature of a black-body that has the surface area and luminosity of the radiant source.

Einstein cross: The multiple image of a gravitational lens produced when a quasar is closely aligned behind an extended mass.

Einstein-de Sitter universe: A flat, Euclidean universe.

Einstein ring: A ring of light produced when a quasar is exactly lined up behind a gravitationally lensing mass of small angular diameter.

Ejecta: The debris thrown out of a crater by the impact that made it.

Ejecta blanket: The layer of debris thrown around an impact crater.

Electric charge: A property of electrons and protons by which they attract or repel one another; a manifestation of the electromagnetic force.

Electric field: The range of influence of an electric charge as it acts over a distance.

Electromagnetic force: The force of nature that combines electricity and magnetism; manifested by electric charge and electric and magnetic fields that act over a distance.

Electromagnetic radiation: Alternating electric and magnetic waves, or photons, that transport energy at the speed of light (c = 299, 792 km/s).

Electromagnetic spectrum: The array of the kinds of electromagnetic waves arranged by wavelength; includes gamma rays, X rays, and ultraviolet, visual, infrared, and radio waves.

Electrons: Negatively charged atomic particles.

Electroweak force: The unified electromagnetic and weak forces.

Elements: The basic constituents of common matter; each element is defined by a different number of nuclear protons.

Ellipse: A curve defined by two foci such that the sum of the distance from any point on the curve to each of the foci is constant; a conic section obtained by slicing a cone with a plane tilted at an angle less than that of the cone's side.

Elliptical galaxies: Spheroidally shaped galaxies without spiral arms.

Emission lines: Bright spectrum lines of radiation at specific wavelengths.

Encke division: A narrow gap in the outer part of Saturn's A ring, produced by a satellite that sweeps up ring material.

Energy: The ability of a body to impress a force on, and to accelerate, another.

Envelope: The thick layer of the Sun that blankets the solar core and transmits solar energy.

Epicycle: In the Ptolemaic system, a secondary planet-carrying orbit centered on a larger orbit that encircles Earth.

Equal areas law: Kepler's second law: the line that connects a planet to the Sun sweeps out equal areas in equal times.

Equator (terrestrial): The great circle on the Earth that is equidistant from the poles.

Equinoxes: The intersections between the ecliptic and the celestial equator.

Escape velocity: The velocity a body needs to achieve a parabolic orbit from another body's surface.

Euclidean: Pertaining to Euclidean geometry; space in which one and only one line can be drawn through a point parallel to another line.

Europa: The second-out and smallest Galilean satellite of Jupiter.

Event horizon: The surface surrounding a black hole where the escape velocity equals the velocity of light.

Evolutionary tracks: Graphs of luminosity against effective temperature (or M_V versus spectral class) on the HR diagram.

Expanding Universe: The steady growth of space that increases distances between clusters of galaxies.

Eyepiece: A lens attached to a telescope that makes light rays parallel, thus allowing an object to be imaged by the eye.

F stars: Yellow-white stars with strong hydrogen lines and metal lines, and temperatures of 6,000 to 7,400 K.

Families of matter: Different groups of atomic particles analogous to the electron-proton pair, each with different kinds of neutrinos.

Farside (of the Moon): The side of the Moon invisible to us, that facing away from the Earth.

Fault: A separation in the planet's crust.

Filaments: Dark ribbons of cool gas in the solar corona seen against the photosphere and confined by magnetic fields; prominences seen against the Sun.

Finder telescope: A small telescope with a wide field of view attached to the side of the main telescope.

Fireball: A brilliant meteor that can produce a meteorite.

First-quarter Moon: A lunar phase in which we see half the illuminated side of the Moon; seen when the Moon makes a right angle with the Sun following new Moon.

Flamsteed numbers: Numbers assigned to stars west-to-east within a constellation.

Flares: Sudden releases of magnetic energy in the solar corona that also brighten the chromosphere.

Flatness problem: The question of why Omega should be so close to 1; resolved by inflation.

Focal Length: The distance from a focusing lens or mirror to the focal point

Focal point/plane: The point on the optical axis, and the plane perpendicular to the axis, where images are focused.

Focus (of an ellipse): One of two points that define an ellipse.

Force (F): That which produces an acceleration of a mass.

Foucault pendulum: A swinging weight that demonstrates the Earth's rotation.

Frequency (v): The number of waves passing a location per second.

Friedmann-Lemaître universes: A set of models for the Universe that consists of open and hyperbolic, and closed and spherical, universes.

Full Moon: A lunar phase in which we see the full illuminated side of the Moon; seen when the Moon is opposite the Sun.

Fusion: A short name for thermonuclear fusion.

G stars: Yellow-white stars with metal lines and particularly strong ionized calcium lines, and temperatures of 4,900 to 6,000 K.

Galactic bulge: The central, thick region of the Galaxy where the disk and the halo come together.

Galactic disk: The flat component of the Galaxy that contains the majority of its stars.

Galactic equator: The center line of the Milky Way, on which galactic longitude is based.

Galactic halo: The sparsely populated, spherical component of the Galaxy that surrounds the Galaxy's disk.

Galactic nucleus: The energetic center of the Galaxy; may be a black hole.

Galactic rotation curve: The rotation velocity of the Galaxy graphed against distance from the galactic center.

Galaxies: Self-contained collections of mass that include stars, interstellar matter, and dark matter.

Galaxy, The: The collection of 200 billion stars, interstellar matter, and dark matter in which we live; it is shaped like a disk surrounded by a sparsely populated halo, all surrounded by a dark matter halo.

Galilean satellites: Jupiter's four largest satellites, discovered by Galileo in 1609.

Gamma-ray bursts: Sudden emissions of gamma rays from unknown sources.

Gamma rays: The shortest-wave electromagnetic radiation, with wavelengths under about 1 Å.

Ganymede: The third-out and largest Galilean satellite of Jupiter.

Gas: A state of matter with neither fixed shape nor volume.

Gas tail: The part of a comet's tail made of ionized gas; it points away from the Sun and is structured by the Sun's wind and magnetic field.

General relativity: That part of relativity that involves accelerations.

Geocentric theory: A theory of the Solar System in which the planets orbit the Earth.

Geosynchronous orbit: An orbit around the Earth in which the orbiting body has a revolution period equal to the Earth's rotation period.

Giant elliptical galaxies: The largest kind of elliptical galaxy, more massive than our Galaxy, and commonly found near the centers of clusters of galaxies.

Giant molecular clouds (GMCs): Massive dusty clouds (typically 200,000 $M_\odot$ and 100 pc across) made mostly of hydrogen molecules.

Giants: Large stars that are brighter than the main sequence for a given temperature; MKK classes II and III.

Gibbous Moon: A lunar phase in which we see more than half the illuminated side of the Moon; seen when the Moon makes more than a right angle with the Sun.

Giotto: The European spacecraft that passed close to and imaged Halley's Comet in 1985; named for the painter Giotto, whose *Adoration of the Magi* included a depiction of Halley's Comet.

Globular clusters: Rich clusters in the galactic halo; found in the halos of other spiral galaxies and in elliptical galaxies.

Gluons: The particles that carry the strong force.

Grand Unified Theories (GUTS): Theories combining the electromagnetic, weak, and strong forces.

Granules: Bright cells in the photosphere, about 700 km across, caused by convection.

Gravitational constant (G): The constant of nature used in the law of gravity; the force between two 1-kg bodies 1 m apart.

Gravitational lens: A body whose gravity causes a relativistic distortion or amplification of the light from a more distant body.

Gravitational redshift: The shift of a photon toward longer wavelengths as a result of its escaping a gravitational field.

Gravitons: The particles that are presumed to carry gravity.

Gravity: The attractive force between masses; the curvature of spacetime caused by the presence of mass.

Great circle: A circle on a sphere whose center is coincident with the center of the sphere.

Great Dark Spot (GDS): A dark, oval, high-pressure storm once seen in Neptune's southern hemisphere; probably caused by a convective plume.

Great Debate: The 1920 debate between Heber Curtis and Harlow Shapley over the nature of spiral nebulae and the existence of external galaxies.

Great Red Spot (GRS): A huge reddish zone between Jupiter's south equatorial belt and south tropical zone.

Great Wall: A dense line of galaxies some 50 to 150 Mpc away, stretching around a third of the sky and containing the Coma cluster.

Great White Spot (GWS): A seasonal northern-hemisphere storm on Saturn that extends around the entire planet.

Greatest brilliancy: The maximum apparent brightness of Venus, which occurs in its crescent phases.

Greatest elongation: For an inferior planet, the maximum possible angular separation from the Sun.

Greenhouse effect: The trapping of radiated heat by atmospheric carbon dioxide and water vapor.

Gregorian calendar: Our modern calendar of 365 days that has an extra day every four years except in century years not divisible by 400; replaced the Julian calendar.

Ground state: The lowest energy level of an atom or ion.

Guest star: The supernova of 1054 in Taurus and source of the Crab Nebula; observed by the Chinese.

Half-life: The time it takes for a specific amount of a radioactive element to decay to half that amount.

Halo (galactic): The sparsely populated, spherical component of a spiral galaxy that surrounds the galaxy's disk.

Harmonic law: Kepler's third law: the sidereal period of a planet in years squared equals the semimajor axis of the planet in astronomical units cubed.

Heat energy: A microscopic form of energy of motion related to the speeds at which particles move in a gas or vibrate in a solid.

Heavy bombardment: The period, early in the history of the Solar System, of a high rate of meteoric impacts on the Earth, Moon, and planets.

Heliocentric theory: A theory of the Solar System in which the planets orbit the Sun.

Helium flash: The explosive ignition of helium in the core of a red giant star of about a solar mass.

Herbig-Haro (HH) **objects:** Knots of gas in the interstellar medium lit by shock waves from bipolar flows.

Hertzsprung-Russell (HR) **diagram:** A plot of stellar magnitudes against spectral types.

High tide: The maximum water line at the shore produced by lunar and solar tides.

Horizon: The line where the land seems to meet the sky; in cosmology, the maximum distance we can look in the Universe, which is determined by the Universe's age and structure.

Horizon problem: Why the cosmic microwave background radiation is the same in all directions (except for a Doppler shift) even though all parts of the Universe could not communicate; resolved by inflation.

Horizontal branch: A branch of roughly constant magnitude extending left from the giant branch in the HR diagrams of globular clusters; horizontal-branch stars fuse helium into carbon in their cores.

Horsehead Nebula: A famous Bok globule located in Orion.

Hot dark matter: Fast-moving dark matter, for example, neutrinos.

Hubble classification: A system of classifying galaxies that branches from ellipticals through the two kinds of spirals to irregulars.

Hubble constant (H_0): The recession velocity of a galaxy divided by distance, a number that defines the rate at which the Universe expands.

Hubble flow: The smooth expansion of space, within which clusters of galaxies flow apart.

Hubble length: The distance at which the speeds of receding cosmological bodies (galaxies or quasars) reach the speed of light.

Hubble relation: The linear correlation between redshifts and distances of galaxies.

Hubble time: The reciprocal of the Hubble constant and the age of an empty universe.

Hydrocarbons: Chemical compounds based on hydrogen and carbon.

Hydrostatic equilibrium: A condition in which the upward push of pressure is balanced by the downward pull of gravity.

Hyperbola: A conic section obtained by slicing a cone with a plane tilted at an angle greater than that of the cone's side; the hyperbola does not close on itself.

Hypothesis: An idea to be tested by experiment or observation.

Image: The depiction of an object in the focal plane.

Impact basins: Large impact craters, commonly filled with dark, solidified lava.

Inequality of the seasons: The fact that northern spring and summer are longer than northern autumn and winter as a result of the eccentricity of the Earth's orbit (reversed in the southern hemisphere).

Inferior conjunction: A conjunction between Venus or Mercury and the Sun in which the planet lies between the Earth and the Sun.

Inferior planets: Mercury and Venus, the two inside the Earth's orbit.

Infinite: Never-ending.

Inflation: A super-rapid expansion of the Universe that may have occurred just after the Big Bang.

Infrared radiation: Radiation with wavelengths longer than can be seen with the human eye (about 7,000 Å) but less than about 0.1 mm.

Interacting galaxies: Galaxies in some state of tidal interaction or collision.

Intercloud medium: A thin, warm, partially ionized gas that lies in the spaces between interstellar clouds.

Intermediate main sequence: The main sequence between 0.8 $M_\odot$ and 8 $M_\odot$; these stars evolve into white dwarfs.

Interstellar absorption: The dimming of starlight by interstellar dust.

Interstellar absorption lines: Narrow absorption lines, superimposed on the spectra of stars, produced by atoms and ions in the interstellar medium.

Interstellar dust: Small solid grains made of silicates or carbon coated with ices and embedded with heavier atoms.

Interstellar medium: The lumpy mixture of gas and dust that pervades the plane of the Galaxy.

Io: The innermost and most active Galilean satellite of Jupiter.

Io torus: A ring of ions and electrons around Jupiter in Io's orbit.

Ionosphere: A layer at the top of a planet's atmosphere in which the gases are ionized.

Ions: Atoms or molecules that are electrically charged because electrons are missing or added.

IRAS: The Infrared Astronomical Satellite, which in 1983 orbited Earth and surveyed the sky at several infrared wavelengths.

Iron peak: The high abundances of the elements around iron.

Irregular galaxies: Small galaxies with little structure.

Ishtar Terra: The northern great rise on Venus.

Isotopes: Variations of a chemical element caused by differences in neutron number.

Isotropic: The same in all directions.

Jovian planets: Those planets with the characteristics of Jupiter: Jupiter, Saturn, Uranus, and Neptune.

Julian calendar: The old calendar dating from Julius Caesar, which had an extra day added every four years; replaced by the Gregorian calendar.

K stars: Orange stars with strong neutral and ionized calcium lines and molecular lines, and temperatures of 3,500 to 4,900 K.

Kelvin (absolute) temperature scale (K): A scale that uses Celsius degrees with zero set at absolute zero; water freezes at 273 K and boils at 373 K.

Kepler's first generalized law: The path of an orbiting body is a conic section with the other body at one focus of the curve.

Kepler's generalized laws of planetary motion: Kepler's laws derived from, and generalized by, Newton's laws of motion and the law of gravity.

Kepler's laws of planetary motion: The law of ellipses, the equal areas law, and the harmonic law, which together describe planetary orbits.

Kepler's second generalized law: In any closed system, angular momentum is conserved.

Kepler's third generalized law: The period of an orbiting body in seconds squared is equal to a constant times the semimajor axis in m (or cm) cubed divided by the sum of the masses in kg (or g).

Kilogram: The basic unit of mass, equal to 1,000 grams.

Kiloparsec (kpc): A thousand parsecs.

Kinetic temperature: Temperature defined by atomic or particle velocities.

Kirchhoff's first law: An incandescent solid or a hot gas under high pressure produces a continuous spectrum.

Kirchhoff's laws of spectral analysis: A set of three laws that define the conditions for the appearance of the spectrum.

Kirchhoff's second law: A hot gas under low pressure produces an emission-line spectrum.

Kirchhoff's third law: A continuous source viewed through a cooler low-density gas produces an absorption-line spectrum.

Kuiper belt: A disk-shaped reservoir of comets outside the orbit of Neptune that produces the short-period comets.

Late heavy bombardment: The last period of meteorite fall before 3.8 billion years ago that produced heavy cratering on planetary and satellite surfaces.

Latitude (ϕ): Arc measurement north and south of the Earth's equator; $\phi = h(NCP)$.

Lava: Liquid rock on the surface of the Earth.

Law of ellipses: Kepler's first law: planets orbit the Sun in elliptical paths with the Sun at one focus.

Leap year: A year in which a day is added; in the Gregorian calendar added as February 29.

Lens: A curved piece of glass that can focus radiation.

Leonids: A famous meteor storm that occurs every 33 years.

Light: Electromagnetic radiation; commonly, radiation visible with the human eye or near the visual domain.

Light curve: A graph, in which the magnitude of a star that varies in brightness is plotted against the time.

Light-gathering power: The degree to which a telescope collects electromagnetic radiation; proportional to the square of the telescope's diameter.

Light-year (ly): The distance a ray of light travels in a year at 299,792 km/s.

Limb: The edge of the solar or lunar disk.

Limb darkening: The darkening of the surface of the photosphere of the Sun toward its edge, caused by the transparency of the photosphere and inwardly increasing temperature.

Liquid: A state of matter that has a fixed volume but not a fixed shape.

Lithosphere: A layer of rock on a terrestrial planet that consists of the crust and the top part of the mantle.

Local Group: The sparse cluster of galaxies that contains our Galaxy.

Local supercluster: The supercluster of galaxies that contains, among others, the Local Group and the Virgo cluster.

Logarithmic: Pertaining to a system in which numbers are represented by powers of ten.

Long-period comets: Comets that have random orbits with periods over 200 years and that come from the Oort cloud.

Long-period variables: Giant pulsating stars with periods over about 100 days and large visual-magnitude ranges.

Look-back time: The time it has taken light to reach us from a distant source.

Low tide: The minimum water line at the shore, produced by lunar and solar tides.

Lower main sequence: The main sequence between 0.08 $M_{\odot}$, and 0.8 M., whose stars have never had time to evolve away from it.

Luminosity (L): The amount of energy radiated by a body per second; measured in watts.

Luminosity (MKK) classes: The categorization of stars by their luminosities and sizes: I, II, III, IV, V for supergiants, bright giants, giants, subgiants, and dwarfs respectively.

Lumpiness problem: Why the Universe is so lumpy with clusters of galaxies while the cosmic microwave background radiation is so smooth.

Lunar highlands: Original lunar crust crushed by heavy cratering during the heavy bombardment.

Lyman series: A series of hydrogen absorption or emission lines that have the ground state as their common bottom level.

M stars: Reddish stars with a strong neutral calcium line and molecular lines, particularly titanium oxide, and temperatures of 2,000 to 3,500 K.

Magellanic Clouds: Two small companion galaxies to our Galaxy, the Large Magellanic Cloud (the LMC) and the Small Magellanic Cloud (the SMC).

Magma: Liquid rock inside the Earth.

Magnetic activity cycle: The 22-year cycle in the amount of solar magnetic activity that produces the 11-year sunspot cycle.

Magnetic field: The range of magnetism as it acts over a distance.

Magnetodisk: An extended disk of ionized gas in Jupiter's magnetosphere.

Magnetosphere: The structure around a planet filled with its magnetic field and particles trapped from the solar wind.

Magnifying power: The amount by which a telescope eyepiece multiplies the apparent angular diameter of an object; equal to the ratio of the focal length of the objective to that of the eyepiece.

Magnitude equation: A relation among absolute magnitude, apparent magnitude, and distance: $M = m + 5 - 5 \log d$ (d is distance in pc).

Magnitudes: Classes of star brightnesses; a measure of stellar brightness in which 5 magnitudes correspond to a brightness factor of 100.

Main belt: The location of the greatest concentration of asteroids, between 2.1 and 3.2 AU from the Sun.

Main (dwarf) sequence: The main band of stars that climbs from lower right to upper left (luminosity increasing with temperature) on the HR diagram; MKK class V.

Main-sequence fitting: The means of deriving cluster distances by comparing their main sequences.

Mantle: The thick layer of rock that surrounds the core of a terrestrial planet.

Mare Imbrium: A major lava-filled basin in the Moon's northern hemisphere.

Marginally open universe: A four-dimensional, flat universe that does not fold back on itself; one and only one line can be drawn through a point parallel to a given line, and the angles of a triangle sum to 180°.

Maria: Dark areas on the Moon made of lava flows that fill many impact basins.

Mariner: The name of a series of planetary spacecraft; *Mariners 2* and *10* went to Venus, *Mariner 10* to Mercury, and *Mariners 4, 6,* and *7* to Mars.

Mass (*M*): The amount of matter in a body, or the degree to which a body's motion resists a force; usually measured in grams or kilograms.

Mass-luminosity relation: $L \propto M^{3.5}$ (average) for main-sequence stars.

Matter-antimatter problem: Why the Universe is constructed of matter and why matter and antimatter did not completely cancel each other out.

Maxwell Montes: The major mountains on Venus, located in Ishtar Terra.

Megaparsec (Mpc): A million parsecs.

Messier Catalogue: A catalogue of 109 nonstellar celestial objects compiled mostly by Charles Messier in the eighteenth century.

Meteor: A bright tube of ionized atmosphere caused by the passage of a meteoroid.

Meteor shower: A display of meteors that appears to emanate from a specific point in the sky (a radiant), caused when the Earth passes near a comet's orbit.

Meteor storm: An intense meteor shower caused by a concentration of cometary debris.

Meteorites: Meteoroids that land on Earth; classified as stones, irons, or stony irons.

Meteoroids: Pieces of interplanetary debris that produce meteors when they enter the Earth's atmosphere and meteorites if they hit the ground.

Metric system: A system of measures in which the units differ by multiples of ten.

Midnight: 0^h or 24^h solar time, when the Sun crosses the celestial meridian below the celestial pole.

Milky Way: The band of light around the sky caused by stars in the disk of our Galaxy.

MKK classes: Luminosity classes of stars.

Models: Mathematical descriptions of physical systems (for example, the changes of temperature, density and pressure change with depth) that allow predictions of observations.

Modern constellations: Constellations generally invented since 1600.

Molecules: Combinations of atoms.

Moon: The Earth's satellite, which orbits Earth under the force of gravity.

Neap tides: Ocean tides with minimum high tide produced at the lunar quarters.

Nearside (of the Moon): The side of the Moon that always faces Earth.

Neutrino: A massless (or nearly massless) particle that carries energy at or near the speed of light.

Neutron star: A collapsed star smaller than about $3 \, M_\odot$, made of degenerate neutrons.

Neutron star limit: The upper limit of about $3 \, M_\odot$, above which stars cannot be supported by degenerate neutrons.

Neutrons: Neutral atomic particles.

New General Catalogue (NGC): A catalogue of over 7,000 nonstellar celestial objects.

New Moon: A lunar phase in which we can see none of the illuminated side of the Moon, the phase occurs when the Moon is aligned with the Sun.

Newtonian focus: A configuration of a primary mirror and a flat secondary mirror that sends the light to the side of a reflecting telescope.

Newton's first law of motion: Left undisturbed, a body will continue in a state of rest or uniform straight-line motion.

Newton's laws of motion: The three laws stated by Isaac Newton that describe how things move.

Newton's second law of motion: The degree of an acceleration is directly proportional to the force applied and inversely proportional to the mass of the body being accelerated.

Newton's third law of motion: For every force applied to a body, there is an equal force exerted in the opposite direction; action equals reaction.

Non-Euclidean geometry: A geometry that does not accept Euclid's parallel postulate.

Nonthermal radiation: Radiation produced by processes that are not the result of heat and that cannot be related to temperature.

Noon: 12^h solar time, when the Sun crosses the upper part of the celestial meridian.

Normal spirals: Spiral galaxies in which the arms come directly from the galaxy's center.

North celestial pole (NCP): The point of rotation of the celestial sphere that lies above the north pole of the Earth.

North Star: Polaris, the second-magnitude star near the North Celestial Pole.

Nova: A stellar brightening or caused by the explosion on the surface of a white dwarf of matter flowing from a main-sequence star.

Nuclear burning: A common term for thermonuclear fusion.

Nucleus: The combined protons and neutrons at the atomic center.

O stars: Bluish stars with neutral and ionized helium lines and temperatures of 28,000 to 50,000 K.

OB associations: Unbound groups of O and B stars found in the disks of spiral galaxies.

Objective: A telescope lens or mirror.

Oblate spheroid: A sphere flattened at its poles.

Obliquity of the ecliptic (ϵ): The tilt of the Earth's axis relative to the perpendicular to the Earth's orbit, or the angle between the celestial equator and the ecliptic, equal to 23°27'.

Oceans: Basins filled with water.

Oceanus Procellorum: The largest recognized impact feature (or combination of such features) on the Moon.

Olbers' paradox: That the night sky is dark when under specified conditions (infinite, uniformly populated, and unchanging) it should not be.

Olympus Mons: The major shield volcano on Mars, located on the slope of the Tharsis bulge.

Omega (Ω): The ratio of the average density of the Universe to the critical density; the Universe is open if Ω is less than 1, marginally and flat if equal to 1, and closed if less than 1.

Oort comet cloud: A reservoir over 100,000 AU in radius that contains a trillion comet nuclei; the origin of the long-period comets.

Open clusters: Loose clusters in the galactic disk.

Open universe: A negatively curved hyperbolic universe that does not close back on itself; there can be more than one line drawn through a point parallel to a given line, and the angles of a triangle sum to less than 180°.

Opposition: The position in which a planet is in the opposite direction from the Sun (or in which two celestial bodies are opposite each other).

Orbit: The path that one body takes about another (or about their center of mass) as a result of their mutual gravitational attraction.

Organic molecules: Molecules that contain carbon.

Orion Nebula: The archetype of diffuse nebulae.

Outflow channels: On Mars, large channels created by eruptions or releases of vast amounts of water.

Ozone: The O_3 molecule.

P wave: A compressional earthquake wave.

Pangaea: The supercontinent that held most of the world's land masses about 200 million years ago.

Parabola: A conic section obtained by slicing a cone with a plane parallel to the side of the cone; the curve that separates the ellipse and the hyperbola.

Parallax: The apparent shift in the position of a body when viewed from different directions; in particular, half the angular shift of a star over the course of the year as a result of the orbital motion of the Earth.

Parallel postulate (of Euclid): One and only one line can be drawn through a point parallel to another line.

Parsec (pc): The distance at which the semimajor axis of the Earth's orbit subtends 1 second of arc; 1 pc = 3.26 light-years = 206,265 AU.

Partial eclipse: The passage of the Moon through only part of the Earth's umbra, or the partial passage of the Moon across the Sun.

Penumbra: A region of partial shadow.

Perfect Cosmological Principle: There are neither special places nor times in the Universe.

Perigee: The point in the lunar orbit closest to the Earth.

Perihelion: The point of closest approach of a planet to the Sun.

Period: The time required for a body to return to a given starting point or position.

Period-luminosity relation: The correlation between the periods of Cepheids and their absolute magnitudes, which allows derivation of distances.

Periodic table: A tabular arrangement of the chemical elements by atomic number and according to their chemical properties.

Perseids: The most famous meteor shower, which takes place on August 12.

Perturbations: Orbital changes induced by outside gravitational forces.

Phases (of the Moon): The different apparent shapes of the Moon caused by our viewing different proportions of the lighted side as the Moon orbits the Earth; crescent, quarter, gibbous, full; similar phases are seen for inferior planets.

Photochemical reactions: Chemical reactions that take place under the action of sunlight.

Photoelectric effect: A phenomenon in which photons knock electrons from matter to produce a flow of electricity.

Photons: Particles of electromagnetic radiation that also incorporate wave motion.

Photosphere: The bright apparent surface of the Sun or a star.

Pixel: Picture element; the individual recording element of a CCD.

Planck time: The smallest age assigned to the known Universe, when the Universe was so dense that our standard theory of gravity, the general theory of relativity, no longer applied.

Planck's constant: The energy carried by a photon with a frequency of 1 Hz; a constant of nature.

Planetary nebula: A shell of illuminated gas around a hot star; the last part of the ejected wind of an asymptotic giant branch star lit by the old nuclear-burning core.

Planetesimals: Primitive bodies of the Solar System that preceded the formation of the planets.

Planets: The Sun's family of major orbiting bodies; small bodies orbiting other stars analogous to the Sun's planets.

Plate tectonics: Deformations of the Earth's crust caused by continental drift.

Plates: The segments of the Earth's crust.

Plumes: Rising columns of hot mantle material that break through a planet's crust, producing shield volcanoes and volcanic floods.

Polarization: An optical process that makes light waves oscillate in a preferential direction.

Poles (celestial): The points of rotation of the celestial sphere that lie above the north and south poles of the Earth.

Poles (terrestrial): The points on the Earth where its rotation axis emerges.

Polycyclic aromatic hydrocarbons (PAHs): Organic interstellar molecules built of benzene rings.

Population I: The galactic disk and its components; stars of high metal content, luminous blue stars, young stars, and interstellar matter.

Population II: The galactic halo and its components; red stars of the lower main sequence, evolved red giants, horizontal branch stars, and stars of lower metal content.

Population III: The original population of the Galaxy made only of hydrogen and helium, that seeded the Galaxy with the first heavy elements; no representatives are known.

Positron: A positive electron.

Post meridiem (P.M.): After noon.

Potential energy: The energy contained by a body as a result of its position or configuration.

Postulate: An unprovable assumption.

Precession: The motion of the celestial poles, celestial equator, and equinoxes caused by the 26,000-year wobble of the Earth's axis.

Pressure: The outward force of compressed matter per unit area.

Primary mirror: The large light-collecting mirror of a reflecting telescope.

Primary wave: A compressional earthquake wave.

Prime focus: The focal position of the primary mirror of a reflecting telescope.

Principle of equivalence: In relativity, that an observer cannot tell the difference between acceleration produced by motion and the acceleration of a gravitational field.

Prograde: In the normal direction of motion; in the context of a planet, in the direction of rotation.

Project Ozma: The first search for extraterrestrial intelligence.

Prominences: Bright arches of cool gas in the solar corona seen against the sky and confined by magnetic fields; filaments seen against the sky.

Proper motion (μ): The angular motion of a star across the line of sight.

Proper (star) names: Individual names assigned to brighter stars, usually reflecting the stars' properties or positions; most are of Arabic origin.

Proton-proton (p-p) chain: A fusion reaction that turns four atoms of hydrogen into one atom of helium.

Protons: Positively charged atomic particles.

Protostars: Stars in the process of formation.

Proxima Centauri: The nearest star and a companion to Alpha Centauri.

Ptolemaic system: A system of geocentric orbits that carry epicycles that carry planets.

Pulsar: A rotating neutron star that beams energy along a tilted magnetic field; it is visible only if a beam hits the Earth.

Quasars: "Quasi-stellar radio sources"; bright, star-like objects with high redshifts; some are strong radio emitters; they are probably distant, young versions of active galaxies.

Quiet Sun: The constant phenomena of the Sun, which do not take large part in the solar magnetic cycle.

Radar: An active observational technique in which radio waves are reflected from a body to determine its distance, speed, and surface features.

Radial velocity (v_r): The relative speed of a body moving along the line of sight.

Radiant: The point in the sky from which the meteors of a meteor shower seem to come.

Radio galaxies: Active elliptical galaxies with powerful central radio sources, bipolar jets, and, sometimes, extended double radio sources.

Radio interferometer: Two or more linked radio telescopes that improve resolving power.

Radio radiation: The longest-wave radiation, with wavelengths longer than about 0.1 mm.

Radio telescope: A telescope that collects and detects radio radiation.

Radioactive dating: Dating of rocks and other objects by establishing the ratio of the amount of the daughter product of radioactive decay to that of the parent.

Radioactive isotopes: Isotopes whose nuclei decay into other nuclei with the release of particles and radiation.

Rays (lunar): Bright lines that emanate from young craters; caused by secondary impacts that expose light-colored rock.

Red giant branch: The stage of stellar evolution in which intermediate-mass stars grow into giants for the first time; these stars have contracting helium cores.

Redshift: The general spectral shift to longer (redder) wavelengths exhibited by galaxies.

Redshift distance: A distance to a galaxy derived from its redshift and the Hubble constant.

Reflecting telescope: A telescope that focuses radiation with a mirror (usually a paraboloid) and that has a variety of focal positions.

Reflection: The return of radiation from a surface at an angle equal to the angle of incidence.

Reflection nebula: A bright cloud created by light scattered by dust from a star too cool to cause ionization (that is, cooler than type B1).

Refracting telescope: A telescope that uses a lens to refract light to a focus.

Refraction: The bending of the path of radiation as it goes from one substance to another.

Refractory elements or compounds: Those that melt or boil only at high temperatures.

Regolith: Lunar or planetary soil, devoid of organic compounds, produced by constant pulverization by meteorites.

Relativistic beaming: Amplification of the radiation from a jet moving near the speed of light toward the observer.

Relativity: The branch of mechanics, developed by Albert Einstein, that lets the speed of light be independent of the speeds of the source or the observer and that describes the action of gravity as a curvature of spacetime.

Resolving power: The ability of a telescope to separate objects; proportional to wavelength divided by the telescope's diameter.

Resonance: A gravitational mechanism in which one orbiting body produces a large gravitational perturbation in another because one has a period that is a simple fraction that of the other.

Retrograde motion: The motion of a planet to the west relative to the stars as a result of the Earth's passing a superior planet in orbit or being passed by an inferior planet; counter to the general revolution of the Solar System; in the context of the planets, opposite the direction of rotation.

Revolution: The movement of one body about another.

Rilles: Channels caused by running lava.

Ring: In the context of the Jovian planets, an encompassing belt of dust and debris.

Ringlets: Small rings only a few hundred km wide that make up Saturn's big rings.

Rise: The appearance of a body from below the horizon.

Roche limit: A limit surrounding a planet within which a fluid body would be torn apart by tides.

Rock and ice: In the context of the Jovian planets, respectively a mixture of heavier materials like silicates and iron and a mixture of volatiles like water, methane, and ammonia.

Rotation: The spinning of a body on its axis.

RR Lyrae stars: Short-period pulsators of classes A and F on the horizontal branch of the HR diagram.

Runaway greenhouse effect: A spiraling effect brought about when high temperature produces more atmospheric carbon dioxide, which produces higher temperature, and so on.

Runoff channels: Ancient dry riverbeds of low-volume water flows.

S-wave: A transverse earthquake wave, one that will not go through a liquid.

Sa galaxy: A normal spiral galaxy with tightly wound arms and a large bulge.

S0 galaxy: A disk galaxy with no spiral arms.

Sagittarius A: The bright radio source at the center of the Galaxy as viewed with a single-dish radio telescope.

Sagittarius A*: The unresolved bright radio source within Sagittarius A; the true nucleus of the Galaxy.

Satellite: A body that orbits a planet.

Sb galaxy: A normal spiral galaxy with intermediately wound spiral arms and an intermediate-sized bulge.

SBa galaxy: A barred spiral galaxy with tightly wound arms and a large bulge.

SBb galaxy: A barred spiral galaxy with intermediately wound spiral arms and an intermediate-sized bulge.

SBc galaxy: A barred spiral galaxy with open spiral arms and a small bulge.

Sc galaxy: A normal spiral galaxy with open spiral arms and a small bulge.

Scale of the Universe: A concept that represents relative distances in the Universe and that increases with time in an expanding universe.

Scarp: A steep cliff or fault.

Scientific method: A method of inquiry whereby observations or experiments are used to establish a theory that, in turn, predicts new experiments or observations for verification.

Scientific (exponential) notation: A means of expressing numbers by using powers of 10.

Search for Extraterrestrial Intelligence (SETI): The radio search for artificial radio signals that would indicate intelligence beyond Earth.

Secondary mirror: A small mirror that intercepts the light from the primary of a reflecting telescope and sends it elsewhere.

Secondary wave: A transverse earthquake wave, one that will not go through a liquid.

Seeing disk: The apparent disk of a star produced by variable refraction in the atmosphere.

Seismograph: A device for recording earthquake waves.

Semimajor axis (a): Half the major axis of an ellipse; the measurement that characterizes the ellipse's size.

Set: The disappearance of a body below the horizon.

Seyfert galaxies: Active spirals with abnormally bright nuclei whose centers radiate strong emission lines.

Shepherd satellites: Small satellites that organize and preserve narrow rings.

Shield volcano: A volcano with runny lava that creates a broad mountain with a low slope.

Shock wave: A wave of sudden increase in pressure.

Short-period comets: Comets with periods under 200 years that come from the Kuiper belt.

Sidereal: Pertaining to the stars.

Sidereal period: The orbital period of the Moon or a planet relative to the stars, or from one orbital point back to the same point.

Silicates: Rocks (including granites and basalts) made of compounds that contain silicon, oxygen, and a variety of metals.

Singularity: A mass that has been shrunk to a point.

Solar core: The inner 30% of the solar radius, in which nuclear reactions are produced.

Solar envelope: The thick layer of the Sun that blankets the core and transports solar energy to the surface.

Solar model: A mathematical description of the Sun that gives its temperature, density, and composition at all points.

Solar nebula: The disk of gas and dust around the forming Sun, out of which grew the planetesimals and the planets.

Solar oscillations: Multiple movements or vibrations of the solar surface.

Solar System: The Sun and its collection of orbiting bodies.

Solar wind: A thin, ionized gas blowing from the Sun.

Solid: A state of matter that has a fixed space and volume.

Solstices: The points on the ecliptic farthest from the celestial equator.

South celestial pole (SCP): The point of rotation of the celestial sphere that lies above the south pole of the Earth.

Space velocity (v_s): The velocity of a star relative to the Sun.

Spacetime: A four-dimensional construction that consists of the three dimensions of space and the one of time.

Seasons: The four periods of the year—spring, summer, autumn, and winter—defined by the position of the Sun relative to the equinoxes and solstices.

Special relativity: That part of relativity that involves constant speed.

Spectral sequence: The categories of stars organized by temperature (OBAFGKM).

Spectrogram: A photographic or graphical rendering of a spectrum.

Spectrograph: A device that separates electromagnetic radiation by wavelength.

Spectroscopic binaries: Binaries with components detected by Doppler shifts.

Spectroscopic distance: Distance estimated from absolute and apparent magnitude as determined by spectral class.

Spectrum: An array of properties; here, the array of electromagnetic waves.

Spectrum lines: Radiation, at specific wavelengths (emission lines) or gaps in a continuous spectrum (absorption lines).

Speed: The rate at which a body changes its distance with time (measured, for example, in m/s, km/s, or km/h).

Spicules: Needlelike projections at the top of the solar chromosphere.

Spiral arms: Arms in a galaxy's disk that wind outward from near its center (or from a bar through its center), contain O and B stars and clouds of interstellar matter, are sites of star formation, and are produced by density waves.

Spiral galaxies: Galaxies displaying spiral arms and a large youthful component.

Spring: The period in the northern hemisphere between March 20 and June 21, when the Sun is moving northward in the northern celestial hemisphere (the period between these dates is autumn in the southern hemisphere).

Spring tides: Ocean tides with maximum high tide produced at new and full Moon.

Standard time: Time that changes in hour units to account for different east-west positions.

Stars: Gaseous, self-luminous bodies similar in nature to the Sun but with diverse properties.

States of matter: The forms matter may take, which include gases, liquids, and solids.

Static universe: A universe that neither expands nor contracts.

Steady-State theory: A theory in which the Universe appears the same from all locations and at all times; this theory satisfies the Perfect Cosmological Principle.

Stefan-Boltzmann law: The flux of radiation from a blackbody is equal to a constant times the body's temperature to the fourth power.

Stellar evolution: The process of change in a star brought about by its aging.

Stellar wind: A flow of mass from the surface of a star.

Stonehenge: A stone monument in southern England aligned with the summer solstice sunrise.

Stratosphere: The layer above the Earth's troposphere in which temperature rises with altitude.

Strong (nuclear) force: The force that binds atomic nuclei.

Subduction: The diving of one crustal plate beneath another.

Subgiants: Stars classified between the giants and the main sequence; MKK class IV.

Summer: The period in the northern hemisphere between June 21 and September 23, when the Sun is moving southward in the northern celestial hemisphere. (The period between these dates is winter in the southern hemisphere.)

Summer solstice: The most northerly point on the ecliptic, north of the celestial equator 23.5°, where the Sun is found about June 21.

Summer Triangle: The triangle defined by the stars Vega, Altair, and Deneb.

Sun: The gaseous, self-luminous body that dominates the Solar System; the nearest star.

Sunspot cycle: The 11-year periodic increase and decrease in the number of sunspots.

Sunspots: Cool, dark regions on the Sun caused by intense magnetic fields that inhibit the flow of energy.

Supercluster: A cluster of clusters of galaxies.

Supergiants: Huge, luminous stars plotted across the top of the HR diagram; MKK class I.

Superior conjunction: A conjunction between Venus or Mercury and the Sun in which the planet lies beyond the Sun.

Superior planets: The planets outside the Earth's orbit, from Mars through Pluto.

Supernova remnant: The expanding cloud of debris caused by the explosion of a supernova.

Supernovae: Outbursts that involve the destruction or near-destruction of stars either by the collapse of their iron cores or by white dwarfs that exceed the Chandrasekhar limit.

Supersymmetry: A relationship in which each kind of atomic particle has a (as yet undetected) counterpart with different, opposing properties.

Symmetry: The unification of the forces of nature at high temperatures.

Synchronous rotation: Rotation with a period equal to the revolutionary period.

Synchrotron radiation: Radiation produced by fast electrons spiraling in a magnetic field.

Synodic period: The orbital period of the Moon or a planet relative to the Sun; for the Moon, the period of the phases.

T Tauri stars: Active, young, variable pre-main-sequence stars.

Tangential velocity (v_t)**:** The relative speed of a body moving perpendicular to the line of sight.

Telescope: A device for gathering and focusing electromagnetic radiation.

Temperature: A measure of the amount of heat energy in a body; also a measure of the average velocities of the particles in a gas.

Terrestrial: Pertaining to Earth or things like Earth.

Terrestrial planets: Earth and those planets like it: Mercury, Venus, and Mars (and perhaps the Moon).

Tharsis bulge: The major volcanic rise on Mars; located in the northern hemisphere, it contains four major volcanoes, including Olympus Mons.

Theory: A model that embraces and explains observational or experimental data.

Theory of Everything (TOE)**:** A theory that combines all the forces of nature.

Thermal radiation: Radiation produced as a result of heat, for example, blackbody radiation.

Thermonuclear fusion: The process of energy generation by the combination of lighter atoms into heavier atoms.

Third-quarter Moon: A lunar phase in which we see half the illuminated side of the Moon; seen when the Moon makes a 270° angle with the Sun, (that is, the right angle before new Moon).

Three-degree background radiation: Another name for the cosmic microwave background radiation.

Tide: A distortion in a body caused by differential gravity; on Earth, a periodic flow of water.

Time dilation: An effect of relativity in which time slows down for a body moving relative to the observer.

Time zones: Zones within which the same time is kept.

Titan: Saturn's large satellite.

Torus: A doughnut-shaped ring.

Total eclipse: An eclipse in which the Moon is entirely immersed in the Earth's umbral shadow, or in which the observer is immersed in the Moon's umbral shadow.

Triple-alpha (3-α) process: The fusion of three atoms of helium into one of carbon.

Tritium: Hydrogen with two neutrons in its nucleus; ^{3}H.

Triton: The large satellite of Neptune.

Trojan asteroids: Asteroids trapped in stable positions along Jupiter's orbit, 60° ahead of and behind the planet.

Tropic of Cancer: The northerly limit, latitude 23.5°N, at which the Sun can be found overhead; the Sun is overhead there about June 21.

Tropic of Capricorn: The southerly limit, latitude 23.5°S, at which the Sun can be found overhead; the Sun is overhead there about December 21.

Tropics: Limits of latitude (23.5°N to 23.5°S) between which the Sun appears overhead at some time during the year.

Troposphere: The layer of atmosphere closest to a solid planet's surface.

Tuning fork diagram: Hubble's classification scheme for galaxies, which shows ellipticals branching to spirals and barred spirals.

21-cm line: A powerful radio line of neutral hydrogen produced in the interstellar medium by an electron in the ground state reversing the direction of its spin from that of the proton to opposite that of the proton.

Tycho: One of the most recent large craters on the Moon.

Tycho's Star: The Supernova of 1572, observed by Tycho.

Ultraviolet radiation: Radiation with wavelengths shorter than the human eye can see but longer than about 100 Å.

Umbra: A region of full shadow.

Uncompressed density: The average density of a planet corrected for the compressing effects of gravity.

Unidentified flying objects (UFOs)**:** Objects in the sky, commonly asserted to be "alien", for which an observer has no ready explanation; caused by misunderstandings of real celestial and terrestrial objects, atmospheric effects, illusions, and hoaxes.

Universal time (UT)**:** Standard time at Greenwich, England, which is used as a worldwide standard.

Universe: The all-encompassing structure that contains everything.

Upper main sequence: The main sequence above $8\,M_\odot$; its stars are too massive to evolve to white dwarfs.

Uraniborg: Tycho's observatory in Denmark.

Valles Marineris: The great fault canyon on Mars, associated with the Tharsis bulge.

Van Allen belts: Doughnut-shaped zones around the Earth filled with high-energy particles.

Variable stars: Stars whose magnitudes change with time.

Velocity (*v*): The combination of speed and direction.

Velocity equation: An equation that relates stellar space velocity, tangential velocity, and radial velocity: $v_s^2 = v_t^2 + v_r^2$.

Venera: Russian spacecraft that landed on Venus.

Vernal equinox (♈): The point on the celestial sphere where the ecliptic crosses the celestial equator as the Sun moves north; the crossing takes place on March 20 or 21.

Very long baseline interferometer (VLBI): An interferometer with separate telescopes synchronized by clocks.

Viking: The name of two spacecraft that imaged, then landed, on Mars.

Virgo cluster: The nearest large cluster of galaxies, in Virgo, 16 Mpc away.

Visual binaries: Binary stars with components that can be separated by eye at the telescope.

Visual magnitudes (*V*): Magnitudes as seen by the human eye or measured with a yellow-green detector.

Visual radiation: Electromagnetic radiation seen with the human eye, with wavelengths between about 4,000 and 7,000 Å.

Voids: Volumes, megaparsecs across, devoid of galaxies.

Volatile elements or compounds: Those that melt or boil at low temperatures.

Volcanoes: Vents in the Earth's crust through which magma, ash, and gas can escape.

Voyager: The name of two spacecraft that probed Jupiter and Saturn; *Voyager 2* also went to Uranus and Neptune.

W and Z particles: The particles that carry the weak force.

Waning phases: The diminishing phases between full and new Moon.

Wavelength (λ): The distance between successive crests of a wave.

Waxing phases: The growing phases between new and full Moon.

Weak force: The third strongest force of nature; involved with radioactive decay.

Weakly interacting massive particles (WIMPs): In supersymmetry theory, massive counterparts to massless (or nearly massless) particles; candidates for cold dark matter.

Weight: The force with which a body is pressed to the surface of another body as a result of gravity.

White dwarfs: Small dim stars about the size of Earth plotted across the bottom of the HR diagram.

Wien law: The wavelength of maximum radiation from a blackbody is a constant divided by the temperature of the body.

Winter: The period in the northern hemisphere between December 22 and March 21, when the Sun is moving northward in the southern celestial hemisphere. (The period between these dates is summer in the southern hemisphere.)

Winter solstice: The most southerly point on the ecliptic, 23.5° south of the celestial equator, where the Sun is found about December 22.

Winter Triangle: The triangle defined by the stars Sirius, Betelgeuse, and Rigel.

X rays: Radiation with wavelengths between the gamma-ray and ultraviolet regions of the spectrum, with wavelengths between about 1 Å and 100 Å.

Young disk: The thin portion of a galaxy's disk that contains its youngest components, the O and B stars, and the bulk of the interstellar medium.

Zenith: The point on the celestial sphere over the observer's head.

Zero-age main sequence (ZAMS): The line on the HR diagram defined by new stars that are just beginning to fuse hydrogen into helium.

Zodiac: The band of constellations that contains the ecliptic.

Zodiacal light: A band of light in the zodiac caused by sunlight scattered from cometary and asteroidal dust.

Zones: In Jupiter's atmosphere, strips of higher-altitude, bright (reflective), lower-temperature clouds.

BIBLIOGRAPHY

ASTRONOMY MAGAZINES

Astronomy, Kalmbach Publishing Co., Waukesha, WI.

Griffith Observer, Griffith Observatory, Los Angeles, CA.

Mercury, the Astronomical Society of the Pacific, San Francisco, CA.

Planetary Report, the Planetary Society, Pasadena, CA.

Sky and Telescope, Sky Publishing Corp., Cambridge, MA.

StarDate, McDonald Observatory, Austin, TX.

GENERAL MAGAZINES WITH ASTRONOMICAL CONTENT

American Scientist, Sigma Xi, New York, NY.

Science News, Science News, New York, NY.

Scientific American, Freeman, New York, NY.

MAPS AND ATLASES

I. Ridpath, ed. 1989. *Norton's Star Atlas.* 19th ed. Cambridge, Mass.: Sky Publishing Corp. (All naked-eye stars and many telescopic objects.)

SC1, SC2, SC3 Star Charts. Cambridge, Mass.: Sky Publishing Corp. (Simple charts for beginning constellation study.)

Tirion, W. 1981. *Sky Atlas 2000.* Cambridge, Mass.: Sky Publishing Corp. (Deep atlas to 8th magnitude includes a large number of nonstellar objects.)

Tirion, W., Rappaport, B., and Lovi, G. *Uranometria 2000.* Richmond, Va.: Willmann-Bell. (Includes stars to 10th magnitude and nonstellar objects.)

SLIDES, VIDEOS, POSTERS, SOFTWARE

Contact the Astronomical Society of the Pacific, 390 Ashton Avenue, San Francisco, CA 94112.

FURTHER READING
General

Abetti, G. 1952. *The History of Astronomy.* New York: Henry Schuman.

Audouze, J., and Israel, G. 1994. *The Cambridge Atlas of Astronomy.* Cambridge, England: Cambridge University Press.

Berry, A. 1961. *A Short History of Astronomy.* Reprint. New York: Dover.

Burnham, R., Jr. 1978. *Burnham's Celestial Handbook.* New York: Dover.

Goldsmith, D. 1990. *The Astronomers.* New York: St. Martin's Press.

Hermann, D. B., and Krisciunas, K. 1984. *The History of Astronomy from Herschel to Hertzsprung.* Cambridge, England: Cambridge University Press.

Illingworth, V. 1985. *The Facts on File Dictionary of Astronomy.* 2nd ed. New York: Facts on File Publications.

Lang, K. R. 1992. *Astrophysical Data: Planets and Stars.* New York: Springer.

Mallas, J. H., and Kreimer, E. 1978. *The Messier Album.* Cambridge, Mass.: Sky Publishing Corp.

Maran, S., ed. 1992. *The Astronomy and Astrophysics Encyclopedia.* New York: Van Nostrand Reinhold.

Moore, P. 1994. *Atlas of the Universe.* New York: Rand McNally.

Pannekoek, A. 1989. *A History of Astronomy.* Reprint. New York: Dover.

Parker, S. P., and Pasachoff, J., eds. 1993. *McGraw-Hill Encyclopedia of Astronomy.* New York: McGraw-Hill.

Sagan, C. 1980. *Cosmos.* New York: Ballantine.

Scientific American Special Issue. 1994. *Life in the Universe.* Vol. 271 (October).

Stott, C., ed. 1991. *Images of the Universe.* Cambridge, England: Cambridge University Press.

Trimble, V. 1992. *Visit to a Small Universe.* Woodbury, New York: American Institute of Physics.

PART I. CLASSICAL ASTRONOMY

Chapter 1. From Earth to Universe

Morrison, P., Morrison, P., and the Office of C. and R. Eames. 1982. *Powers of Ten.* Scientific American Library. New York: Freeman.

Chapter 2. Earth, Sun, and Sky

Aveni, A. F. 1986. Archaeoastronomy: Past, Present, and Future. *Sky and Telescope.* 72:456 (November).

Aveni, A. F. 1989. *Empires of Time: Calendars, Clocks, and Cultures.* New York: Basic Books.

Daniel, G. 1980. Megalithic Monuments. *Scientific American.* 243:80 (July).

Gingerich, O. 1986. Islamic Astronomy. *Scientific American.* 254: 74 (April).

Hawkins, G. S. 1965. *Stonehenge Decoded*. Garden City, New York: Doubleday.

Kaler, J. B. 1996. *The Ever-Changing Sky*. Cambridge, England: Cambridge University Press.

Moyer, G. 1982. The Gregorian Calendar. *Scientific American*. 246:144 (May).

Chapter 3. The Face of the Sky

Allen, R. H. 1963. *Star Names, Their Lore and Meaning*. Reprint. New York: Dover.

Cozens, G. 1986. Discover the Southern Skies–I. *Sky and Telescope*. 71:126 (February).

Cozens, G. 1986. Discover the Southern Skies–II. *Sky and Telescope*. 71:237 (March).

Fraknoi, A. 1989. Your Astrology Defense Kit. *Sky and Telescope*. 78:146 (August).

Gingerich, O. 1984. The Origin of the Zodiac. *Sky and Telescope*. 67:218 (March).

Krupp, E. C. 1991. *Beyond the Blue Horizon*. New York: HarperCollins.

Kunitzsch, P. 1983. How We Got Our "Arabic" Star Names. *Sky and Telescope*. 65:20 (January).

Kunitzsch, P., and Smart, T. 1986. *Short Guide to Modern Star Names and their Derivations*. Wiesbaden, Germany: Otto Harrassowitz.

Menzel, D. H., and Pasachoff, J. 1982. *A Field Guide to the Stars and Planets*. 3rd. ed. New York: Houghton-Mifflin.

Ridpath, I. 1988. *Star Tales*. New York: Universe Books.

Urton, G. 1981. *At the Crossroads of the Earth and the Sky: An Andean Cosmology*. Austin: University of Texas Press.

Villard, R. 1989. The World's Biggest Star Catalogue. *Sky and Telescope*. 78:583 (December).

Williamson, R. A. 1984. *Living the Sky: The Cosmos of the American Indian*. Boston: Houghton-Mifflin.

Chapter 4. Earth, Moon, and Planets

Gingerich, O. 1982. The Galileo Affair. *Scientific American*. 247:131 (August).

Gingerich, O. 1993. *The Eye of Heaven: Ptolemy, Copernicus, Kepler*. New York: American Institute of Physics.

Gingerich, O. 1993. How Galileo Changed the Rules of Science. *Sky and Telescope*. 85:32 (March).

Hodson, D. G., ed. 1974. *The Place of Astronomy in the Ancient World*. Oxford: Oxford University Press.

McPeak, W. J. 1990. Tycho Brahe Lights Up the Universe. *Astronomy*. 18:29 (December).

Meeus, J., Grosjean, C. C., and Vanderleen, W. 1966. *Canon of Solar Eclipses*. New York: Pergamon.

Oppolzer, T. 1962. *Canon of Eclipses*. Reprint. New York: Dover.

Pedersen, O. 1974. *A Survey of the Almagest*. Odense, Denmark: Odense University Press.

Schaeffer, B. E. 1992. Lunar Eclipses that Changed the World. *Sky and Telescope*. 84:639 (December).

Stephenson, F. R. 1982. Historical Eclipses. *Scientific American*. 247:170 (October).

Zirker, J. B. 1984. *Total Eclipses of the Sun*. New York: Prentice-Hall.

Chapter 5. Gravity

Christianson, G. E. 1987. Newton's *Principia*: A Retrospective. *Sky and Telescope*. 74:18 (July).

Cohen, I. B. 1981. Newton's Discovery of Gravity. *Scientific American*. 244:165 (March).

Drake, S. 1980. Newton's Apple and Galileo's Dialogue. *Scientific American*. 243:151 (August).

Einstein, A. 1961. *Relativity: The Special and General Theory*. Reprint. New York: Crown.

Sagan, C. 1995. The First New Planet. *Astronomy*. 23:34 (March).

Wheeler, J. A. 1990. *A Journey into Gravity and Spacetime*. Scientific American Library. New York: Freeman.

Will, C. F. 1986. *Was Einstein Right? Putting General Relativity to the Test*. New York: Basic Books.

Chapter 6. Atoms, Light, and Telescopes

Atkins, P. W. 1991. *Atoms, Electrons, and Change*. Scientific American Library. New York: Freeman.

Davis, J. 1991. Measuring the Stars. *Sky and Telescope*. 82:361 (October).

King, H. C. 1979. *The History of the Telescope*. Reprint. New York: Dover.

O'Dell, C. R. 1989. Building the Hubble Space Telescope. *Sky and Telescope*. 78:31 (July).

Peterson, C. C., and Brandt, J. C. 1995. *Hubble Vision*. Cambridge, England: Cambridge University Press.

Powell, C. S. 1991. Mirroring the Cosmos. *Scientific American*. 265:113 (November).

Readhead, A. C. S. 1982. Radio Astronomy by Very Long Baseline Interferometry. *Scientific American*. 246:52 (June).

Ressmeyer, R. H. 1992. Keck's Giant Eye. *Sky and Telescope*. 84:623 (December).

Sinnott, R. W., and Nyren, K. 1993. The World's Largest Telescopes. *Sky and Telescope*. 86:27 (July).

Strom, S. E. 1991. New Frontiers in Ground-based Optical Astronomy. *Sky and Telescope*. 82:18 (July).

Svec, M. T. 1992. The Birth of Electronic Astronomy. *Sky and Telescope*. 83:496 (May).

Tucker, W., and Tucker, K. 1986. *The Cosmic Enquirers*. Cambridge, Mass.: Harvard University Press.

Verschuur, G. 1987. *The Invisible Universe Revealed: The Story of Radio Astronomy.* New York: Springer.

Wearner, R. 1992. The Birth of Radio Astronomy. *Astronomy.* 20:49 (June).

Weinberg, S. 1983. *The Discovery of Subatomic Particles.* New York: Freeman.

Williamson, S. J., and Cummins, H. Z. *Light and Color in Nature and Art.* New York: Wiley.

PART II.   PLANETARY ASTRONOMY

General

Beatty, J. K., and Chaikin, A., eds. 1990. *The New Solar System.* 3rd ed. Cambridge, England: Cambridge University Press.

Kaufmann, W. J. 1979. *Planets and Moons.* New York: Freeman.

McLaughlin, W. I. 1989. Voyager's Decade of Wonder. *Sky and Telescope.* 78:16 (July).

Miner, E. D. 1990. Voyager 2's Encounter with the Gas Giants. *Physics Today.* 43(no. 7):40 (July).

Morrison, D. 1993. *Exploring Planetary Worlds.* Scientific American Library. New York: Freeman.

Morrison, D., and Owen, T. 1996. *The Planetary System.* 2d ed. New York: Addison-Wesley.

Rothery, D. A. 1992. *The Satellites of the Outer Planets.* Oxford: Oxford University Press.

Chapter 7. Home: The Double World

Akasofu, S.-I. 1989. The Dynamic Aurora. *Scientific American.* 260:90 (May).

Alvarez, W., and Asaro, F. 1990. An Extraterrestrial Impact. *Scientific American.* 263:78 (October).

Anderson, D. L., and Dziewonski, A. D. 1984. Seismic Tomography. *Scientific American.* 251:60 (October).

Badash, L. 1989. The Age-of-the-Earth Debate. *Scientific American.* 261:90 (August).

Beatty, J. K. 1995. New Measures of the Moon. *Sky and Telescope.* 90:32 (July).

Bonatti, E. 1987. The Rifting of Continents. *Scientific American.* 256:97 (March).

Broecker, W. S. 1983. The Ocean. *Scientific American.* 249:146 (September).

Chaikin, A. 1994. *A Man on the Moon.* New York: Viking.

Davis, N. 1992. *The Aurora Watcher's Handbook.* College, Alaska: The University of Alaska Press.

Foxworthy, B. L., and Hill, M. 1982. *Volcanic Eruptions of 1980 at Mount St. Helens.* U.S. Geological Survey Professional Paper 1249. Washington, D.C.: U.S. Government Printing Office.

Frencheteau, J. 1983. The Oceanic Crust. *Scientific American.* 249:114 (September).

Gehrels, T. 1996. Collisions with Comets and Asteroids. *Scientific American.* 274:54 (March).

Grieve, R. A. F. 1990. Impact Cratering on the Earth. *Scientific American.* 262:66 (April).

Grove, N. 1992. Crucibles of Creation. *National Geographic.* 182:5.

Ingersoll, A. P. 1983. The Atmosphere. *Scientific American.* 249:162 (September).

Kosovsky, L. J., and Farouk, E.-B. 1970. *The Moon as Viewed by Lunar Orbiter.* Washington, D.C.: NASA.

Masursky, H., Colton, G. W., and Farouk, E.-B. 1978. *Apollo Over the Moon: A View from Orbit.* Washington, D.C.: NASA.

Moore, P. 1981. *The Moon.* New York: Rand McNally.

Morrison, D. 1995. Target: Earth. *Astronomy.* 23:34 (October).

Musgrove, R. G. 1971. *Lunar Photographs from Apollos 8, 9, and 10.* Washington, D.C.: NASA.

Siever, R. 1983. The Dynamic Earth. *Scientific American.* 249:46 (September).

Spudis, P. 1996. The Giant Holes of the Moon. *Astronomy.* 24:50 (May).

Taylor, G. J. 1994. The Scientific Legacy of Apollo. *Scientific American.* 271:40 (July).

White, R. M. 1990. The Great Climate Debate. *Scientific American.* 263:36 (July).

White, R. S., and McKenzie, D. P. 1989. Volcanism at Rifts. *Scientific American.* 261:62 (July).

Wilhelms, D. E. 1987. *The Geologic History of the Moon.* U.S. Geological Survey Professional Paper 1348. Washington, D.C.: U.S. Government Printing Office.

Chapter 8. Rocky Worlds

Barlow, N. 1995. The Prodigal Sister. *Mercury.* 24:23 (Sept./Oct.).

Beatty, J. K. 1996. Life From Ancient Mars? *Sky and Telescope.* 92:18 (October).

Burnham, R. 1993. What Makes Venus Go. *Astronomy.* 21:40 (January).

Cattermole, P. 1994. *Venus: The Geological Story.* Baltimore: Johns Hopkins University Press.

Davies, M. E. et al. 1978. *Atlas of Mercury.* Washington, D.C.: NASA.

Edgett, K., Geissler, P., and Herkenhoff, K. 1993. The Sands of Mars. *Astronomy.* 21:26 (June).

French, B. M. 1977. *Mars: The Viking Discoveries.* Washington, D.C.: NASA.

Naeye, R. 1996. Was There Life on Mars? *Astronomy.* 24:46 (November).

Robinson, C. 1995. Magellan Reveals Venus. *Astronomy.* 23:32 (February).

Saunders, R. S. et al. 1991. Magellan at Venus. *Science.* 252:185–312.

Saunders, R. S. 1991. Magellan at Venus: A Magellan Progress Report. *Mercury.* 20:131 (Sept./Oct.).

Sheehan, W. 1988. Mars 1909: Lessons Learned. *Sky and Telescope*. 76:247 (September).

Solomon, S. C. 1993. The Geophysics of Venus. *Physics Today*. 46(no. 7):48–55 (July).

Solomon, S. C. et al. 1992. Venus Tectonics: An Overview of Magellan Observations. *Journal of Geophysical Research*. 97:13, 199.

Stofan, E. R. 1993. The New Face of Venus. *Sky and Telescope*. 86:22 (August).

Strom, R. G. 1990. Mercury: The Forgotten Planet. *Sky and Telescope*. 80:256 (September).

Chapter 9. Great Worlds

Beatty, J. K. 1981. Voyager at Saturn. *Sky and Telescope*. 62:430 (November).

Beatty, J. K. 1996. Galileo: An Image Gallery. *Sky and Telescope*. 92:24 (November).

Beebe, R. F. 1990. Queen of the Giant Storms. *Sky and Telescope*. 80:339 (October).

Eliott, J., and Kerr, R. 1987. *Rings*. Cambridge, Mass.: MIT Press.

Hunt, G., and Moore, P. 1981. *Jupiter*. New York: Rand McNally.

Hunt, G., and Moore, P. 1982. *Saturn*. New York: Rand McNally.

Ingersoll, A. P. 1981. Jupiter and Saturn. *Scientific American*. 245:90 (December).

Morrison, D. 1980. *Voyage to Jupiter*. Washington, D.C.: NASA.

Morrison, D. 1982. *Voyages to Saturn*. Washington, D.C.: NASA.

O'Meara, S. J. 1991. Saturn's Great White Spot Spectacular. *Sky and Telescope*. 81:144 (February).

Rogers, J. H. 1995. *The Giant Planet Jupiter*. Cambridge, England: Cambridge University Press.

Soderblom, L. A. 1980. *Scientific American*. 243:88 (November).

Soderblom, L. A., and Johnson, T. V. 1982. The Moons of Saturn. *Scientific American*. 246:101 (January).

Chapter 10. Outer Worlds

Berry, R. 1989. Neptune Revealed. *Astronomy*. 17:22 (December).

Binzel, R. P. 1990. Pluto. *Scientific American*. 262:50 (June).

Cuzzi, J. N., and Esposito, L. W. 1987. The Rings of Uranus. *Scientific American*. 257:52 (July).

Dowling, T. 1990. Big, Blue: The Twin Worlds of Uranus and Neptune. *Astronomy*. 18:42 (October).

Hunt, G., and Moore, P. 1989. *Atlas of Uranus*. Cambridge, England: Cambridge University Press.

Ingersoll, A. P. 1987. Uranus. *Scientific American*. 256:38 (January).

Johnson, T. V., Brown, R. H., and Soderblom, L. A. 1987. The Moons of Uranus. *Scientific American*. 256:48 (April).

Miner, E. D. 1989. Voyager's Last Encounter. *Sky and Telescope*. 78:26 (July).

Moore, P. 1989. The Discovery of Neptune. *Mercury*. 18:98 (July/August).

Tombaugh, C. 1991. Plates, Pluto, and Planet X. *Sky and Telescope*. 81:360 (April).

No author. 1990. Neptune and Triton: Worlds Apart. *Sky and Telescope*. 79:136 (February).

Chapter 11. Creation and Its Debris

Balsiger, H., Fechtig, H., and Geiss, J. 1988. A Close Look at Halley's Comet. *Scientific American*. 259:96 (September).

Binzel, R. P., Barucci, M. A., and Fulchignoni, M. 1991. The Origins of the Asteroids. *Scientific American*. 265:88 (October).

Binzel, R. P., Gehrels, T., and Matthews, M. S. 1989. *Asteroids II*. Tucson: University of Arizona Press.

Brandt, J. C. 1992. *Rendezvous in Space: The Science of Comets*. New York: Freeman.

Brandt, J. C., and Niedner, M. B., Jr. 1986. The Structure of Comet Tails. *Scientific American*. 254:49 (January).

Delsemme, A. H. 1989. Whence Come Comets? *Sky and Telescope*. 77:260 (March).

Dodd, R. T. 1981. *Meteorites*. Cambridge, England: Cambridge University Press.

Gingrich, O. 1986. Newton, Halley, and the Comet. *Sky and Telescope*. 71:230 (March).

Gropman, D. 1985. *Comet Fever: A Popular History of Halley's Comet*. New York: Simon & Schuster.

Kronk, G. W. 1988. Meteor Showers. *Mercury*. 17:162 (Nov./Dec.).

Kronk, G. W. 1988. *Meteor Showers: A Descriptive Catalogue*. Hillside, New Jersey: Enslow.

Levy, D. H. 1994. *The Quest for Comets*. New York: Plenum Press.

Levy, D. H., Shoemaker, E. M., and Shoemaker, C. S. 1995. *Scientific American*. 273:84 (August).

Luu, J. X., and Jewitt, D. C. 1996. The Kuiper Belt. *Scientific American*. 274:46 (May).

McFadden, L.-A., and Chapman, C. R. 1992. Interplanetary Fugitives. *Astronomy*. 20:30 (August).

Musser, G. 1994. The Big Hit. *Mercury*. 23:13 (July/Aug.).

Norton, O. R. 1994. *Rocks from Space*. Missoula, Montana: Mountain Press Publishing Company.

Spratt, C., and Stephens, S. 1992. Against All Odds: Meteorites That Have Struck Home. *Mercury*. 21:50 (March/April).

Stern, A. 1992. Where Has Pluto's Family Gone? *Astronomy*. 20:41 (September).

Weissman, P. 1995. Making Sense of Shoemaker-Levy 9. *Astronomy*. 23:48 (May).

Wetherill, G. W. 1981. The Formation of the Earth from Planetesimals. *Scientific American.* 244:162 (June).

Whipple, F. L. 1985. *The Mystery of Comets.* Washington, D.C.: Smithsonian Institution Press.

Whipple, F. L. 1987. The Black Heart of Comet Halley. *Sky and Telescope.* 73:242 (March).

PART III.   STELLAR ASTRONOMY

General

Aller, L. H. 1971. *Atoms, Stars, and Nebulae.* Cambridge, Mass.: Harvard University Press.

Hoskin, M. 1986. William Herschel and the Making of Modern Astronomy. *Scientific American.* 254:106 (April).

Kaler, J. B. 1997. *Cosmic Clouds: Birth, Death, and Recycling in the Galaxy.* Scientific American Library. New York: Freeman.

Kaler, J. B. 1992. *Stars.* Scientific American Library. New York: Freeman.

Chapter 12. The Sun

Bahcall, J. N. 1990. The Solar Neutrino Problem. *Scientific American.* 262:54 (May).

Fischer, D. 1992. Closing In on the Solar-Neutrino Problem. *Sky and Telescope.* 84:378 (October).

Hathaway, D. H. 1995. Journey to the Heart of the Sun. *Astronomy.* 23:38 (January).

Kennedy, J. R. 1996. GONG: Probing the Sun's Hidden Heart. *Sky and Telescope.* 92:20 (October).

Kippenhahn, R. 1994. *Discovering the Secrets of the Sun.* New York: Wiley.

Lang, K. R. 1995. *Sun, Earth, and Sky.* New York: Springer.

Lang, K. R. 1996. Unsolved Mysteries of the Sun–Part I. *Sky and Telescope.* 92:38 (August).

Lang, K. R. 1996. Unsolved Mysteries of the Sun–Part II. *Sky and Telescope.* 92:24 (September).

McCrea, W. 1991. Arthur Stanley Eddington. *Scientific American.* 264:92 (June).

Phillips, K. J. H. 1992. *Guide to the Sun.* Cambridge, England: Cambridge University Press.

Wentzel, D. G. 1989. *The Restless Sun.* Washington, D.C.: Smithsonian Institution Press.

Chapter 13. The Stars

Baliunas, S., and Sarr, S. 1992. Unfolding the Mysteries of Stellar Cycles. *Astronomy.* 20:42 (May).

Kaler, J. B. 1989. *Stars and their Spectra.* Cambridge, England: Cambridge University Press.

Kaler, J. B. 1991. The Faintest Stars in the Galaxy. *Astronomy.* 19:26 (August).

King, I. R. 1985. Globular Clusters. *Scientific American.* 252:79 (June).

Kopal, Zdenek. 1990. Eclipsing Binary Stars. *Mercury.* 19:88 (May/June).

McAlister, H. A. 1996. Twenty Years of Seeing Double. *Sky and Telescope.* 92:28 (November).

Spradley, J. L. 1990. The Industrious Mrs. Fleming. *Astronomy.* 18:48 (July).

Steffey, P. C. 1992. The Truth About Star Colors. *Sky and Telescope.* 84:266 (September).

Terrell, D. 1992. Close Binary Stars. *Astronomy.* 20:34 (October).

Trimble, V. 1984. A Field Guide to Close Binary Stars. *Sky and Telescope.* 68:306 (October).

van den Bergh, S. 1992. Star Clusters: Enigmas in Our Backyard. *Sky and Telescope.* 83:508 (May).

White, R. E. 1991. Globular Clusters: Fads and Fallacies. *Sky and Telescope.* 81:24 (January).

Chapter 14. Star Formation

Black, D. C. 1996. Other Suns, Other Planets? *Sky and Telescope.* 92:20 (August).

Blitz, L. 1982. Giant Molecular-Cloud Complexes in the Galaxy. *Scientific American.* 246:84 (April).

Boss, A. P. 1995. Companions to Young Stars. *Scientific American.* 273:38 (October).

Boss, A. P. 1991. The Genesis of Binary Stars. *Astronomy.* 19:34 (June).

Chyba, C. 1992. The Cosmic Origins of Life on Earth. *Astronomy.* 20:29 (November).

Cohen, M. 1988. *In Darkness Born.* Cambridge, England: Cambridge University Press.

Dame, T. M. 1988. The Molecular Milky Way. *Sky and Telescope.* 76:22 (July).

Drake, F., and Sobel, D. 1992. *Is Anyone Out There? The Scientific Search for Extraterrestrial Intelligence.* New York: Delacorte.

Frank, A. 1996. Starmaker. *Astronomy.* 24:52 (July).

Goldsmith, D., and Owen, T. 1992. *The Search for Life in the Universe.* 2nd ed. New York: Addison-Wesley.

Goldstein, A. 1990. Magnificent Orion. *Astronomy.* 18:79 (November).

Greenberg, J. M. 1984. The Structure and Evolution of Interstellar Grains. *Scientific American.* 250:124 (June).

Horgan, J. 1991. In the Beginning. *Scientific American.* 264:116 (February).

Knapp, G. 1995. The Stuff Between the Stars. *Sky and Telescope.* 89:20 (May).

Lada, C. J. 1993. Deciphering the Mysteries of Stellar Origins. *Sky and Telescope.* 85:18 (May).

Lada, C. J., and Shu, F. H. 1990. The Formation of Sunlike Stars. *Science.* 248:564.

Malin, D. 1982. The Dust Clouds of Sagittarius. *Sky and Telescope.* 63:254 (March).

Malin, D. 1987. In the Shadow of the Horsehead. *Sky and Telescope*. 74:253 (September).

Reipurth, B., and Heathcote, S. 1995. Herbig-Haro Objects and the Birth of Stars. *Sky and Telescope*. 90:38 (October).

Schorn, R. A. 1981. Extraterrestrial Beings Don't Exist. *Sky and Telescope*. 62:207 (September).

Smith, D. H. 1985. Reflection Nebulae: Celestial Veils. *Astronomy*. 13:207 (September).

Stahler, S. W. 1991. The Early Life of Stars. *Scientific American*. 265:48 (July).

Stephens, S. 1996. The Excesses of Youth. *Astronomy*. 24:36 (September).

Tipler, F. J. 1982. The Most Advanced Civilization in the Galaxy Is Ours. *Mercury*. 11:5 (Jan./Feb.).

Verschuur, G. 1989. *Interstellar Matters*. New York: Springer.

Verschuur, G. 1992. Star Dust. *Astronomy*. 20:46 (March).

Verschuur, G. 1992. Interstellar Molecules. *Sky and Telescope*. 83:379 (April).

Chapter 15. The Life and Death of Stars

Balick, B. 1987. The Shaping of Planetary Nebulae. *Sky and Telescope*. 73:125 (February).

Bethe, H. A. 1990. Supernovae. *Physics Today*. 43(no. 9):24 (September).

Bethe, H. A., and Brown, G. 1985. How a Supernova Explodes. *Scientific American*. 252:60 (May).

Croswell, K. 1992. The Best Black Hole in the Galaxy. *Astronomy*. 20:30 (March).

de Duve, C. 1996. Searching for Life on Other Planets. *Scientific American*. 274:60 (April).

Gingerich, O., and Welther, B. 1985. Harlow Shapley and the Cepheids. *Sky and Telescope*. 70:540 (December).

Hoffmeister, C., Richter, G., and Wenzel, W. 1985. *Variable Stars*. 1985. New York: Springer.

Graham-Smith, F. 1990. Pulsars Today. *Sky and Telescope*. 80:240 (September).

Greenstein, G. 1983. *Frozen Star*. New York: Freundlich.

Hayes, J. C., and Burrows, A. 1995. A New Dimension to Supernovae. *Sky and Telescope*. 90:30 (August).

Kaler, J. B. 1982. Bubbles from Dying Stars. *Sky and Telescope*. 63:129 (February).

Kaler, J. B. 1988. Journeys on the HR Diagram. *Sky and Telescope*. 75:483 (May).

Kaler, J. B. 1990. Realm of the Hottest Stars. *Astronomy*. 18:32 (February).

Kaler, J. B. 1990. The Coolest Stars. *Astronomy*. 18:20 (May).

Kaler, J. B. 1990. The Largest Stars in the Galaxy. *Astronomy*. 18:30 (October).

Kaler, J. B. 1991. The Smallest Stars in the Universe. *Astronomy*. 19:50 (November).

Kaler, J. B. 1991. The Brightest Stars. *Astronomy*. 19:30 (May).

Kaler, J. B. 1994. Hypergiants. *Astronomy*. 22:32 (March).

Kaufmann, W. 1979. *Black Holes and Warped Spacetime*. New York: Freeman.

Kawaler, S. D., and Winget, W. E. 1987. White Dwarfs: Fossil Stars. *Sky and Telescope*. 74:132 (August).

Kwok, S. 1996. A Modern View of Planetary Nebulae. *Sky and Telescope*. 92:38 (July).

Luminet, J.-P. 1987. *Black Holes*. Cambridge, England: Cambridge University Press.

Marshall, L. A. 1988. *The Supernova Story*. New York: Plenum.

Merrill, P. 1940. *The Spectra of Long Period Variable Stars*. Chicago: University of Chicago Press.

Nather, R. E., and Winget, D. E. 1992. Taking the Pulse of White Dwarfs. *Sky and Telescope*. 83:374 (April).

Petit, M. 1987. *Variable Stars*. New York: Wiley.

Shipman, H. 1980. *Black Holes, Quasars, and the Universe*. New York: Houghton Mifflin.

Starrfield, S., and Shore, S. N. 1995. The Birth and Death of Nova V1974 Cygni. *Scientific American*. 272:76 (January).

Soker, N. 1992. Planetary Nebulae. *Scientific American*. 266:78 (May).

Trimble, V. 1986. White Dwarfs: The Once and Future Suns. *Sky and Telescope*. 72:348 (October).

Williams, R. E. 1981. The Shells of Novas. *Scientific American*. 244:120 (April).

Woosley, S., and Weaver, T. 1989. The Great Supernova of 1987. *Scientific American*. 261:32 (August).

Chapter 16. The Galaxy

Binney, J. 1995. The Evolution of Our Galaxy. *Sky and Telescope*. 89:20 (March).

Bok, B. J. 1981. The Milky Way Galaxy. *Scientific American*. 244:93 (March).

Bok, B., and Bok, P. 1981. *The Milky Way*. Cambridge, Mass.: Harvard University Press.

Burnham, R. 1990. Strange Doings at the Milky Way's Core. *Astronomy*. 18:39 (October).

Croswell, K. 1992. Galactic Archaeology. *Astronomy*. 20:29 (July).

Croswell, K. 1993. Intruder Galaxies. *Astronomy*. 21:28 (November).

Friedlander, M. 1990. Cosmic Rays: "A Thin Rain of Charged Particles." *Mercury*. 19:130 (Sept./Oct.).

Helfrand, D. J. 1988. Fleet Messengers from the Cosmos. *Sky and Telescope*. 75:263 (March).

Jayawardhana, R. 1995. Destination: Galactic Center. *Sky and Telescope*. 89:26 (June).

Oort, J. 1992. Exploring the Nuclei of Galaxies (Including Our Own). *Mercury*. 21:57 (March/April).

Robinson, L. J. 1982. The Black Heart of the Milky Way. *Sky and Telescope.* 64:133 (August).

Smith, D. H. 1990. Seeking the Origin of Cosmic Rays. *Sky and Telescope.* 79:479 (May).

Townes, C. H., and Genzel, R. 1990. What Is Happening at the Center of Our Galaxy? *Scientific American.* 262:46 (April).

Trimble, V., and Parker, S. 1995. Meet the Milky Way. *Sky and Telescope.* 89:26 (January).

Twarog, B. A. 1985. Chemical Evolution of the Galaxy. *Mercury.* 14:107 (July/August).

Verschuur, G. L. 1990. The Magnetic Milky Way. *Astronomy.* 18:32 (June).

Verschuur, G. L. 1993. Journey into the Galaxy. *Astronomy.* 21:33 (January).

PART IV. GALAXIES AND THE UNIVERSE

General

Jones, B. 1989. The Legacy of Edwin Hubble. *Astronomy.* 17:38 (December).

Osterbrock, D., and Brashear, R. 1990. Young Edwin Hubble. *Mercury.* 19:2 (Jan./Feb.).

Sandage, A., Sandage, M., and Kristian, J., eds. 1975. *Galaxies and the Universe.* Chicago: University of Chicago Press.

Tucker, W., and Tucker, K. 1988. *Dark Matter.* New York: Morrow.

Chapter 17. Galaxies

Barnes, J., Hernquist, L., and Schweizer, F. 1991. Colliding Galaxies. *Scientific American.* 265:40 (August).

Benningfield, D. 1996. Galaxies Colliding in the Night. *Astronomy.* 24:36 (November).

Burns, J. O. 1990. Chasing the Monster's Tail: New Views of Cosmic Jets. *Astronomy.* 18:29 (August).

Croswell, K. 1995. What Lies at the Milky Way's Center? *Astronomy.* 23:32 (May).

de Vaucouleurs, G. 1983. The Distance Scale of the Universe. *Sky and Telescope.* 66:511 (December).

Dressler, A. 1993. Galaxies Far Away and Long Ago. *Sky and Telescope.* 85:22 (April).

Elmegreen, D. M., and Elmegreen, B. 1993. What Puts the Spiral in Spiral Galaxies? *Astronomy.* 21:34 (September).

Finkbeiner, A. 1992. Active Galactic Nuclei: Sorting Out the Mess. *Sky and Telescope.* 84:138 (August).

Ferris, T. 1980. *Galaxies.* New York: Stewart, Tabori, and Chang.

Ford, H., and Tsvetanov, Z. I. 1996. Massive Black Holes in the Hearts of Galaxies. *Sky and Telescope.* 91:28 (June).

Hodge, P. 1986. *Galaxies.* Cambridge, Mass.: Harvard University Press.

Hodge, P. 1987. The Local Group: Our Galactic Neighborhood. *Mercury.* 16:2 (Jan./Feb.).

Keel, W. W. 1989. Crashing Galaxies, Cosmic Fireworks. *Sky and Telescope.* 77:18 (January).

Lake, G. 1992. Understanding the Hubble Sequence. *Sky and Telescope.* 83:513 (May).

Lake, G. 1992. Cosmology of the Local Group. *Sky and Telescope.* 84:613 (December).

Miley, G. K., and Chambers, K. C. 1993. The Most Distant Radio Galaxies. *Scientific American.* 268:54 (June).

Parker, B. 1990. *Colliding Galaxies: The Universe in Turmoil.* New York: Plenum.

Percy, J. R. 1984. Cepheids: Cosmic Yardsticks, Celestial Mysteries? *Sky and Telescope.* 68:517 (December).

Price, J. S., and Caldwell, K. 1995. Galaxies that Go Bump in the Night. *Mercury.* 24:23 (July/August).

Rubin, V. C. 1983. Dark Matter in Spiral Galaxies. *Scientific American.* 248:96 (June).

Sandage, A. 1961. *The Hubble Atlas of Galaxies.* Washington, D.C.: Carnegie Institution of Washington.

Silk, J. 1986. Formation of the Galaxies. *Sky and Telescope.* 72:582 (December).

Smith, D. H. 1985. Mysteries of Galactic Jets. *Sky and Telescope.* 69:213 (March).

Smith, R. W. 1983. The Great Debate Revisited. *Sky and Telescope.* 65:28 (January).

Struble, M. F. 1988. Diversity Among Galaxy Clusters. *Sky and Telescope.* 75:16 (January).

Tremain, S. 1992. The Dynamical Evidence for Dark Matter. *Physics Today.* 82:28 (February).

Veilleux, S., Cecil, G., and Bland-Hawthorne, J. 1996. Colossal Galactic Explosions. *Scientific American.* 274:98 (February).

Chapter 18. The Expansion and Structure of the Universe

Arp, H. 1987. *Quasars, Redshifts, and Controversies.* Berkeley, Calif.: Interstellar Media.

Burns, J. O. 1986. Very Large Structures in the Universe. *Scientific American.* 255:38 (July).

Burstein, D., and Manly, P. L. 1993. Cosmic Tug of War. *Astronomy.* 21:40 (July).

Courvoisier, T. J.-L., and Robson, E. I. 1991. The Quasar 3C 273. *Scientific American.* 264:50 (June).

Croswell, K. 1993. Have Astronomers Solved the Quasar Enigma? *Astronomy.* 21:29 (February).

Dressler, A. 1987. The Large-Scale Streaming of Galaxies. *Scientific American.* 257:46 (September).

Freedman, W. 1992. The Expansion Rate and Size of the Universe. *Scientific American.* 267:54 (November).

Geller, M. J. 1990. Mapping the Universe: Slices and Bubbles. *Mercury.* 19:66 (May/June).

Geller, M. J., and Huchra, J. P. 1991. Mapping the Universe. *Sky and Telescope.* 82:134 (August).

Gregory, S. A. 1988. Active Galaxies and Quasars: A Unified View. *Mercury.* 17:111 (July/August).

Hodge, P. 1993. The Extragalactic Distance Scale: Agreement at Last? *Sky and Telescope.* 86:24 (August).

Jacoby, G. H. et al. 1992. *A Critical Review of Selected Techniques for Measuring Extragalactic Distances.* Publ. Astronomical Society of the Pacific. 104:599.

Osterbrock, D. E., and Brashear, R. S. 1993. Edwin Hubble and the Expanding Universe. *Scientific American.* 269:84 (July).

Preston, R. 1988. Beacons in Time: Maarten Schmidt and the Discovery of Quasars. *Mercury.* 17:2 (Jan./Feb.).

Spergel, D. N., and Turok, N. G. 1992. Textures and Cosmic Structures. *Scientific American.* 266:52 (March).

Tully, R. B. 1982. Unscrambling the Local Supercluster. *Sky and Telescope.* 63:550 (June).

Turner, E. L. 1988. Gravitational Lenses. *Scientific American.* 259:54 (July).

Tyson, A. 1992. Mapping Dark Matter with Gravitational Lenses. *Physics Today.* 82(no. 6):24 (June).

Weedman, D. W. 1986. *Quasar Astronomy.* Cambridge, England: Cambridge University Press.

Wilkes, B. J. 1991. The Emerging Picture of Quasars. *Astronomy.* 19:35 (December).

Chapter 19. The Universe

Brush, S. G. 1992. How Cosmology Became a Science. *Scientific American.* 267:62 (August).

Collins, G. P. 1992. COBE Measures Anisotropy in Cosmic Microwave Background Radiation. *Physics Today.* 82(no. 6):17 (June).

Cowen, R. 1994. The Debut of Galaxies. *Astronomy.* 22:44 (December).

Davies, P. 1985. Relics of Creation. *Sky and Telescope.* 69:112 (February).

Davies, P. 1985. New Physics and the Big Bang. *Sky and Telescope.* 70:406 (November).

Davies, P. 1990. Matter-Antimatter. *Sky and Telescope.* 79:257 (March).

Davies, P. 1991. Everyone's Guide to Cosmology. *Sky and Telescope.* 81:250 (March).

Davies, P. 1992. The First One Second of the Universe. *Mercury.* 21:82 (May/June).

Davies, P. 1994. *The Last Three Minutes.* New York: Basic Books.

Djorgovski, S. G. 1995. Fires at Cosmic Dawn. *Astronomy.* 23:36 (September).

Dressler, A. 1991. Observing Galaxies Through Time. *Sky and Telescope.* 82:126 (August).

Eicher, D. 1995. Our Strange, Scrappy Ancestors. *Astronomy.* 23:62 (December).

Guth, A. H., and Steinhardt, P. J. 1984. The Inflationary Universe. *Scientific American.* 250:116 (May).

Harrison, E. R. 1981. *Cosmology.* Cambridge, England: Cambridge University Press.

Hawking, S. 1988. *A Brief History of Time.* Bantam: New York.

Jayawardhana, R. 1993. The Age Paradox. *Astronomy.* 21:39 (June).

Kanipe, J. 1996. Dark Matter and the Fate of the Universe. *Astronomy.* 24:34 (October).

Kippenhahn, R. 1987. Light From the Depths of Time. *Sky and Telescope.* 73:140 (February).

Lederman, L. M., and Schramm, D. N. 1989. *From Quarks to the Cosmos.* Scientific American Library. New York: Freeman.

Linde, A. 1994. The Self-Reproducing Inflationary Universe. *Scientific American.* 271:48 (November).

Odenwalf, S., and Fienberg, R. T. 1993. Galaxy Redshifts Reconsidered. *Sky and Telescope.* 85:31 (February).

Overbye, D. 1991. *Lonely Hearts of the Cosmos.* New York: HarperCollins.

Parker, B. 1986. Discovery of the Expanding Universe. *Sky and Telescope.* 72:227 (September).

Powell, C. S. 1991. Star Bursts: The Deepening Mystery of the Gamma-Ray Sky. *Scientific American.* 265:32 (December).

Riordan, M., and Schramm, D. N. 1991. *The Shadows of Creation.* New York: Freeman.

Schramm, D. N. 1991. The Origin of Cosmic Structure. *Sky and Telescope.* 82:140 (August).

Schramm, D. N. Dark Matter and Cosmic Structure. *Sky and Telescope.* 88:28 (October).

Silk, J. 1989. *The Big Bang.* 2nd ed. New York: Freeman.

Silk, J. 1994. *A Short History of the Universe.* Scientific American Library. New York: Freeman.

Talcott, R. 1994. Everything You Wanted to Know About the Big Bang. *Astronomy.* 22:28 (January).

Talcott, R. 1995. Gamma-Ray Bursters: Near or Far? *Astronomy.* 23:56 (December).

Trimble, V., and Mussur, G. 1995. Clusters, Lensing, and the Future of the Universe. *Mercury.* 24:6 (May/June).

Weinberg, S. 1988. *The First Three Minutes.* Updated Ed. New York: HarperCollins.

Wesson, P. S. 1989. Olbers' Paradox Solved at Last. *Sky and Telescope.* 77:594 (June).

ACKNOWLEDGMENTS

LITERARY PERMISSIONS

Figure 2.19(b) Illustration by Snowden Hodges from *Living the Sky* by Ray A. Williamson. Line illustrations copyright © 1984 by Snowden Hodges. Reprinted by permission of Houghton Mifflin Company. All rights reserved.

Figure 2.25 Adapted from *Stars* by James B. Kaler, page 30. Copyright © 1992 by Scientific American Library. Reprinted by permission of W.H. Freeman and Company.

Figure 4.2(a) From *Stars* by James B. Kaler, page 9. Copyright © 1992 by Scientific American Library. Reprinted by permission of W.H. Freeman and Company.

Figure 4.9 U.S. Naval Observatory.

Figure 7.20(b) Adaptation of Figure 3.12 from *The Cambridge Encyclopedia of Earth Sciences* edited by David G. Smith, page 47. Copyright © 1981 by Cambridge University Press. Reprinted by permission of Cambridge University Press and the author.

Figure 8.7 R. G. Strom

Figure 8.21 Bottom From *Sky and Telescope*, November 1982, page 421. Copyright © 1982 Sky Publishing Corporation. Courtesy Sky Publishing Corporation.

Figure 9.4 Adapted from *The New Solar System*, Third Edition, edited by J. Kelly Beatty and Andrew Chaikin, page 143. Copyright © 1990 Sky Publishing Corporation. Reprinted by permission.

Figure 9.11(a) Illustration of model of Callisto reprinted by permission of the Jet Propulsion Laboratory.

Figure 9.11(c) Illustration of model of Europa reprinted by permission of the Jet Propulsion Laboratory.

Figure 9.11(d) Illustration of model of Io reprinted by permission of the Jet Propulsion Laboratory.

Figure 9.15(a) Adapted from *The New Solar System*, Third Edition, edited by J. Kelly Beatty and Andrew Chaikin, page 143. Copyright © 1990 Sky Publishing Corporation. Reprinted by permission.

Figure 10.4 Adapted from *Sky and Telescope*, February 1990, page 148. Copyright © 1990 Sky Publishing Corporation. Reprinted by permission.

Figure 10.5 From "Voyager II Encounter with the Gas Plants" by Ellis D. Miner, from *Physics Today*, July 1990, page 40. Reprinted by permission of the American Institute of Physics and the author.

Figure 11.4(a) Adapted from *Sky and Telescope*, June 1992, p. 609. Reprinted by permission of Lucy McFadden.

Figure 11.12(b) From Figure 19 by J.C. Brandt from *The New Solar System*, Third Edition, edited by J. Kelly Beatty and Andrew Chaikin. Reprinted by permission.

Figure 11.13(b) From *The New Solar System*, Third Edition, edited by J. Kelly Beatty and Andrew Chaikin, page 225. Copyright © 1990 Sky Publishing Corporation. Courtesy Sky Publishing Corporation.

Figure 11.13(c) From *The New Solar System*, Third Edition, edited by J. Kelly Beatty and Andrew Chaikin, page 227. Copyright © 1990 Sky Publishing Corporation. Courtesy Sky Publishing Corporation.

Figure 11.14 From *Sky and Telescope*, January 1993, page 26. Artwork by Steven Simpson. Copyright © 1993 Sky Publishing Corporation. Courtesy Sky Publishing Corporation.

Figure 11.18(a) From *Sky and Telescope*, October 1985, page 317. Copyright © 1985 Sky Publishing Corporation. Courtesy Sky Publishing Corporation.

Figure 12.8(b) Adapted from *The Restless Sun* by Donat G. Wentzel, page 76. Copyright © 1989 by the Smithsonian Institution. Reprinted by permission.

Figure 12.13(a) Adapted from *The Sun* by Iain Nicolson, page 41. Copyright © 1982 by Iain Nicolson. Reprinted by permission of Reed Book Services Ltd. Data originally compiled by J.A. Eddy, University Corporation for Atmospheric Research.

Figure 12.13(b) From "A Search from Rhenium Lines in the Fraunhofer Spectrum" by J.W. Swensson from *Solar Physics*, Volume 13, No. 1, July 1970. Copyright © 1970 by D. Reidel Publishing Company. Reprinted by permission of Kluwer Academic Publishers.

Figure 12.19 Adapted from *Stars* by James B. Kaler, page 114. Copyright © 1992 by Scientific American Library. Reprinted by permission of W.H. Freeman and Company.

Figure 12.22 From *Stars* by James B. Kaler, page 125. Copyright © 1992 by Scientific American Library. Reprinted by permission of W.H. Freeman and Company.

Figure 13.12 From *Landolt-Börnstein, Group VI: Astronomy Astrophysics and Space Research, Volume I: Astronomy and Astrophysics* edited by H.H. Voigt, page 284. Copyright © 1965 by Springer-Verlag. Reprinted by permission.

Figure 13.24(a, b) From figures on pages 51 (Hyades), 26 (Pleiades), and 14 & 15 (Double Cluster) from *An Atlas of Open*

Cluster Colour-Magnitude Diagrams by Gretchen L. Hagen, Volume 4, 1970. Reprinted by permission.

Figure 13.24(c) Data from figure *The Astrophysical Journal,* 105, 492, 1947. Reprinted by permission of the author.

Figure 13.25 From figures on pages 51 (Hyades) and 26 (Pleiades) from *An Atlas of Open Cluster Colour-Magnitude Diagrams* by Gretchen L. Hagen, Volume 4, 1970. Reprinted by permission.

Figure 14.13 From figure by K. Akabane, M. Morimoto, and M. Ishiguro, Nobeyama Radio Observatory.

Figure 14.24 Adapted figure from Steven W. Stahler, with tracks by Icko Iben, Jr.. Reprinted by permission.

Figure 15.3 Figure, "HR diagrams & evolution of open clusters" from "Observational Approach To Evolution II. A Computed Luminosity Function For K0–K2 Stars From $M_V = +5$ To $M_V = -4.5$" by Allan Sandage from *The Astrophysical Journal,* March 1957, Volume 125, Number 2, page 435. Reprinted by permission of the author.

Figure 15.4 Adaptation of figure by A. Maeder and G. Meynet from *Astronomy & Astrophysics Supplements,* Volume 76, page 411. Reprinted by permission of the authors.

Figure 15.7 Graph Reprinted by permission of the publishers from *The Story of Variable Stars* by Leon Campbell and Luigi Jacchia, Cambridge, Mass.: Harvard University Press. Copyright © 1941 by the Blakiston Company.

Figures 15.8 and 15.11 Adapted from Icko Iben.

Figure 15.10 Inset From *Variable Stars* by Michel Petit, page 54. Copyright © 1987 by John Wiley & Sons, Ltd. Reprinted by permission of John Wiley & Sons, Ltd.

Figure 15.14 Adaptation of figure by A. Maeder and G. Meynet from *Astronomy & Astrophysics Supplements,* Volume 76, page 411. Reprinted by permission of the author. Observed supergiants on HR diagram from *The Astrophysical Journal,* 284, 565, 1984; 232, 409, 1979. Reprinted by permission of the author.

Figure 15.16 Adapted from *Stars* by James B. Kaler, pages 179 and 189. Copyright © 1992 by Scientific American Library. Reprinted by permission of W.H. Freeman and Company.

Figure 15.21 (a) Figure by A. Hewish et al., in *Nature,* Vol. 217.

Figure 15.22 Adapted from *Stars* by James B. Kaler, page 199. Reprinted by permission of George Greenstein.

Figure 16.1 Figure from *Stars* by James B. Kaler, page 101. Reprinted by permission.

Figure 16.11 Inset From *Sky and Telescope,* May 1990. Copyright © 1990 Sky Publishing Corporation. Courtesy Sky Publishing Corporation.

Figure 18.2(b) Adapted from *Galaxies and the Universe,* edited by Allan Sandage, Mary Sandage and Jerome Kristian, page 782. Copyright © 1975 by The University of Chicago. Reprinted by permission of The University of Chicago Press.

Figure 18.6 Figure by Rob Hess from *Sky and Telescope,* June 1982, page 554. Copyright © 1982 Sky Publishing Corporation. Courtesy Sky Publishing Corporation.

Figure 19.4(b) Adapted from *Sky and Telescope,* March 1991, page 253. Copyright © 1991 Sky Publishing Corporation. Reprinted by permission.

Figure 19.7 Adapted from *Sky and Telescope,* March 1991, page 255. Copyright © 1991 Sky Publishing Corporation. Reprinted by permission.

Figure 19.9 Adapted from *The Big Bang: The Creation and Evolution of the Universe* by Joseph Silk, page 95. Copyright © 1980 by W.H. Freeman and Company. Reprinted by permission.

Figure 19.10 Adaptation of Figure 6 from "Observational Tests of World Models" by Allan Sandage from *Annual Review of Astronomy and Astrophysics,* Volume 26, 1988, page 599. Copyright © 1988 by Annual Review Inc. Adapted, with permission, from Annual Reviews Inc. and the author.

Figure 19.12 Adapted from *From Quarks to the Cosmos: Tools of Discovery* by Leon M. Lederman and David N. Schramm, page 152. Copyright © 1989 by Scientific American Library. Reprinted by permission of W.H. Freeman and Company.

Figure 19.13 From *From Quarks to the Cosmos: Tools of Discovery* by Leon M. Lederman and David N. Schramm, page 144. Copyright © 1989 by Scientific American Library. Reprinted by permission of W.H. Freeman and Company.

Figure 19.15 "Looking Back to the Big Bang" from "Astronomers Detect Proof of 'Big Bang'" from *The New York Times,* Friday, April 24, 1992, page A16. Copyright © 1992 by The New York Times Company. Reprinted by permission.

PHOTO CREDITS

Page 1 Bibliotheque Nationale, Paris
Figure 1.2 NASA
Figure 1.3 NASA
Figure 1.4 Mt. Wilson and Las Campanas Observatories
Figure 1.6 NASA
Figure 1.7 University of Illinois Prairie Observatory
Figure 1.8 James B. Kaler, Smithsonian Institution.
Figure 1.9(a) Rudolph Schild, Smithsonian Astrophysical Observatory
Figure 1.9(b) Laird Thompson, University of Hawaii
Figure 1.10 Dennis DiCicco
Figure 1.12 California Institute of Technology
Figure 1.13 ROE/Anglo-Australian Telescope Board, photo by David F. Malin

Figure 2.1(a) James B. Kaler
Figure 2.2 Akira Fujii
Figure 2.7 James B. Kaler
Figure 2.8 National Optical Astronomy Observatories
Figure 2.10(a) Gregory G. Dimijian
Figure 2.10(b) Anglo-Australian Telescope Board, photo by David F. Malin
Figure 2.19(a) Department of Environment, Crown Copyright
Figure 2.22 Mario Grassi

Figure 3.1 University of Illinois Library
Figure 3.2 Akira Fujii
Figure 3.3(b) Rick Olson
Figure 3.4(a) Akira Fujii
Figure 3.5 University of Illinois Library
Figure 3.6 Akira Fujii
Figure 3.7 University of Illinois Library

Figure 3.8 Istanbul University Library
Figure 3.9 Dennis DiCicco

Figure 4.1 Lick Observatory
Figure 4.3 (7 photos) Lick Observatory
Figure 4.8 Mark E. Killion
Figure 4.10(a) NASA
Figure 4.10(b) Yerkes Observatory
Figure 4.10(c) Richard Berry
Figure 4.10(d) Tersch Enterprises
Figure 4.10(e) National Optical Astronomy Observatories
Figure 4.11 James B. Kaler
Figure 4.19 (4 photos) E. C. Slipher, Lowell Observatory
Figure 4.21 "Copernicus Concerning the Revolutions of the Heavenly Bodies," 1543
Figure 4.23(a) From Jan Blaeu, *Grand Atlas au Cosmographie Blaviane,* vol. I, Amsterdam, 1667 Bibliotheque Publique et Universitaire, Geneva
Figure 4.23(b) New York Public Library; Astor, Lenox and Tilden Foundations
Figure 4.24 Scripta Mathematica, Yeshiva University
Figure 4.25 Scala/Art Resource, NY

Figure 5.1 Brown Brothers
Figure 5.7(a, b) Lowell Observatory
Figure 5.8(a, b) NASA
Figure 5.9 Corbis-Bettmann

Figure 6.4 The Exploratorium
Figure 6.10 Deutsches Museum
Figure 6.19 Ressmeyer/Starlight/Corbis
Figure 6.20 California Institute of Technology
Figure 6.23 California Institute of Technology
Figure 6.24 James B. Kaler
Figure 6.26 James B. Kaler
Figure 6.27(a) Palomar Observatory
Figure 6.28 NRAO/AUI
Figure 6.29 Roger Ressmeyer/Starlight
Figure 6.30(a–c) NASA

Page 151 U.S.G.S., Flagstaff, NASA
Figure 7.1 NASA
Figure 7.3(a, b) Maine Department of Sea and Shore Fisheries
Figure 7.5(a) James B. Kaler
Figure 7.5(b, c) David Weintraub/Photo Researchers, Inc.
Figure 7.6 W. Haxby, NOAA/NESDIS/NGDC
Figure 7.7 NASA
Figure 7.8(a) Lick Observatory
Figure 7.11(a) Ewan Whitaker, Lunar and Planetary Lab
Figure 7.11(b, c) NASA
Figure 7.12(a) Lick Observatory
Figure 7.12(b) NASA
Figure 7.13 NASA
Figure 7.14(a) Ewan Whitaker, Lunar and Planetary Lab
Figure 7.14(b, c) NASA
Figure 7.15 Paul Lucey/NASA
Figure 7.17(a, c) NASA
Figure 7.17(b) Astrogeology Team, U.S.G.S., Flagstaff, AZ
Figure 7.18 NASA
Figure 7.19(a) Courtesy Meteor Crater Enterprises

Figure 7.19(b) NASA
Figure 7.23 Scientific American
Figure 7.25 Phil Degginger/H. Armstrong Roberts
Figure 7.27 Inset Louis Frank, University of Iowa
Figure 7.28 Dyballa/Zefa/H. Armstrong Roberts
Figure 7.29(a–d) Center for Astrophysics

Figure 8.1(a) NASA
Figure 8.1(a) Inset Lick Observatory
Figure 8.1(b) L. Esposito (University of CO, Boulder)/NASA
Figure 8.1(c) Hubble Space Telescope, NASA
Figure 8.1(d) Philip James (University of Toledo), Steven Lee (University of CO, Boulder), NASA
Figure 8.5(a–c) NASA
Figure 8.6 (a, b) NASA
Figure 8.8 NASA
Figure 8.10 NASA/JPL
Figure 8.11(a–c) NASA/JPL
Figure 8.12(a, b) Vernadsky Institute, USSR Academy of Sciences
Figure 8.13 NASA/JPL
Figure 8.14(a, b) NASA/JPL
Figure 8.15(a, d) Lowell Observatory
Figure 8.16(a, b) Lowell Observatory
Figure 8.19 NASA/JPL
Figure 8.20 U.S.G.S., Flagstaff, Arizona
Figure 8.21 NASA
Figure 8.22(a, b) NASA/JPL
Figure 8.23(a, b) NASA/JPL
Figure 8.24 NASA
Figure 8.25 NASA/JPL
Figure 8.26(a, b) NASA/JPL
Figure 8.27 NASA
Figure 8.28(a, b) NASA

Figure 9.1 National Optical Astronomy Observatories
Figure 9.3 NASA/JPL
Figure 9.5 NASA/JPL
Figure 9.9 Courtesy Nick Schneider (University of CO, Boulder)
Figure 9.11(a–d) NASA/JPL
Figure 9.11 Inset NASA/JPL
Figure 9.12 (a, b) NASA/JPL
Figure 9.13 NASA/JPL
Figure 9.14 NASA/JPL
Figure 9.15(b) NASA/JPL
Figure 9.16 NASA
Figure 9.18(a, b) NASA/JPL
Figure 9.19 Amanda Bosh (Lowell Observatory), Andrew Rivkin (University of AZ), HST High Speed Photometer Instrument Team (R.C. Bless, PI), NASA, STScI
Figure 9.20(a) NASA/JPL
Figure 9.20(b) NASA/JPL
Figure 9.21 NASA/JPL
Figure 9.23(a–e) NASA/JPL
Figure 9.24(a) NASA/JPL
Figure 9.24(b) Peter H. Smith (University of AZ Lunar & Planetary La), NASA
Figure 9.24(c) Don Davis

Figure 10.1(a) Mauna Kea Observatory

Figure 10.1(b, c)   Lick Observatory
Figure 10.2(a–c)   NASA/JPL
Figure 10.3(a)   NASA
Figure 10.3 Inset   David Crisp (JPL), Heidi Hammel (MIT), WFPC2 Science Team, NASA
Figure 10.6   Hubble Space Telescope, STScI, WFPC2
Figure 10.7(a, b)   NASA/JPL
Figure 10.9(a–d)   NASA/JPL
Figure 10.10   NASA/JPL
Figure 10.10 Inset   NASA/JPL
Figure 10.11(a)   U.S. Naval Observatory
Figure 10.11(b)   Space Telescope Science Institute
Figure 10.12 (a, b)   A. Stern (SWRI), M. Buie (Lowell Observatory), NASA, ESA
Figure 10.13   NASA/JPL

Figure 11.1   Astrophysical Station, Haute-Provence
Figure 11.2(a)   Smithsonian Institution
Figure 11.2(b–d)   John A. Wood
Figure 11.3   Yerkes Observatory
Figure 11.4(b)   Richard P. Binzel (MIT)
Figure 11.5(a)   Courtesy Sky Publishing
Figure 11.5(b)   Sara Eichmiller Ruck
Figure 11.5(c)   W. Menke
Figure 11.6(a)   B. Zellner (Georgia Southern University), NASA
Figure 11.6(b)   NASA
Figure 11.6(c)   S. J. Ostro, California Institute of Technology
Figure 11.7   Akira Fujii
Figure 11.8(a)   Lowell Observatory
Figure 11.8(b)   Scientific American
Figure 11.9   Bernard, Bennett, and Rice
Figure 11.10   Gary Goodman
Figure 11.11(a)   Giraudon/Art Resource, NY
Figure 11.11(b)   Scala/Art Resource, NY
Figure 11.12(c)   NASA, G.S.F.S., Max Planck Institute, Univ. of MD, Dept. of Astronomy, C. Lisse, M. Mumma, K. Dennerl, J. Schmidt, J. Englhauser, ROSAT
Figure 11.13(a, b)   Halley Multicolor Camera Team
Figure 11.15(a)   P. Jewitt, University of Hawaii
Figure 11.15(b)   A. Cochran (Univ. of TX), NASA
Figure 11.16(a)   HST, Comet Team
Figure 11.16(b)   MSSSO, ANU, SPL, Photo Researchers, Inc.
Figure 11.16(c)   HST Comet Team
Figure 11.16(d, e)   H. Hammel, MIT/NASA
Figure 11.17   Maroshi Hayashi
Figure 11.18(b)   Union Pacific Railroad
Figure 11.20   A. and M. Meinel, *Sunsets, Twilights, and Evening Skies,* Cambridge University Press, 1983

Page 309   H. Yang (Univ. of IL), J. Hester (Univ. of AZ), NASA, Ground-based image courtesy of Palomar, Caltech and STScI Digitized Sky Survey
Figure 12.1   Mt. Wilson and Las Campanas Observatory
Figure 12.3(a)   Royal Swedish Academy of Sciences
Figure 12.3(b)   NASA
Figure 12.7(a)   Gary McDonald
Figure 12.7(b)   Richard Berry
Figure 12.8(a)   National Optical Astronomy Observatories

Figure 12.9   NASA
Figure 12.10(a)   Perkin-Elmer Corporation
Figure 12.11   Mt. Wilson and Las Campanas Observatory
Figure 12.12(a, b)   Kitt Peak/National Optical Astronomy Observatories
Figure 12.13(a)   John Eddy, UCAR
Figure 12.13(b)   Royal Greenwich Observatory
Figure 12.16(a)   W. P. Sterne, Jr.
Figure 12.16(b)   NASA
Figure 12.17   NASA
Figure 12.18(a)   Big Bear Solar Observatory
Figure 12.18(b)   Courtesy Satoshi Matsuda
Figure 12.20   NASA
Figure 12.23   National Optical Astronomy Observatories

Figure 13.1(a)   University of Chicago
Figure 13.3(a, b)   Lowell Observatory
Figure 13.5(b)   Palomar Observatory
Figure 13.6   Akira Fujii
Figure 13.8(a, b)   *Atlas of Representative Spectra,* Y. Yamashita, K. Nairai, and Y. Norimoto, University of Tokyo Press
Figure 13.9   University of Michigan
Figure 13.10(a, b)   UPI/Corbis-Bettmann
Figure 13.14   Sproul Observatory
Figure 13.15(a–c)   *Atlas of Representative Spectra,* Y. Yamashita, K. Nairai, and Y. Norimoto, University of Tokyo Press
Figure 13.17(a–c)   Yerkes Observatory
Figure 13.18(b)   Yerkes Observatory
Figure 13.20   T. Schmidt-Kaler
Figure 13.22   Akira Fujii
Figure 13.23   California Institute of Technology
Figure 13.24(a)   James B. Kaler
Figure 13.24(b)   David F. Malin/Royal Observatory, Edinburgh
Figure 13.24(c)   Mark E. Killion
Figure 13.26(a)   National Optical Astronomy Observatories
Figure 13.26(b)   Space Telescope Science Institute
Figure 13.26(c)   *Dudley Observatory Report No. 11,* A. G. D. Phil., M. F. Cullen, and R. E. White

Figure 14.1   C. R. O'Dell (Rice University), STScI, NASA
Figure 14.2   Lick Observatory
Figure 14.4(a)   Palomar Observatory
Figure 14.4(c)   NASA
Figure 14.7(a)   Anglo-Australian Telescope Board
Figure 14.8   Royal Observatory, Edinburgh/AATB/SPL/ Photo Researchers, Inc.
Figure 14.9   Dennis DiCicco
Figure 14.10   NASA/JPL
Figure 14.15   Columbia and Palomar Observatory
Figure 14.16(a)   Mt. Wilson Observatory
Figure 14.16(b)   Palomar Observatory
Figure 14.18(a)   Max Planck Society for the Advancement of Sciences
Figure 14.18(b)   J. Morse/STScI/NASA
Figure 14.20   R. Schild
Figure 14.21   Palomar Sky Survey
Figure 14.23   Royal Observatory, Edinburgh/AATB/SPL/ Photo Researchers, Inc.

Figure 14.23 Inset NASA/IPAC
Figure 14.25 Ian McLean, UCLA
Figure 14.26(a) Anglo-Australian Telescope Board
Figure 14.26(b) Space Telescope Science Institute
Figure 14.26(c) Jeff Hester, Paul Scowen (AZ State University), NASA, STScI
Figure 14.27 S. Kuklari (Caltech), D. Golimowsky, JHU, NASA
Figure 14.28(a–c) M. J. McCoughrean (MPIA), C. R. O'Dell (Rice University), NASA
Figure 14.29 Top Backman, CTIO
Figure 14.29 Bottom R. Terrile, JPL
Figure 14.30 Research by Geoff Marcy, Paul Butler, artwork by Leigh Anne McConnaughy
Figure 14.31(a, b) Robert Schaeffer

Figure 15.9(a, b) Lowell Observatory
Figure 15.10 S. Ridgway, National Optical Astronomy Observatories
Figure 15.12(a) P. Plait and N. Soker, University of Virginia
Figure 15.12(b) Anglo-Australian Telescope Board
Figure 15.13(a) A. E. Morton
Figure 15.13(b) Steward Observatory, University of Arizona
Figure 15.15(a) Royal Observatory Edinburgh
Figure 15.15(b) Jon Morse (University of CO), NASA
Figure 15.17 Bill Fletcher
Figure 15.18(c) John C. Hayes (University of AZ)
Figure 15.19(a, b) National Optical Astronomy Observatories
Figure 15.20(a) California Institute of Technology
Figure 15.20(b) Royal Observatory, Edinburgh/AATB/SPL/Photo Researchers, Inc.
Figure 15.21(b) National Optical Astronomy Observatories
Figure 15.26 Dennis DiCicco
Figure 15.26 Inset ROSAT

Figure 16.2 David F. Malin, Anglo-Australian Telescope Board
Figure 16.5(a) University of Bochum, Germany
Figure 16.5(b) Mt. Wilson and Palomar Observatories, P. C. van der Kruit
Figure 16.9 NOAO
Figure 16.10(a) G. Westerhout
Figure 16.10(b) P. Solomon
Figure 16.11 © Jon Lomberg and The National Air & Space Museum
Figure 16.12 National Optical Astronomy Observatories
Figure 16.13 Main Image Ronald Royer
Figure 16.13 Upper Inset Robert Catchpole, South African Astronomical Observatory
Figure 16.13 Lower Inset NRAO by Farhad Yusef-Zadeh, D. R. Chance, and M. Morris
Figure 16.14 Farhad Yusef-Zadeh, Northwestern University
Figure 16.17 J. Bachall, Institute for Advanced Study, Princeton, NASA

Page 457 Robert Williams, Hubble Deep Field Team (STScI), NASA
Figure 17.1 1845 drawing by Lord Rosse, Ireland

Figure 17.2 LIFE/J. R. Eyerman
Figure 17.3 Mt. Wilson and Las Campanas Observatories
Figure 17.3 Inset Mt. Wilson and Las Campanas Observatories
Figure 17.3(b) Palomar Observatory
Figure 17.4(a, b) Palomar Observatory
Figure 17.5(a) Left and Center Palomar Observatory
Figure 17.5(a) Right Mt. Wilson and Las Campanas Observatories
Figure 17.5(b) Left, Center, and Right Palomar Observatory
Figure 17.7(a, b) ROE/Anglo-Australian Telescope Board
Figure 17.9 Dr. Wendy L. Freedman Observatories of the Carnegie Institution of Washington/NASA
Figure 17.10(a) Mt. Wilson and Palomar Observatories
Figure 17.10(b) Palomar
Figure 17.10(c) Anglo-Australian Telescope Board
Figure 17.11 National Optical Astronomy Observatories
Figure 17.12(a) G. Bothun, University of Oregon
Figure 17.12(b) California Institute of Technology
Figure 17.14 NRAO/AUI
Figure 17.15 NOAO
Figure 17.16 NASA, H. Ford et al.
Figure 17.17(a) European Southern Observatory
Figure 17.17 Inset David F. Malin/Anglo-Australian Telescope Board
Figure 17.17(b) Brad Whitmore, Francois Schweizer, STScI, NASA
Figure 17.18 European Southern Observatory
Figure 17.19 T. Van Albada
Figure 17.20 Inset STScI/Hubble Space Telescope
Figure 17.21 Space Telescope Science Institute
Figure 17.22(a) HST
Figure 17.22(b) NRAO
Figure 17.24(a) Tom Kinman, Kitt Peak National Observatory
Figure 17.24(b) Smithsonian Institution
Figure 17.25 NRAO, R. Perley, J. Dreher, J. Cowan
Figure 17.25 Inset Laird Thompson, University of Hawaii

Figure 18.1 Mt. Wilson and Las Campanas Observatories
Figure 18.7 Emilio Falco/Center for Astrophysics
Figure 18.8 T. Tyson, National Optical Astronomy Observatories
Figure 18.9 R. Williams/Hubble Deepfield Team, NASA
Figure 18.10(a) Palomar Observatory
Figure 18.10(b) Bachall, Kirhakos, Schneider
Figure 18.11(a) Palomar Observatory
Figure 18.11(b) NOAO
Figure 18.12(a) Based on L. Eachus and William Liller and *The Astrophysical Journal*, published by the University of Chicago Press; copyright 1975 The American Astronomical Society
Figure 18.12(b) T. J.-L. Courvoisier and E. I. Robson
Figure 18.13 Alan Bridle, NRAO
Figure 18.14 ROSAT
Figure 18.15(a) Palomar Observatory
Figure 18.15(b) A. N. Stockton, University of Hawaii
Figure 18.16(b) Hewitt and Turner, NRAO

Figure 18.16(c)   Space Telescope Science Institute
Figure 18.17   W. Couch (University of NSW), R. Ellis (Cambridge University), NASA, STScI

Figure 19.14   NASA
Figure 19.17   NCSA, Greg Bryan, Michael Norman (University of IL)

Figure 19.18   A. Dressler (Carnegie Institutions, Washington), M. Dickinson (STScI), D. Macchetto (ESA/STScI), M. Giavalisco (STScI), NASA
Figure 19.19 Clockwise from Top   California Institute of Technology; David F. Malin, Anglo-Australian Telescope Board; Akira Fujii; Mt. Wilson and Las Campanas Observatory, Carnegie Institution; NASA; James B. Kaler

INDEX*

---

*Photographs of named objects are indicated by boldfaced numbers.